Microorganisms for Sustainability

Volume 8

Series editor
Naveen Kumar Arora, Environmental Microbiology, School for Environmental Science, Babasaheb Bhimrao Ambedkar University, Lucknow, Uttar Pradesh, India

More information about this series at http://www.springer.com/series/14379

Dilfuza Egamberdieva • Nils-Kåre Birkeland
Hovik Panosyan • Wen-Jun Li
Editors

Extremophiles in Eurasian Ecosystems: Ecology, Diversity, and Applications

Springer

Editors
Dilfuza Egamberdieva
Faculty of Biology
National University of Uzbekistan
Tashkent, Uzbekistan

Nils-Kåre Birkeland
Department of Biology
University of Bergen
Bergen, Norway

Hovik Panosyan
Faculty of Biology
Yerevan State University
Yerevan, Armenia

Wen-Jun Li
School of Life Sciences
Sun Yat-Sen University
Guangzhou, China

ISSN 2512-1901 ISSN 2512-1898 (electronic)
Microorganisms for Sustainability
ISBN 978-981-13-0328-9 ISBN 978-981-13-0329-6 (eBook)
https://doi.org/10.1007/978-981-13-0329-6

Library of Congress Control Number: 2018949042

Printed on acid-free paper

This Springer imprint is published by the registered company Springer Nature Singapore Pte Ltd.
The registered company address is: 152 Beach Road, #21-01/04 Gateway East, Singapore 189721, Singapore

Foreword

The extremophiles are those microorganisms thriving under extreme conditions where no other living being will have any chance to survive. The Eurasian zone examples of extreme habitats are those presenting high temperatures, such as hot springs, or those with high salt concentrations, such as salines and hypersaline lakes. The study of extremophiles brings great interest from both the physiological or environmental and industrial perspectives. For instance, they represent key organisms to understand the limits of life on Earth and the potential existence of life somewhere else, and the study of their biological characteristics forcing them to thrive under extreme conditions is a basic milestone to understand the origin of life and its development through a variety of environments so that microbes have been able to spread all over our planet. At present, it is known that microorganisms present a huge diversity, and among them the extremophiles are of particular interest because the extreme conditions where they inhabit create relatively restricted environments which, a priori, should facilitate the analysis of complex cellular interactions within those ecosystems. Besides, the extreme nature of those interactions and the physiological capabilities developed in the extremophiles are attracting the attention of industrial applications. This is easy to understand because the biotechnological industry is looking for unique products and applications and the use of highly durable and stable biocatalysts so that higher production efficiencies and lower costs can be achieved.

Large numbers of publications from American researchers and from extreme environments in America are reported. However, one should not forget the contributions from Eurasian systems and investigators. This book promotes that research being carried out by scientists on the topics of ecosystems, diversity, and applications of extremophiles in Eurasia. Thus, this highly significant work adds to the worldwide advancement in the field and sums up to earlier contributions on the discovery of novel extremophiles, their ecology, and biotechnological applications made by numerous highly relevant scientists from the Eurasian continent.

Institute of Natural Resources
and Agrobiology, Spanish Council
for Research, IRNAS-CSIC
Sevilla, Spain

Juan M. Gonzalez
jmgrau@irnase.csic.es

Contents

1 **Insights into the Thermophile Diversity in Hot Springs of Pakistan** 1
Arshia Amin, Iftikhar Ahmed, Nauman Khalid, Yao Zhang, Min Xiao, and Wen-Jun Li

2 **Hot Springs of India: Occurrence and Microbial Diversity** 29
Manik Prabhu Narsing Rao, Lan Liu, Jian-Yu Jiao, Min Xiao, and Wen-Jun Li

3 **Diversity of Thermophiles in Terrestrial Hot Springs of Yunnan and Tibet, China** 57
Wen-Dong Xian, Manik Prabhu Narsing Rao, En-Min Zhou, Lan Liu, Min Xiao, and Wen-Jun Li

4 **Microbial Diversity of Terrestrial Geothermal Springs in Lesser Caucasus** 81
Hovik Panosyan, Armine Margaryan, Liana Poghosyan, Ani Saghatelyan, Ekaterine Gabashvili, Ekaterine Jaiani, and Nils-Kåre Birkeland

5 ***Geobacillus* and *Anoxybacillus* spp. from Terrestrial Geothermal Springs Worldwide: Diversity and Biotechnological Applications** 119
Armine Margaryan, Grigor Shahinyan, Pargev Hovhannisyan, Hovik Panosyan, Nils-Kåre Birkeland, and Armen Trchounian

6 **Thermophiles and Their Exploration for Thermostable Enzyme Production** 167
Nikoleta Boteva and Margarita Kambourova

7 **Thermophilic Chemolithotrophic Bacteria in Mining Sites** 187
Narine S. Vardanyan and Arevik K. Vardanyan

8 **Thermophilic and Halophilic Microorganisms Isolated from Extreme Environments of Turkey, with Potential Biotechnological Applications** 219
Kemal Guven, Fatma Matpan Bekler, and Reyhan Gul Guven

9 **Hypersaline Environments of Iran: Prokaryotic Biodiversity and Their Potentials in Microbial Biotechnology** 265
Atefeh Safarpour, Mohammad Ali Amoozegar, and Antonio Ventosa

10 **Halotolerant and Halophilic Microbes and Their Environmental Implications in Saline and Hypersaline Lakes in Qinghai Province, China** 299
Hongchen Jiang, Jianrong Huang, and Jiang Yang

11 **Soil Salinity and Microbes: Diversity, Ecology, and Biotechnological Potential** 317
Dilfuza Egamberdieva, Kakhramon Davranov, and Stephan Wirth

12 **Halophilic *Actinobacteria* Biological Activity and Potential Applications** 333
Osama Abdalla Abdelshafy Mohamad, Li Li, Jin-Biao Ma, Shaimaa Hatab, Bakhtiyor A. Rasulov, Zulpiya Musa, Yong-Hong Liu, and Wen-Jun Li

13 **Microbial Diversity in Asian Deserts: Distribution, Biotechnological Importance, and Environmental Impacts** 365
Nimaichand Salam, Zi-Wen Yang, Mipeshwaree Devi Asem, Wael N. Hozzein, and Wen-Jun Li

14 **Trichoderma from Extreme Environments: Physiology, Diversity, and Antagonistic Activity** 389
Laith Khalil Tawfeeq Al-Ani

15 **Exopolysaccharide-Producing Microorganisms from Extreme Areas: Chemistry and Application** 405
Annarita Poli, Paola Di Donato, Giuseppina Tommonaro, Gennaro Roberto Abbamondi, Ilaria Finore, and Barbara Nicolaus

16 **Why Settle for Mediocre, When Extremophiles Exist?** 435
Shivanshi Vashist and Rohit Sharma

17 **Use of Acidophilic or Acidotolerant *Actinobacteria* for Sustainable Agricultural Production in Acidic Soils** 453
K. Tamreihao, Nimaichand Salam, and Debananda S. Ningthoujam

About the Editors

Dilfuza Egamberdieva graduated with a degree in Biology from the National University of Uzbekistan in 1993. She received her Ph.D. in Agricultural Sciences from the Humboldt University of Berlin, Germany, in 2000. She pursued her postdoctoral studies at the Helsinki University of Finland (2001, 2008), University of Florence, Italy (2003), Manchester Metropolitan University, UK (2006), Leiden University of Netherlands (2007), and Leibniz Centre for Agricultural Landscape Research, Germany (2014–2017). In 2009, she established the research laboratory "Plant Microbe Interactions" at the National University of Uzbekistan. For her outstanding academic achievements she received numerous fellowship and awards including the UNESCO-L'OREAL Fellowship for Women in Science (2006), IUBMB Young Scientist Award (2006), ASM Morrison Rogosa Award (2006), UNESCO-MAB Award (2005), Alexander Von Humboldt Fellowship (2014), and in 2012 the TWAS Prize in Agricultural Sciences for her innovative contributions to the study of plant-microbe interactions in stressed environments. She has produced over 80 peer-reviewed publications in international journals and 40 chapters.

Nils-Kåre Birkeland graduated with a degree in Microbiology from the University of Bergen (Norway), and received his Dr. philos. degree in Molecular Microbiology from the University of Oslo (Norway) in 1992. He also has work experience at the NTNU (Norway), University of Tromsø (Norway), Norwegian University of Life Sciences, and from sabbaticals in USA. From 1996 he has been a Professor of Microbiology at the University of Bergen. His main research areas during the last 20 years have been anaerobic microbiology, extremophiles, and microbial biotechnology. He has isolated and described a number of novel taxa of extremophiles and analyzed the molecular mechanisms for high-temperature adaptations and diversity in extreme environments. He has published 97 papers in peer-reviewed journals and 3 book chapters, and has coordinated a number of international higher educational/research networks and projects.

Hovik Panosyan graduated with a degree in Biology from Yerevan State University (YSU) in 1999. He received his Ph.D. in Microbiology from the Institute of Botany of NAS of Armenia in 2003. He has been a faculty member at YSU since 2002 and was promoted to Associate Professor in 2011. His main research areas are Microbial Ecology and Biology of Extremophilic Microbes. He has received numerous research fellowships and awards including the FEBS Short-Term Fellowship (2009 and 2004), FEMS Research Fellowship (2009), NFSAT (2011), DAAD (2013) and has participated in international research together with American, European, and Asian partners in the context of CRDF and Norwegian SIU grants. He is currently coordinator and leader of international research and educational programs, as well as ISME ambassador of Armenia. He has work experience at the University of Bergen (Norway), LMU Munich (Germany), University of Nevada, Las Vegas (USA), and Institute of Biomolecular Chemistry Naples (Italy). He has published more than 45 research papers in peer-reviewed journals, 2 books, and 18 chapters.

Wen-Jun Li received his Ph.D. in Microbiology from the Institute of Applied Ecology, Chinese Academy of Sciences in 2002. He has been Head of the Actinobacterial Research Group and Vice Director of Yunnan Institute of Microbiology, Yunnan University, China, since January of 2009, and Vice Dean of School of Life Sciences, Yunnan University, China, since January of 2012, and currently works as a Distinguished Professor at the School of Life Sciences, Sun Yat-Sen University, Guangzhou, China. He has written 3 monographs and more than 610 publications and holds more than 15 authorized patents. He has been appointed an editorial board member of 7 peer-reviewed journals and was invited to be a reviewer for over 20 international journals. He was awarded the WFCC (The World Federation for Culture Collections) Skerman Award for Microbial Taxonomy in 2007, and five other provincial level awards in 2005, 2007, 2009, 2010, and 2013, respectively, for his outstanding research achievements in microbial systematics. Recently, he was also appointed as member of the Executive Board (EB) of International Committee on Systematics of Prokaryotes (ICSP) during its meeting at Valencia, Spain (2017), and also Secretary in General of Bergey's International Society for Microbial Systematics (BISMiS) in 2018.

Insights into the Thermophile Diversity in Hot Springs of Pakistan

1

Arshia Amin, Iftikhar Ahmed, Nauman Khalid, Yao Zhang, Min Xiao, and Wen-Jun Li

Abstract

The hot springs are populated by mesophilic, thermotolerant, and hyperthermophilic bacteria. These populations are diverse, and some of them show combinations of other extreme conditions, for example, acidic, alkaline, high pressure, and high concentrations of salts and heavy metals. Anaerobes inhabiting hot springs are considered to be the closest living descendants of the earliest living forms on earth, and their study offers understandings about the origin and evolution of life. In this chapter, thermal spring bacterial diversity from Pakistani ecology is reviewed. The bacterial populations in Pakistani hydrothermal vent

A. Amin
National Culture Collection of Pakistan (NCCP), Bioresources Conservation Institute (BCI), National Agricultural Research Centre (NARC), Islamabad, Pakistan

State Key Laboratory of Biocontrol and Guangdong Provincial Key Laboratory of Plant Resources, School of Life Sciences, Sun Yat-Sen University, Guangzhou, People's Republic of China

Department of Biosciences, Capital University of Science and Technology, Islamabad, Pakistan

I. Ahmed (✉)
National Culture Collection of Pakistan (NCCP), Bioresources Conservation Institute (BCI), National Agricultural Research Centre (NARC), Islamabad, Pakistan
e-mail: iftikharnarc@hotmail.com

N. Khalid
School of Food and Agricultural Sciences, University of Management and Technology, Lahore, Pakistan

Y. Zhang · M. Xiao
State Key Laboratory of Biocontrol and Guangdong Provincial Key Laboratory of Plant Resources, School of Life Sciences, Sun Yat-Sen University, Guangzhou, People's Republic of China

D. Egamberdieva et al. (eds.), *Extremophiles in Eurasian Ecosystems: Ecology, Diversity, and Applications*, Microorganisms for Sustainability 8,
https://doi.org/10.1007/978-981-13-0329-6_1

environments showed a great genetic diversity, and most members of these populations appeared to be uncultivated and unidentified organisms. Analysis suggested that some microorganisms of novel phylotypes predicted by molecular phylogenetic analysis were likely present in thermal spring environments. Libraries were predominantly composed of rare phylotypes that appeared to be unclassified, and the number and type of phylotypes observed were correlated with biogeography as well as biogeochemistry. These findings broaden our opinion of the genetic diversity of bacteria in hot water spring environments. The global-scale bacterial diversity of other hot water spring environments, on the other hand, may be beyond present proficiencies for authentic study.

Keywords

Thermophiles · Thermal springs · Bacterial diversity · Taxonomy · Biogeography · Biogeochemistry · Unculturable methods

1.1 Introduction

Temperature as an environmental factor compels all living microorganisms. In contrast to the upper temperature boundaries, the lower temperature boundaries for growth among microorganisms are not well defined (Russell et al. 1990). Thermophiles are the microorganisms that "love" heat. A word of caution is necessary regarding the use of the term "thermophilic." The term means different temperature ranges for different groups of microorganisms. For example, *Candida thermophile* is described as a thermophilic yeast with a maximum growth temperature of 51 °C. The optimal growth temperature for this microorganism is 30–35 °C (Shin et al. 2001). Among bacteria, this would be a thermotolerant species. The record for the widest temperature span for growth is held by *Methanothermobacter thermautotrophicus* that able to grow from 22 to 75 °C (Gerday and Glansdorff 2007).

Strain 121, a Fe(III)-reducing archaea isolated from a hydrothermal vent along the Juan de Fuca Ridge, is reported to have a doubling time of 24 h at 121 °C and remains viable after exposure to temperatures as high as 130 °C (Kashefi and Lovley 2003). The most heat-resistant spore is held by *Moorella thermoacetica* strain JW/DB-4. Under autotrophic conditions at 60 °C, this bacterium forms spores with a decimal reduction time of 2 h at 121 °C. A subpopulation of spores apparently requires 1 h at 100 °C to become fully activated before germinating (Byrer et al. 2000).

W.-J. Li (✉)
State Key Laboratory of Biocontrol and Guangdong Provincial Key Laboratory of Plant Resources, School of Life Sciences, Sun Yat-Sen University, Guangzhou, People's Republic of China

Key Laboratory of Biogeography and Bioresource in Arid Land, Xinjiang Institute of Ecology and Geography, Chinese Academy of Sciences, Ürűmqi, People's Republic of China
e-mail: liwenjun3@mail.sysu.edu.cn

Thermus aquaticus, an aerobic, thermophilic bacterium, was isolated from Yellowstone National Park, Wyoming, in the late 1960s (Brock and Freeze 1969), and the microorganism's Taq DNA polymerase has become an essential component of molecular biology. Brock (1997) stated that among different types of research methods, the one done by individual scientist usually returns late. *Thermus aquaticus*, an anaerobic bacterium was isolated from Yellowstone National Park, Wyoming, in the late 1960s by Thomas Brock (Brock and Freeze 1969). That lead to the discovery of Taq polymerase which become a breakthrough in molecular biology. Cosmopolitan microorganisms from thermal environments include *Methanothermobacter thermautotrophicus*, *Thermoanaerobacter thermohydrosulfuricus*, *Thermoanaerobacterium thermosaccharolyticum*, and *Geobacillus stearothermophilus*.

The interaction correlation between biogeography and biogeochemistry in thermal environments is also worthy. As an example, three combinations can be defined by Engle et al. (1995):

Relaxedbiogeographyandbiogeochemistry:Forexample,*Thermoanaerobacterium thermosaccharolyticum* and *Thermobrachium celere* have a relaxed biogeography and biogeochemistry. They have been isolated from a variety of environments from several locations including thermobiotic, mesobiotic, slightly alkaline, and acidic environments.

Relaxed biogeography and restricted biogeochemistry: For example, *Clostridium paradoxum* and *Clostridium thermoalcaliphilum* are isolated from sewage sludge on four different continents, but only from sewage sludge.

Restricted biogeography and relaxed biogeochemistry: That is, *Anaerobranca horikoshii* has only been isolated from a specific area behind the old faithful ranger station in Yellowstone National Park, but from several pools in that area, representing a spectrum of pH values from acidic (pH 5) to alkaline (pH 8.5). Although relatively easy to isolate, strains of *A. horikoshii* have not been obtained from other areas of Yellowstone National Park or other countries, nor has its sequence been found in environmental 16S rRNA gene libraries.

Strazzulli et al. (2017) studied that spots of volcanic activity exist all over the Earth's surface and under the sea. They offer a variety of different environments for extremophilic microorganisms. Hot springs are always full of hyperthermophiles, the majority of which belong to the domain of archaea. Combination of extreme temperature with other physicochemical parameters, i.e., acidic, alkaline, high pressure, and high concentrations of salts and heavy metals, also selects specific classes of bacteria and discourages remaining classes for which these conditions are not favorable (Cowan et al. 2015). Archaea which resides in hot springs are claimed to be the closest living descendants of the primitive living forms on earth and are considered as models to study origin and evolution of life (Olsen et al. 1994).

1.2 History

It is well established that a number of archaeal and bacterial species live under extreme environmental conditions, which include pressure, high temperature, UV light, ionizing radiation, very low levels of nutrients, and low or high levels of pH

(Gerday and Glansdorff 2007). Cavicchioli (2002) suggested the possibility that while considering these extremophiles as models, we can get insights into the lifestyle at celestial habitat.

It would be highly significant to establish identification between ancient and primitive organisms. It has been observed that cladistically ancient organisms are located near the root of universal rRNA-based trees, but they do not own primitive molecular genetic apparatus, nor they are more basic in their metabolic abilities than their aerobic equivalents (Islas et al. 2003). Pre-RNA worlds are the foundation of primitive living systems, in which life may have been based on polymers using backbones other than ribose phosphate and possibly the bases different from guanine, adenine, uracil, and cytosine (Levy and Miller 1998), followed by a stage in which life was based on RNA both as the genetic material and as catalysts (Joyce 2002). Only very few facts support hyperthermophilic origin of life. Firstly, the deepest, branches of rRNA-based molecular phylogenies are full of hyperthermophiles (Pace 1991, 1997). Secondly, immediately after earth formation, the surface of the earth was extremely hot and planet remained molten for some time after its formation. About 4.6×10^9 years ago, life on earth and only hyperthermophilic life were possible (Wiegel 1998). The biphasic temperature-growth curves of many thermophiles growing at elevated temperatures and the existence of cryptic thermophiles are considered as additional arguments for the start of life in the range of 60–90 °C and that hyperthermophiles as well as mesophiles and psychrophiles are adaptations to changed environments.

While some antagonists say that the earth's surface speedily lost temperature to provide mesophilic origin of life (Wilde et al. 2001). Chemical decomposition of recognized biochemical compounds, i.e., amino acids, nucleobases, RNA, and thermolabile molecules, has half-lives for decomposition at temperatures between 250 and 350 °C at the most a few minutes (Miller and Bada 1988). Another theory that supports mesophilic origin of life came from Gulen et al. (2016). Petrov et al. (2015) believe that the property of ribosome that shields it in high temperature, e.g., RNA foldings, evolved slowly during evolution. Hyperthermophilic microbial lifestyles are the product of secondary adaptations that developed during early stages of cell evolution, but we do not have an information on the composition of the terrestrial atmosphere during the period of the origin of life or on the temperature, ocean pH values, and other general and local environmental conditions that were important for the emergence of living systems (Lazcano and Bada 2003). Delaye et al. (2005) believe that the origin of the mutant sequences ancestral to those found in all existing species and the divergence of the bacteria, archaea, and eukarya were not synchronous events, i.e., the separation of the primary domains took place later, perhaps even much later, and then the appearance of the genetic components of their least common ancestors. The cenancestor is thus one of the last evolutionary outcomes of a series of ancestral events, including lateral gene transfer, gene losses, and paralogous duplications that took place before the separation of bacteria, archaea, and eukarya. Dworkin et al. (2002) and Forterre et al. (2002) believes that if hyperthermophile is not truly primordial, then heat-loving lifestyles may be remainders of a secondary adaptation that evolved after the origin of life and before or soon after the separation of the major lineages. Forterre et al. (2002) believe in adaptation of bacteria to extreme environments by lateral transfer of reverse gyrase and other thermo-adaptive traits from heat-loving archaea. Dworkin et al. (2002) believe that outcompetition of older mesophiles by hyperthermophiles originally adapted to stress-inducing conditions other than high temperatures.

Wilson (1992) created the term biodiversity and wrote *The Diversity of Life*. At that time, there were 4800 species described in the "kingdom" Monera. Currently more than 30,000 whole genomes have been submitted from all three domains of life bacteria, archaea, and eukarya and are available in the Joint Genome Institute's Integrated Microbial Genomes database (Hug et al. 2016). Recently, Hug et al. (2016) gave new tree of life (Fig. 1.1) in which 92 phyla which are representing total bacterial eukaryotic and archaeal diversity and includes 92 phyla belonging to bacteria, 26 of archaea five of the eukaryotes. Genome-resolved metagenomics and single-cell genomics of hundreds of genomes revealed that all members have comparatively small genomes and most of them have restricted metabolic capacities or are symbionts. Therefore, all cells either lack thorough citric acid cycles or respiratory chains, and furthermost few have limited or no ability to synthesize nucleotides and amino acids. It is presumed that these reduced metabolisms resulted from either super phylum-wide damages or inherited characteristics. If its result of inherited characteristics, then symbiotic lifestyles were secondary adoptions from once more complex organisms appeared.

1.3 Thermal Environments and Biodiversity

What makes thermal environment a popular model to test biogeographical hypotheses is its island-like nature. Using similar strains of the thermophilic archaeon *Sulfolobus* originating from hot springs in Yellowstone National Park and Italy, Zillig et al. (1980) formulated the hypothesis that "geographical barriers between habitats of the same type do not exist for microorganisms." This hypothesis also corresponds to the oft-quoted hypothesis that "everything is everywhere and the environment selects" (Beijerinck 1913). However, Whitaker et al. (2003) attribute genetic divergence detected by multilocus sequence analysis of strains of *Sulfolobus solfataricus* from five sites to geographic isolation.

Recent reports on bacterial diversity of hot water springs revealed that it is difficult to propose the reasons for the presence of specific bacterial species in a thermal spring because these ecosystems are always deviate when influenced by an outside influence, for example, Hu et al. (2017) reported that in acidic thermal springs in New Zealand temperature (range 30–80 °C) was the only significant variable associated with community turnover. Near 40 °C, chemolithoautotrophs were dominant, whereas, at temperatures >65 °C, the microbial community was dominated approximately solely by sulfur-oxidizing archaea. At mesophilic temperatures, the community structure was diverse, encompassing both archaea and bacteria but dominated mainly by chemolithotrophic sulfur and hydrogen oxidizers. In another report, Jiang and Takacs-Vesbach (2017) revealed that despite similar pH of all studied sites of Yellow Stone National Park, bacterial diversity varied a lot.

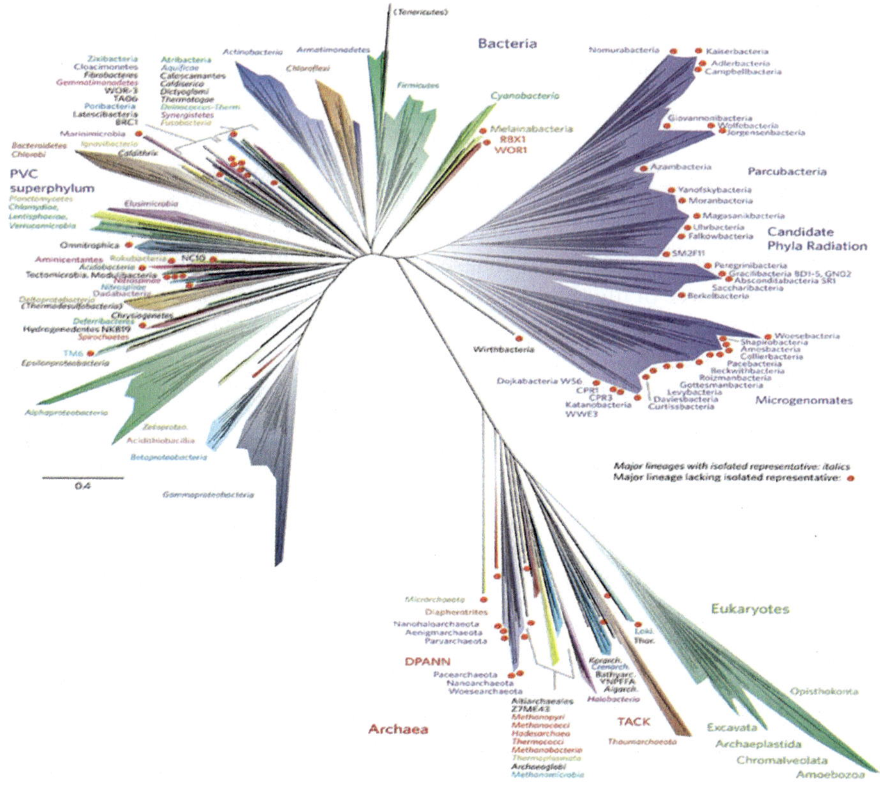

Fig. 1.1 Figure adopted from Hug et al. (2016) is showing tree of life. All major lineages are highlighted with genome-wide depiction, but mostly are phylum-level branches. Major lineages are allocated random colors and named, with the published and described names, in italics. Uncultured lineages are highlighted with red dots and are nonitalic. Brackets around *Tenericutes* and *Thermodesulfobacteria* show that these are subbranches of *Firmicutes* and *Deltaproteobacteria*, respectively. Phylum like *Proteobacteria* which is not monophyletic (*Alphaproteobacteria*, *Betaproteobacteria*, *Deltaproteobacteria*, *Gammaproteobacteria*) is shown separately. The candidate phyla radiation (CPR) is assigned a single color because all are uncultured and unclassified and are still in the process of description at lower taxonomic levels. Further analysis by ribosomal proteins as well as by primitive genetic code showed that there is vast difference in composition of three domains, i.e., thermophilic, mesophilic, and halophilic domains (®Macmillan Publishers Limited, 2017)

Lowest-temperature site predominant phyla were *Chloroflexi*, *Bacteroidetes*, *Proteobacteria*, and *Firmicutes*. Metagenome study revealed that all genes related to energy production were present, i.e., transcription, carbohydrate transport, genes related to sulfate reduction, dissimilatory nitrogen reduction, and H_2S. In another study, Merkel et al. (2017) on Kamchatka Peninsula hot springs revealed dominance of sulfur-oxidizing bacteria of genus *Sulfurihydrogenibium* which were followed by the second most dominant anaerobic bacteria of genus *Caldimicrobium*. At high-temperature sites, archaea of the genus *Vulcanisaeta* were abundant, and at acidic springs *Nanoarchaeota* and uncultured *Thermoplasmataceae* A10 were also present.

1.4 General Features and Geography of Pakistan Hot Water Springs

The Main Mantle Thrust and the Main Karakoram Thrust (MKT) in Chilas and Hunza areas of Northern Pakistan are host to many hydrothermal activity with numerous thermal springs distributed between latitude 30°–37° N and longitude 73°–77° E (Bakht 2000). One of the ways to address the woes of energy crisis effectively in the developing world is through the use of geothermal energy resources (Gondal et al. 2017). There are seven hot springs in Murtazaabad which lie along the Main Karakoram Thrust in Northern Areas of Pakistan with the surface temperature range of 47–92 °C. All the thermal waters of Pakistan are formulated from $NaHCO_3$. Tattapani and Tato thermal springs along the Main Mantle Thrust have a surface temperature from 48 to 92 °C. These are also $NaHCO_3$ type. Geothermal springs of Chagai are related to the youngest volcano (Koh-I-Sultan) of Pakistan. The northern areas having geothermal fields at Tattapani, Tato, and Murtazaabad are located between the latitudes 35°20′ N–36°30′ N and longitudes 74°E–76°E with sheer topography and U-shaped valleys, which are drained by the rivers Indus, Gilgit, and Hunza, while the rivers Shigar, Shyok, Ishkuman, and Yasin form the major branches to these rivers.

Other important mountain ranges of the area are the Kailas, Rakaposhi, and Masherbrum. Rainfall in these areas is light, and the geotectonic development of the northern areas of Pakistan occurs during late Cretaceous to Cenozoic era. The creation involved three tectonic elements, i.e., the Indo-Pakistan shield and its northern sedimentary cover (Indian Mass), the rocks deposited on the southern part of the Eurasian Mass, and the Kohistan island arc sequence (Ahmad et al. 2015).

Amin et al. (2017b) reported bacterial diversity and ecological interactions with physicochemical parameters in 9 hot water springs scattered along Himalayan geothermal region where temperature ranges from 60 to 95 °C and pH from 6.2 to 9.4, and in mineralogy from HCO_3^- (Tato field), sulfates (Tattapani) to mixed type (Fig. 1.2) (Murtazaabad).

Among various hot water springs present in Pakistan (Table 1.1), Chang et al. (2013) have reported on the chemical composition of Manghopir thermal spring for the year 2008. In our published report by Amin et al. (2017b), we revealed that among the studied sites (Table 1.1) were Tato field thermal springs that were bicarbonated in nature and had higher bicarbonates (525–610 mg L^{-1}) than sulfates (410–460 mg L^{-1}) and the Tattapani hot springs that were sulfate type considering bicarbonate level (133–159 mg L^{-1}) was significantly lower than sulfates (545–684 mg L^{-1}). The Murtazaabad thermal springs were mixed type with high level of both sulfates and bicarbonates (710–940 mg L^{-1}). The levels of sulfates and bicarbonates in the Murtazaabad springs were also significantly higher than the other two sites.

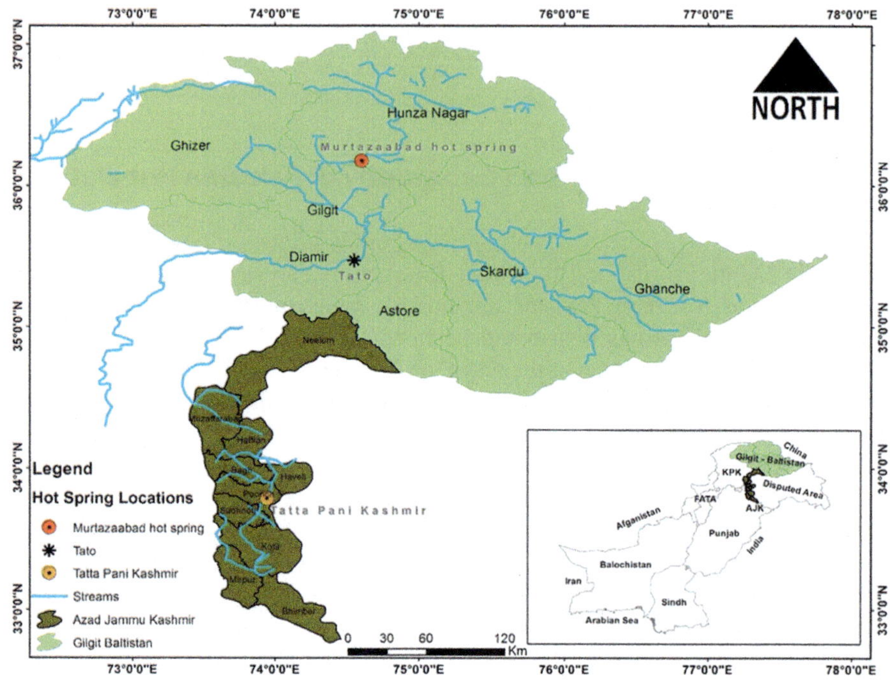

Fig. 1.2 Geographical locations of thermal springs of Pakistan

1.5 Bacterial Diversity in Hot Water Springs of Pakistan

In order to scratch the surface in revealing a new vista, we designed a study to discover unique and undiscovered microbiota of hot water springs present in Pakistan by second-generation pyrosequencing and compared the gap between cultured and uncultured microbiota by assessing the relationships between microbial community compositions and environmental conditions (e.g., water geochemistry). Two new species of bacteria were characterized by polyphasic taxonomic approach for validation at species level *Nocardioides pakistanensis* sp. nov. and *Streptomyces caldifontis* sp. nov. (Amin et al. 2016, 2017a).

According to unculturable method, at 97% OTU (operational taxonomic unit) level, 5535 quality reads were distributed into 972 microbial genera and 53 phyla. OTUs of the phyla *Proteobacteria* and *Chloroflexi* were found to be dominantly present in all the sampling sites. Distinct phyla which seemed to outcompete were *Proteobacteria*, *Chloroflexi*, *Thermotoga*, *Bacteriodetes*, *Deinococcus-Thermus*, *Nitrospirae*, and *Acidobacteria*, and other well-reported thermal spring phyla were UT06, OP11, BRC1, OD1, OP8, OP1, OP3, OP9, OMAN, and NKB19. Environmental properties like pH, temperature, and sulfur influenced the community structure at most that was depicted by the presence of sulfur- and nitrate-reducing bacteria, but the influence of other factors on microbial community

Table 1.1 Chemical composition of hot water springs of Pakistan

Thermal spring	pH	EC µS cm^{-1}	Temp. (°C)	Ca^{2+}	Na^{+}	K^{+}	HCO_3^-	SiO_2	SO_4^{-2}	NH4	N2	CO (ppm)
				mg L^{-1}								
Tato field[a]	7.9–8.8	1119–1569	70–90	2.5–3.5	200–300	20–26	525–610	10–91	410–460	200–300	8.4–9.7	20.3–25.3
Tattapani[a]	6.2–9.4	1032–1060	75–90	0.1–0.2	150–170	2.5–4.9	133–159	45–58	545–690	250–270	11.3–13.8	40.0–42.4
Murtazaabad[a]	6.7–9.2	1730–1742	90–95	2.6–9	400–500	35–60	710–940	5–10	710–940	300–400	2.3–3.7	29.8–32.6
Crocodile's pool-S1 Manghopir Karachi[b]	7.0–7.9	2465–4790	22–33	64–400	58–855	14–24	335–960	ND	240–860	ND	ND	ND
Crocodile's pool springs S2 Manghopir Karachi[b]	7.02–7.8	2770–3730	19–29	60–508	43–635	12–18	340–490	ND	242–812	ND	ND	ND
Chilas[c]	ND	ND	20	75.3	215.5	215.5	ND	79.4–83.0	ND	ND	ND	ND
Jaglot[c]	ND	ND	20–65	127.7–160.1	56.9–85.0	56.9–85.0	ND	82.1–88.7	ND	ND	ND	ND
Budelas[c]	7.85	776–1060	39–40	104.5–116.6	153.8–159.1	153.8–159.1	ND	113.4–121.0	ND	ND	ND	ND
Hakuchar[c]	ND	ND	49–50	0.0–97.7	251.2–252.3	251.2–252.3	ND	115.6–118.2	ND	ND	ND	ND

EC electrical conductivity, *Temp.* temperature
Values are given as means of three readings:
[a]Amin et al. (2017b)
[b]Chang et al. (2013)
[c]Bakht (2000)

assemblage like anaerobic stress in deep water; methane, ammonia and presence of planktonic material were supported by the presence of *Chloroflexi* and *haloanaerobia*. Dominance of *Chloroflexus* and low number of order *Aquificales* were also studied by Skirnisdottir et al. (2000) at upper temperature of 88–90 °C that matched our results. High-temperature and low-oxygen site of Tattapani spring (TP-H3-c) had the largest OTUs for sulfur bacteria. *Deltaproteobacteria* purple sulfur bacteria were most dominant in sulfur-rich (Tattapani hot water spring) TP-H3 sites. The presence of *Cyanobacteria* at (Tato field hot water spring) TF-H2-a and (Tato field hot water spring) TF-H2-b affected number and diversity of purple sulfur bacteria even at high temperature. The highest numbers of OTUs for purple sulfur bacteria, that is, 217, were present in (Tattapani hot water spring samples) TP-H3-c which was a sulfur-rich site, and the second highest number was observed in (Tato field hot water spring) TF-H2 where sulfur level was much lower, but phylum *Cyanobacteria* population was present to support their growth. Another unique phylum UT06 was present with rich diversity in (Tattapani hot water spring) TP-H3 and (Murtazaabad hot water spring) MA-H4 and low in Tato field samples. Its diversity and number were evenly high in all samples of MA-H4, but in TP-H3 its distribution was not even. Similar physiochemical properties of hot water springs located at far distances and varying geographical locations were responsible for linked microbial community. The presence of closely related microbial species in neighboring hot water springs indicated that movement of water and soil was also responsible for designing microbial community structure in adjacent environments. These geothermal sites should be considered to explore natural biogeochemical cycles and role of specific microorganisms in energizing these cycles to exploit these potentials in the near future. In these hot water springs, enhanced conditions of high temperature, alkaline pH, and methanogenesis all met automatically, and genes for methanogenesis were switched in methanogens. Study suggested that methanotrophy in these thermal sites was not restricted to only one type of methanotroph, but members of type I methanotroph, aerobic methanotroph, *Betaproteobacteria* methanotroph, and type II methanotroph all collectively were responsible for methane cycle in thermal systems (Amin et al. 2017b).

1.6 Survival Mechanisms at Thermophilic Environment

Survival of thermophilic bacteria (Fig. 1.3), at high temperatures, is because of various adaptations in physiological systems and genetics as stress response to stabilize homeostasis (Wang et al. 2015). Few examples are production of DNA-binding proteins, activation of heat shock proteins, activation of reactive oxygen species, and efficient repair damage (Ranawat and Rawat 2017).

Other mechanisms involve amino acid substitutions (Arnórsdóttir et al. 2009), hydrophobic cores (Bezsudnova et al. 2012), interactions among subunits (Pang and Allemann 2007), and inactivation of spores by high hydrostatic pressure (Sarker et al. 2015) and by adjusting membrane fluidity after adjusting membrane fatty acid

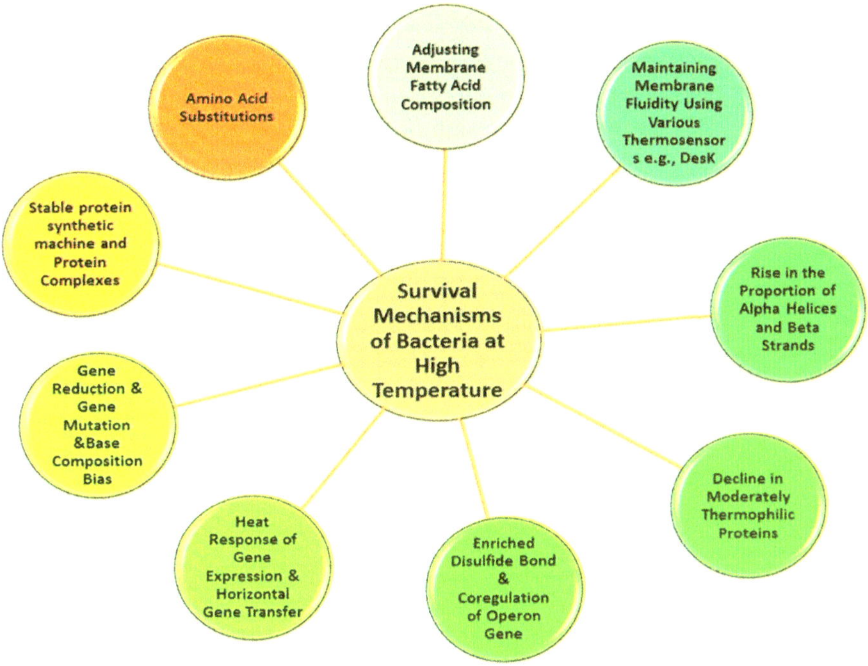

Fig. 1.3 Survival mechanisms related to transcriptome, genome, proteome, and other adaptive features of bacteria for survival at high temperatures

composition (Yoon et al. 2015) and also by maintaining membrane fluidity using various thermosensors, e.g., DesK (Cybulski et al. 2015).

Another worth-mentioning fact of thermophilic proteins is substantial rise in the proportion of alpha helices and beta strands, with a decline in irregular region; with moderately thermophilic proteins, alpha helical increase is dominant, whereas in extremely thermophilic ones, beta strand rise is more extensive (Chakravorty et al. 2017; Szilágyi and Závodszky 2000).

Another distinguishing factor among extreme thermophilic proteins and moderately thermotolerant proteins is amino acid composition. In moderately thermophilic proteins, lysine content is less than arginine content, but in extremely thermophilic proteins, lysine residues are more in number because of their requirement of stronger electrostatic interactions and lysine is a charged residue. Amino acid residues which are not high temperature tolerant, i.e., methionine and asparagine, are less in extreme thermophilic proteins (Mrabet et al. 1992). However, the important phenomena, i.e., lysine succinylation and lysine propionylation which are important for protein function, are not different in extremely thermophilic and moderately thermophilic proteins. These are common protein functions which are not dependent on temperature tolerance (Okanishi et al. 2017).

High temperature is also related to strong association between ions. The despairs for the ion pair decreases with an increase in temperatures. In thermostable proteins, internal water molecules (bridging water molecules) are present which ensures that ions are not fully desolated at high temperature. Hence it was established that salt bridges are very important in the design of thermostable proteins (Bikkina et al. 2017). Upregulation of proteins which are responsible for protection against heat stress, i.e., heat shock proteins (*HSPs*), is common in all thermophiles (Mizobata et al. 2000), i.e., upregulation of *HSP60, HSP20, groEL-growES, hrcA-grpE-dnaJ*, and *dnaK-sHSP* in *Thermotoga maritima* and *sHSPs* in *Sulfolobus solfataricus* during temperature rise (Shockley et al. 2003; Johnson et al. 2006). Other chaperon proteins which are upregulated in thermophilic strains *Thermoanaerobacter tengcongensis* and *Thermotoga maritime* in response to heat stress include *GroEL, GroES, DnaK*, and *GrpE* (Chen et al. 2012a, b; Wang et al. 2012).

1.7 Types of Thermal Environments

Various types of thermal environments exist including terrestrial, solar-heated, marine, subsurface, anthropogenic, temporary, and mesobiotic environments; a brief overview of these environments was reviewed below.

1.7.1 Terrestrial Thermal Environments

The northern areas of Pakistan have a large number of hot springs in the Gilgit, Hunza, and Yasin valleys. The Tattapani hot springs are located on Karakoram Highway at the right bank of Indus River. These springs are located at the altitude of 1200 m. There are two hot springs in Murtazaabad, located in the Hunza valley, downward near the bank of the Khunjerab River: Murtazaabad Zairen and Murtazaabad Balai hot springs. Murtazaabad Balai hot spring is located somewhat upper as compared to the Murtazaabad Zairen hot spring. Other hot springs are located 3.0 km earlier from Darkot Pass in Yasin valley upper to the Rawat base camp. It is situated at the height of about 4650 m from the sea level. Two hot springs are oozing out here, which seem to have the same origin (Ahmad et al. 2013). Shuja (1986) and Bakht (2000) have also found numerous hot springs along the Main Mantle Thrust and Main Karakoram Thrust in Chilas and Hunza areas, respectively. The geothermal system here is the result of the collision of the Indian and Eurasian plates. Hot springs are scattered and their temperature ranges up to 91 °C. Three parts of Pakistan, i.e., Kashmir, Khyber Pakhtunkhwa, and Baluchistan, are the potential zones where geothermal resources are located. Major tectonic elements during the Cenozoic and Mesozoic era have shaped the geological structures that are observed in Pakistan today. These structural elements are indicators for delineating and developing the potential geothermal resources of the country. Worldwide these environments are found at geysers, solfataras (mud or paint pots), and mud or paint pots in volcanically active regions throughout the world, including Iceland,

Western North America, New Zealand, Japan, Eastern Russia, and the rest of the so-called Pacific Ring of Fire; major examples include Yellowstone National Park at North America which is being studied dating back to 1897 (Reysenbach and Shock 2002). Neutral to alkaline areas richer in chloride salts or carbonate were observed in areas of terrestrial environments (Zhao et al. 2005).

1.7.2 Solar-Heated Environments

Solar-heated environments may occur anywhere on earth receiving solar energy inputs. Such environments are likely inhabited by mesophilic, thermotolerant, and thermophilic microorganisms because solar energy can heat some soils to 60 °C and shallow waters to 50 °C at certain times of the day or year, as pointed out by Brock (2012). Thermal environments on the earth's surface also experience evaporation, and thus many environments have elevated salinity and, therefore, halophilic inhabitants. For example, *Thermohalobacter berrensis*, a thermophilic and halophilic bacterium, was isolated from a solar slattern in France (Cayol et al. 2000). Halo-alkali-thermophiles and halophilic (up to 25% NaCl 4.5 M sodium ion as NaCl/Na_2CO_3), thermophilic (up to 75 °C), and alkaliphilic (up to pH 10.5) triple extremophiles, coined, have been isolated from dry salts from salt flats in Nevada and from sediments of athalassohaline lakes in Wadi El Natrun, Egypt (Mesbah and Wiegel 2005).

1.7.3 Marine Environments

Marine thermal environments may occur at Beaches: Hot Water Beach (Whitianga, New Zealand), Pozzuoli (Italy), Savusavu (Fiji Island) (Burgess 2009). Under 8 m of water: Vents off the coast of Mílos Island, Greece (Sievert et al. 2000). Under 2500 m: Abysmal of water, deep-sea hydrothermal vents first discovered in 1977 near the Galápagos Islands (Corliss et al. 1979).

Organisms inhabiting such environments face multiple challenges, i.e., venting water can exceed 300 °C, but in deep-sea vents, it cools quickly upon mixing with cold, deep-sea water; habitat types range from those preferred by hyperthermophiles to temperatures habitable by psychrophiles (Kelley et al. 2002), i.e., black smoker chimneys, associated with volcanic psychrophiles activity; and plate spreading zones generally are fueled by high concentrations of sulfides (Kelley et al. 2002). Serpentinite-hosted systems, like the Lost City hydrothermal field, are enriched in hydrogen and methane as energy sources (Kelley et al. 2005). *Thermococcus barophilus*, obtained from the snake pit region of the Mid-Atlantic Ridge, requires elevated pressure for growth at or above 95 °C (Marteinsson et al. 1999). *Pyrococcus* strain ES4 shows an extension of T_{max} under increased pressure (Pledger et al. 1994; Summit et al. 1998).

Jolivet et al. (2003, 2004) reported that at hydrothermal vents, the level of natural radioactivity can be 100 times greater than that at the earth's surface because of increased occurrence of elements such as Pb, Po, and Rn. For example, archaea *Thermococcus gammatolerans* was isolated from a hydrothermal site in Guaymas Basin, *Thermococcus marinus* from the snake pit hydrothermal site on the Mid-Atlantic Ridge, and *Thermococcus radiotolerans* from a hydrothermal site in the Guaymas Basin. Additionally, all organisms existing in marine environments also have some tolerance for moderate (around 3%) salinity.

1.7.4 Subsurface Environment

Subsurface thermal environments include petroleum reservoirs and geothermally heated lakes and aquifers. Activity in subsurface environments varies with the availability of nutrients, water, energy, depth, surrounding matrix, and source materials. Lethal temperatures may not occur until as much as 10,000 m below the surface (Pedersen 2000) with few exceptions, e.g., Uzon Caldera; temperatures well above 100 °C can occur at depths of only a few meters (Burgess 2009). A depth record for culturable life has been established at 5278 m (Szewzyk et al. 1994). Elevated temperatures found within petroleum reservoirs can be up to 130 °C (Grassia et al. 1996). The geochemical conditions in reservoirs are variable because of age, source material, and surrounding geology and prokaryote communities (Orphan et al. 2003). Takahata et al. (2000) have proposed that microorganisms in these environments may face oligotrophic conditions. Subsurface geothermal aquifers such as the well- known and expensive Great Artesian Basin of Australia are non-volcanically heated but experience temperatures up to nearly 103 °C.

1.7.5 Anthropogenic Environments

Anthropogenic habitats include household and water heaters and industrial process environments and thermal effluent from power plants (Brock 2012). One of the earliest well-known anaerobic thermophiles, *Thermoanaerobacter* (*basonym Clostridium*) *thermohydrosulfuricus*, was isolated from an Austrian sugar factory (Lee et al. 1993). Other thermophiles have been isolated from thermally polluted effluent from a carpet factory (Carreto et al. 1996), the smoldering slag heap of a uranium mine (Fuchs et al. 1996), and mushroom compost (Korn-Wendisch et al. 1995). Strains of *Thermus aquaticus* have been isolated from various anthropogenic thermal environments including hot tap water and greenhouse soil (Brock and Freeze 1969).

1.7.6 Temporary Environments and Mesobiotic Environments

Thermophiles can be isolated from various environments, such as animal droppings, manure piles, and compost, temporarily heated by biodegradation of organic material, sun-heated soils, and sediments at the edges of lakes and puddles which can have temperatures up to 50 °C but are frequently around 35–45 °C, whereas most of the thermophiles isolated from these environments are *Firmicutes*. One example is the archaeon *Methanothermobacter thermautotrophicus*. This species can be easily isolated from sun-heated black sediments of lakes and mesobiotic sewage plants, but it also has been isolated from sun-heated wood stumps in Georgia, United States (Luo et al. 2013), and mesobiotic environments such as cold stream sediments in Germany (Wiegel et al. 1981) or sediments of Lake Mendota, Wisconsin, for which temperatures have never reached 16 °C. In contrast, thermophiles, living in steady thermal environments such as thermal spring and sediments even if substrate concentration is low, do not have that selection pressure for very rapid growth as long as their residence time in the pool is longer than their doubling time (Fig. 1.4).

1.8 Diversity of Thermophiles

1.8.1 Cultural Diversity

Most of the microorganisms from nearly all environments inhabit are presently uncultured (Hugenholtz 2002). Considering the extreme conditions in which most thermophiles thrive, some require special handling or novel approaches for their enrichment, culturing, and isolation (Mesbah and Wiegel 2005). During our study on thermal springs of Pakistan, based on the 16S rRNA gene sequence similarities, we observed that 248 isolated strains belonged to 37 genera and 3 major phyla which were *Proteobacteria*, *Firmicutes*, and *Actinobacteria*. Of the potentially novel species of *Actinobacteria* and *Bacteria*, two were also characterized by polyphasic taxonomy. These strains were characterized as novel species of the genera *Nocardioides* and *Streptomyces* (Amin et al. 2016, 2017a).

Until now many species of thermophilic anaerobic bacteria have been isolated and described, and few anaerobic bacteria isolated and described include *Thermoanaerobacter tengcongensis* sp. nov. (Xue et al. 2001), *Chlorobium tepidium* sp. nov. (Wahlund et al. 1991), *Pyrobaculum igneiluti* sp. nov. (Lee et al. 2017), and *Desulfuribacillus stibiiarsenatis* sp. nov. (Abin and Hollibaugh 2017) (others are mentioned in Table 1.2). Their habitats include geothermal areas (Wiegel and Ljungdahl 1981) (Jessen and Orlygsson 2012) and deep-sea vents (Slobodkin et al. 1999). Low-oxygen concentrations are usually present in the habitats of anaerobes; hence most known thermophilic species are obligate or facultative anaerobes (Amend and Shock 2001).

1.8.2 Phylogenetic and Genetic Diversity

Amplification of 16S rRNA genes directly from environmental DNA has shown intense variation in amount of diversity among prokaryotes and novel lineages of thermophilic bacteria and archaea (Kimura et al. 2005; Burgess 2009; Burgess et al. 2007). Sequences from deep-sea hydrothermal vents led to the identification of

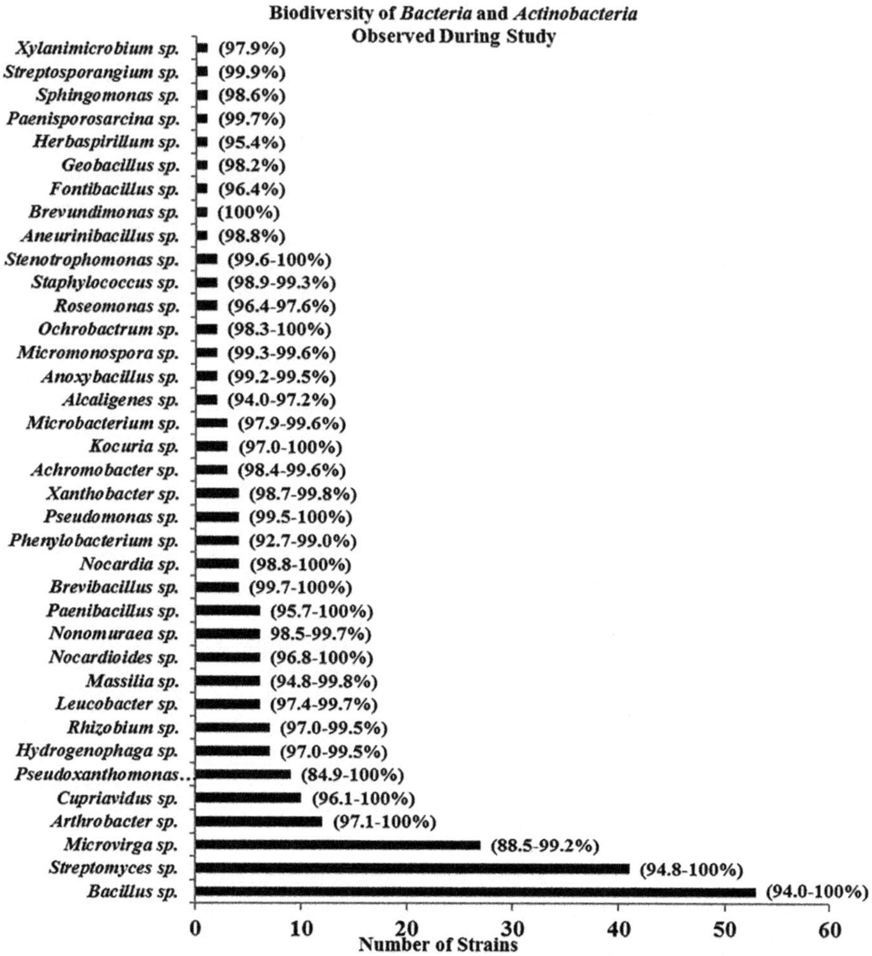

Fig. 1.4 Diversity of bacteria and *Actinobacteria* in hot water springs of Pakistan. Percentage in brackets is showing the 16S rRNA gene sequence similarity

novel lineages among archaea and bacteria (Reysenbach et al. 2000), but this is not an absolute fact because some thermal environment communities may contain only a few phylogenetic types; e.g., Reysenbach and Shock (2002) identified only three major phylogenetic groups out of 35 clones analyzed during a study on Yellowstone National Park. During a study on Pakistan hot water springs by unculturable technique, major phyla observed were *Proteobacteria*, *Chloroflexi*, *Thermotoga*, *Bacteriodetes*, *Deinococcus-Thermus*, *Nitrospirae*, and *Acidobacteria*, and other well-reported thermal spring phyla which are still unclassified were UT06, OP11, BRC1, OD1, OP8, OP1, OP3, OP9, OMAN, and NKB19 (Fig. 1.5). Presence of 40.1% unclassified OTUs clearly suggest the presence of many undiscovered and unexplored unique microbiota within these sites (Amin et al. 2017b).

Table 1.2 Few of the new species of thermophilic and mesophilic, aerobic and anaerobic bacteria isolated from hot water springs and identified by polyphasic taxonomic approach from 2016 till date

Isolated strains	Temperature	References
Lampropedia cohaerens sp. nov.	45	Tripathi et al. (2016)
Bacillus licheniformis RBS 5 sp. nov.	65	Salem et al. (2016)
Caldimicrobium thiodismutans sp. nov.	75	Kojima et al. (2016)
Inmirania thermothiophila gen. nov.	35–68	Slobodkina et al. (2016)
Tepidibacillus decaturensis sp. nov.	20–60	Dong et al. (2016)
Chelatococcus thermostellatus sp. nov.	50	Ibrahim et al. (2016)
Deferrisoma palaeochoriense sp. nov.	30–70	Pérez-Rodríguez et al. (2016)
Streptomyces sp. Al-Dhabi-1 sp. nov.	55	Al-Dhabi et al. (2016)
Thermostilla marina gen. nov., sp. nov.	30–68	Slobodkina et al. (2016)
Brevibacillus gelatini sp. nov.	45	Inan et al. (2016)
Cyanobacterial strains	26–58	Bravakos et al. (2016)
Athalassotoga saccharophila gen. nov., sp. nov. *Mesoaciditogales* ord. nov. *Mesoaciditogaceae* fam. nov.	30–60	Itoh et al. (2016)
Brevibacillus borstelensis cifa_chp40	37–50	Tripathy et al. (2016b)
Brevibacillus sediminis sp. nov.	50–55	Xian et al. (2016)
Sulfuritortus calidifontis gen. nov., sp. nov.	15–48	Kojima et al. (2017)
Nocardioides pakistanensis sp. nov.	20–40	Amin et al. (2016)
Caldimicrobium thiodismutans sp. nov.	40–77	Kojima et al. (2016)
Streptomyces caldifontis sp. nov.	18–40	Amin et al. (2017a)
Pyrobaculum igneiluti sp. nov.	90	Lee et al. (2017)
Desulfuribacillus stibiiarsenatis sp. nov.	37	Abin and Hollibaugh (2017)
Tibeticola sediminis gen. nov., sp. nov.	37–43	Khan et al. (2017)
Balneicella halophila gen. nov., sp. nov.	20–50	Fadhlaoui et al. (2016)
Thermoanaerobacterium butyriciformans sp. nov.	50–55	López et al. (2017)

While limited cultivation-based study of the geothermal springs in this region has been reported (Javed et al. 2012), a cultivation-independent study which provides a more comprehensive assessment of microbial diversity was still lacking before our studies. Great plate count anomaly illustrates that less than 1% of existing microorganisms are culturable. Under such conditions, culture-independent approaches facilitate the exploration of microbial diversity from diverse habitats (Hou et al. 2013). In Pakistan thermal springs, higher species richness and abundance in sediments of Tattapani than in sediments of Tato field and Murtazaabad were reported to be due to moderate temperature, high silicates, and high sulfate contents of Tattapani springs. Lau et al. (2009) and Yim et al. (2006) also reported the influence of temperature (<70 °C) for the presence of *Cyanobacteria* and *Chloroflexi* in hot springs.

Thermophilic and hyperthermophilic bacteria had been predominantly isolated from streamers with temperature above 75 °C and mainly comprised of the phyla *Aquificae*, *Deinococcus-Thermus*, *Thermodesulfobacteria*, and *Thermotogae* and

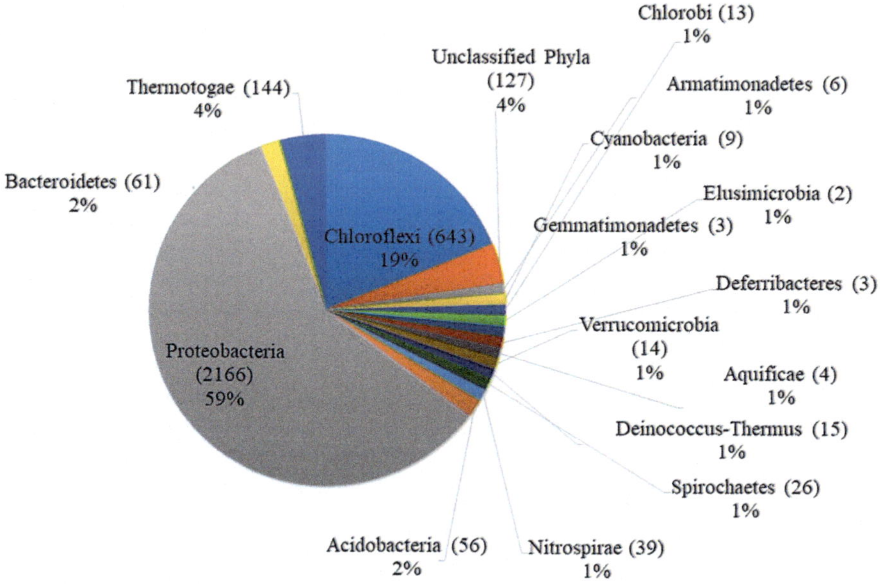

Fig. 1.5 Pie chart showing distribution of bacterial phyla and *Actinobacteria* in hot water springs of Pakistan by pyrosequencing approach of 16S rRNA gene. Values in brackets are showing an average number of OTUs (operational taxonomic units). Unclassified OTUs belong to UT06, Caldithrix_p, BRC1, OP8, OP11, MATCR, OMAN, Bacteria_uc, NKB19, JX105615_p, WS1, and O

some members of the phyla *Proteobacteria* and *Firmicutes* (Miller et al. 2009; Wang et al. 2013). Murtazaabad hot springs with relatively higher temperature (90–95 °C) favored the growth of thermophilic bacterial phylum *Thermotogae*. These domains were also detected in higher proportion in sites of Tata field and Tattapani, where average temperature is above 85 °C. However, OTUs of phyla *Aquificae* and *Deinococcus-Thermus* were more dominant in sites ranging in temperature from 70 to 85 °C. At sites with low silica and high temperature, OTUs belonging to phylum *Chloroflexi* were dominant. Kambura et al. (2016) believed that the existence of phyla A*ctinobacteria* and *Firmicutes* was an adaptation in low-nutrient conditions of the hot springs.

1.8.3 Metabolic Diversity

Among all types of microbial metabolism from thermal environments, chemolithotrophy either autotrophy or heterotrophy is a foundation of hyperthermophilic communities in sunless and too hot environments which are not suitable for photoautotrophic production. Few chemolithoautotrophs, e.g., bacteria of the order *Aquificales*, are considered as primary producers in these ecosystems (Blank et al. 2002). Among bacteria are anaerobic *Firmicutes* such as the facultative chemolithoautotrophs *Moorella thermoacetica* that undergoes homoacetogenic fermentations from carbohydrates and the anaerobe *Ammonifex degensii*, capable of forming ammonium from nitrate via chemolithoautotrophic growth (Huber et al. 1996), Fe(III)-reducing *Thermolithobacter ferrireducens*, and *Thermolithobacter carboxydivorans*, a hydrogenic CO utilizer (Wiegel et al. 2003). Photoheterotrophic *Chloroflexus aggregans*, *Chloroflexus aurantiacus*, *Heliobacterium modesticaldum*, and *Roseiflexus castenholzii* (Hanada et al. 2002). In some examples, in situ geochemistry of thermal environments may be shaping the dominant metabolisms or perhaps is shaped by the dominant metabolisms (Orphan et al. 2003). Many thermal environments are enriched in elements that are toxic to humans, such as arsenic and selenium, and some microorganisms in these habitats use toxic, redox-active elements to gain energy, via either oxidation or reduction (Donahoe-Christiansen et al. 2004).

During a study by Amin et al. (2017b), Pakistan thermal springs were explored for bacterial diversity and it was reported that among Murtazaabad hot water spring, sulfur-reducing bacteria was extensively present in deep waters and Physiological functions revealed that in sulpur rich grothermal springs with anoxic waters, methane is produced by consortium of methanotophs and sulfur reducing bacteria (Tripathy et al. 2016a, b; Delgado-Serrano et al. 2014). Type I and II methanotrophs and SRB were major constituents among phylum *Proteobacteria* and likely involved in the mineral recycling under the low-oxygen conditions of hot springs, which in turn helped in energy production. In acidic hot springs, this metabolism of energy recycling was reported to be initiated by methane-oxidizing phylum *Verrucomicrobia* (Islam et al. 2008; Sharp et al. 2012).

Few strains which are dependent on varied metabolic classes isolated from the studied thermal springs included phylum *Clostridia*, which is obligatorily dependent on methanogens or on the presence of an external electron acceptor; thermophilic anaerobic, Mn(IV)- and Fe(III) reducing *Carboxydocella* species *Carboxydocella_uc* and *Carboxydocella manganica*; and CO-assimilating chemolithoautotroph which survived under aerobic conditions by using CO dehydrogenases under anaerobic conditions. Genus *Ammoniphilus* were present in the three sites of Tattapani which were obligatory oxalotrophic and haloalkalitolerant bacteria and required a high concentration of ammonium ions and pH of 6.8–9.5. Unclassified species belonging to genera *Anaerosporobacter* and *Nitratireductor* were also detected which were in accordance to the study made by Stackebrandt (2014) who also reported oxidation of NO_2 to nitrate by *Nitrospira* at high temperature and subsequent reduction of nitrate to nitrous oxide or complete oxidation to N_2 by members of the order *Thermales*, *Aquificales*, and *Bacillales* (Nakagawa and Fukui 2002).

1.8.4 Ecological Diversity

The thermophilic prokaryotes have introduced us to novel modes of life because of biological interactions in geothermally heated environments. The discovery of deep-sea hydrothermal vent communities demonstrated that life can exist at temperatures 100 °C as well as at 2 °C on the basis of associated microbial vent community (Corliss et al. 1979). Novel symbioses between eukaryotes and prokaryotes have been identified at deep-sea vents, such as the association between the tube worm *Riftia pachyptila* and chemosynthetic, sulfur-oxidizing bacteria (Cavanaugh et al. 1981) or the thermotolerant Pompeii worm, which utilizes eurythermal enzymes of a community of prokaryotes living on its back (Chevaldonné et al. 2000). Fledgling field of microecology is rapidly expanding, and thermal environments are exemplary systems for it (Magurran 2013). In hot water springs, the diversity of microorganisms within mats of *Cyanobacteria* has been examined, and the importance of a prokaryote species is determined based on its role in their environments (Ward et al. 1998) and the effect of temperature on structuring community of prokaryotes through genetic parameters and the distribution of different metabolic types (Norris et al. 2002). FISH is also a helpful method that enables us to examine the structural distribution of microorganism of known phylogenetic affiliations (Nübel et al. 2002). Lipids present within the membranes of prokaryotes can be diagnostic for various types of microorganisms and have provided insight into the distribution of microorganisms among different environments. For example, analysis of glycerol dialkyl glycerol tetraethers (GDGTs) from selected hot springs in Nevada exposed the presence of the archaeal lipid crenarchaeol, which was believed to be present in low-temperature and marine environments. The second evidence came from the presence of DGGE band sequences of 16S rRNA genes from these springs which were related to thermophilic *Crenarchaeota* and confirmed that the presence of crenarchaeol is not exclusive to the cold-adapted, marine branch of the *Crenarchaeota* (Pearson et al. 2004).

1.9 Conclusion

In Pakistan to date, not a single study has been reported for bacterial diversity of hot water springs except studies by our group. Various other studies in which selective bacteria were isolated from hot water springs of Pakistan include isolation of strain *Ralstonia* sp. MRL-TL from hot water spring to check its ability to degrade poly(ε-caprolactone) (PCL) (Shah et al. 2015), Analysis of power generation from geothermal resources (Ahsan Mustaqeem et al. 2015), Euthermal hot water spring Mango Pir was studied for physicochemical and biological studies and Cyanophyta, Zooplankton, Bacillarophyta and Nematoda were isolated (Jahangir et al. 2001), freshwater spring was studied from Kohat, Pakistan, and the quality assessments of the drinking water were carried out by determining total coliform bacteria, total plate count, total fecal coliform and *E. coli* (Ahmad et al. 2013). Another study from Pakistan strain AK9 was isolated from hot water spring of Tattapani Azad Kashmir, Pakistan; cellulase enzyme was extracted and purified which reserved its activity from 50 to 70 °C and 3–7pH. They reported that *B. amyloliquefaciens* AK9 can be used in bioconversion of lignocellulosic biomass to fermentable sugar (Irfan et al. 2017).

Further we suggest that in the future we should focus to unravel all ignored geochemically important sites specially to fill gap of limited culturing techniques and substrates so far known to study these non-cultured bacteria. The taxonomic results obtained will provide information for exploration in this regime and thus to discover new whole area for further research in the subject of "extremophiles" which will lead to studies on controlled experiments which would help identify which factors influence species distribution at most. For example, Kumar et al. (2015), Boucher et al. (2011), and Porter et al. (2017) reported that core genomes of *Rhizobium leguminosarum* and *Vibrio cholerae* have high similarity, but the accessory genome is much varied (Porter et al. 2017). Specific classes of bacteria can be studied along with their biogeochemical cycles, i.e., methane cycle-related microorganisms, anaerobes, nitrogen cycle-related bacteria, sulfate-reducing bacteria, and archaea. Novel enzymes can be extracted from novel genomes by metagenomic studies and whole-genome analysis. Pasternak et al. (2013) stated that that impact of bacterial diversity and their abundance on nature can be explained only by using full-genome proteomic comparisons.

We have tried to highlight that microbial community composition varies with change in biogeography, biogeochemistry, temperature, pH, physicochemical parameters, and many other factors. Globally our results highlight the Pakistan thermal springs, in part the effects of changes in bacterial population in specific set of conditions, and particularly observed the microbial community differences, novel microbiota, and the need to further investigate them to cover all metabolic and genome-wide aspects. It will help researchers to extend this research for narrowing down habitat in genomic interconnected populations, discrete, particular bacterial lineages can be found in contrasting soil types, in case if genomic interconnections will be low, variations in the core and accessory genomes could be found to solve distribution of distinct biogeographical patterns.

References

Abin CA, Hollibaugh JT (2017) *Desulfuribacillus stibiiarsenatis* sp. nov., an obligately anaerobic, dissimilatory antimonate- and arsenate-reducing bacterium isolated from anoxic sediments, and emended description of the genus *Desulfuribacillus*. Int J Syst Evol Microbiol 67:1011–1017

Ahmad S, Ali A, Ullah I, Naz N, Ali A, Ali N (2013) Bacteriological and biochemical evaluation of the spring's water of district Buner Khyber Pakhtunkhwa Pakistan. Int J Adv Res Technol 2:452–460

Ahmad M, Tasneem M, Rafique M, Akram W, Iqbal N, Rafiq M, Fazil M (2015) Identification of recharge sources and dating of groundwater using isotope and CFC techniques. In: Proceedings of the first international conference on environmentally sustainable development. Islamabad, Pakistan. pp 1125–1135

Ahsan Mustaqeem S, Qureshi SR, Waqar A (2015) Exergetic and energetic analysis of power generation from geothermal resources in Pakistan. In: Proceedings of the World Geothermal Congress, Melbourne, Australia. pp 1–11

Al-Dhabi NA, Esmail GA, Duraipandiyan V, Arasu MV, Salem-Bekhit MM (2016) Isolation, identification and screening of antimicrobial thermophilic Streptomyces sp. Al-Dhabi-1 isolated from Tharban hot spring, Saudi Arabia. Extremophiles 20(1):79–90

Amend JP, Shock EL (2001) Energetics of overall metabolic reactions of thermophilic and hyperthermophilic Archaea and Bacteria. FEMS Microbiol Rev 25:175–243

Amin A, Ahmed I, Habib N, Abbas S, Xiao M, Hozzein WN, Li W-J (2016) *Nocardioides pakistanensis* sp. nov., isolated from a hot water spring of Tatta Pani in Pakistan. Antonie Van Leeuwenhoek 109:1101–1109

Amin A, Ahmed I, Khalid N, Osman G, Khan IU, Xiao M, Li W-J (2017a) *Streptomyces caldifontis* sp. nov., isolated from a hot water spring of Tatta Pani, Kotli, Pakistan. Antonie Van Leeuwenhoek 110:77–86

Amin A, Ahmed I, Salam N, Kim B-Y, Singh D, Zhi X-Y, Xiao M, Li W-J (2017b) Diversity and distribution of thermophilic bacteria in hot springs of Pakistan. Microbial Ecol 74:116–127

Arnórsdóttir J, Sigtryggsdóttir ÁR, Thorbjarnardóttir SH, Kristjánsson MM (2009) Effect of proline substitutions on stability and kinetic properties of a cold adapted subtilase. J Biochem 145:325–329

Bakht MS (2000) An overview of geothermal resources of Pakistan. In: Proceedings of the World Geothermal Congress, Kyushu, Japan. pp 77–83

Beijerinck M (1913) Jaarboek van de Koninklijke Akademie v. Wetenschoppen Muller, Amsterdam, The Netherlands. In Dutch

Bezsudnova EY, Boyko KM, Polyakov KM, Dorovatovskiy PV, Stekhanova TN, Gumerov VM, Ravin NV, Skryabin KG, Kovalchuk MV, Popov VO (2012) Structural insight into the molecular basis of polyextremophilicity of short-chain alcohol dehydrogenase from the hyperthermophilic archaeon *Thermococcus sibiricus*. Biochimie 94:2628–2638

Bikkina S, Bhati AP, Padhi S, Priyakumar UD (2017) Temperature dependence of the stability of ion pair interactions, and its implications on the thermostability of proteins from thermophiles. J Chem Sci 129:405–414

Blank CE, Cady SL, Pace NR (2002) Microbial composition of near-boiling silica-depositing thermal springs throughout Yellowstone National Park. Appl Environ Microbiol 68:5123–5135

Boucher Y, Cordero OX, Takemura A, Hunt DE, Schliep K, Bapteste E, Lopez P, Tarr CL, Polz MF (2011) Local mobile gene pools rapidly cross species boundaries to create endemicity within global Vibrio cholerae populations. MBio 2:e00335–e00310

Bravakos P, Kotoulas G, Skaraki K, Pantazidou A, Economou-Amilli A (2016) A polyphasic taxonomic approach in isolated strains of Cyanobacteria from thermal springs of Greece. Mol Phylogenet Evol 98:147–160

Brock TD (1997) The value of basic research: discovery of *Thermus aquaticus* and other extreme thermophiles. Genetics 146:1207–1210

Brock TD (2012) Thermophilic microorganisms and life at high temperatures. Springer, New York

Brock TD, Freeze H (1969) *Thermus aquaticus* gen. n. and sp. n., a nonsporulating extreme thermophile. J Bacteriol 98:289–297

Burgess EA (2009) Geomicrobiological description of two contemporary hydrothermal pools in Uzon Caldera, Kamchatka, Russia, as models for sulfur biogeochemistry. PhD Thesis, University of Georgia, USA

Burgess EA, Wagner ID, Wiegel J (2007) Thermal environments and biodiversity. In: Physiology and biochemistry of extremophiles. American Society of Microbiology, Washington, DC

Byrer DE, Rainey FA, Wiegel J (2000) Novel strains of *Moorella thermoacetica* form unusually heat-resistant spores. Arch Microbiol 174:334–339

Carreto L, Moore E, Nobre MF, Wait R, Riley PW, Sharp RJ, DA COSTA MS (1996) *Rubrobacter xylanophilus* sp. nov., a new thermophilic species isolated from a thermally polluted effluent. Int J Syst Evol Microbiol 46:460–465

Cavanaugh CM, Gardiner SL, Jones ML, Jannasch HW, Waterbury JB (1981) Prokaryotic cells in the hydrothermal vent tube worm Riftia pachyptila Jones: possible chemoautotrophic symbionts. Science 213:340–342

Cavicchioli R (2002) Extremophiles and the search for extraterrestrial life. Astrobiology 2:281–292

Cayol J-L, Ducerf S, Patel B, Garcia J-L, Thomas P, Ollivier B (2000) *Thermohalobacter berrensis* gen. nov., sp. nov., a thermophilic, strictly halophilic bacterium from a solar saltern. Int J Syst Evol Microbiol 50:559–564

Chakravorty D, Faheem Khan M, Patra S (2017) Thermostability of proteins revisited through machine learning methodologies: from nucleotide sequence to structure. Curr Biotechnol 6:39–49
Chang MS, Gachal GS, Qadri AH, Sheikh MY (2013) Ecological impacts on the status of Marsh Crocodiles in Manghopir Karachi. Int J Adv Res 1:42–46
Chen W, Wang BJ, Hong HZ, Yang H, Liu SJ (2012a) *Deinococcus reticulitermitis* sp. nov., isolated from a termite gut. Int J Syst Evol Microbiol 62:78–83
Chen Z, Wang Q, Lin L, Tang Q, Edwards JL, Li S, Liu S (2012b) Comparative evaluation of two isobaric labeling tags, DiART and iTRAQ. Anal Chem 84:2908–2915
Chevaldonné P, Fisher C, Childress J, Desbruyères D, Jollivet D, Zal F, Toulmond A (2000) Thermotolerance and the 'Pompeii worms'. Mar Ecol Prog Ser 208:293–295
Corliss JB, Dymond J, Gordon LI, Edmond JM (1979) On the Galapagos Rift. Science 203:1073–1083
Cowan D, Ramond J, Makhalanyane T, De Maayer P (2015) Metagenomics of extreme environments. Curr Opin Microbiol 25:97–102
Cybulski LE, Ballering J, Moussatova A, Inda ME, Vazquez DB, Wassenaar TA, de Mendoza D, Tieleman DP, Killian JA (2015) Activation of the bacterial thermosensor DesK involves a serine zipper dimerization motif that is modulated by bilayer thickness. Proc Natl Acad Sci 112:6353–6358
Delaye L, Becerra A, Lazcano A (2005) The last common ancestor: what's in a name? Orig Life Evol Biosph 35:537–554
Delgado-Serrano L, López G, Bohorquez LC, Bustos JR, Rubiano C, Osorio-Forero C, Junca H, Baena S, Zambrano MM (2014) Neotropical Andes hot springs harbor diverse and distinct planktonic microbial communities. FEMS Microbiol Ecol 89:56–66
Donahoe-Christiansen J, D'Imperio S, Jackson CR, Inskeep WP, McDermott TR (2004) Arsenite-oxidizing Hydrogenobaculum strain isolated from an acid-sulfate-chloride geothermal spring in Yellowstone National Park. Appl Environ Microbiol 70:1865–1868
Dong Y, Sanford RA, Boyanov MI, Kemner KM, Flynn TM, O'Loughlin EJ, Locke RA II, Weber JR, Egan SM, Fouke BW (2016) Tepidibacillus decaturensis sp. nov.: a microaerophilic, moderately thermophilic iron-reducing bacterium isolated from a depth of 1.7 km in the Illinois Basin, USA. Int J Syst Evol Microbiol 66(10):3964–3971
Dworkin JP, Lazcano A, Miller SL (2002) The roads to and from the RNA world. J Theor Biol 222:127–134
Engle M, Li Y, Woese C, Wiegel J (1995) Isolation and characterization of a novel Alkalitolerant Thermophile, *Anaerobranca horikoshii* gen. nov., sp. nov. Int J Syst Evol Microbiol 45:454–461
Fadhlaoui K, Hania WB, Postec A, Hamdi M, Ollivier B, Fardeau M-L (2016) Balneicella halophila gen. nov., sp. nov., an anaerobic bacterium, isolated from a thermal spring and description of Balneicellaceae fam. nov. Int J Syst Evol Microbiol 66(11):4692–4696
Forterre P, Brochier C, Philippe H (2002) Evolution of the Archaea. Theor Popul Biol 61:409–422
Fuchs T, Huber H, Teiner K, Burggraf S, Stetter K (1996) *Metallosphaera prunae*, sp. nov., a novel metal-mobilizing, thermoacidophilic Archaeum, isolated from a uranium mine in Germany. Syst Appl Microbiol 18:560–566
Gerday C, Glansdorff N (2007) Physiology and biochemistry of extremophiles. ASM Press, Washington, DC
Gondal IA, Masood SA, Amjad M (2017) Review of geothermal energy development efforts in Pakistan and way forward. Renew Sust Energ Rev 71:687–696
Grassia GS, McLean KM, Glénat P, Bauld J, Sheehy AJ (1996) A systematic survey for thermophilic fermentative bacteria and archaea in high temperature petroleum reservoirs. FEMS Microbiol Ecol 21:47–58
Gulen B, Petrov AS, Okafor CD, Vander Wood D, O'Neill EB, Hud NV, Williams LD (2016) Ribosomal small subunit domains radiate from a central core. Sci Rep 6:20885
Hanada S, Takaichi S, Matsuura K, Nakamura K (2002) *Roseiflexus castenholzii* gen. nov., sp. nov., a thermophilic, filamentous, photosynthetic bacterium that lacks chlorosomes. Int J Syst Evol Microbiol 52:187–193

Hou W, Wang S, Dong H, Jiang H, Briggs BR, Peacock JP, Huang Q, Huang L, Wu G, Zhi X (2013) A comprehensive census of microbial diversity in hot springs of Tengchong, Yunnan Province China using 16S rRNA gene pyrosequencing. PLoS One 8:e53350

Hu Z, Xu C, McDowell NG, Johnson DJ, Wang M, Luo Y, Zhou X, Huang Z (2017) Linking microbial community composition to C loss rates during wood decomposition. Soil Biol Biochem 104:108–116

Huber R, Rossnagel P, Woese C, Rachel R, Langworthy TA, Stetter KO (1996) Formation of ammonium from nitrate during chemolithoautotrophic growth of the extremely thermophilic bacterium *Ammonifex degensii* gen. nov. sp. nov. Syst Appl Microbiol 19:40–49

Hug LA, Baker BJ, Anantharaman K, Brown CT, Probst AJ, Castelle CJ, Butterfield CN, Hernsdorf AW, Amano Y, Ise K (2016) A new view of the tree of life. Nat Microbiol 1:16048

Hugenholtz P (2002) Exploring prokaryotic diversity in the genomic era. Genome Biol 3:REVIEWS0003

Ibrahim MH, Lebbe L, Willems A, Steinbüchel A (2016) Chelatococcusthermostellatus sp. nov., a new thermophile for bioplastic synthesis: comparative phylogenetic and physiological study. AMB Exp 6(1):1–9

Inan K, Ozer A, Guler HI, Belduz AO, Canakci S (2016) Brevibacillus gelatini sp. nov., isolated from a hot spring. Int J Syst Evol Microbiol 66(2):712–718

Islam T, Jensen S, Reigstad LJ, Larsen Ø, Birkeland N-K (2008) Methane oxidation at 55 C and pH 2 by a thermoacidophilic bacterium belonging to the Verrucomicrobia phylum. Proc Natl Acad Sci 105:300–304

Islas S, Velasco AM, Becerra A, Delaye L, Lazcano A (2003) Hyperthermophile and the origin and earliest evolution of life. Int Microbiol 6:87–94

Irfan M, Tayyab A, Hasan F, Khan S, Badshah M, Shah AA (2017) Production and characterization of organic solvent-tolerant cellulase from Bacillus amyloliquefaciens AK9 isolated from hot spring. Appl Biochem Biotechnol 182(4):1390–1402

Itoh T, Onishi M, Kato S, Iino T, Sakamoto M, Kudo T, Takashina T, Ohkuma M (2016) Athalassotoga saccharophila gen. nov., sp. nov., isolated from an acidic terrestrial hot spring, and proposal of Mesoaciditogales ord. nov. and Mesoaciditogaceae fam. nov. in the phylum Thermotogae. Int J Syst Evol Microbiol 66(2):1045–1051

Jahangir T, Khuhawar M, Leghari S, Leghari A (2001) Physico-chemical and biological study of Mangho Pir euthermal springs Karachi, Sindh, Pakistan. Online J Biol Sci 1:636–639

Javed M, Zahoor S, Sabar H, Haq I, Babar M (2012) Thermophilic bacteria from the hot springs of gilgit (Pakistan). J Anim Plant Sci 22:83–87

Jessen JE, Orlygsson J (2012) Production of ethanol from sugars and lignocellulosic biomass by Thermoanaerobacter J1 isolated from a hot spring in Iceland. Biomed Res Int 2012:86982

Jiang X, Takacs-Vesbach CD (2017) Microbial community analysis of pH 4 thermal springs in Yellowstone National Park. Extremophiles 21:135–152

Johnson M, Conners S, Montero C, Chou C, Shockley K, Kelly R (2006) The Thermotoga maritima phenotype is impacted by syntrophic interaction with *Methanococcus jannaschii* in hyperthermophilic coculture. Appl Environ Microbiol 72:811–818

Jolivet E, L'Haridon S, Corre E, Forterre P, Prieur D (2003) *Thermococcus gammatolerans* sp. nov., a hyperthermophilic archaeon from a deep-sea hydrothermal vent that resists ionizing radiation. Int J Syst Evol Microbiol 53:847–851

Jolivet E, Corre E, L'Haridon S, Forterre P, Prieur D (2004) *Thermococcus marinus* sp. nov. and *Thermococcus radiotolerans* sp. nov., two hyperthermophilic archaea from deep-sea hydrothermal vents that resist ionizing radiation. Extremophiles 8:219–227

Joyce GF (2002) The antiquity of RNA-based evolution. Nature 418:214–221

Kambura AK, Mwirichia RK, Kasili RW, Karanja EN, Makonde HM, Boga HI (2016) Bacteria and Archaea diversity within the hot springs of Lake Magadi and Little Magadi in Kenya. BMC Microbiol 16:136

Kashefi K, Lovley DR (2003) Extending the upper temperature limit for life. Science 301:934–934

Kelley DS, Baross JA, Delaney JR (2002) Volcanoes, fluids, and life at mid-ocean ridge spreading centers. Annu Rev Earth Planet Sci 30:385–491

Kelley DS, Karson JA, Früh-Green GL, Yoerger DR, Shank TM, Butterfield DA, Hayes JM, Schrenk MO, Olson EJ, Proskurowski G (2005) A serpentinite-hosted ecosystem: the Lost City hydrothermal field. Science 307:1428–1434

Kimura H, Sugihara M, Yamamoto H, Patel BK, Kato K, Hanada S (2005) Microbial community in a geothermal aquifer associated with the subsurface of the Great Artesian Basin, Australia. Extremophiles 9:407–414

Korn-Wendisch F, Rainey F, Kroppenstedt R, Kempf A, Majazza A, Kutzner H, Stackebrandt E (1995) *Thermocrispum* gen. nov., a new genus of the order Actinomycetales, and description of *Thermocrispum municipale* sp. nov. and *Thermocrispum agreste* sp. nov. Int J Syst Evol Microbiol 45:67–77

Kojima H, Umezawa K, Fukui M (2016) Caldimicrobium thiodismutans sp. nov., a sulfur-disproportionating bacterium isolated from a hot spring, and emended description of the genus Caldimicrobium. Int J Syst Evol Microbiol 66(4):1828–1831

Kojima H, Watanabe M, Fukui M (2017) Sulfuritortus calidifontis gen. nov., sp. nov., a novel sulfur oxidizer isolated from a hot spring microbial mat. Int J Syst Evol Microbiol 67(5):1355–1358

Khan IU, Hussain F, Tian Y, Habib N, Xian W-D, Jiang Z, Amin A, Yuan C-G, Zhou E-M, Zhi X-Y (2017) Tibeticola sediminis gen. nov., sp. nov., a thermophilic bacterium isolated from hot spring. Int J Syst Evol Microbiol 67(5):1133–1139

Kumar N, Lad G, Giuntini E, Kaye ME, Udomwong P, Shamsani NJ, Young JPW, Bailly X (2015) Bacterial genospecies that are not ecologically coherent: population genomics of *Rhizobium leguminosarum*. Open Biol 5:140133

Lau MC, Aitchison JC, Pointing SB (2009) Bacterial community composition in thermophilic microbial mats from five hot springs in Central Tibet. Extremophiles 13:139–149

Lazcano A, Bada JL (2003) The 1953 Stanley L. Miller experiment: fifty years of prebiotic organic chemistry. Orig Life Evol Biosph 33:235–242

Lee Y-E, Jain MK, Lee C, Zeikus JG (1993) Taxonomic distinction of saccharolytic thermophilic anaerobes: description of *Thermoanaerobacterium xylanolyticum* gen. nov., sp. nov., and *Thermoanaerobacterium saccharolyticum* gen. nov., sp. nov.; reclassification of *Thermoanaerobium brockii*, *Clostridium thermosulfurogenes*, and *Clostridium thermohydrosulfuricum* E100-69 as *Thermoanaerobacter brockii* comb. nov., *Thermoanaerobacterium thermosulfurigenes* comb. nov., and *Thermoanaerobacter thermohydrosulfuricus* comb. nov., respectively; and transfer of *Clostridium thermohydrosulfuricum* 39E to *Thermoanaerobacter ethanolicus*. Int J Syst Evol Microbiol 43:41–51

Lee JY, Iglesias B, Chu CE, Lawrence DJ, Crane EJ III (2017) *Pyrobaculum igneiluti* sp. nov., a novel anaerobic hyperthermophilic archaeon that reduces thiosulfate and ferric iron. Int J Syst Evol Microbiol 67:1714–1719

Levy M, Miller SL (1998) The stability of the RNA bases: implications for the origin of life. Proc Natl Acad Sci 95:7933–7938

López G, Cañas-Duarte S, Pinzón-Velasco A, Vega-Vela N, Rodríguez M, Restrepo S, Baena S (2017) Description of a new anaerobic thermophilic bacterium thermoanaerobacterium butyriciformans sp nov. Syst Appl Microbiol 40(2):86–91

Luo G, Wang W, Angelidaki I (2013) Anaerobic digestion for simultaneous sewage sludge treatment and CO biomethanation: process performance and microbial ecology. Environ Sci Technol 47:10685–10693

Magurran AE (2013) Measuring biological diversity. Wiley, Malden

Marteinsson VT, Birrien J-L, Reysenbach A-L, Vernet M, Marie D, Gambacorta A, Messner P, Sleytr UB, Prieur D (1999) *Thermococcus barophilus* sp. nov., a new barophilic and hyperthermophilic archaeon isolated under high hydrostatic pressure from a deep-sea hydrothermal vent. Int J Syst Evol Microbiol 49:351–359

Merkel AY, Pimenov NV, Rusanov II, Slobodkin AI, Slobodkina GB, Tarnovetckii IY, Frolov EN, Dubin AV, Perevalova AA, Bonch-Osmolovskaya EA (2017) Microbial diversity and autotrophic activity in Kamchatka hot springs. Extremophiles 21:307–317

Mesbah NM, Wiegel J (2005) Halophilic thermophiles: a novel group of extremophiles. Microbial diversity: current perspectives and potential applications. IK Publishing House, New Delhi, pp 91–118
Miller SL, Bada JL (1988) Submarine hot springs and the origin of life. Nature 334:609–611
Miller SR, Strong AL, Jones KL, Ungerer MC (2009) Bar-coded pyrosequencing reveals shared bacterial community properties along the temperature gradients of two alkaline hot springs in Yellowstone National Park. Appl Environ Microbiol 75:4565–4572
Mizobata T, Kagawa M, Murakoshi N, Kusaka E, Kameo K, Kawata Y, Nagai J (2000) Overproduction of Thermus sp. YS 8-13 manganese catalase in Escherichia coli. Eur J Biochem 267:4264–4271
Mrabet NT, Van den Broeck A, Stanssens P, Laroche Y, Lambeir A, Matthijssens G, Jenkins J, Chiadmi M, Van Tilbeurgh H (1992) Arginine residues as stabilizing elements in proteins. Biochemistry 31:2239–2253
Nakagawa T, Fukui M (2002) Phylogenetic characterization of microbial mats and streamers from a Japanese alkaline hot spring with a thermal gradient. J Gen Appl Microbiol 48:211–222
Norris TB, Wraith JM, Castenholz RW, McDermott TR (2002) Soil microbial community structure across a thermal gradient following a geothermal heating event. Appl Environ Microbiol 68:6300–6309
Nübel U, Bateson MM, Vandieken V, Wieland A, Kühl M, Ward DM (2002) Microscopic examination of distribution and phenotypic properties of phylogenetically diverse Chloroflexaceae-related bacteria in hot spring microbial mats. Appl Environ Microbiol 68:4593–4603
Okanishi H, Kim K, Fukui K, Yano T, Kuramitsu S, Masui R (2017) Proteome-wide identification of lysine succinylation in thermophilic and mesophilic bacteria. Biochim Biophys Acta 1865:232–242
Olsen GJ, Woese CR, Overbeek R (1994) The winds of (evolutionary) change: breathing new life into microbiology. J Bacteriol 176:1–6
Orphan V, Goffredi SK, Delong E, Boles J (2003) Geochemical influence on diversity and microbial processes in high temperature oil reservoirs. Geomicrobiol J 20:295–311
Pace NR (1991) Origin of life-facing up to the physical setting. Cell 65:531–533
Pace NR (1997) A molecular view of microbial diversity and the biosphere. Science 276:734–740
Pang J, Allemann RK (2007) Molecular dynamics simulation of thermal unfolding of *Thermotoga maritima* DHFR. Phys Chem Chem Phys 9:711–718
Pasternak Z, Pietrokovski S, Rotem O, Gophna U, Lurie-Weinberger MN, Jurkevitch E (2013) By their genes ye shall know them: genomic signatures of predatory bacteria. ISME J 7:756–769
Pearson A, Huang Z, Ingalls A, Romanek C, Wiegel J, Freeman K, Smittenberg R, Zhang C (2004) Nonmarine crenarchaeol in Nevada hot springs. Appl Environ Microbiol 70:5229–5237
Pedersen K (2000) Exploration of deep intraterrestrial microbial life: current perspectives. FEMS Microbiol Lett 185:9–16
Petrov AS, Gulen B, Norris AM, Kovacs NA, Bernier CR, Lanier KA, Fox GE, Harvey SC, Wartell RM, Hud NV (2015) History of the ribosome and the origin of translation. Proc Natl Acad Sci 112:15396–15401
Pérez-Rodríguez I, Rawls M, Coykendall DK, Foustoukos DI (2016) Deferrisoma palaeochoriense sp. nov., a thermophilic, iron (III)-reducing bacterium from a shallow-water hydrothermal vent in the Mediterranean Sea. Int J Syst Evol Microbiol 66(2):830–836
Pledger RJ, Crump BC, Baross JA (1994) A barophilic response by two hyperthermophilic, hydrothermal vent Archaea: an upward shift in the optimal temperature and acceleration of growth rate at supra-optimal temperatures by elevated pressure. FEMS Microbiol Ecol 14:233–241
Porter SS, Chang PL, Conow CA, Dunham JP, Friesen ML (2017) Association mapping reveals novel serpentine adaptation gene clusters in a population of symbiotic Mesorhizobium. ISME J 11:248–262
Ranawat P, Rawat S (2017) Stress response physiology of thermophiles. Arch Microbiol 199:391–414
Reysenbach A-L, Shock E (2002) Merging genomes with geochemistry in hydrothermal ecosystems. Science 296:1077–1082

Reysenbach A-L, Longnecker K, Kirshtein J (2000) Novel bacterial and archaeal lineages from an in situ growth chamber deployed at a Mid-Atlantic Ridge hydrothermal vent. Appl Environ Microbiol 66:3798–3806

Russell N, Harrisson P, Johnston I, Jaenicke R, Zuber M, Franks F, Wynn-Williams D (1990) Cold adaptation of microorganisms. Philos Trans R Soc Lond B 326:595–611

Salem RB, Abbassi MS, Cayol J-L, Bourouis A, Mahrouki S, Fardeau M-L, Belhadj O (2016) Thermophilic *Bacillus licheniformis* rbs 5 isolated from hot tunisian spring co-producing alkaline and thermostable [alpha]-amylase and protease enzymes. J Microbiol Biotechnol Food Sci 5(6):557

Sarker MR, Akhtar S, Torres JA, Paredes-Sabja D (2015) High hydrostatic pressure-induced inactivation of bacterial spores. Crit Rev Microbiol 41:18–26

Shah AA, Nawaz A, Kanwal L, Hasan F, Khan S, Badshah M (2015) Degradation of poly (ε-caprolactone) by a thermophilic bacterium Ralstonia sp. strain MRL-TL isolated from hot spring. Int Biodeterior Biodegrad 98:35–42

Sharp C, Stott M, Dunfield P (2012) Detection of autotrophic verrucomicrobial methanotrophs in a geothermal environment using stable isotope probing. Front Microbiol 3:303

Shin K-S, Shin YK, Yoon J-H, Park Y-H (2001) *Candida thermophila* sp. nov., a novel thermophilic yeast isolated from soil. Int J Syst Evol Microbiol 51:2167–2170

Shockley KR, Ward DE, Chhabra SR, Conners SB, Montero CI, Kelly RM (2003) Heat shock response by the hyperthermophilic archaeon *Pyrococcus furiosus*. Appl Environ Microbiol 69:2365–2371

Shuja TA (1986) Geothermal areas in Pakistan. Geothermics 15:719–723

Sievert SM, Kuever J, Muyzer G (2000) Identification of 16S ribosomal DNA-defined bacterial populations at a shallow submarine hydrothermal vent near Milos Island (Greece). Appl Environ Microbiol 66:3102–3109

Skirnisdottir S, Hreggvidsson GO, Hjorleifsdottir S, Marteinsson VT, Petursdottir SK, Holst O, Kristjansson JK (2000) Influence of sulfide and temperature on species composition and community structure of hot spring microbial mats. Appl Environ Microbiol 66:2835–2841

Slobodkin A, Tourova T, Kuznetsov B, Kostrikina N, Chernyh N, Bonch-Osmolovskaya E (1999) *Thermoanaerobacter siderophilus* sp. nov., a novel dissimilatory Fe (III)-reducing, anaerobic, thermophilic bacterium. Int J Syst Evol Microbiol 49:1471–1478

Slobodkina GB, Baslerov RV, Novikov AA, Viryasov MB, Bonch-Osmolovskaya EA, Slobodkin AI (2016) Inmirania thermothiophila gen. nov., sp. nov., a thermophilic, facultatively autotrophic, sulfur-oxidizing gammaproteobacterium isolated from a shallow-sea hydrothermal vent. Int J Syst Evol Microbiol 66(2):701–706

Stackebrandt E (2014) The family Gracilibacteraceae and transfer of the genus Lutispora into Gracilibacteraceae. In: The prokaryotes. Springer, Berlin, Heidelberg, pp 149–151

Strazzulli A, Iacono R, Giglio R, Moracci M, Cobucci-Ponzano B (2017) Metagenomics of hyperthermophilic environments: biodiversity and biotechnology. In: Microbial ecology of extreme environments. Springer, Cham, pp 103–135

Summit M, Scott B, Nielson K, Mathur E, Baross J (1998) Pressure enhances thermal stability of DNA polymerase from three thermophilic organisms. Extremophiles 2:339–345

Szewzyk U, Szewzyk R, Stenström T (1994) Thermophilic, anaerobic bacteria isolated from a deep borehole in granite in Sweden. Proc Natl Acad Sci 91:1810–1813

Szilágyi A, Závodszky P (2000) Structural differences between mesophilic, moderately thermophilic and extremely thermophilic protein subunits: results of a comprehensive survey. Structure 8:493–504

Takahata Y, Nishijima M, Hoaki T, Maruyama T (2000) Distribution and physiological characteristics of hyperthermophiles in the Kubiki oil reservoir in Niigata, Japan. Appl Environ Microbiol 66:73–79

Tripathi C, Mahato NK, Singh AK, Kamra K, Korpole S, Lal R (2016) Lampropedia cohaerens sp. nov., a biofilm-forming bacterium isolated from microbial mats of a hot water spring, and emended description of the genus Lampropedia. Int J Syst Evol Microbiol 66(3):1156–1162

Tripathy S, Padhi SK, Mohanty S, Samanta M, Maiti NK (2016a) Analysis of the metatranscriptome of microbial communities of an alkaline hot sulfur spring revealed different gene encoding pathway enzymes associated with energy metabolism. Extremophiles:1–12
Tripathy S, Padhi SK, Sen R, Maji U, Samanta M, Mohanty S, Maiti NK (2016b) Draft genome sequence of Brevibacillus borstelensis cifa_chp40, a Thermophilic Strain Having Biotechnological Importance. J Gen 4:4
Wahlund TM, Woese CR, Castenholz RW, Madigan MT (1991) A thermophilic green sulfur bacterium from New Zealand hot springs, *Chlorobium tepidum* sp. nov. Arch Microbiol 156(2):81–90
Wang Z, Tong W, Wang Q, Bai X, Chen Z, Zhao J, Xu N, Liu S (2012) The temperature dependent proteomic analysis of *Thermotoga maritima*. PLoS One 7:e46463
Wang S, Hou W, Dong H, Jiang H, Huang L, Wu G, Zhang C, Song Z, Zhang Y, Ren H (2013) Control of temperature on microbial community structure in hot springs of the Tibetan Plateau. PLoS One 8:e62901
Wang Q, Cen Z, Zhao J (2015) The survival mechanisms of thermophiles at high temperatures: an angle of omics. Physiology 30:97–106
Ward DM, Ferris MJ, Nold SC, Bateson MM (1998) A natural view of microbial biodiversity within hot spring cyanobacterial mat communities. Microbiol Mol Biol Rev 62:1353–1370
Whitaker RJ, Grogan DW, Taylor JW (2003) Geographic barriers isolate endemic populations of hyperthermophilic archaea. Science 301:976–978
Wiegel J (1998) Anaerobic alkalithermophiles, a novel group of extremophiles. Extremophiles 2:257–267
Wiegel J, Ljungdahl LG (1981) *Thermoanaerobacter ethanolicus* gen. nov., spec. nov., a new, extreme thermophilic, anaerobic bacterium. Arch Microbiol 128:343–348
Wiegel J, Braun M, Gottschalk G (1981) *Clostridium thermoautotrophicum* species novum, a thermophile producing acetate from molecular hydrogen and carbon dioxide. Curr Microbiol 5:255–260
Wiegel J, Hanel J, Aygen K (2003) *Chemolithoautotrophic thermophilic* iron (III)-reducer. In: Biochemistry and physiology of anaerobic Bacteria. Springer, New York, NY, pp 235–251
Wilde SA, Valley JW, Peck WH, Graham CM (2001) Evidence from detrital zircons for the existence of continental crust and oceans on the Earth 4.4 Gyr ago. Nature 409:175–178
Wilson EO (1992) The diversity of life. WW Norton & Company, New York
Xian W-D, Yin Y-R, Liu L, Yuan C-G, Hussain F, Khan I, Habib N, Zhou E-M, Li W-J (2016) Brevibacillus sediminis sp. nov., isolated from a hot spring. Int J Syst Evol Microbiol 66(2):548–553
Xue Y, Xu Y, Liu Y, Ma Y, Zhou P (2001) *Thermoanaerobacter tengcongensis* sp. nov., a novel anaerobic, saccharolytic, thermophilic bacterium isolated from a hot spring in Tengcong, China. Int J Syst Evol Microbiol 51:1335–1341
Yim LC, Hongmei J, Aitchison JC, Pointing SB (2006) Highly diverse community structure in a remote central Tibetan geothermal spring does not display monotonic variation to thermal stress. FEMS Microbiol Ecol 57:80–91
Yoon Y, Lee H, Lee S, Kim S, Choi K-H (2015) Membrane fluidity-related adaptive response mechanisms of foodborne bacterial pathogens under environmental stresses. Food Res Int 72:25–36
Zhao W-D, Romanek CS, Mills G, Wiegel J, Zhang C-L (2005) Geochemistry and microbiology of hot springs in Kamchatka. Russ Geol J China Univ 11:217–223
Zillig W, Stetter KO, Wunderl S, Schulz W, Priess H, Scholz I (1980) The Sulfolobus-"Caldariella" group: taxonomy on the basis of the structure of DNA-dependent RNA polymerases. Arch Microbiol 125(3):259–269

2 Hot Springs of India: Occurrence and Microbial Diversity

Manik Prabhu Narsing Rao, Lan Liu, Jian-Yu Jiao, Min Xiao, and Wen-Jun Li

Abstract

Hot springs indicate where hot water is emitted from the Earth; they are situated in many parts of the world, and most of them are known for their medical importance. Extensive research has been conducted to understand their chemical composition and microbial diversity. This chapter focuses on some important Indian hot springs, their locations, components, and importance. Much emphasis is placed on understanding their microbial diversity, including culture dependent, as well as independent methods. This chapter also sheds light on various Indian hot spring's uniqueness, novel strains that have been reported, and information regarding the genome sequence for strains that have been isolated from Indian hot springs. The bioactive molecules, such as enzymes and antibiotics obtained from hot springs, are also listed here.

Keywords

Indian hot springs · Culture-dependent study · Culture-independent study · Novel strains

2.1 Introduction

Hot springs are places where warm or hot groundwater comes out from the Earth. Their water contains a lot of dissolved solids and minerals and everything that is necessary for the creation of life. It is believed that life on Earth evolved in such an

M. P. Narsing Rao · L. Liu · J.-Y. Jiao · M. Xiao · W.-J. Li (✉)
State Key Laboratory of Biocontrol and Guangdong Key Laboratory of Plant Resources, School of Life Sciences, Sun Yat-Sen University, Guangzhou, People's Republic of China
e-mail: liwenjun3@mail.sysu.edu.cn

D. Egamberdieva et al. (eds.), *Extremophiles in Eurasian Ecosystems: Ecology, Diversity, and Applications*, Microorganisms for Sustainability 8,
https://doi.org/10.1007/978-981-13-0329-6_2

environment (Corliss et al. 1981; Bisht et al. 2011). Even archaea, the third domain of life, was discovered in hot springs (Barns et al. 1994). The hot water rising from hot springs is either heated by geothermal heat (i.e., heat from the Earth's interior) or by coming in contact with magma (i.e., the hot water is generated by the heat produced from the Earth's mantle). Generally, the temperature of rocks within the Earth increases with depth. The rate of temperature increase with depth is known as the geothermal gradient. When water percolates deeply enough into the crust, it will be heated as it comes into contact with hot stones. When the percolating fluid reaches a sufficiently high temperature, the pressure that is generated forces the fluid through the pores and fissures back to the surface of the Earth. In active volcanic zones, water is heated when comes into contact with magma. The high temperature gradient near magma may cause water to be heated sufficiently that it boils, creating pressure and pushing it to the Earth's surface (Mehta and Satyanarayana 2013).

Hot springs are found in many parts of the world (Fig. 2.1) such as China (Hedlund et al. 2012), India (Verma et al. 2015), Japan (Nishiyama et al. 2013), the Philippines (Huang et al. 2013), Indonesia (Aditiawati et al. 2009), Iceland, Russia (Reigstad et al. 2010), etc. Since ancient times, hot springs have been used for medicinal purposes. There has been a long history of Japanese people bathing in hot

Fig. 2.1 Some hot springs around the globe

springs for sanitation (Liang et al. 2015). In India, people bathe in hot springs to treat skin diseases and stomach and rheumatic disorders. It is also believed that women taking a bath in hot springs during festival periods will be cured of infertility (Bisht et al. 2011). Many reports have shown that the water of hot springs may have therapeutic effects for treating a number of diseases (e.g., cardiovascular disease, atopic dermatitis, ankylosing spondylitis, asthma, inflammatory arthritis, rheumatic disease, and rhinosinusitis) (Liang et al. 2015). Apart from their therapeutic value, hot springs can also act as a model system for studying extraterrestrial life (Sharma et al. 2013).

The microbial diversity study of hot springs increased rapidly after Thomas Brock's discovery of *Thermus aquaticus* in the thermal vents of Yellowstone National Park (Brock 1997). Researchers started exploring similar environments in different parts of the world, such as North America (Costa et al. 2009), China, the Philippines (Yang et al. 2015), Japan (Masaki et al. 2016), India (Kumar et al. 2004), Russia (Merkel et al. 2017), and other countries. In terms of microbial diversity, temperature is one of the most important factors that govern species abundance and distribution. High temperature exerts pressure on microbial species, which leads to the selection of specific flora. The microbes that grow at high temperatures have bioactive molecules that are commercially important because of their thermostability and thermoactivity. The best example is Taq DNA polymerase obtained from *Thermus aquaticus* (Sen and Maiti 2014; Chien et al. 1976). Although there is a large body of literature available about microbial diversity studies in hot springs, only a few studies focused on Indian hot springs and their microbial diversity analysis. In this chapter, we discuss several major Indian hot springs, their physicochemical properties, and their microbial diversity analysis and bioactive molecules reported.

In India, there are approximately 400 geothermal springs (Zimik et al. 2017) found either solitary or in groups (Verma et al. 2015; Mangrola et al. 2015a). The temperature of Indian hot springs ranges from 30 to 100 °C; the majority of the hot springs are not volcanic in origin (Pandey and Negi, 1995). In India, geothermal studies of hot springs started in 1864 with a study by Schlagintweit (1865). This author documented 99 hot springs throughout the country. Oldham (1888) published an inventory of 300 thermal springs in India (Bisht et al. 2011). The microbial diversity analysis of hot springs in India was first noted in the nineteenth century. Drouet (1938) described 12 thermal species obtained from Yale, North India. Gonzalves (1947) studied the algal flora of the hot springs of Vajreshwari near Bombay. Thomas and Gonzalves (1965a, b, c, d, e) studied the algal flora of the Akloli, Ganeshpuri, Palli, Sav, Aravali, Tooral, Rajewadi, Unai, Lasundra, and Unapdeo hot springs.

Hot springs in India (Fig. 2.2) are located in various provinces, namely, Himalayan geothermal province, Naga Lushai geothermal province, Andaman-Nicobar Islands province, West Coast geothermal province, Cambay Graben geothermal province, Aravalli province, Son-Narmada-Tapti geothermal province, Godavari geothermal province, Mahanadi geothermal province, and South Indian Cratonic province (Sharma 2010).

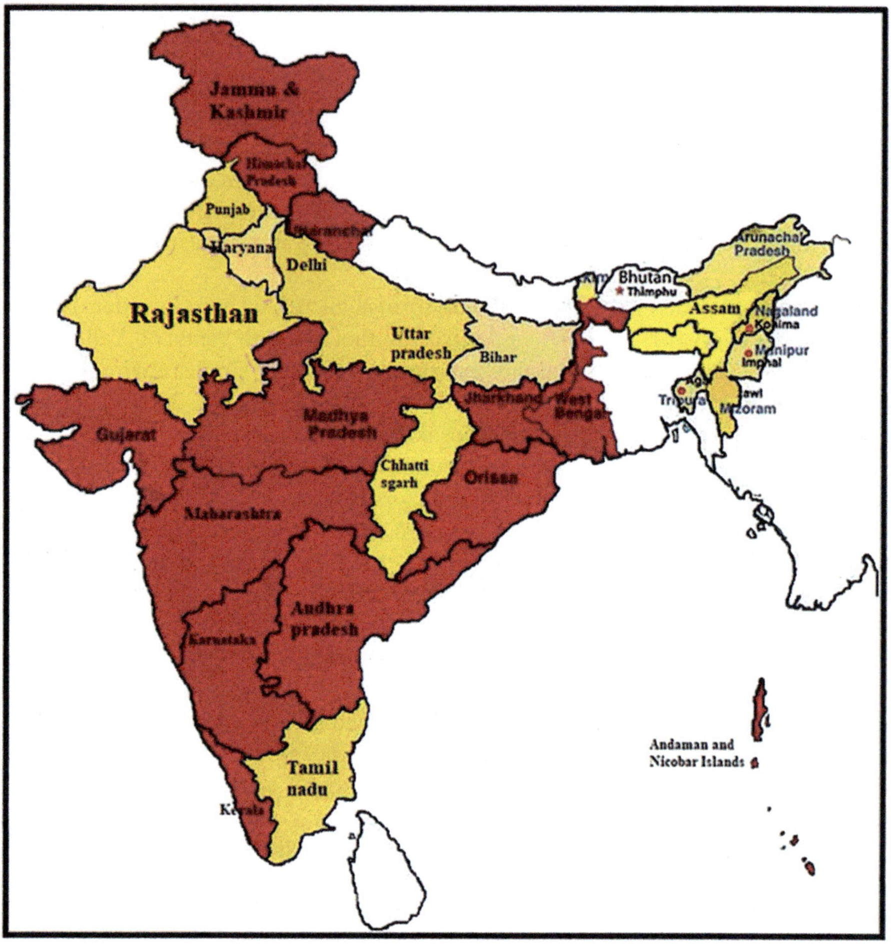

Fig. 2.2 The red highlighted provinces are some of the important hot springs of India

2.2 Some Important Indian Hot Springs and Their Physicochemical Parameters

2.2.1 Jammu and Kashmir

In Jammu and Kashmir, geothermal are found in three main areas, the Chenab Valley/Himalayan foothills, the Kashmir Valley, and Ladakh (Craig et al. 2013; Thussu 2002). The thermal discharge from Chenab Valley/Himalayan foothills and in the Kashmir Valley ranges from 40 to 65 °C, while in Ladakh the thermal discharge has been recorded up to 130 °C (Craig et al. 2013).

Table 2.1 Physicochemical properties of Jammu and Kashmir hot springs

Hot springs	Temperature (°C)	pH	Total dissolved solids	Some important elements detected	Reference
Kurah	53	7.6	475	Ca, Mg, Na, K, F, B, SiO_2, HCO_3, Cl, and SO_4	Craig et al. (2013)
Tattapani	45	7.5	500	Ca, Mg, Na, K, F, B, SiO_2, HCO_3, Cl, and SO_4	
Mahogala	42	7.6	275	Ca, Mg, Na, K, F, SiO_2, HCO_3, Cl, and SO_4	
Gul	40	7.25	2557	Ca, Mg, Na, K, F, B, and SiO_2, HCO_3, Cl, and SO_4	
Chinkah	51	7.1	860	Ca, Mg, Na, K, F, B, SiO_2, HCO_3 and Cl	
Sidhu	65	7.3	1578	Ca, Mg, Na, K, F, SiO_2, HCO_3, Cl, and SO_4	
Kiar	56	6.8	256	Ca, Mg, Na, K, F, B, SiO_2, HCO_3, Cl, and SO_4	
Yurdu	46	7	208	Ca, Mg, Na, K, F, B, SiO_2, HCO_3, Cl, and SO_4	
Sweed	28	7.2	219	Ca, Mg, Na, K, F, B, SiO_2, HCO_3, and Cl	
Atholi	55	8	290	Ca, Mg, Na, K, F, B, SiO_2, HCO_3, Cl, and SO_4	
Tatwain	40	7.1	538	Ca, Na, K, F, B, SiO_2, HCO_3, Cl, and SO_4	
Galhar	60	6.95	235	Ca, Mg, Na, K, F, B, SiO_2, HCO_3, Cl, and SO_4	
Puga	84	8.9	2278	Ca, Mg, Na, K, F, B, SiO_2, HCO_3, Cl, and SO_4	
Chamuthang	83	8.8	Not determined	Ca, Mg, Na, K, F, B, SiO_2, HCO_3, Cl, and SO_4	
Nubra (Panamik)	76	7.7	570	Ca, Mg, Na, K, F, B, SiO_2, HCO_3, Cl, and SO_4	

In Chenab Valley, there are 12 hot springs located at Kurah, Tattapani, Mahogala, Gul, Chinkah, Sidhu, Kiar, Yurdu, Sweed, Atholi, Tatwain, and Galhar with thermal discharge ranges from 28 to 65 °C. The least thermal discharge has been recorded at Sweed while the highest at Sidhu (Table 2.1). The thermal discharge in these places has neutral to alkaline pH with one exception at Kiar which has slight acidic pH (Table 2.1). The highest total dissolved solids have been recorded at Gul while the least at Yurdu (Table 2.1). The thermal discharge at Tattapani has high content of dissolved minerals particularly phosphorus. Tattapani hot spring is of special importance due to its medicinal qualities, particularly for treating skin diseases and bone and joint ailments (Craig et al. 2013; Thussu 2002). The thermal discharge from Chenab Valley also harbors many elements which are listed in Table 2.1.

Ladakh is the most remote region of Jammu and Kashmir. The most famous hot springs in this region are located at Puga, Chumathang, and Panamik (Fig. 2.3). The

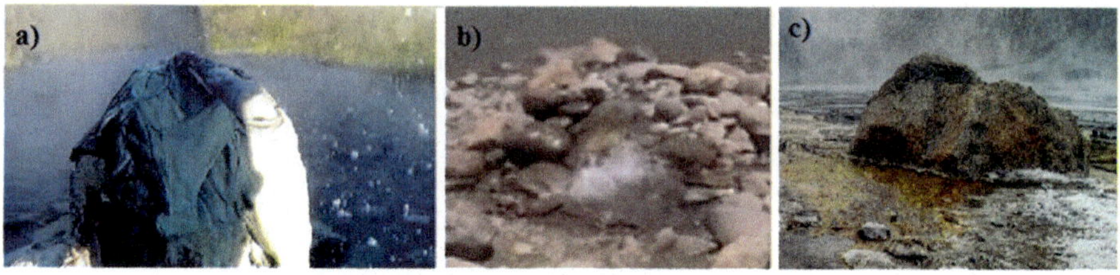

Fig. 2.3 Hot springs (**a**) Puga, (**b**) Chumathang, and (**c**) Panamik

surface temperature of the thermal springs at Panamik ranges from 65 to 76 °C. The hot sulfurous springs at Panamik are particularly famous because during the days of the caravan trade between India and Central Asia along the Silk Route, men bound for Yarkand (now in China's Xinjian province) took their last bath here as they prepared for the treacherous and grueling journey across the Saser Muztagh range (Craig et al. 2013).

The Puga and Chumathang geothermal areas are located in the Indus Valley in eastern Ladakh region of the North-West Himalaya. Puga exhibits vigorous activity in the form of hot springs, mud pools, and sulfur and borax deposits, while at Chumathang thermal activity is in the form of hot springs (Craig et al. 2013; Thussu 2002). The thermal discharges at Puga range from 30 to 84 °C, while those at Chumathang range from 85 to 87 °C. Both these areas have alkaline pH (Table 2.1) with sodium bicarbonate-chloride water type. The Chumathang thermal discharge has similar chemical composition as that of Puga thermal discharge, but Chumathang thermal discharge has slightly higher sulfate level (Craig et al. 2013; Thussu 2002). Oxygen and hydrogen isotope studies indicate that the Chumathang thermal waters, like the Puga waters, are predominantly of meteoric origin (Craig et al. 2013; Thussu 2002).

2.2.2 Himachal Pradesh

The hot springs of Himachal Pradesh are among the hottest springs in India (Dwivedi et al. 2012). There are many hot springs in Himachal Pradesh; some of them are listed below:

(a) Beas Valley: Geothermal system extends for about 45 km between Bashist in the North and Takoli in the South. It consists of eight hot springs, and the temperature of those springs ranges from 30 to 57 °C (Sharma 2010).
(b) Parvati Valley: 40 km area of the Parvati Valley contains six thermal springs, and the temperature of these hot springs ranges from 21 to 150 °C. The important springs in this region are Mannikaran (temperature ranges from 86 to 94 °C), Khirganga (with temperature 150 °C), Kasol (with temperature 100 °C), and Awas (with temperature 58 °C). The most intensively studied hot springs in Himachal Pradesh is Mannikaran (Fig. 2.4). The geothermal field of this hot

Fig. 2.4 Manikaran Gurudwara hot spring

spring lies in the Parvati Valley and extends in a linear zone of 1.5 km, where, sporadically, thermal springs emerge as spouts with temperatures reaches up to 96 °C (Chandrasekharam et al. 2005). Besides sulfur, the water contains a high dose of uranium and radioactive minerals (Bisht et al. 2011). The temperature of the thermal discharge from Mannikaran bore wells is about 86 °C to the maximum temperature recorded 101 °C.

(c) Satluj and Spiti Valley: The temperature of Satluj and Spiti Valley springs ranges from 23 to 73 °C and includes Tapri, Chuza-Sumdo (temperature of the hot spring ranges from 23 to 59 °C), Tattapani (temperature of the hot spring ranges from 32 to 61 °C; it is a hot sulfur spring located on the banks of the River Satluj), Garam Kund, and Vasisht (Pandey and Negi 1995; Bisht et al. 2011).

2.2.3 Uttarakhand

The hot springs of Uttarakhand are situated at Yamunotri, Surya Kund (near Rudraprayag), Tapt Kund (on the bank of the river Alaknanda), Bhagirathi Valley in Uttarkashi District (Gangnani, Bhukki, Songarh), Darma Valley in Pithoragarh District, and Madhya Maheshwar Valley (Bisht et al. 2011).

2.2.4 Jharkhand

The hot springs in Jharkhand contain high concentrations of fluoride; water flowing in shallow streams reaches a temperature as high as 70 °C. In some hot springs, layers of reddish precipitate (due to Fe^{3+}) form on the sides. Beneath the reddish layer, the soil is darker containing iron in the reduced form. Some of the important hot springs are Surajkund (Fig. 2.5), Belkapi, Tattapani, Lugaratha, Siddhpur, Pindarkund, Dwari, Nunbil, Siddhapur Tataloi, Panchvati, Rameswar Kund, Shringirishi, Rishi Kund, Sita Kund, Lakshmi Kund, Janmakund, Bhimbandh, Jhurka, Rajgir, and Tapovan with a temperature ranging from 45 to 88 °C (Jain et al. 2014).

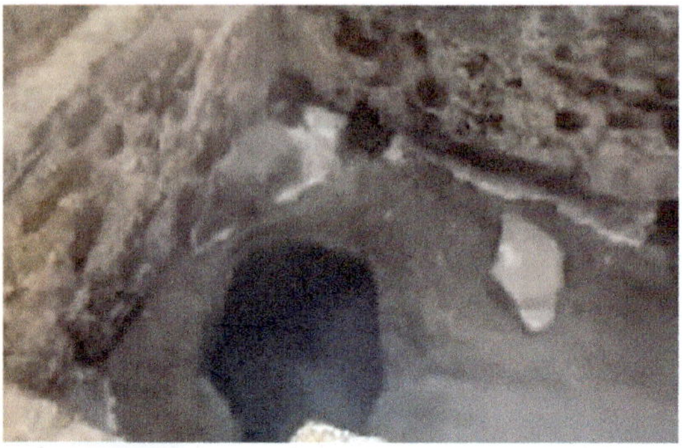

Fig. 2.5 Surajkund hot spring, Jharkhand

Fig. 2.6 Bakreshwar hot spring, West Bengal

2.2.5 West Bengal

The hot springs in West Bengal include Bakreshwar, Tantloi, Kendughata, Bholeghata, and Tantni. The most extensively studied hot spring is Bakreshwar (Fig. 2.6). The water of this hot spring is alkaline being charged with Ca^{++} and HCO_3^- with profuse gaseous activity and temperature ranging from 35 to 66.5 °C (Jana 1973; Bisht et al. 2011).

2.2.6 Odisha

Hot springs in Odisha are located at Atri, Badaberena, Taptapani, Tarobalo, Deulajhari, Athmallik, Magarmuhan, Bankhol, and Boden (Sahoo et al. 2015; Panday and Das 2010). The thermal water discharging from these springs ranges from 28 to 69 °C having moderate acidic to alkaline pH (5.05–8.93) with variation in the chemical characteristics (Table 2.2). Most of the Odisha hot springs are sulfur springs and famous for their medicinal properties (Bisht et al. 2011). The thermal discharge from Attri, Magarmuhan, Bankhol, Taptapani, and Boden is from a single

Table 2.2 Physicochemical properties of Odisha hot springs

Hot springs	Temperature (°C)	pH	Dissolved solids (mg/L)	Some important elements detected	References
Attri	56–57	7.42–8.93	499–534	Cl, SO_4, Ca, Na, F, and K	Zimik et al. (2017)
Tarabalo	55–58	7.96–8.89	214–381	Cl, SO_4, Ca, Na, F, and K	Zimik et al. (2017)
Deulajhari	56–69	6.68–8.36	563–595	Cl, SO_4, Ca, Na, F, and K	Zimik et al. (2017); Singh and Subudhi (2016)
Magarmuhan	36–37	6.37–6.67	18.8–20.9	Cl, SO_4, Na, and Ca	Zimik et al. (2017)
Bankhol	42–45	5.05–6.74	16.9–17.6	Cl, SO_4, Na, and Ca	Zimik et al. (2017)
Badaberena	36–40	8.64–8.8	165–186	Cl, SO_4, Ca, Na, F and K	Zimik et al. (2017)
Taptapani	41	6.84–7.71	188–194	Cl, SO_4, Ca, Na, F, and K	Zimik et al. (2017)
Boden	28–29	6.73–6.96	201–246	Cl, SO_4, Ca, and Na	Zimik et al. (2017)
Athamallik	56	7.4	780.4	Cl, SO_4, Ca, Na, F, and K	Badhai et al. (2015)

spot, whereas Tarabalo, Deulajhari, and Badaberena discharge is from multiple spots (Zimik et al. 2017). The highest temperature (69 °C) has been observed in Deulajhari while the least in Boden (29 °C) (Zimik et al. 2017). The thermal discharge at Deulajhari originates below Shivalinga which attaches a distinct religious attribute. The most important and unique thing about this hot spring is that two ponds are attached with water flowing into each other and one pond is having hot water whereas the other is cold (Singh and Subudhi 2016; Bisht et al. 2011). The water of the Deulajhari hot springs is believed to contain medicinal properties that can heal a number of diseases (Bisht et al. 2011). The active hot springs in this place are Agnikunda, Taptakunda, Himakunda, Amrutakunda, and Labakusakunda (Bisht et al. 2011).

Athamallik hot springs are located in Angul district. This hot spring has a series of outlets within a radius of 70 m. The water temperature at the main outlet was 56 °C and in the surrounding areas ranged from 43 to 50 °C with pH of 7.4 (Panday and Das 2010). Thermal discharge from Taptapani has mild acidic to neutral pH with 41 °C temperature. The hot springs at Atri and Badaberena are located in Khurda district having a temperature ranging from 36 to 57 °C (Zimik et al. 2017). Tarabalo hot spring is situated close to Atri hot spring, which has thermal water discharge (55–58 °C) similar to that of Atri (Badhai et al. 2015; Zimik et al. 2017). The dominant ions discharging from Attri, Tarabalo, and Deulajhari are sodium and chloride, whereas other locations have bicarbonate. Among these thermal springs, higher sulfate concentration has been observed at Attri, Tarabalo, and Deulajhari. In addition to major ions, lithium concentrations were found in some hot springs (Zimik et al. 2017).

2.2.7 Madhya Pradesh

The Indian state of Madhya Pradesh has several hot springs located at Dhuni Pani (Amarkantak), Tattapani (Surguja district), Salbardi region (Betul district), Chavalpani (Pachmarhi), and Anhoni (Chhindwara-Hoshangabad) having thermal discharge ranging from 30 to 98 °C (Bisht et al. 2011; Saxena et al. 2017). Recently, hot springs of Anhoni and Tattapani have been extensively explored. The thermal discharge at Anhoni has a temperature range of 43.5–55 °C, while those at Tattapani has thermal discharge range of 61.5–98 °C. These two springs have neutral to alkaline pH (Table 2.3). The total dissolved solids were high in Tattapani when compared to Anhoni (Table 2.3). These two hot springs also harbor many elements like lithium, lead, mercury, and many more (Table 2.3).

2.2.8 Gujarat

The hot springs at Gujarat are Tulsishyam (Fig. 2.7a) (Junagarh District), Tuwa (Fig. 2.7b) (Panchmahal District), Lasundra (Kheda District), and Unani (Surat District) (Ghelani et al. 2015). Tulsishyam hot spring is arranged under the deep sedimentary basin of tertiary age. The water temperature and pH of this spring range

Table 2.3 Physicochemical properties of Madhya Pradesh hot springs

Hot springs	Temperature (°C)	pH	Total dissolved solids (ppm)	Elements detected	Reference
Anhoni	43.5–55	7.5–7.8	590–690	Li, B, Mg, Al, Si, K, Ca, V, Cr, Mn, Fe, Co, Ni, Cu, Zn, Se, Sr, Mo, Cd, Cs, Ba, La, Ce, Pb, S, and Hg	Saxena et al. (2017)
Tattapani	61.5–98	7–7.8	700–880	Li, B, Mg, Al, Si, K, Ca, V, Cr, Mn, Fe, Co, Ni, Cu, Zn, Se, Sr, Mo, Cd, Cs, Ba, La, Ce, Pb, S, and Hg	

Fig. 2.7 Hot springs of Gujarat: (**a**) Tulsishyam, (**b**) Tuwa

from 58 to 67 °C and 6.8 to 7.7, respectively. The water is clear with no visible turbidity with typical odor of hydrogen sulfide (Ghelani et al. 2015).

Lasundra hot spring is a sulfuric hot water spring. The temperature of the hot spring ranges from 42 to 52 °C. The base of the reservoir contains stones and algal growth along with soil sediments (Mangrola et al. 2015a).

Tuwa hot spring is one of the unexplored hot water springs bearing the temperature of 54–65 °C. The water is alkaline with high salt and mineral content. The bottom of the reservoir has hard rocks and photosynthetic microbes thrive in the water column (Mangrola et al. 2015b).

2.2.9 Maharashtra

The hot springs in Maharashtra are located at Vajreshwari (Fig. 2.8a), Ganeshpuri (Fig. 2.8b), Sativali (Fig. 2.8c), Unhavare (Fig. 2.8d), Akloli Kund, Nimboli, Banganga, Pimplas, Shahapur, Unhere, Unhala, and Unkeshwar.

Vajreshwari hot spring, which is 75 km away from Mumbai, was named after Goddess Vajreshwari. There are around 21 hot water springs in a 5-kilometer radius of the temple, and the temperature of these hot springs ranges from 43 to 49 °C. In some hot springs, the water appears blackish due to accumulation of minerals (Bisht et al. 2011).

The hot springs at Akloli (near Thane District) are located on the left bank of Tansi River. There are ten hot springs. A concrete enclosure has been constructed at each hot spring location. The temperature of these hot springs varies from 45 to 48 °C with sporadic gaseous activity (Sarolkar 2005).

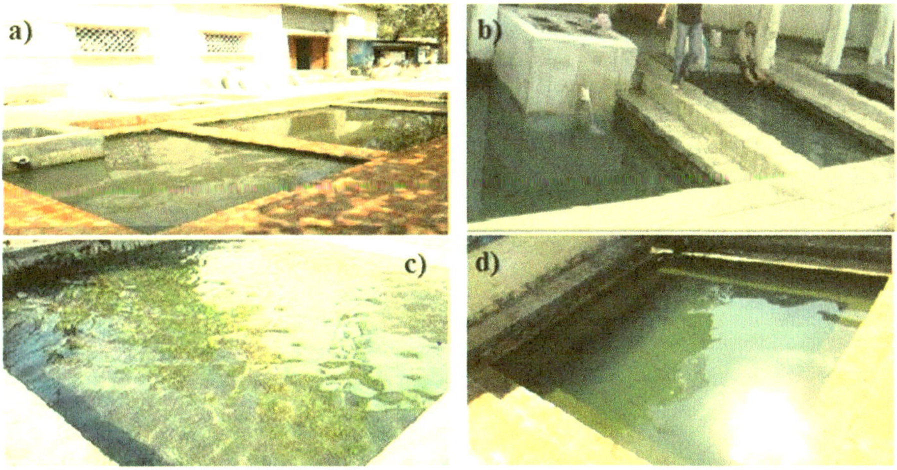

Fig. 2.8 Hot springs of Maharashtra: (**a**) Vajreshwari, (**b**) Ganeshpuri, (**c**) Sativali, and (**d**) Unhavare

Ganeshpuri is located in the Bhiwandi taluka of Thane District and is about 2 km away from Vajreshwari. Ganeshpuri hot springs discharge through a concrete enclosure into tanks, and the temperature of the water is about 52 °C with flows out at 15 lpm (Sarolkar 2005).

There are six hot springs located on the banks of Vandri stream flowing near Sativli. The main hot spring is located on the south bank, and small sprouts are located on the north bank of the stream. The main spring shows profuse gas emission (Sarolkar 2005).

Unkeshwar is located at Nanded District, the temperature of water ranges between 50 and 60 °C with the pH 7.3. The main Unkeshwar hot spring is within the Unkeshwar temple in the form of kund. The water has a sulfurous smell with feeble gaseous activity and discharge through the jointed Deccan basalts. The chemical stress like availability of high phosphorus and sulfur concentrations and slightly higher organic content enriches the microbial diversity of this hot spring (Mehetre et al. 2016).

2.2.10 Karnataka, Andhra Pradesh, and Telangana

The hot spring located at Karnataka is called Bendru Theertha (Fig. 2.9a), is a sulfur spring, which is 13 km away from Mangalore (Bisht et al. 2011). The water is moderately alkaline and categorized as Na-HCO_3 major ion facies. The low concentration of trace elements and minerals indicates that meteoric water is the main source of these thermal fluids. The water consists of high total dissolved solid residues which may be due to rock-water interaction in elevated temperature (Ramanathan and Chandrasekharan 1997; Gurumurthy and Neelagund 2010).

Gundala, a village in Bhadrachalam Mandal in Khammam District of Telangana state, and Mahanandi temple, (Fig. 2.9b) situated in Kurnool District, Andhra

Fig. 2.9 (**a**) Bendru Theertha hot spring, (**b**) Hot spring of Mahanandi temple

Pradesh, consist of hot springs. The hot spring in Gundala located on the bed of the river and pit dug will produce hot water, while Mahanandi is famous for its hot water pool (Bisht et al. 2011).

2.2.11 Tamil Nadu, Kerala, and Andaman and Nicobar

In Tamil Nadu, Godavari Valley contains nearly about 30 hot springs mainly in Mannargudi Thiruthuraipoondi area and along the coastal tract of Aranthangi. In Kerala, hot springs are located in Varkala, while in Andaman and Nicobar various hot water springs are present with a temperature range of more than 200 °C located near active volcanoes (Bisht et al. 2011).

2.3 Culture-Dependent Microbial Diversity Analysis and Bioactive Molecules from Indian Hot Springs

The pioneer work of Robert Koch and Louis Pasteur in culture-dependent method has played a crucial role in the field of microbiology for isolating microbes. The culture-dependent method has several advantages such as exploration of microbes for potential biotechnological applications and discovery of novel isolates for future studies (Kumar et al. 2014; Piterina and Pembroke 2010).

Microorganisms capable of growing at high temperature were first noticed by Miquel (1888). Microbial inhabitant studies of the hot springs by culture-dependent method started with the isolation of thermophilic bacteria from Yellowstone National Park by Marsh and Larsen (1953). Since then, all around the globe, many researchers put their effort to study the microbial diversity based on culture-dependent method. In India, many studies have been carried out to study microbial diversity of the various hot springs.

Culture-dependent microbial diversity study of Himachal Pradesh hot springs (using water samples), namely, Tattapani (District Mandi), Manikaran, and Vashist (District Kullu), has been evaluated. A total of 101 microbial strains have been isolated; among them only two were fungi and the remaining were bacteria. All the isolates can grow at or above 50 °C. Gram staining results showed that only four isolates were Gram-negative. Most of the bacterial shapes were rods and the rest were either coccus or coccobacillus. These isolates have the ability to produce amylase, cellulase, and xylanase (Sharma et al. 2013). Microbial diversity of two hot springs, Soldhar and Ringigad (located in the Garhwal region of Uttaranchal Himalaya), has been carried out. Serially diluted soil samples have been placed on tryptone-yeast extract agar, potato dextrose agar, actinomycetes isolation agar, thiosulfate agar, sulfate-reducing medium, sulfur medium, and Pikovskaya medium. The highest aerobic bacterial colony-forming unit (CFU) has been observed on actinomycetes isolation agar for Soldhar hot spring (50 CFU × 10^4 g^{-1}), while the highest bacterial CFU has been observed on tryptone-yeast extract agar in case of Ringigad hot spring (49 CFU × 10^4 g^{-1}). A temperature of 50 °C has been considered as optimum for the isolation of microorganisms for both hot springs, and up to 80 °C bacterial growth has been observed. Media such as thiosulfate agar, sulfate-reducing medium, and sulfur medium were devoid of any bacterial growth, indicating the absence of sulfur-metabolizing populations. However, a zone of clearance was observed on Pikovskaya medium, indicating the presence of phosphate-solubilizing populations. The highest anaerobic CFU has been observed in Soldhar hot spring when compared to Ringigad hot spring. The majority of the bacteria isolated were *Bacillus* and most of them have endospores; this property may be the reason that these bacteria could withstand high temperature. Apart from bacteria, yeasts were also obtained; however, only a few can grow at high temperature (60 °C). The different bacterial isolation pattern has been observed in these two hot springs. The water holding capacity of the Soldhar soil was almost three times that of Ringigad hot spring. The Soldhar hot spring sample contained high amounts of Cu, Fe, and Mn, while the Ringigad hot spring sample was devoid of Cu but had high phosphate (Kumar et al. 2004).

Odisha harbors many hot springs, and their culture-dependent microbial diversity studies have been evaluated. The microbial diversity of three hot springs of Odisha, namely, Atri, Taptapani, and Tarobalo, has been studied by conventional culture-dependent approach using water samples (temperature and pH of the water ranged from 48 to 58 °C and 7–8). A total of 48 isolates have been obtained belonging to the family *Bacillaceae*, *Paenibacillaceae*, *Planococcaceae*, *Pseudomonadaceae*, and *Enterobacteriaceae*. The majority of the bacterial isolates were affiliated to the genus *Bacillus*. The optimum temperature for growth for these isolates varied from 37 °C to 50 °C. Though the genus *Bacillus* was predominant in all three hot springs, but these sites did not have much overlap or similarity. The genus *Kurthia* has been isolated in Atri, but not in Taptapani and Tarobalo. *Klebsiella* has been isolated in Taptapani, but not in Atri and Tarobalo. *Brevibacillus* has been isolated in Atri and Tarobalo, but not in Taptapani (Sen and Maiti 2014). These data suggest that, though these hot springs are situated in the same state, their microbial diversity didn't have much overlap.

The microbial diversity of Maharashtra hot springs (Vajreshwari and Ganeshpuri) has been evaluated using water samples. The temperature of the water in these springs ranged from 40 to 65 °C. A total of 73 bacteria strains encompassing 8 *Actinobacteria* and 65 *Eubacteria* have been isolated using actinomycetes isolation agar and soybean casein digest agar. A total of 46 and 27 bacteria were isolated from Vajreshwari and Ganeshpuri, respectively. Vajreshwari hot spring had 19 cocci and 11 rod-shaped Gram-positive bacteria and 7 rods and 5 coccobacilli-shaped Gram-negative bacteria. Ganeshpuri hot spring had 8 cocci and 13 rod-shaped Gram-positive bacteria and 8 rods and 4 coccobacilli Gram-negative bacteria; remaining were *Actinobacteria*. The isolated strains showed good antimicrobial activity against both Gram-positive and Gram-negative bacteria (Pednekar et al. 2011). Unkeshwar (Nanded District) hot spring's microbial diversity has been evaluated using water samples. Microbes were isolated using nutrient agar, tryptone-yeast extract agar, tryptone-yeast glucose agar, Vogel-Johnson agar, glucose sodium azide glycerol agar, thiosulfate agar, J agar, brain heart infusion agar, Gram-negative agar, and *Bacillus* agar plates. Among various media used, nutrient agar (95 CFU) had luxuriant bacteria growth followed by tryptone-yeast glucose agar (90 CFU), brain heart infusion agar (88 CFU), tryptone-yeast extract agar (70 CFU), J agar (68 CFU), *Bacillus* agar (62 CFU), thiosulfate agar (54 CFU), Gram-negative agar (51 CFU), Vogel-Johnson agar (23 CFU), and glucose sodium azide glycerol agar (13 CFU). A total of ten pure colonies have been obtained and identified. Out of the ten identified isolates, seven isolates belong to class *Firmicutes* and three belong to class *Gammaproteobacteria*. Six isolates were Gram-positive, three were Gram-negative, and one isolate was Gram-variable. Out of the six Gram-positive isolates, five were spore formers. These isolates produce various bioactive molecules such as caseinase, urease, amylase, oxidase, gelatinase, and lipase (Pathak and Rathod 2014).

Besides having high microbial diversity and being a potential source for stable bioactive molecules, Indian hot springs serve as an important niche for novel strains. Many novel species and even novel genera (Table 2.4) have been reported.

Manikaran hot water spring (Himachal Pradesh) served as a potential source for novel strains. Novel species, namely, *Thermus parvatiensis* (Dwivedi et al. 2015), *Lampropedia cohaerens* (Tripathi et al. 2016b), and *Fictibacillus halophilus* (Sharma et al. 2016), have been recently reported. *Thermus parvatiensis*, a Gram-negative, yellow-pigmented bacterium, has been isolated from the water sample, which exhibits protease activity up to 70 °C. This novel species reported for its growth at temperature ranging from 60 to 80 °C with optimum growth at 70 °C (Dwivedi et al. 2015). Novel species *Fictibacillus halophilus* (Gram-positive) and *Lampropedia cohaerens* (Gram-negative) have been isolated from microbial mats of Manikaran hot water spring (Sharma et al. 2016; Tripathi et al. 2016b). *Fictibacillus halophilus* reported for its growth at high salt concentration (up to 12%) and had a temperature range for growth from 28 to 45 °C with optimum growth at 37 °C (Sharma et al. 2016). *Lampropedia cohaerens* had the ability to form biofilm and grow up to 55 °C (Tripathi et al. 2016b).

Novel species, namely, *Comamonas thiooxidans* (Narayan et al. 2010), *Gulbenkiania indica* (Jyoti et al. 2010), *Chelatococcus sambhunathii* (Panday and Das 2010), and *Pannonibacter indica* (Bandyopadhyay et al. 2013), were isolated from Athamallik (Orissa); sulfur spring sediment. Novel species *Comamonas thiooxidans* had the ability to oxidize thiosulfate under mixotrophic growth condition (Narayan et al. 2010), and *Pannonibacter indicus* exhibited remarkable arsenic tolerance (Bandyopadhyay et al. 2013). Warm spring located in Assam forest served as potential reservoir for novel strains. Six novel candidates, namely, *Aquimonas voraii* (Saha et al. 2005a), *Paenibacillus assamensis* (Saha et al. 2005b), *Aeromonas sharmana* (Saha and Chakrabarti 2006a), *Flavobacterium indicum* (Saha and Chakrabarti 2006b), *Emticicia oligotrophica* (Saha and Chakrabarti 2006c), and *Fontibacillus aquaticus* (Saha et al. 2010), have been reported.

Apart from Himachal Pradesh, Orissa, and Assam hot springs, novel species were also reported from different places such as Meghalaya (Jakrem hot spring, *Caldimonas meghalayensis*, Rakshak et al. 2013), Bhubaneswar (Atri hot spring, *Thiomonas bhubaneswarensis*, Panda et al. 2009), and Jharkhand (Suryakund hot spring, *Anoxybacillus suryakundensis*, Deep et al. 2013; *Tepidiphilus thermophilus*,

Table 2.4 Novel strains from Indian hot springs

Location	Novel species/Genus	Source	Gram staining	Temperature range/ optimum growth	GenBank accession number	Type of strains/ deposited no	References
Manikaran, Himachal Pradesh	*Thermus parvatiensis* sp. nov.	Water	−ve	60–80 °C, optimum at 70 °C	EU017402	RL^T (=MTCC 8932^T= DSM 21745^T)	Dwivedi et al. (2015)
	Lampropedia cohaerens sp. nov.	Microbial mat	−ve	20–55 °C, optimum at 37 °C	KP265299	$CT6^T$ (=DSM 100029^T =KCTC 42939^T =MCC 2711^T	Tripathi et al. (2016b)
	Fictibacillus halophilus sp. nov.	Microbial mat	+ve	28–45 °C, optimum at 37 °C	KP265300	$AS8^T$ (=MCC 2765^T=DSM 100124^T=KCTC 33758^T)	Sharma et al. (2016)
Athamallik, Orissa	*Comamonas thiooxidans* sp. nov.	Sediment	−ve	12–40 °C, optimum at 37 °C	DQ322069	$S23^T$ (=DSM17888^T = JCM 14801^T).	Narayan et al. (2010)
	Gulbenkiania indica sp. nov.	Sediment	−ve	15–45 °C, optimum at 30–37 °C	DQ415656	$HT27^T$ (=DSM 17901^T =JCM 15969^T)	Jyoti et al. (2010)
	Chelatococcus sambhunathii sp. nov.	Sediment	−ve	20–50 °C, optimum at 37–42 °C	DQ322070	$HT4^T$ (=DSM 18167^T=JCM 14988^T)	Panday and Das (2010)
	Pannonibacter indica sp. nov.	Sediment	−ve	20–45 °C, optimum at 37 °C	EF608175	$HT23^T$ (=JCM 16851^T = DSM 23407^T = LMG 25769^T	Bandyopadhyay et al. (2013)
Jakrem hot spring, Meghalaya	*Caldimonas meghalayensis* sp. nov.	Water	−ve	25–55 °C, optimum at 45 °C	HF562216	$AK31^T$ (= MTCC 11703^T = JCM 18786^T	Rakshak et al. (2013)
Atri hot spring, Bhubaneswar	*Thiomonas bhubaneswarensis* sp. nov.	Sediment	−ve	25–45 °C, optimum at 30–37 °C	DQ092334	$S10^T$ (=DSM 18181^T =JCM 14806^T)	Panda et al. (2009)

Suryakund hot spring Jharkhand	*Anoxybacillus suryakundensis* sp. nov.	Sediment	+ve	40–60 °C, optimum at 55 °C	KC958552	JS1^T (= DSM 27374^T = LMG 27616^T =JCM19211^T)	Deep et al. (2013)
	Tepidiphilus thermophilus JHK30^T sp. nov.	Sediment	−ve	30–60; optimum at 50–55	HM543264	JCM 19170^T =LMG 27587^T =5DSM 27220^T	Poddar et al. (2014)
Assam	*Aquimonas voraii* gen. nov. sp. nov	Water	−ve	25–42 °C,	AY544768	GPTSA 20^T (=MTCC 6713^T=JCM 12896^T)	Saha et al. (2005a)
	Paenibacillus assamensis sp. nov.	Water	Gram-variable	20–37 °C, optimum temperature not determined	AY884046	GPTSA 11^T (=MTCC 6934^T=JCM 13186^T)	Saha et al. (2005b)
	Aeromonas sharmana sp. nov.	Water	−ve	15–42 °C, optimum temperature not determined	DQ013306	GPTSA-6^T (=MTCC 7090^T=DSM 17445^T	Saha and Chakrabarti (2006a)
	Flavobacterium indicum sp. nov.,	Water	−ve	15–42 °C, optimum at 37 °C	AY904351	GPTSA100-9^T (=MTCC 6936^T=DSM 17447^T	Saha and Chakrabarti (2006b)
	Emticicia oligotrophica gen. nov., sp. nov.,	Water	−ve	15–42 °C, optimum temperature not determined	AY904352	GPTSA100-15^T (=MTCC 6937^T=DSM 17448^T	Saha and Chakrabarti (2006c)
	Fontibacillus aquaticus gen. nov., sp. nov.	Water	+ve	20–42 °C, optimum temperature not determined	DQ023221	GPTSA 19^T (=MTCC 7155^T=DSM 17643^T)	Saha et al. (2010)
Irde, Karnataka	*Calidifontibacter indicus.* gen. nov., sp. nov	Water	+ve	20–37; optimum at 30	EF187228	MTCC 8338^T =DSM 22967^T =JCM 16038^T	Ruckmani et al. (2011)

Poddar et al. 2014). This indicates that novel species are widely distributed in Indian hot springs. Not only novel species, Indian hot springs were also reported for having novel genera such as *Calidifontibacter* reported from Karnataka (Puttur hot spring, Ruckmani et al. 2011) and *Aquimonas* and *Fontibacillus* reported from Assam (Saha et al. 2005a, 2010). From the above, it can be said that Indian hot springs serve as a potential source for novel microorganisms.

2.4 Whole-Genome Sequence of Bacterial Strains Isolated from Indian Hot Springs

Since the first two complete bacterial genome sequences were published, the science of bacteria has dramatically changed. Using third-generation DNA sequencing, it is now possible to completely sequence a bacterial genome in a few hours. Sequencing of bacterial genome sequences is now a standard procedure, and the information obtained has provided a major impact on our understanding of the bacterial world (Land et al. 2015). Whole-genome sequence of bacteria isolated from Indian hot springs (Table 2.5) has been determined for their insight into the metabolic capabilities and functions (Dwivedi et al. 2012; Mahato et al. 2014; Sharma et al. 2014). Whole-genome analyses of four bacteria, namely, *Thermus* sp. strain RL (Dwivedi et al. 2012), *Deinococcus* sp. strain RL (Mahato et al. 2014), *Cellulosimicrobium* sp. strain MM (Sharma et al. 2014), and *Lampropedia cohaerens* strain CT6^{T} sp. nov. (Tripathi et al. 2016a), isolated from Manikaran hot spring (Himachal Pradesh) have been carried out. Protease-producing *Thermus* sp. strain RL represented a genome size of 20,36,600 bp with 68.77% G+C content. The genome annotation of this strain has predicted 1986 protein-coding genes and 710 hypothetical proteins (Dwivedi et al. 2012). *Deinococcus* sp. strain RL genome has been sequenced for comparative genome analysis with closely related radioresistant members. The total length of the genome has been estimated to be 2,792,068 bp, with 69.4% G+C content (Mahato et al. 2014).

The genome of strain CT6^{T}, which is the type strain for *Lampropedia cohaerens*, has been sequenced in order to supplement the phenotypic taxonomical observations with genetic data and obtain genomic insights into heavy metal resistance and metabolic potential of gene complements of this microbial mat dweller. Strain CT6^{T} represented a genome size of 3,158,922 bp with 41 contigs, 63.5% G+C content, and 282,3 coding sequences. Strain CT6^{T} contained genes responsible for imparting resistance to arsenic, copper, cobalt, zinc, cadmium, and magnesium, providing survival advantages at a thermal location (Tripathi et al. 2016a). The genome sequence of *Cellulosimicrobium* sp. strain MM has been carried to understand the mechanism for its growth at high arsenic-rich environment. The draft genome (3.85 Mb) of *Cellulosimicrobium* sp. strain MM consists of 3718 coding sequence with 74.4% G+C content. *Cellulosimicrobium* sp. strain MM also encodes for mannanase endo-β-1, 3glucanases, endo-β-1,4-xylanases, and chitinases (Sharma et al. 2014). *Brevibacillus borstelensis* cifa_chp40 isolated from Attri hot spring (Bhubaneswar) has been reported to degrade low-density polythene and produces essential enzymes

Table 2.5 Genome sequence of bacterial strains isolated from Indian hot springs

Hot springs	Bacterial strains	Genome size; G+C content (mol%)	Accession numbers	References
Manikaran, Himachal Pradesh	*Thermus* sp. strain RL	20,36,600 bp; 68.77	AIJQ00000000	Dwivedi et al. (2012)
	Deinococcus sp. strain RL	2,792,068 bp; 69.4	JMQF00000000	Mahato et al. (2014)
	Cellulosimicrobium sp. strain MM	3.85 Mbp; 74.4	JPQW00000000	Sharma et al. (2014)
	Lampropedia cohaerens strain CT6^T sp. nov.	3,158,922 bp; 63.5	LBNQ00000000	Tripathi et al. (2016a)
Atri, Bhubaneswar	*Brevibacillus borstelensis* cifa_chp40	5,196,578 bp; 51.90	JPRB00000000	Tripathy et al. (2016)
	Thiomonas bhubaneswarensis S10^{T}sp. nov.	3.2 Mb; 65.0	LIPV00000000.1	Narayan et al. (2016a)
Athamallik, Orissa	*Gulbenkiania indica* HT27^{T}sp. nov.	2.8 Mb; 63.0	LION00000000	Badhai et al. (2016a)
	Chelatococcus sambhunathii HT4^{T}sp. nov.	4.4 Mb; 67.8	LIOL00000000	Badhai et al. (2016b)
	Comamonas thiooxidans S23^{T}sp. nov.	5.3 Mb; 62.0	LIOM00000000	Narayan et al. (2016b)
	Pannonibacter indicus HT23^{T}sp. nov.	4.2 Mb; 63.5	LIPT01000000	Bandyopadhyay et al. (2017)
Surajkund, Jharkhand	*Tepidiphilus thermophilus* JHK30^T sp. nov.	2.3 Mb; 66.1	LIPU00000000	Poddar et al. (2016)
	Anoxybacillus suryakundensis JS1^T sp. nov.	2.6 Mb; 42.0	LIOK00000000	Deep et al. (2016)
Tattapani, Chhattisgarh	*Anoxybacillus mongoliensis* strain MB4	30,188.3 bp; 58.3	MRZM00000000	Mittal et al. (2017)
Assam	*Flavobacterium indicum* GPSTA100-9^T sp. nov.	2,993,089 bp; 31.8	HE774682	Barbier et al. (2012)

like protease, lipase, esterase, and amidase. To understand the insight into the metabolic capabilities, function, and evolution of this important bacterium, genome sequencing has been carried out. The total genome length of *Brevibacillus borstelensis* cifa_chp40 was 5,196,578 bp assembled into 38 scaffolds with 51.90% G+C content (Tripathy et al. 2016). An obligate mixotrophic bacterium, *Thiomonas bhubaneswarensis* S10^{T}sp. nov., isolated from Atri hot spring (Bhubaneswar) has been investigated to elucidate the pathway(s) and mechanism of electron transport during thiosulfate oxidation. The total genome size of this strain was 3.2 Mb with a G+C content of 65.0%. Whole-genome sequence analysis revealed the presence of complete *sox* (sulfur oxidation) gene cluster (*soxCDYZAXB*) including the sulfur oxygenase reductase (SOR), sulfide quinone reductase (SQR), sulfide dehydrogenase (flavocytochrome *c* (fcc)), thiosulfate dehydrogenase (Tsd), sulfite dehydrogenase (SorAB), and intracellular sulfur oxidation protein (DsrE/DsrF). In addition, genes encoding respiratory electron transport chain components, viz., complex I (NADH dehydrogenase), complex II (succinate dehydrogenase), and complex III (ubiquinone-cytochrome *c* reductase), and various types of terminal oxidases (cytochrome *c* and quinol oxidase) have been identified in the genome (Narayan et al. 2016a). The genome sequence of bacteria isolated from Orissa (Badhai et al. 2016a, b; Narayan et al. 2016b; Bandyopadhyay et al. 2017), Jharkhand (Poddar et al. 2016; Deep et al. 2016), Chhattisgarh (Mittal et al. 2017), and Assam (Barbier et al. 2012) has also been carried out, and their details are mentioned in Table 2.5.

2.5 Culture-Independent Microbial Diversity analysis of Indian Hot Springs

Though culture-dependent method was used for studying the microbial diversity, however, this method has several disadvantages. In culture-dependent method, most of the microorganisms remain hidden or difficult to grow because essential nutrients for growth or optimal environmental conditions such as temperature, pH, and essential mixtures of gases may not be present. Aerobic and anaerobic organisms cannot be cultured together; hence, majority of them remain unknown, thereby limiting information at the genomic and phenotypic level (Kumar et al. 2014; Piterina and Pembroke 2010).

In the past few years, the application of culture-independent genomics or metagenomics approaches coupled with high-throughput DNA sequencing has proved a promising tool to investigate the population diversity, gene content, function, and ecological significance of microbial communities living in diverse hot spring environments (Badhai et al. 2015). Thus, with the advances in these methods, all the hidden facts about the microbial ecology have been revealed.

The bacterial and archaeal diversity of Manikaran hot spring has been studied through metagenomic analysis using the water samples. Gram-positive, endospore-forming *Firmicutes* has dominated the hot spring, followed by *Aquificae* and the *Deinococcus-Thermus* group. *Bacillus megaterium*, *Bacillus sporothermodurans*, *Hydrogenobacter* sp. GV4-1, *Thermus thermophiles*, and *Thermus brockianus* were the main bacterial species. In archaea, phylum *Crenarchaeota* has dominated with *Pyrobaculum aerophilum* and *Pyrobaculum calidifontis.* Further, several bacterial and archaeal sequences remained taxonomically unresolved, indicating potentially novel microorganisms in this geothermal ecosystem (Bhatia et al. 2015).

Metagenomic analysis of water sample from Lasundra hot spring (Gujarat) has been carried out. The majority of the sequences were of bacterial origin (99.21%) followed by eukaryotes (0.43%) and archaea (0.11%). A total of 33 prokaryotic phyla, including 29 bacterial and 4 archaeal phyla along with 20 eukaryotic phyla, have been detected. *Firmicutes* (95.5%) was the most abundant prokaryotic phyla followed by *Proteobacteria* (2.0%), *Actinobacteria* (0.8%), *Bacteroidetes* (0.1%), *Cyanobacteria* (0.1%), and *Euryarchaeota* (0.09%). At the family level, *Bacillaceae* dominated (90.1%) followed by *Paenibacillaceae* (1.3%), *Clostridiaceae* (0.8%), *Listeriaceae* (0.5%), and *Staphylococcaceae* (0.5%). The most leading genera were *Bacillus* (86.7%), *Geobacillus* (2.4%), *Paenibacillus* (1.0%), *Clostridium* (0.7%), and *Listeria* (0.5%). The stress-associated genes such as oxidative stress, periplasmic stress, osmotic stress, heat shock, cold shock, acid stress, and detoxification account for 3.0% (Mangrola et al. 2015b). The presence of photosynthetic bacteria, heterotrophs, and autotrophs in the hot spring metagenome suggested the nutritive interaction among the microorganisms. Metagenomic analysis of this hot spring showed the presence of a total of 183,408 predicted protein-coding regions, of which 33.3% features have no significant similarities to the protein database. A total of 104,110 features have been assigned to functional categories with the COG approach, of which 45.4% was metabolism connected and 19.6% falls in poorly characterized group, indicating the possibilities of having a novel gene (Mangrola et al. 2015b).

Community analysis of Tuwa hot spring, using the shotgun sequencing approach, presented that 99.1% of sequences belong to bacteria, 0.3% to eukaryotes, 0.2% to virus, and 0.05% to archaea, while 0.4% belong to unclassified and 0.07% to unidentified sequences. A total of 22 bacterial phyla include 90 families and 201 species with *Firmicutes* (97.0%) as dominant bacterial phylum. *Proteobacteria* and *Actinobacteria* account for 1.3% and 0.4%, respectively. 4.0% of genes were assigned for stress responses, and 3% of genes were fit into the metabolism of aromatic compounds (Mangrola et al. 2015a).

Metagenome sequences from Tulsishyam hot spring suggest the dominance of bacteria (98.2%), followed by eukaryotes (1.5%). About 0.3% of metagenomes were unidentified suggesting the wealth of uncultivable bacteria. A total of 16 bacterial phyla, 97 families, and 287 species have been predicted. *Firmicutes* (65.38%) has been the abundant phylum followed by *Proteobacteria* (21.21%). The dominating family was *Peptostreptococcaceae* (37.33%) followed by *Clostridiaceae* (23.36%) and *Enterobacteriaceae* (16.37%). The dominant species were *Clostridium bifermentans* and *Clostridium lituseburense* (Ghelani et al. 2015).

The microbial community of Jakrem hot spring (Meghalaya) has been evaluated through 16S rRNA sequence analysis targeting V3 region. The bacterial community that dominated was *Firmicutes* (61.60%), followed by *Chloroflexi* (21.37%) and *Cyanobacteria* (12.96%). The unclassified bacteria accounts for 1.2%. Prominent families reported were *Clostridiaceae* (60.92%), *Chloroflexaceae* (21.26%), and *Pseudanabaenaceae* (12.77%) (Panda et al. 2015).

The microbial diversity of hot springs of Odisha, Atri, and Taptapani has been studied by 16S rRNA deep sequencing analysis, targeting V3 region using the sediment samples. Atri and Taptapani metagenomes were classified into 50 and 51 bacterial phyla, respectively. *Proteobacteria* (45.17%) dominated the Taptapani sample followed by *Bacteroidetes* (23.43%) and *Cyanobacteria* (10.48%), while *Chloroflexi* (52.39%), *Nitrospirae* (10.93%), and *Proteobacteria* (9.98%) dominated the Atri sample. A large number of sequences remained taxonomically unresolved in both hot springs, indicating the presence of potentially novel microbes (Sahoo et al. 2015).

Comparative microbial community analysis of the four hot springs, Athamallik (located in the district of Angul), Taptapani (located in the district of Ganjam), Tarabalo (located in the district of Nayagarh), and Atri (located in the district of Khorda), has been carried out which showed that 56.2% of the total sequence reads in each metagenomic data sets were taxonomically classified, while 43.8% were of unknown taxa. The majority of the sequence reads were classified as bacteria (54.5%) followed by archaea (1.7%) and eukarya (<0.1%). A total of 30, 30, 32, and 36 phyla have been identified in the samples of Athamallik, Taptapani, Tarabalo, and Atri, respectively. Within the bacterial domain, only 66.5% of the assigned sequences were classified at the phylum level, and there was high abundance of the phyla *Chloroflexi* and *Proteobacteria* with little variation. *Acetothermia* was abundant in Tarabalo, whereas *Verrucomicrobia*, *Ignavibacteriae*, and *Cyanobacteria* were abundant in Taptapani. Within the Archaea domain about 79% of the assigned sequences were classified at the phylum level. In all samples, the phylum *Euryarchaeota* was found to be the most abundant followed by phylum *Crenarchaeota* (with exception of Taptapani). A high taxonomic diversity has been observed in these samples; however, there were only little variations in the overall functional profiles of the microbial communities. Genes involved in the metabolism of carbohydrates and carbon fixation were the most abundant functional class of genes present in these hot springs. The distribution of genes involved in carbon fixation predicted the presence of all the six known autotrophic pathways in the metagenomes (Badhai et al. 2015). 16S rRNA gene sequence analysis of Tantloi hot spring microbial community revealed a significant bacterial diversity represented by at least ten taxonomic divisions of bacteria with the majority belonging to the division *Deinococcus-Thermus*; there were representatives of the divisions *Proteobacteria*, *Firmicutes*, *Nitrospira*, *Chloroflexi*, *Aquificae*, *Cyanobacteria*, *Thermotogae*, and *Verrucomicrobia*. A significant metabolic diversity was represented by at least ten taxonomic divisions of bacteria with a clear predominance of *Thermus* in Tantloi hot spring. Approximately 80% of the sequences obtained in this study represented novel phylotypes that had less than 97% similarity with known sequences (Jain et al. 2014).

Metagenome sequencing of Unkeshwar hot springs revealed 41 phyla (including bacteria and archaea) with 719 different species. The dominant phyla were *Actinobacteria* (56%), *Verrucomicrobia* (24%), *Bacteroides* (13%),

Deinococcus-Thermus (3%), and *Firmicutes* (2%) and viruses (2%). At the phylum level, dominant bacterial phyla were *Actinobacteria*, *Bacteroides*, *Deinococcus-Thermus*, *Firmicutes*, and *Planctomycetes*. Bacterial genera like *Rhodococcus*, *Microbacterium*, *Propionibacterium*, *Flavobacterium*, *Deinococcus*, *Caulobacter*, *Brevundimonas*, *Methylobacterium*, *Paracoccus*, *Roseomonas*, *Novosphingobium*, Sphingomonas, Achromobacter, Acidovorax, and Aquabacterium were dominant (Mehetre et al. 2016).

2.6 Conclusion

India is endowed with more than 400 hot water springs. Since ancient times, it is believed that these hot springs have medical importance. The culture-dependent studies showed that Indian hot springs are diverse with respect to their content and microbial diversity. The microbes isolated from these have special qualities and have thermostable bioactive molecules. Indian hot springs have been reported to have many novel strains, and culture-independent studies suggest that a large number of sequences remained taxonomically unresolved, indicating the presence of potential novel microbes. Although a lot of literature are available on Indian hot springs, still many of them remain unexplored. In this regard, researches have to put their step forward of studying these hot springs and discovering useful things which will be beneficial for mankind.

Acknowledgments This work was supported by Science and Technology Infrastructure Project (No. 2015FY110100) and Key Project of International Cooperation of Ministry of Science & Technology (MOST) (No. 2013DFA31980), Natural Science Foundation of China (Nos. 31470139 and 31600103), and China Postdoctoral Science Foundation No. 2017M612796. W J Li was also supported by Guangdong Province Higher Vocational Colleges & Schools Pearl River Scholar Funded Scheme (2014).

References

Aditiawati P, Yohandini H, Madayanti F, Akhmaloka (2009) Microbial diversity of acidic hot spring (Kawah Hujan B) in geothermal field of Kamojang area, West Java-Indonesia. Open Microbiol J 3:58–66

Badhai J, Ghosh TS, Das SK (2015) Taxonomic and functional characteristics of microbial communities and their correlation with physicochemical properties of four geothermal springs in Odisha, India. Front Microbiol 6:1166. https://doi.org/10.3389/fmicb.2015.01166

Badhai J, Narayan KD, Whitman WB, Das SK (2016a) Draft genome sequence of *Gulbenkiania indica* strain HT27[T] (DSM 17901[T]) isolated from a sulfur spring in India. Genome Announc 4:e00830–e00816. https://doi.org/10.1128/genomeA.00830-16

Badhai J, Whitman WB, Das SK (2016b) Draft genome sequence of *Chelatococcus sambhunathii* strain HT4[T] (DSM 18167[T]) isolated from a hot spring in India. Genome Announc 4(4):e00825–e00816. https://doi.org/10.1128/genomeA.00825-16

Bandyopadhyay S, Schumann P, Das SK (2013) *Pannonibacter indica* sp. nov., a highly arsenate-tolerant bacterium isolated from a hot spring in India. Arch Microbiol 195(1):1–8

Bandyopadhyay S, Whitman WB, Das SK (2017) Draft genome sequence of *Pannonibacter indicus* strain HT23^T (DSM 23407^T), a highly arsenate tolerant bacterium isolated from a hot spring in India. Genome Announc 5(18):e00283–e00217. https://doi.org/10.1128/genomeA.00283-17

Barbier P, Houel A, Loux V, Poulain J, Bernardet JF, Touchon M, Duchaud E (2012) Complete genome sequence of *Flavobacterium indicum* GPSTA100-9^T, isolated from warm spring water. J Bacteriol 194:3024–3025

Barns SM, Fundyga RE, Jeffries MW, Pace NR (1994) Remarkable archaeal diversity detected in a Yellowstone national Park hot spring environment. Proc Natl Acad Sci U S A 91(5):1609–1613

Bhatia S, Batra N, Pathak A, Green SJ, Joshi A, Chsauhan A (2015) Metagenomic evaluation of bacterial and archaeal diversity in the geothermal hot springs of Manikaran, India. Genome Announc 3(1):e01544–e01514. https://doi.org/10.1128/genomeA.01544-14

Bisht SS, Das NN, Tripathy NK (2011) Indian hot- water springs: a bird's eye view. JoEECC 1(1):1–15

Brock TD (1997) The value of basic research: discovery of *Thermus aquaticus* and other extreme thermophiles. Genetics 146:1207–1210

Chandrasekharam D, Alam MA, Minissale A (2005) Thermal discharges at Manikaran, Himachal Pradesh, India. In: Proceedings World Geothermal Congress Antalya, Turkey 24–29 April 2005

Chien A, Edgar DB, Trela JM (1976) Deoxyribonucleic acid polymerase from the extreme thermophile *Thermus aquaticus*. J Bacteriol 127:1550–1557

Corliss JB, Baross JA, Hoffman SE (1981) An hypothesis concerning the relationship between submarine hot springs and the origin of life on Earth. Oceanol Acta suppl:59–69

Costa KC, Navarro JB, Shock EL, Zhang CL, Soukup D, Hedlund BP (2009) Microbiology and geochemistry of great boiling and mud hot springs in the United States Great Basin. Extremophiles 13:447–459

Craig J, Absar A, Bhat G, Cadel G, Hafiz M, Hakhoo N, Kashkari R, Moore J, Ricchiuto TE, Thurow J, Thusu B (2013) Hot springs and the geothermal energy potential of Jammu & Kashmir State, N.W. Himalaya, India. Earth-Sci Rev 126:156–177

Deep K, Poddar A, Das SK (2013) *Anoxybacillus suryakundensis* sp. nov, a moderately thermophilic, alkalitolerant bacterium isolated from hot spring at Jharkhand, India. PLoS One 8(12):e85493. https://doi.org/10.1371/journal.pone.0085493

Deep K, Poddar A, Whitman WB, Das SK (2016) Draft genome sequence of *Anoxybacillus suryakundensis* strain JS1^T (DSM27374^T) isolated from a hot spring in Jharkhand, India. Genome Announc 4(4):e00824–e00816. https://doi.org/10.1128/genomeA.00824-16

Drouet F (1938) Myxophyceae of the Yale North India expedition, collected by G. E. Hutchinson. Trans Am Microsc Soc 57(2):127–131

Dwivedi V, Sangwan N, Nigam A, Garg N, Niharika N, Khurana P, Khurana JP, Lal R (2012) Draft genome sequence of *Thermus* sp. strain RL, isolated from a hot water spring located atop the Himalayan ranges at Manikaran, India. J Bacteriol 194(13):3534

Dwivedi V, Kumari K, Gupta SK, Kumari R, Tripathi C, Lata P, Niharika N, Singh AK, Kumar R, Nigam A, Garg N, Lal R (2015) *Thermus parvatiensis* RL^T sp. nov., isolated from a hot water spring, located a top the Himalayan ranges at Manikaran, India. Indian J Microbiol 55:357–365

Ghelani A, Patel R, Mangrola A, Dudhagara P (2015) Cultivation-independent comprehensive survey of bacterial diversity in Tulsi Shyam hot springs, India. Genom Data 4:54–56

Gonzalves E (1947) The algal flora of the hot springs of Vajreswari near Bombay. Jour Univ Bombay 16:22–27

Gurumurthy DM, Neelagund SE (2010) *Geobacillus* sp. Iso 5, a novel amylase-producing thermophile from thermal springs in Konkan region of Southern India. J Earth Sci 21(1):319–322

Hedlund BP, Cole JK, Williams AJ, Hou W, Zhou E, Li WJ, Dong H (2012) A review of the microbiology of the Rehai geothermal field in Tengchong, Yunnan Province, China. Geo Sci Front 3(3):273–288

Huang Q, Jiang H, Briggs BR, Wang S, Hou W, Li G, Wu G, Solis R, Arcilla CA, Abrajano T, Dong H (2013) Archaeal and bacterial diversity in acidic to circum neutral hot springs in the Philippines. FEMS Microbiol Ecol 85:452–464

Jain P, Reza HM, Pa S (2014) Molecular phylogenetic analysis of bacterial community and characterization of Cr(VI) reducers from the sediments of Tantloi hot spring, India. Aquat Biosyst 10:7. https://doi.org/10.1186/2046-9063-10-7

Jana BB (1973) The thermal springs of Bakreswar, India physico-chemical conditions, flora and fauna. Hydrobiologia 41:291–307

Jyoti V, Narayan KD, Das SK (2010) *Gulbenkiania indica* sp. nov., isolated from a sulfur spring. Int J Syst Evol Microbiol 60:1052–1055

Kumar B, Trivedi P, Mishra AK, Pandey A, Palni LM (2004) Microbial diversity of soil from two hot springs in Uttaranchal Himalaya. Microbiol Res 159:141–146

Kumar M, Yadav AN, Tiwari R, Prasanna R, Saxena AK (2014) Deciphering the diversity of culturable thermotolerant bacteria from Manikaran hot springs. Ann Microbiol 64:741–751

Land M, Hauser L, Jun SR, Nookaew I, Leuze MR, Ahn TH, Karpinets T, Lund O, Kora G, Wassenaar T, Poudel S, Ussery DW (2015) Insights from 20 years of bacterial genome sequencing. Funct Integr Genomics 15:141–161

Liang J, Kang D, Wang Y, Yu Y, Fan J, Takashi E (2015) Carbonate ion-enriched hot spring water promotes skin wound healing in nude rats. PLoS One 10(2):e0117106. https://doi.org/10.1371/journal.pone.0117106

Mahato NK, Tripathi C, Verma H, Singh N, Lal R (2014) Draft genome sequence of *Deinococcus* sp. strain RL isolated from sediments of a hot water spring. Genome Announc 2(4):e00703–e00714. https://doi.org/10.1128/genomeA.00703-14

Mangrola A, Dudhagara P, Koringa P, Joshi CG, Parmar M, Patel R (2015a) Deciphering the microbiota of Tuwa hot spring, India using shotgun metagenomic sequencing approach. Genom Data 4:153–155

Mangrola AV, Dudhagara P, Koringa P, Joshi CG, Patel RK (2015b) Shotgun metagenomic sequencing based microbial diversity assessment of Lasundra hot spring, India. Genom Data 4:73–75

Marsh CL, Larsen DH (1953) Characterization of some thermophilic bacteria from the hot springs of Yellowstone National Park. J Bacteriol 65:193–197

Masaki Y, Tsutsumi K, Hirano S, Okibe N (2016) Microbial community profiling of the Chinoike Jigoku ("Blood Pond Hell") hot spring in Beppu, Japan: isolation and characterization of Fe(III)-reducing *Sulfolobus* sp. strain GA1. Res Microbiol 167:595–603

Mehetre GT, Paranjpe AS, Dastager SG, Dharne MS (2016) Complete metagenome sequencing based bacterial diversity and functional insights from basaltic hot spring of Unkeshwar, Maharashtra, India. Genom Data 7:140–143

Mehta D, Satyanarayana T (2013) Diversity of hot environments and thermophilic microbes. In: Satyanarayana T, Littlechild J, Kawarabayasi Y (eds) Thermophilic microbes in environmental and industrial biotechnology. Springer, Dordrecht, pp 3–60

Merkel AY, Pimenov NV, Rusanov II, Slobodkin AI, Slobodkina GB, Tarnovetckii IY, Frolov EN, Dubin AV, Perevalova AA, Bonch-Osmolovskaya EA (2017) Microbial diversity and autotrophic activity in Kamchatka hot springs. Extremophiles 21(2):307–317

Miquel P (1888) Monographie d'un bacille vivant au-dela de 70 °C. Ann Micrographic 1:4

Mittal P, Saxena R, Sharma VK (2017) Draft genome sequence of *Anoxybacillus mongoliensis* strain MB4, a sulfur-utilizing aerobic thermophile isolated from a hot spring in Tattapani, Central India. Genome Announc 5:e01709–e01716. https://doi.org/10.1128/genomeA.01709-16

Narayan KD, Pandey SK, Das SK (2010) Characterization of *Comamonas thiooxidans* sp. nov and the comparison of thiosulfate oxidation with *Comamonas testosteroni* and *Comamonas composti*. Curr Microbiol 61:248–253

Narayan KD, Sabat SC, Das SK (2016a) Mechanism of electron transport during thiosulfate oxidation in an obligately mixotrophic bacterium *Thiomonas bhubaneswarensis* strain S10 (DSM 18181^T). Appl Microbiol Biotechnol 101:1239–1252

Narayan KD, Badhai J, Whitman WB, Das SK (2016b) Draft genome sequence of *Comamonas thiooxydans* strain $S23^T$ ($DSM17888^T$), a thiosulfate-oxidizing bacterium isolated from a sulfur spring in India. Genome Announc 4(4):e00834–e00816. https://doi.org/10.1128/genomeA.00834-16

Nishiyama M, Yamamoto S, Kurosawa N (2013) Microbial community analysis of a coastal hot spring in Kagoshima, Japan, using molecular and culture-based approaches. J Microbiol 51(4):413–422
Oldham RD (1888) Some notes on the geology of the North-West Himalayas. Rec Geol Surv India 21:149–157
Panda SK, Jyoti V, Bhadra B, Nayak KC, Shivaji S, Rainey FA, Das SK (2009) *Thiomonas bhubaneswarensis* sp. nov., a novel obligately mixotrophic, moderately thermophilic, thiosulfate oxidizing bacterium. Int J Syst Evol Microbiol 59:2171–2175
Panda AK, Bisht SS, Kumar NS, Mandal SD (2015) Investigations on microbial diversity of Jakrem hot spring, Meghalaya, India using cultivation-independent approach. Genom Data 4:156–157
Panday D, Das SK (2010) *Chelatococcus sambhunathii* sp. nov., a moderately thermophilic alphaproteobacterium isolated from a hot spring sediment. Int J Syst Evol Microbiol 60:861–865
Pandey OP, Negi JG (1995) Geothermal fields of India: a latest update. In: Proceedings of World Geothermal Congress International Geothermal Association, Bochum, Germany, pp 163–171
Pathak AP, Rathod MG (2014) Cultivable bacterial diversity of terrestrial thermal spring of Unkeshwar, India. J Biochem Tech 5:814–818
Pednekar P, Jain R, Mahajan G (2011) Anti-infective potential of hotspring bacteria. J Glob Infect Dis 3:241–245
Piterina AV, Pembroke JT (2010) Preparation and analysis of environmental DNA: optimisation of techniques for phylogenetic analysis of ATAD sludge. In: Mendez-Vilaz A (ed) Current research, technology and education topics in applied microbiology and microbial biotechnology. Formatex, Badajoz, pp 1533–1538
Poddar A, Lepcha RT, Das SK (2014) Taxonomic study of the genus *Tepidiphilus*: transfer of *Petrobacter succinatimandens* to the genus *Tepidiphilus* as *Tepidiphilus succinatimandens* comb. nov., emended description of the genus *Tepidiphilus* and description of *Tepidiphilus thermophilus* sp. nov., isolated from a terrestrial hot spring. Int J Syst Evol Microbiol 64:228–235
Poddar A, Lepcha RT, Whitman WB, Das SK (2016) Draft genome sequence of *Tepidiphilus thermophilus* strain JHK30^{T} (JCM 19170^{T}) isolated from a terrestrial hot spring in India. Genome Announc 4(4):e00832–e00816. https://doi.org/10.1128/genomeA.00832-16
Rakshak K, Ravinder K, Nupur STN, Kumar PA (2013) *Caldimonas meghalayensis* sp. nov., a novel thermophilic betaproteobacterium isolated from a hot spring of Meghalaya in Northeast India. Antonie Van Leeuwenhoek 104(6):1217–1225
Ramanathan A, Chandrasekharam D (1997) Geochemistry of Rajpur and Puttur thermal springs of the West Cost, India. J Geol Soc India 49:559–556
Reigstad LJ, Jorgensen SL, Schleper C (2010) Diversity and abundance of *Korarchaeota* in terrestrial hot springs of Iceland and Kamchatka. ISME J 4:346–356
Ruckmani A, Kaur I, Schumann P, Klenk HP, Mayilraj S (2011) *Calidifontibacter indicus* gen. nov., sp. nov., a member of the family *Dermacoccaceae* isolated from a hot spring, and emended description of the family *Dermacoccaceae*. Int J Syst Evol Microbiol 61:2419–2424
Saha P, Chakrabarti T (2006a) *Aeromonas sharmana* sp. nov., isolated from a warm spring. Int J Syst Evol Microbiol 56:1905–1909
Saha P, Chakrabarti T (2006b) *Flavobacterium indicum* sp. nov., isolated from warm spring water in Assam. India. Int J Syst Evol Microbiol 56:2617–2621
Saha P, Chakrabarti T (2006c) *Emticicia oligotrophica* gen. nov., sp. nov., a new member of the family '*Flexibacteraceae*', phylum Bacteroidetes. Int J Syst Evol Microbiol 56:991–995
Saha P, Krishnamurthi S, Mayilraj S, Prasad GS, Bora TC, Chakrabarti T (2005a) *Aquimonas voraii* gen. nov., sp. nov., a novel gammaproteobacterium isolated from a warm spring of Assam, India. Int J Syst Evol Microbiol 55:1491–1495
Saha P, Mondal AK, Mayilraj S, Krishnamurthi S, Bhattacharya A, Chakrabarti T (2005b) *Paenibacillus assamensis* sp. nov., a novel bacterium isolated from a warm spring in Assam, India. Int J Syst Evol Microbiol 55:2577–2581
Saha P, Krishnamurthi S, Bhattacharya A, Sharma R, Chakrabarti T (2010) *Fontibacillus aquaticus* gen. nov., sp. nov., isolated from a warm spring. Int J Syst Evol Microbiol 60:422–428

Sahoo RK, Subudhi E, Kumar M (2015) Investigation of bacterial diversity of hot springs of Odisha, India. Genom Data 6:188–190

Sarolkar PB (2005) Geochemical characters of hot springs of West Coast, Maharashtra State, India. In: Proceedings World Geothermal Congress, Antalya, Turkey, 24–29 April 2005

Saxena R, Dhakan DB, Mittal P, Waiker P, Chowdhury A, Ghatak A, Sharma VK (2017) Metagenomic analysis of hot springs in Central India reveals hydrocarbon degrading thermophiles and pathways essential for survival in extreme environments. Front Microbiol 7:2123. https://doi.org/10.3389/fmicb.2016.02123

Schlagintweit RD (1865) Enumeration of hot springs of India and high Asia. J Asiat Soc Bengal 33:51–73

Sen R, Maiti NK (2014) Genomic and functional diversity of bacteria isolated from hot springs in Odisha, India. Geomicrobiol J 31:541–550

Sharma SK (2010) Geothermal: a sustainable energy outlook for India. In: Proceedings World Geothermal Congress, Bali, Indonesia, 25–29 April 2010

Sharma N, Vyas G, Pathania S (2013) Culturable diversity of thermophilic microorganisms found in hot springs of northern Himalayas and to explore their potential for production of industrially important enzymes. Sch Acad J Biosci 1:165–178

Sharma A, Hira P, Shakarad M, Lal R (2014) Draft genome sequence of *Cellulosimicrobium* sp. strain MM, isolated from arsenic-rich microbial mats of a Himalayan hot spring. Genome Announc 2(5):e01020–e01014. https://doi.org/10.1128/genomeA.01020-14

Sharma A, Kohli P, Singh Y, Schumann P, Lal R (2016) *Fictibacillus halophilus* sp. nov., from a microbial mat of a hot spring atop the Himalayan range. Int J Syst Evol Microbiol 66:2409–2416

Singh A, Subudhi E (2016) Structural insights of microbial community of Deulajhari (India) hot spring using 16s-rRNA based metagenomic sequencing. Genom Data 7:101–102

Thomas J, Gonzalves EA (1965a) Thermal algae of Western India. I. Algae of the hot springs at Akloli and Ganeshpuri. Hydrobiologia 25:330–340

Thomas J, Gonzalves EA (1965b) Thermal algae of Western India. II. Algae of the hot springs at Palli. Hydrobiologia 25:340–351

Thomas J, Gonzalves EA (1965c) Thermal algae of Western India. III. Algae of the hot springs at Sav. Hydrobiologia 26:21–28

Thomas J, Gonzalves EA (1965d) Thermal algae of Western India. IV. Algae of the hot springs at Aravali, Tooral and Rajewadi. Hydrobiologia 26:29–40

Thomas J, Gonzalves EA (1965e) Thermal algae of Western India, VI. Algae of the hot springs at unai, Lasundra and unapdeo. Hydrobiologia 26:55–65

Thussu JL (2002) Geothermal energy resources of India. Geol Surv India Spec Publ 69:210

Tripathi C, Mahato NK, Rani P, Singh Y, Kamra K, Lal R (2016a) Draft genome sequence of *Lampropedia cohaerens* strain CT6^T isolated from arsenic rich microbial mats of a Himalayan hot water spring. Stand Genom Sci 11(1):64. https://doi.org/10.1186/s40793-016-0179-1

Tripathi C, Mahato NK, Singh AK, Kamra K, Korpole S, Lal R (2016b) *Lampropedia cohaerens* sp. nov., a biofilm-forming bacterium isolated from microbial mats of a hot water spring, and emended description of the genus *Lampropedia*. Int J Syst Evol Microbiol 66:1156–1162

Tripathy S, Padhi SK, Sen R, Maji U, Samanta M, Mohanty S, Maiti NK (2016) Draft genome sequence of *Brevibacillus borstelensis* cifa_chp40, a thermophilic strain having biotechnological importance. J Genom 4:4–6

Verma A, Dhiman K, Gupta M, Shirkot P (2015) Bioprospecting of thermotolerant bacteria from Hot Water Springs of Himachal Pradesh for the production of Taq DNA polymerase. Proc Natl Acad Sci India Sect B Biol Sci 85:739–749

Yang J, Zhou E, Jiang H, Li WJ, Wu G, Huang L, Hedlund BP, Dong H (2015) Distribution and diversity of aerobic carbon monoxide-oxidizing bacteria in geothermal springs of China, the Philippines, and the United States. Geomicrobiol J 32:903–913

Zimik HV, Farooq SH, Prusty P (2017) Geochemical evaluation of thermal springs in Odisha, India. Environ Earth Sci 76(593). https://doi.org/10.1007/s12665-017-6925-x

3 Diversity of Thermophiles in Terrestrial Hot Springs of Yunnan and Tibet, China

Wen-Dong Xian, Manik Prabhu Narsing Rao, En-Min Zhou, Lan Liu, Min Xiao, and Wen-Jun Li

Abstract

The Yunnan-Tibetan geothermal zone (YTGZ), located in western and southwestern China, harbors hundreds of hot springs with a wide range of temperatures and pH. These hot springs provide large and diverse niche for thermophilic microorganisms. In this chapter, we will discuss culture-dependent and culture-independent studies that have been conducted to understand Yunnan-Tibetan hot spring microbial diversity. Several novel taxa isolated and their uniqueness are listed. This chapter also sheds light on various physicochemical factors that structure the microbial diversity. The bioactive molecules and functional genes reported from these hot springs are also listed here.

Keywords

Terrestrial hot springs · Thermophiles · Yunnan · Tibet · Culture-dependent and culture-independent analysis · Novel taxa

3.1 Introduction

Terrestrial hot springs, a type of extreme environments, existed on Earth for billions of years (Gold 1992). The physical and chemical characteristics of the hot springs make it a unique habitat which is quite different from the surrounding environments (Bisht et al. 2011). Hot springs are once perceived to be sterile environment (Chaudhuri et al. 2017), but the pioneer work of Thomas Brock's in discovering

---

W.-D. Xian · M. P. Narsing Rao · E.-M. Zhou · L. Liu · M. Xiao · W.-J. Li (✉)
State Key Laboratory of Biocontrol, School of Life Sciences, Sun Yat-Sen University, Guangzhou, China
e-mail: liwenjun3@mail.sysu.edu.cn

D. Egamberdieva et al. (eds.), *Extremophiles in Eurasian Ecosystems: Ecology, Diversity, and Applications*, Microorganisms for Sustainability 8,
https://doi.org/10.1007/978-981-13-0329-6_3

Thermus aquaticus from thermal vents of Yellowstone National Park (Brock 1997) has completely changed our understanding of hot spring microbial diversity. Hot springs harbor unique microbial diversity that could be the source of commercially important products (Satyanarayana et al. 2005). Understanding how living communities survive and structured in hot spring conditions is important because hot springs are similar to the postulated early chemical environment on Earth (Li et al. 2015) and thus hot springs become a model ecosystems for research on the origin and evolution of life (Farmer 1998; Whitaker et al. 2003). Culture-dependent microbial analysis of hot springs reveals the presence of many new taxonomic and functional lineages (Xue et al. 2001; Huang et al. 2010). Although culture-dependent analysis is regarded as an effective method to understand the microbial diversity, these are not sufficient to explore the microbial diversity as it does not reveal a clear picture of the community diversity due to lack of cultivation of microorganisms in laboratory conditions (Kikani et al. 2017). Recent advances in culture-independent microbial diversity analysis have showed a remarkable progress in understanding community diversity. The application of culture-independent analysis has proved to be a promising tool to investigate detailed insight of hot spring microbial habitats in terms of diversity, adaptation, functions, and ecological significance (Badhai et al. 2015).

Geothermal Professional Committee of China Energy Research Society in 1986 has identified nearly 3398 hot springs distributed across China. The most concentrated regions of thermal springs are located in Yunnan and Tibet (Liao 2018). Extensive studies have been carried out in understanding microbial diversity of Yunnan and Tibet hot springs (Liu et al. 2016; Wang et al. 2014; Hou et al. 2013). In this chapter, we discuss several major hot springs of Yunnan and Tibet, their physicochemical properties, and their microbial diversity and bioactive molecules.

3.2 Terrestrial Hot Springs in Yunnan-Tibetan Geothermal Zone (YTGZ)

The YTGZ (Fig. 3.1) located between the Indian Plate and Eurasian Plate is well known for its volcanic activity and geothermal features (Wang et al. 2013). The northeastern edge of the YTGZ belongs to Himalayan Geothermal Belt (HGB) which resulted from the collision of the Indian Plate with the Eurasian Plate. HGB is more than 50 km wide and extends for 3000 km, distributed throughout India, Tibet, Yunnan, Myanmar, and Thailand, associated with at least 600 geothermal systems (Hochstein and Regenauer-Lieb 1998). The total amount of thermal springs in YTGB accounts for half of the total number of thermal springs in China with wide ranges of temperature and pH (Wang et al. 2013; Wu et al. 2015). Some of the important Yunnan and Tibet hot springs and their physicochemical parameters are listed in Table 3.1.

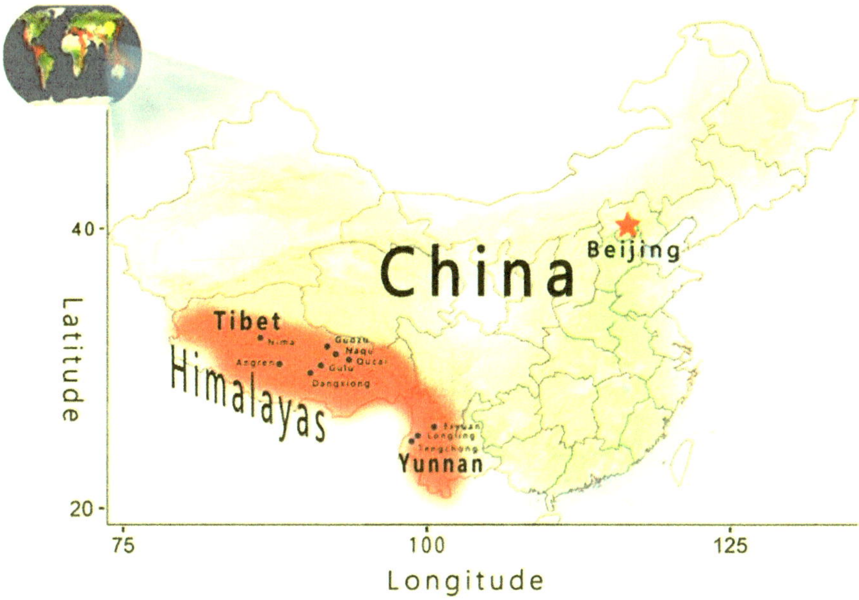

Fig. 3.1 The red highlighted portion shows some of the important Yunnan and Tibet hot springs

A total number of thermal springs in Tibet are 645, in which Nagqu Prefecture accounts for highest hot springs (187), followed by Qamdo Prefecture (126), Xigaze (117), Ngari Prefecture (88), Nyingchi Prefecture (49), Shangnan Prefecture (44), and Lhasa City (34). Tibet has 48 boiling springs (having temperature greater than 86 °C), 179 hot springs (having temperature less than 86 °C), 294 warm springs (having temperature less than 45 °C), and 127 tepid springs (having temperature less than 35 °C) (Liao 2018). Most of boiling springs found in Tibet discharge sodium chloride type of water which is boron rich. Some boiling springs discharge Cl–HCO_3–Na-type water or HCO_3–Cl–Na-type water due to blending of different degrees of cold water. Thermal springs in Tibet have high salinity (Liao 2018).

Yunnan Province is the largest province in terms of number of thermal springs in China with unbelievable physical and chemical features. More than 862 hot springs have been reported in Yunnan Province; among these springs, 20 are boiling springs, 314 are hot springs, 208 are warm springs, and 321 are tepid springs. All these high-temperature boiling springs have high fluorine and low boron content. In lower-temperature hot springs and warm springs, especially tepid springs, their water chemistry type is basically HCO_3–Ca or HCO_3–Mg type. TDS value of Yunnan springs is very low, except the Rehai geothermal field of Tengchong (Liao 2018).

Table 3.1 Descriptions of some important hot springs of Yunnan and Tibet

Hot springs	Geographical location	Source temperature (°C)	Source pH	Description	References
Rehai National Park hot springs	Rehai,Tengchong,Yunnan	69–97	2.58–9.39	A wide physicochemical diversity of springs up to 97 °C temperature	Hedlund et al. (2012)
Eryuan Niujie hot springs	Niujie, Eryuan, Yunnan	73–78	7.3–7.4	Hot springs with black mat or black sandy sediments	Song et al. (2013a)
Ruidian geothermal fields	Diantan, Eryuan, Yunnan	73.8–81.6	6.71–7.29	Two large neutral springs constructed into tetragonal (Jinze) or octagonal (Gongxiaoshe) pool and several small pools	Hou et al. (2013)
Jiwa town hot springs	Nima County, Tibet	41.7–56	6.5–6.8	A geothermal field with a large amount of nameless hot springs, elevation(m): 4618-4630	Huang et al. (2011)
Rongma town hot springs	Nima County, Tibet	26.2–64.1	6.8–7.2	A geothermal field with a large amount of nameless hot springs, elevation(m): 4718-4725	Huang et al. (2011)
Gulu town hot springs	Naqu County, Tibet	52–81.2	8.2	A geothermal field with dozens of nameless hot springs, elevation(m): 4711-4735	Huang et al. (2011)
Guozu town hot springs	Naqu County, Tibet	22.1	7.2	A geothermal field with moderate hot springs	Wang et al. (2013)
Nima town hot springs	Naqu County, Tibet	43–48	7	A geothermal field with a large amount of nameless hot springs	Wang et al. (2013)
Naqu County hot springs	Naqu County, Tibet	49	7.8	A geothermal field with a large amount of nameless hot springs	Wang et al. (2013)
Qucai town hot springs	Naqu County, Tibet	60–75	7.6–8.1	A geothermal field with a large amount of nameless hot springs	Wang et al. (2013)
Yangbajing hot springs	Dangxiong County, Tibet	67–85	6	Geothermal power plant,with some abandoned thermal vent, elevation(m): 4360	Chen et al. (2009)

3.3 Culture-Dependent Microbial Diversity Analysis

A large number of cultivation-dependent studies have been performed in Yunnan and Tibet hot springs (Hedlund et al. 2015; Xian et al. 2016; Khan et al. 2017a, b). Several new taxa (Table 3.2) and their bioactive potential have been reported from Yunnan and Tibet hot springs (Duan et al. 2014; Xian et al. 2016; Khan et al. 2017a, b; Chen et al. 2012).

3.3.1 Physicochemical Factors Structuring Microbial Diversity

The seasonal dynamics of both the physicochemical conditions and the microbial communities inhabiting hot springs in Tengchong County, Yunnan Province, China, have been evaluated. The seasonal variation, especially the rainy season changed the physicochemical conditions and microbial communities. Monsoon samples showed increased concentrations of potassium, total organic carbon, ammonium, calcium, sodium, and total nitrogen with decreased ferrous iron relative to the dry season. High mesophilic community has been observed after monsoon which may be flushed into springs due to enhanced rain influx (Briggs et al. 2014).

Wang and co-workers have evaluated temporal changes of sediment and water microbial community in Tengchong hot springs, Yunnan Province, China. The authors suggest that the microbial communities were not transported into hot springs from the surroundings by increased surface runoff, but rather their occurrence or even dominance was due to large temporal variations of physicochemical conditions such as pH, temperature, and dissolved organic carbon. Water and sediment communities responded differently to temporal physicochemical changes. Water communities were found stable, while sediment communities were more responsive to temporal geochemical changes. Greater temporal variations were observed in individual taxa than at the whole community structure level (Wang et al. 2014).

3.3.2 *Proteobacteria*, *Firmicutes*, and *Aquificae*

Culture-dependent analysis showed that *Proteobacteria*, *Firmicutes*, and *Aquificae* were the prominent groups residing in the hot springs in India (Kumar et al. 2004; Sen and Maiti 2014; Pathak and Rathod 2014). Culture-dependent analysis of the samples from Yunnan and Tibetan hot springs also showed the presence of *Proteobacteria*, *Firmicutes*, and *Aquificae* groups.

Several novel genera such as *Caldalkalibacillus* (*Caldalkalibacillus thermarum* as type species) (Xue et al. 2015), *Crenobacter* (*Crenobacter luteus* as type species) (Dong et al. 2015), *Crenalkalicoccus* (*Crenalkalicoccus roseus* as type species) (Ming et al. 2016), *Caldovatus* (*Caldovatus sediminis* as type species) (Habib et al. 2017a), and *Tibeticola* (*Tibeticola sediminis* as type species) (Khan et al. 2017b) have been reported from Yunnan and Tibet hot springs. Genera, namely, *Caldalkalibacillus*, *Crenobacter*, *Crenalkalicoccus*, and *Caldovatus* have been

Table 3.2 Novel bacterial and archaeal strains isolated from Yunnan and Tibet

Novel bacterial strains	Isolation sources	Gram staining	Type strain/ deposition number	GenBank accession number	Oxygen requirement	Growth requirements			References
						Temperature °C range ($T_{opt.}$ °C)	pH range ($pH_{opt.}$)	NaCl range (optimum) (w/v)	
Bacteria									
Crenobacter luteus	Sediment from hot spring of Hehua, Tengchong, Yunnan Province	Negative	YIM 78141^T (=BCRC 80650^T = KCTC 32558^T = DSM 27258^T)	KF771276	Aerobic	10–55 (40–50)	6.0–10.0 (8.0–9.0)	0–3% (0–1%)	Dong et al. (2015)
Crenalkalicoccus roseus	Sediment sample collected from an alkaline hot spring in Tengchong County, Yunnan Province	Negative	YIM (78023^T (=JCM 19657^T = KACC 17825^T)	KJ361470	Aerobic	20–60 (40–50)	6.0–9.0 (8.0)	0–2.5 (0–2.0%)	Ming et al. (2016)
Caldovatus sediminis	Sediment sample from Hamazui hot spring in Tengchong County, Yunnan Province	Negative	YIM 72346^T (=KCTC 52714^T = CGMCC 1.16330^T)	MF446885	Aerobic	37–50 (45)	6.0–9.0 (6.5–7.0)	0.5–1.0% (0.5%)	Habib et al. (2017a)
Tibeticola sediminis	Sediment sample collected from a hot spring in Tibet Province	Negative	YIM 73013^T (=DSM 101684^T = KCTC 42873^T)	KX161735	Aerobic	30–55 (37–45)	6.0–8.0 (7.0)	0–1% (ND)	Khan et al. (2017b)

Laceyella sediminis	Sediment sample collected from a hot spring in Tengchong County, Yunnan Province	Positive	RHA1^T (=DSM 45263^T = CCTCC AA 208058^T)	FJ422144	Aerobic	28–65 (55)	5.0–9.0 (7.0)	0–1% (0%)	Chen et al. (2012)
Brevibacillus sediminis	Sediment of a hot spring in the Tagejia geothermal field, Angren, Tibet Province	Positive	YIM 78300^T (= DSM 29928^T = CPCC 100738^T	KP985221	Aerobic	37–65 (50–55)	6.0–8.0 (7.0)	0–1.5% (ND)	Xian et al. (2016)
Altererythrobacter lauratis	Sediment collected from the Tagejia hot spring in Tibet	Negative	YIM 75003^T = KCTC 52606^T = CCTCC AB2016258^T)	KX808673.	Aerobic	10–55 (37–45)	5.0–11.0 (8.0)	0–11% (1–6%)	Yuan et al. (2017)
Altererythrobacter palmitatis		Negative	YIM 75004^T = KCTC 52607^T = CCTCC AB2016270^T)	KX808674	Aerobic	10–55 (37–45)	5.0–9.0 (8.0)	0–11% (1–6%).	
Thermoanaerobacter tengcongensis	Mixed sediment and water sample taken from a hot spring in Tengcong, located in Yunnan Province	Negative	MB4^T (= Chinese collection of microorganisms AS 1.2430^T = JCM 11007^T)	AF209708	Obligate anaerobic	50–80 (75)	5.5–9.0 (7.0–7.5)	0–2.5% (0.2%)	Xue et al. (2001)

(continued)

Table 3.2 (continued)

Novel bacterial strains	Isolation sources	Gram staining	Type strain/ deposition number	GenBank accession number	Oxygen requirement	Growth requirements			References
						Temperature °C range (T_{opt}. °C)	pH range ($pH_{opt.}$)	NaCl range (optimum) (w/v)	
Themoanaerobacterium calidifontis	Mixed sediment and water sample taken from a hot spring in Baoshan of Yunnan Province	Positive	Rx1 ^{T} (=JCM 18270 ^{T} = CCTCC M 2011109^T)	AB544080	Obligate anaerobe	38–68 (50–55)	4.5–8.0 (7.0)	0–3% (ND)	Shang et al. (2013)
Thermus rehai	Water sample hot springs in Rehai of Tengchong, Yunnan Province	Negative	RH99-GF7504^T (=CCTCC-AB200292^T)	AF331969	Aerobic	40–80 (65–70)	5.5–10(7.5–8.5.)	ND	Lin et al. (2002)
Thermus caliditerrae	Sediment sample from hydrothermal explosion (Shuirebaozhaqu) area in Tengchong County, Yunnan Province	Negative	YIM 77925^T (=DSM 25901^T = CCTCC 2012061^T)	KC852874	Aerobic	50–70 (65)	6.0–8.0 (7.0)	0–0.5% (0%)	Ming et al. (2014)
Thermus amyloliquefaciens	Sediment sample collected from Eryuan of Dali, Yunnan Province	Negative	YIM 77409^T (= DSM 25898^T = KCTC 32024^T)	KP284528	Aerobic	50–70 (60–65)	6.0–8.0 (7.0)	In absence of NaCl	Yu et al. (2015)
Thermus caldifontis	Sediment sample collected from Tibet hot spring	Negative	YIM 73026^T (=NBRC 112415^T = CCTCC AB 2016305^T)	KX580314	Aerobic	50–70 (60)	6.0–8.0 (7.0)	0.5–1% (ND)	Khan et al. (2017a)

Meiothermus rosaceus	Tengchong hot springs, Yunnan, PR China	Negative	RH9901^T (=CCTCC-AB200291^T)	AF312766	Aerobic	40–70(55)	4.5–10.5(8–9)	ND	Chen et al. (2002)
Meiothermus roseus	Sediment from a hot spring in Tengchong County, Yunnan Province	Negative	YIM (71031^T (=KCTC 42495^T = NBRC 110900^T)	KP232922	Aerobic	37–65 (50)	6.0–9.0 (7.0–7.5)	1.0%(ND)	Ming et al. (2015)
Meiothermus luteus	Sediment sample from a hot spring in Tengchong County, Yunnan Province	Negative	YIM 72257^T (=KCTC 52599^T =CCTCC AB 2017100^T)	KY608080	Aerobic	50–65 (60)	6.0–9.0 (7.0–8.0)	0–1% (0–0.5%)	Habib et al. (2017b)
Streptomyces calidiresistens	Sediment sample collected from Hehua hot spring in Tengchong, Yunnan Province	Positive	YIM 78087^T (=BCRC 16955^T = DSM 42108^T = JCM 19629^T)	KJ361473	Aerobic	28–50 (37–45)	4.0–10.0 (8.0–9.0)	0–9% (1–3%)	Duan et al. (2014)
Caldalkalibacillus thermarum	Water sample from Drumbeat in Rehai Park, Tengchong, Yunnan Province	Positive	HA6^T (=CGMCC 1.4242^T = JCM 13486^T)	AY753654	Aerobic	45–65 (60)	7.5–10 (8.5)	0–6% (1.5%)	Xue et al. (2015)

(continued)

Table 3.2 (continued)

Novel bacterial strains	Isolation sources	Gram staining	Type strain/ deposition number	GenBank accession number	Oxygen requirement	Growth requirements: Temperature °C range (T_{opt}. °C)	pH range ($pH_{opt.}$)	NaCl range (optimum) (w/v)	References
Archaea									
Metallosphaera tengchongensis	Muddy water sample of a sulfuric hot spring located in Tengchong County, Yunnan Province	Negative	Ric-A^T (=NBRC 109472^T = CGMCC 1.12287^T)	KJ735100	Facultatively aerobic	55–75 (70)	1.5–6.5 (3.5)	0–1% (ND)	Peng et al. (2015)
Metallosphaera cuprina	Muddy water sample collected from the edge of the hot springs of Tengchong County, Yunnan Province	Negative	Ar-4^T (=JCM 15769^T = CGMCC 1.7082^T)	FN796482	Facultatively aerobic	55–75 (65)	2.5–5.5 (3.5)	0–1% (ND)	Liu et al. (2011a)

ND Not determined

reported from Tengchong (Yunnan province) hot springs, while *Tibeticola* has been reported from Tibet hot spring. The optimum temperature for growth of these novel genera varies; *Caldalkalibacillus thermarum* and *Crenalkalicoccus roseus* have been reported for optimum growth at 60 °C. *Crenobacter luteus* had optimum growth at 40–50 °C; *Caldovatus sediminis* and *Tibeticola sediminis* had optimum growth at 45 °C. *Crenobacter luteus*, *Crenalkalicoccus roseus*, *Caldovatus sediminis*, and *Tibeticola sediminis* were reported for their growth from slightly acidic to alkaline pH with optimum pH at 8.0–9.0, 8.0, 6.5–7.0, and 7.0, respectively (Xue et al. 2015; Dong et al. 2015; Ming et al. 2016; Habib et al. 2017a, b). Further, several novel aerobic species, namely, *Laceyella sediminis* (Chen et al. 2012), *Brevibacillus sediminis* (Xian et al. 2016), *Altererythrobacter lauratis*, and *Altererythrobacter palmitatis* (Yuan et al. 2017), and obligate anaerobes such as *Thermoanaerobacter tengcongensis* (Xue et al. 2001), and *Thermoanaerobacterium calidifontis* (Shang et al. 2013) have also been reported. In-depth analysis of novel species, *Thermoanaerobacterium calidifontis*, showed the ability to produce ethanol and ability to convert thiosulfate to elemental sulfur and reduce sulfite to hydrogen sulfide (Shang et al. 2013). *Thermoanaerobacter tengcongensis* reported to reduce thiosulfate and sulfur to hydrogen sulfide (Xue et al. 2001). Further, complete sequence of *Thermoanaerobacter tengcongensis* has been carried out which suggests *Thermoanaerobacter tengcongensis* has a genome size of 2,689,445 bp. The genome encodes 2588 predicted coding sequences. Among them, 1764 (68.2%) were similar to documented proteins, and the rest, 824 coding sequences (31.8%), were functionally unknown (Bao et al. 2002).

Many culture-independent analyses of microbial diversity in hot springs of Tengchong, Yunnan Province, were carried out suggesting *Aquificales* populations as the abundant group (Hou et al. 2013; Song et al. 2013a), but attempts to isolate *Aquificales* from Tengchong hot springs have never been made. Hedlund and his co-workers made an attempt to isolate diverse members of the *Aquificales* from geothermal springs in Tengchong, China. The authors have isolated five strains of *Aquificales* from diverse springs (temperature 45.2–83.3 °C and pH 2.6–9.1). Phylogenetic analysis showed that four of the strains belong to the genera *Hydrogenobacter*, *Hydrogenobaculum*, and *Sulfurihydrogenibium*, including some strains distinct enough to likely justify as new species of *Hydrogenobacter* and *Hydrogenobaculum*. They also suggested that one strain was distinct enough to represent as new genus in the *Hydrogenothermaceae* family. All strains were capable of aerobic respiration under microaerophilic conditions; however, they had variable capacity for chemolithotrophic oxidation of hydrogen and sulfur compounds and nitrate reduction (Hedlund et al. 2015).

Efforts to isolate acidophilic bacteria from Yunnan Province hot springs have been made. It was noticed that acidophilic mesophiles in these regions were more diverse, and several ferrous iron and sulfur-oxidizing genera such as *Acidiphilum* (Liu et al. 2007), *Acidothiobacillus*, *Alicyclobacillus*, *Leptospirillum*, and *Sulfobacillus* (Jiang et al. 2009) were present.

3.3.3 *Deinococcus-Thermus*

The genus *Thermus* has been regarded as models to investigate the mechanism of thermostability of thermophiles (Saiki et al. 1972). The diversity of *Thermus* has been evaluated in 15 hot springs of Rehai geothermal area, Tengchong, China. The isolation was carried out using *Thermus* and YIM14 medium. A total of 57 *Thermus* strains have been recovered. Strains from YIM14 medium were physiologically more diverse than strains from *Thermus* medium (Guo et al. 2003).

Several novel species, namely, *Thermus rehai* (Lin et al. 2002), *Thermus caliditerrae* (Ming et al. 2014), *Thermus amyloliquefaciens* (Yu et al. 2015), and *Thermus caldifontis* (Khan et al. 2017a), have been reported from Yunnan and Tibetan hot springs. All the above-described species reported for the growth till 70 °C (Lin et al. 2002; Ming et al. 2014; Yu et al. 2015; Khan et al. 2017a).

Thermus spp. isolated from Yunnan and Tibetan hot springs have been reported as a potential source for bioactive molecules. The novel species *Thermus amyloliquefaciens* was reported for its ability to liquify starch (Yu et al. 2015).

Gong and co-worker evaluated the diversity of thermostable alkaline phosphatase-producing bacteria in Tengchong (Yunnan province) hot springs. Sixty strains belonging to three genera have been isolated. Among them, one strain designated RHY12-2 had highest phosphatase activity. The 16S rRNA gene sequence of the strain RHY12-2 showed the strain was a member of the genus *Thermus* (highest similarity with *Thermus scotoductus* 96.3%) and probably new species. The enzyme had a single peptide with a molecular mass of about 52 kDa with highly specific activity and thermal resistance. The optimum enzyme activity is observed at pH 8.0–10.0 and temperature 70–80 °C (Gong et al. 2005). *Thermus* play a significant role in the cesium assembly. The bacterium *Thermus* sp. TibetanG7, isolated from hot springs in Tibet, China, has been examined for the ability to accumulate cesium from solutions. The accumulation of cesium by this microorganism was rapid with 40%–50% accumulation within the first 5 min (Wang et al. 2007).

The genus *Meiothermus* which was reclassified from the genus *Thermus* (Nobre et al. 1996) is the most common genus isolated from hot spring (Chan et al. 2015). Tengchong hot springs (Yunnan, China) served as potential source for harboring novel species of the genus *Meiothermus*. Several novel species such as *Meiothermus rosaceus* (Chen et al. 2002), *Meiothermus roseus* (Ming et al. 2015), and *Meiothermus luteus* (Habib et al. 2017b) have been reported.

3.3.4 *Actinobacteria*

Actinobacteria have been widely concerned due to their ability of producing various kinds of antibiotics and bioactive molecules (Liu et al. 2016). *Actinobacteria* residing in hot springs exhibits unique metabolic activity. (Xu et al. 1998). Recently, our group has evaluated actinobacterial diversity in ten hot springs distributed over three geothermal fields, namely, Hehua, Rehai, and Ruidian. A total of 58 thermophilic actinobacterial strains have been isolated, and the 16S rRNA gene sequence result showed that these strains shared high similarities with actinobacterial genera, namely, *Actinomadura*, *Micromonospora*, *Microbispora*, *Micrococcus*, *Nocardiopsis*, *Nonomuraea*, *Promicromonospora*, *Pseudonocardia*, *Streptomyces*, and *Verrucosispora*. Some of the strains had low sequence similarity which suggested that these hot springs harbor many novel strains. The isolated strains showed good antimicrobial activity. Fifty-three strains exhibited antimicrobial activities against *Acinetobacter baumannii*. Eighteen and three strains showed inhibitory activities against *Micrococcus luteus* and *Staphylococcus aureus*, respectively. Further, 22 strains were positive for PCR amplification of at least 1 of the 3 biosynthetic gene clusters (PKS-I, PKS-II, and NRPS) (Liu et al. 2016).

Many novel strains have been reported from Yunnan province. During our investigation on thermophilic actinobacterial diversity from hot springs, strain YIM 78087 has been isolated from a sediment sample collected from the Hehua hot spring in Yunnan Province, southwest China. The strain can grow up to 50 °C temperature and tolerate NaCl up to 9% (w/v). The strain was reported for its growth in acidic as well as basic conditions (pH 4.0–10.0). The 16S rRNA gene sequence result showed that strain YIM 78087 belonged to the genus *Streptomyces* and was closely related to *Streptomyces fimbriatus* DSM 40942 (97.18%), *Streptomyces marinus* DSM 41968 (97.05%), and *Streptomyces qinglanensis* DSM 42035 (97.1%). When the taxonomic position of strain YIM 78087 was evaluated, it represented as a novel species of genus *Streptomyces*, for which the name *Streptomyces calidiresistens* has been proposed (Duan et al. 2014). Similarly, strain Y-14046 isolated from a hot spring in Eryuan, Yunnan, China, has been described as new species (*Streptomyces thermogriseus*) of the genus *Streptomyces* (Xu et al. 1998).

3.3.5 Thermophilic Archaea

Compared with bacteria, studies on cultivation of archaea in Yunnan and Tibetan hot springs were rather limited. Xiang and co-workers isolated and characterized novel species *Sulfolobus tengchongensis* RT8-4 from acidic hot spring located in Tengchong, Yunnan Province, China (Xiang et al. 2003). *Sulfolobus tengchongensis* RT8-4 had long and curved peritrichous flagella with aerobic growth either a lithotrophic or heterotrophic mode. It grew fastest at 85–90 °C and was capable of slow growth at 95 °C. Growth has been observed at various pH values ranging from 1.7 to 6.5 with optimum growth at pH 3.5 (Xiang et al. 2003).

A facultatively aerobic novel species, *Acidianus tengchongensis*, has been isolated from a Tengchong acidothermal spring. The optimal pH and temperature for growth reported were 2.5 and 70 °C, respectively. *Acidianus tengchongensis* cells were non-motile, and under anaerobic conditions *Acidianus tengchongensis* reduces elemental sulfur with molecular hydrogen, producing hydrogen sulfide. Under aerobic conditions, it oxidizes elemental sulfur and produces sulfuric acid. No growth reported when cultivated in iron medium, indicating that ferrous iron not be used as an energy source (He et al. 2004).

The diversity of *Sulfolobus* in acidic hot springs of Tengchong of Yunnan, China, has been evaluated by Jian and co-workers. Eleven thermoacidophilic strains from six acidic hot springs have been isolated. The 16S rRNA gene sequence result showed that these strains belong to the genus *Sulfolobus* but are distinct enough to designate as new species (Jian et al. 2010).

Many studies have been carried out to use acidothermophilic archaeal species isolated Yunnan hot springs for bioleaching. Zou and co-workers have isolated thermoacidophilic archaea (*Sulfolobus acidocaldarius*) from hot sulfur spring in the Yunnan Province and conducted bioleaching activity in both laboratory batch bioreactors and leaching columns on low-grade chalcopyrite ore. They reported that bioreactor experiments showed 97% rate of copper bioleaching (in 12 days). In the case of column leaching, tests of a two-phase leaching have been conducted. In the first phase, *Thiobacillus ferrooxidans* has been used, followed by a 140-day thermoacidophilic archaeal leaching in the second phase. The average leaching rate of copper achieved by thermoacidophilic archaea has found to be 195 mg/(L·d), while for the control experiments (for the *Thiobacillus ferrooxidans*), it was 78 mg/(L·d), indicating thermoacidophilic archaea possesses a more powerful oxidizing ability than *Thiobacillus ferrooxidans* (Zou et al. 2006). Two novel acidothermophilic archaeal species *Metallosphaera tengchongensis* (Peng et al. 2015) and *Metallosphaera cuprina* (Liu et al. 2011a) reported for oxidizing metal sulfide ores, showing their potential in bioleaching. These two species were isolated from muddy water samples collected at the edge of the hot springs of Tengchong, Yunnan Province, China. *Metallosphaera tengchongensis* and *Metallosphaera cuprina* were aerobic and facultatively chemolithoautotrophic with growth at temperature ranging from 55 to 75 °C (Peng et al. 2015; Liu et al. 2011a). For better understanding of bioleaching potential, complete genome sequence of *Metallosphaera cuprina* has been carried out. The genome in total carried 2029 open reading frames (ORFs). Genome annotation and metabolic reconstruction supported the idea that *Metallosphaera cuprina* lived a facultative life. *Metallosphaera cuprina* strain fixed CO_2 via the 3-hydroxypropionate/4-hydroxybutyrate cycle, and this strain assimilated carbohydrates via the nonphosphorylated Entner-Doudoroff pathway. It had a complete tricarboxylic acid (TCA) cycle and an incomplete phosphate pentose pathway. Oxidation of RISCs by the heterodisulfide reductase complex, sulfide/quinone oxidoreductase, thiosulfate/quinone oxidoreductase, tetrathionate hydrolase, and sulfite/acceptor oxidoreductase in *Metallosphaera cuprina* has been proposed (Liu et al. 2011b).

3.3.6 Thermophilic Virus

Viruses are the most abundant biological entities in every ecosystem, even in hot springs (Lopez-Lopez et al. 2013). They are probably the only predators in these communities and may be involved in the control of host mortality (Lopez-Lopez et al. 2013). In this section, we provided insights into the latest study being carried out to understand viruses in Yunnan and Tibetan hot springs.

Thermus bacteriophage named TSP4 has been isolated from Tengchong hot springs, China. TSP4 belonged to the *Siphoviridae* family and had a hexagonal head of 73 nm in diameter, an extremely long and flexible tail of 785 nm in length and 10 nm in width (Lin et al. 2010). The first reported *Meiothermus* phage, MMP17 (*Meiothermus Myoviridae* phage 17), has been isolated from hot spring in Eryuan County, Yunnan. MMP17 was a typical myovirus with an icosahedral head (42 nm in diameter) and a tail of 120 nm in length and 17 nm in width. Its DNA was about 33.5–39.5 kb in size. MMP17 was very stable at 55–60 °C and pH 6–7. An average of 15 phages were released from each infected cell (Lin et al. 2011). A virus, denoted STSV1 infecting the hyperthermophilic archaeon *Sulfolobus tengchongensis*, has been isolated from acidic hot springs located in Tengchong, China. The virus STSV1 was spindle (230 by 107 nm) with a tail of variable length (68 nm on average) at one end. STSV1 shape was similar to the members of the family *Fuselloviridae* but much larger than known fuselloviruses. After infecting its host, STSV1 multiplied rapidly to high titers (>10^{10} PFU/ml). Replication of the virus retards host growth, but does not lyse host cells. STSV1 do not integrate into the host chromosome and existed in a carrier state. The STSV1 DNA modifies in an unusual fashion, presumably by virally encoded modification systems. STSV1 harbors a double-stranded DNA genome of 75,294 bp, which shares no significant sequence similarity to those of fuselloviruses. The viral genome contains a total of 74 open reading frames (ORFs), among which 14 have a putative function. Five ORFs that encode viral structural proteins, including a putative coat protein of high abundance, have been noticed. The products of the other nine ORFs have been mentioned to be involved in polysaccharide biosynthesis, nucleotide metabolism, and DNA modification (Xiang et al. 2005).

3.4 Culture-Independent Microbial Diversity Analysis

Culture-dependent method revealed immense limitation for addressing microbial diversities. Majority of microbes in various environments, including hot springs, are still not isolated using traditional cultivation methods (Streit and Schmitz 2004). Hence, culture-dependent microbial analysis does not give clear idea about the microbial diversity residing in a particular environment. In the past few years, the application of culture-independent analysis has proved to be a promising tool to investigate the population diversity, gene content, function, and ecological significance of microbial communities living in diverse hot spring environments (Hou et al. 2013; Badhai et al. 2015). Several culture-independent analyses have been carried out in Yunnan and Tibet hot springs (Song et al. 2009, 2010, 2013a, b; Hou et al. 2013).

A comprehensive cultivation-independent census of microbial communities in 37 samples collected from Rehai and Ruidian geothermal fields, located in Tengchong County, Yunnan Province, has been evaluated using 16S rRNA gene pyrosequencing to understand microbial diversity. The temperature and pH of the samples sites ranged from 55.1 to 93.6 °C and 2.5 to 9.4, respectively. Richness found low in all samples, with 21–123 species-level operational taxonomic units (OTUs). The bacterial phylum *Aquificae* and archaeal phylum *Crenarchaeota* dominated in Rehai samples, yet the dominant taxa within these phyla depended on temperature, pH, and geochemistry. Rehai springs with low pH (2.5–2.6), high temperature (85.1–89.1 °C), and high sulfur contents favored *Sulfolobales*, whereas lower temperature (55.1–64.5 °C) with low pH (2.6–4.8) favored the *Aquificae* (genus *Hydrogenobaculum*). Rehai springs with neutral-alkaline pH (7.2–9.4) and high temperature (80 °C) with high concentrations of silica and salt ions (Na, K, and Cl) favored the *Aquificae* (genus *Hydrogenobacter*). Ruidian water samples harbored a single *Aquificae* (genus *Hydrogenobacter*), whereas microbial communities in Ruidian sediment samples were more diverse at the phylum level and distinctly different from those in Rehai and Ruidian water samples, with high abundance of uncultivated lineages, close relatives of the ammonia-oxidizing archaeon "*Candidatus* Nitrosocaldus yellowstonii," and candidate division O1aA90 and OP1. These differences between Ruidian sediments and Rehai samples were likely caused by temperature, pH, and sediment mineralogy (Hou et al. 2013).

Diversity of *Crenarchaeota* has been investigated in eight terrestrial hot springs (pH 2.8–7.7; temperature 44–96 °C) located in Tengchong, China, using 16S rRNA gene phylogenetic analysis. A total of 826 crenarchaeotal clones were sequenced, and a total of 47 OTUs were identified. About 93% of the OTUs were identical to those retrieved from hot springs and other thermal environments. The result suggests that temperature predominates over pH in affecting crenarchaeotal diversity in Tengchong hot springs. Crenarchaeotal diversity in moderate-temperature (59–77 °C) hot springs was the highest, indicating that the moderately hot-temperature springs may provide optimal conditions for speciation of *Crenarchaeota* (Song et al. 2010).

Investigation on the community diversity and composition in Yunnan and Tibetan hot springs using a barcoded 16S rRNA gene-pyrosequencing approach has been carried out. 16 hot spring samples from five thermal fields, namely, Tengchong, Longling, and Eryuan in Yunnan Province and Gulu and Qucai in Tibet, have been collected. Hot spring samples had a range of temperature (47–96 °C) and pH (3.2–8.6) conditions. *Proteobacteria*, *Aquificae*, *Firmicutes*, *Deinococcus-Thermus*, and *Bacteroidetes* comprised the large portion of the bacterial communities in acidic hot springs (in Yunnan). Nonacidic hot springs (both Yunnan and Tibet) harbor more and variable bacterial phyla than acidic springs; the major phyla of Tibetan hot springs were similar to the Yunnan nonacidic samples but showed different relative abundances. For example, *Bacteroidetes* in Tibetan nonacidic hot springs shows higher abundance than Yunnan. *Desulfurococcales* and unclassified *Crenarchaeota* were the dominated groups in archaeal populations from most of the nonacidic hot springs, whereas the archaeal community structure in acidic hot springs was simpler and dominated by *Sulfolobales* and *Thermoplasmata*. The phylogenetic analyses showed that *Aquificae* and *Crenarchaeota* were predominant in the investigated springs and possessed many phylogenetic lineages that have never been detected in other hot springs in the world (Song et al. 2013a).

Culture-independent approach that combines CARD-FISH, qPCR, and 16S rRNA gene clone library has been carried out to investigate the abundance, community structure, and diversity of microbes along a steep thermal gradient in the Tengchong geothermal field named Shuirebaozhaqu. The authors observed a remarkable change in bacterial and archaeal abundance with temperature changes. Under low-temperature conditions (52.3–74.6 °C), the microbial community that dominated was bacteria. The community was dominated by five phyla, namely, *Proteobacteria*, *Firmicutes*, *Nitrospirae*, *Thermotogae*, and *Cyanobacteria*. The greatest diversity was observed in the phylum *Proteobacteria*, with 11 genera belonging to the classes *Alphaproteobacteria*, *Betaproteobacteria*, and *Gammaproteobacteria*. Archaea dominant at 74.6 °C and 90.8 °C, the number of cells was lowest, but the archaea accounted more than 90% of the total number of cells. Additionally, the microbial communities at high temperatures (74.6–90.8 °C) were substantially simpler than those at the low-temperature sites. Only a few bacterial genera, namely, *Caldisericum*, *Thermotoga*, and *Thermoanaerobacter*, and archaeal genera *Vulcanisaeta* and *Hyperthermus* dominated at high temperature. Some bacteria were observed at both low temperature and high temperature but with different abundance. Genera such as *Hippea*, *Syntrophus*, and *Geobacter* were more adapted to hyperthermal environments, whereas genera such as *Methylobacterium*, *Novosphingobium*, *Achromobacter*, *Desulfomonile*, *Rubrivivax*, *Haemophilus*, *Sorangium*, and *Thauera* were only detected at low temperatures (Li et al. 2015).

Microbial community composition and diversity in hot springs of the Tibetan Plateau across a wide range of temperatures have been evaluated. Thirteen hot spring samples from Nima, Gulu, Naqu, Guozu, and Qucai in Naqu County have been collected and evaluated its microbial diversity using the 16S rRNA gene-pyrosequencing approach. The temperature of these springs ranged from 22.1 to 75 °C. The results suggested that bacteria (42 bacterial phyla) in Tibetan hot springs were more abundant and far more diverse than archaea (5 archaeal phyla). The dominant bacterial phyla systematically varied with temperature. Moderate temperatures (75–66 °C) favored *Aquificae*, whereas low temperatures (60–22.1 °C) favored *Deinococcus-Thermus*, *Cyanobacteria*, and *Chloroflexi*. The relative abundance of *Aquificae* was correlated positively with temperature, but the abundances of *Deinococcus-Thermus*, *Cyanobacteria*, and *Chloroflexi* were negatively correlated with temperature. *Cyanobacteria* and *Chloroflexi* were abundant in Tibetan hot springs, and their abundances were positively correlated at low temperatures (55–43 °C) but negatively correlated at moderate temperatures (75–55 °C). Most archaeal sequences were related to *Crenarchaeota* with only a few related to *Euryarchaeota* and *Thaumarchaeota* (Wang et al. 2013).

The first culture-independent report specifically to actinobacterial diversity in three hot springs located in Tengchong (Frog Mouth hot spring) in China, Kamchatka (Robb Flag hot spring) in Russia, and Nevada boiling spring in the USA has been carried out using denaturing gradient gel electrophoresis (DGGE), restriction fragment length polymorphism (RFLP), and actinobacterial 16S rRNA gene phylogenetic analysis. The authors noticed very diverse actinobacterial populations, and most of the retrieved actinobacterial 16S rRNA gene sequences were affiliated with uncultured *Actinobacteria*. The actinobacterial clone sequences retrieved were affiliated to *Actinomycetales*, *Rubrobacterales*, uncultured *Candidatus* Microthrix, and unclassified *Actinobacteria.* The actinobacterial diversity was noticed at high temperature. Unexpected high actinobacterial diversity was observed in Tengchong hot spring where temperature was 81 °C suggesting these *Actinobacteria* might have an extraordinary capability to adapt to hot spring environments. In this study, authors for the first time were able to retrieve sequences affiliated to *Frankineae* and uncultured *Candidatus* Microthrix in hot spring with temperature as high as 81 °C (Song et al. 2009).

During the global metagenomic survey in geothermal springs, our group found a new bacterial candidate phylum, *Candidatus* Kryptonia with two genera *Candidatus* Chrysopegis kryptomonas and *Candidatus* Kryptobacter tengchongensis from Yunnan hot springs. This lineage had remained hidden as a taxonomic blind spot because of mismatches in the primers commonly used for ribosomal gene surveys. The discovery of a new candidate phylum from the Yunnan hot springs emphasizes that extraordinary microbial novelty is still waiting for the discovery (Eloe-Fadrosh et al. 2016).

3.5 Function Genes and Ecology

Microbial diversity in the hot springs plays a major role in controlling the cycling of organic and inorganic compounds, thereby directly affecting characteristics of environments (Huang et al. 2010; Wu et al. 2015). Various studies have been carried out to understand microbial diversity controlling the cycling of organic and inorganic compounds in Yunnan and Tibetan hot springs (Wu et al. 2015; Song et al. 2013b).

3.5.1 Ammonia-Oxidizing Microorganisms

Ammonium is considered as the major source of inorganic nitrogen in most geothermal springs (Zhang et al. 2008). The abundance of ammonia-oxidizing microorganisms (AOM) and effect of environmental variables in 13 hot springs located in Yunnan Province, China, have been studied. Ammonia-oxidizing archaeal (AOA) abundance ranged 0.02–1.32%, whereas no ammonia-oxidizing bacteria were detected. AOA abundance was significantly correlated with concentrations of NH_3, NO^{2-}, NO^{3-}, pH, and temperature, but not related to salinity and concentrations of Fe^{2+} and salinity (Huang et al. 2010). Studies in Tengchong hot springs showed that the *amoA* gene of aerobic ammonia-oxidizing archaea can be transcribed at temperatures higher than 74 °C and up to 94 °C, suggesting that archaeal nitrification can potentially occur at near boiling temperatures (Jiang et al. 2010).

3.5.2 Archaeal *accA* Gene Genes

Archaea carrying the *accA* gene, encoding the alpha subunit of the acetyl CoA carboxylase, autotrophically fix CO_2 using the 3-hydroxypropionate/4-hydroxybutyrate pathway (Berg et al. 2007; Song et al. 2013b). The abundance and diversity of archaeal *accA* gene in Yunnan hot springs have been studied using DNA- and RNA-based phylogenetic analyses and quantitative polymerase chain reaction. The results showed that archaeal *accA* genes were present and expressed in the investigated Yunnan hot springs with a wide range of temperatures (66–96 °C) and pH (4.3–9.0). The majority of the amplified archaeal *accA* gene sequences were affiliated with the ThAOA/HWCG III [thermophilic ammonia-oxidizing archaea (AOA)/hot water crenarchaeotic group III]. The archaeal *accA* gene abundance was very close to that of AOA *amoA* gene, encoding the alpha subunit of ammonia monooxygenase. These data suggest that AOA in terrestrial hot springs might acquire energy from ammonia oxidation coupled with CO_2 fixation using the 3-hydroxypropionate/4-hydroxybutyrate (Song et al. 2013b).

3.5.3 Arsenite-Oxidizing Microorganisms

Arsenic is widely distributed in nature and can exist in four oxidation states, As(III), As(0), As(III), and As(V). Arsenic oxyanions could be used for energy generation of prokaryotes, either by oxidizing arsenite or by respiring arsenate (Oremland and Stolz 2003). There are certain microbes like arsenite-oxidizing microorganisms containing arsenite oxidase that catalyzes the transformation of arsenite [As(III)] to arsenate [As(V)] (Lett et al. 2012). *aioA* gene is a molecular biomarker for studying the distribution and activity of arsenite-oxidizing bacteria in various environments.

The abundance and diversity of arsenite-oxidizing bacteria in the geothermal features of Tengchong County of Yunnan Province, Dachaidan County of Qinghai Province, and Tibet have been investigated. The results showed that the *aioA* gene abundance increased as temperature decreased, whereas its diversity at the OTU level (97% cutoff) increased with increase in temperature. This suggests that temperature played an important role in affecting *aioA* gene distribution and thus arsenic speciation. The *aioA* gene population (at OTU level) differed among the studied regions, indicating geographic isolation may be an important factor controlling *aioA* gene distribution in hot springs (Wu et al. 2015).

3.6 Conclusion and Future Perspectives

The investigations have demonstrated that huge diverse and novel thermophilic bacterial and archaeal communities with bioactive potentials thrive in Yunnan and Tibet hot springs. Most of the microbial communities are still unclassified or unknown, which awaits further exploration. Environmental factors play an important role in structuring microbial communities, and hence these factors should be considered in future analysis. Only few reports have been described in studying thermophilic virus in Yunnan and Tibet hot springs. Hence, attempts to identify the distribution pattern and host-virus interaction in these hot springs have to be conducted.

Acknowledgment This work was supported by the Key Project of International Cooperation of Ministry of Science and Technology (MOST) (No. 2013DFA31980), Science and Technology Infrastructure Work Project (No. 2015FY110100), Natural Science Foundation of China (Nos. 31470139 and 31600103), and China Postdoctoral Science Foundation No. 2017 M612796. W-J.L. was also supported by Guangdong Province Higher Vocational Colleges and Schools Pearl River Scholar Funded Scheme (2014).

References

Badhai J, Ghosh TS, Das SK (2015) Taxonomic and functional characteristics of microbial communities and their correlation with physicochemical properties of four geothermal springs in Odisha, India. Front Microbiol 6:1166. https://doi.org/10.3389/fmicb.2015.01166

Bao Q, Tian Y, Li W, Xu Z, Xuan Z, Hu S, Dong W, Yang J, Chen Y, Xue Y, Xu Y, Lai X, Huang L, Dong X, Ma Y, Ling L, Tan H, Chen R, Wang J, Yu J, Yang H (2002) A complete sequence of the *T. tengcongensis* genome. Genome Res 12(5):689–700

Berg IA, Kockelkorn D, Buckel W, Fuchs G (2007) A 3-hydroxypropionate/4-hydroxybutyrate autotrophic carbon dioxide assimilation pathway in Archaea. Science 318:1782–1786

Bisht SS, Das NN, Tripathy NK (2011) Indian hot- water springs: a bird's eye view. JoEECC 1(1):1–15

Briggs BR, Brodie EL, Tom LM, Dong H, Jiang H, Huang Q, Wang S, Hou W, Wu G, Huang L, Hedlund BP, Zhang C, Dijkstra P, Hungate BA (2014) Seasonal patterns in microbial communities inhabiting the hot springs of Tengchong, Yunnan Province, China. Environ Microbiol 16(6):1579–1591

Brock TD (1997) The value of basic research: discovery of *Thermus aquaticus* and other extreme thermophiles. Genetics 146:1207–1210

Chan CS, Chan KG, Tay YL, Chua YH, Goh KM (2015) Diversity of thermophiles in a Malaysian hot spring determined using 16S rRNA and shotgun metagenome sequencing. Front Microbiol 6:177. https://doi.org/10.3389/fmicb.2015.00177

Chaudhuri B, Chowdhury T, Chattopadhyay B (2017) Comparative analysis of microbial diversity in two hot springs of Bakreshwar, West Bengal, India. Genom Data 12:122–129

Chen C, Lin L, Peng Q, Ben K, Zhou Z (2002) *Meiothermus rosaceus* sp. nov. isolated from Tengchong hot spring in Yunnan, China. FEMS Microbiol Lett 216(2):263–268

Chen B, Ma C, Zhou S, Bu H, Jiang S, Li H, Nizhen (2009) Bacterial diversity of disused thermal vents in Tibetan Yangbajing Region. Wei Sheng Wu Xue Bao 49(2):217–226

Chen JJ, Lin LB, Zhang LL, Zhang J, Tang SK, Wei YL, Li WJ (2012) *Laceyella sediminis* sp. nov., a thermophilic bacterium isolated from a hot spring. Int J Syst Evol Microbiol 62:38–42

Dong L, Ming H, Zhou EM, Yin YR, Liu L, Feng HG, Xian WD, Nie GX, Li WJ (2015) *Crenobacter luteus* gen. nov., sp. nov., isolated from a hot spring. Int J Syst Evol Microbiol 65:214–219

Duan YY, Ming H, Dong L, Yin YR, Zhang Y, Zhou EM, Liu L, Nie GX, Li WJ (2014) *Streptomyces calidiresistens* sp. nov., isolated from a hot spring sediment. Antonie Van Leeuwenhoek 106(2):189–196

Eloe-Fadrosh EA, Paez-Espino D, Jarett J, Dunfield PF, Hedlund BP, Dekas AE, Grasby SE, Brady AL, Dong H, Briggs BR, Li WJ, Goudeau D, Malmstrom R, Pati A, Pett-Ridge J, Rubin EM, Woyke T, Kyrpides NC, Ivanova NN (2016) Global metagenomic survey reveals a new bacterial candidate phylum in geothermal springs. Nat Commun 7:10476. https://doi.org/10.1038/ncomms10476

Farmer J (1998) Thermophiles, early biosphere evolution, and the origin of life on Earth: implications for the exobiological exploration of Mars. J Geophys Res 103(E12):28457–28461

Gold T (1992) The deep, hot biosphere. Proc Natl Acad Sci U S A 89(13):6045–6049

Gong N, Chen C, Xie L, Chen H, Lin X, Zhang R (2005) Characterization of a thermostable alkaline phosphatase from a novel species *Thermus yunnanensis* sp. nov. and investigation of its cobalt activation at high temperature. Biochim Biophys Acta 1750(2):103–111

Guo CL, Wang T, Zhu W, Zhang DH, Cui XL, Xu LH, Peng QA (2003) The phylotype of *Thermus* from the Rehai geothermal area, Tengchong, China. J Microbiol 41(2):152–156

Habib N, Khan IU, Hussain F, Zhou EM, Xiao M, Ahmed I, Zhi XY, Li WJ (2017a) *Caldovatus sediminis* gen. nov., sp. nov., a moderately thermophilic bacterium isolated from a hot spring. Int J Syst Evol Microbiol 67(11):4716–4721

Habib N, Khan IU, Hussain F, Zhou EM, Xiao M, Dong L, Zhi XY, Li WJ (2017b) *Meiothermus luteus* sp. nov., a slightly thermophilic bacterium isolated from a hot spring. Int J Syst Evol Microbiol 67(8):2910–2914

He ZG, Zhong H, Li Y (2004) *Acidianus tengchongensis* sp. nov., a new species of acidothermophilic archaeon isolated from an acidothermal spring. Curr Microbiol 48(2):159–163

Hedlund BP, Cole JK, Williams AJ, Hou W, Zhou E, Li WJ, Dong H (2012) A review of the microbiology of the Rehai geothermal field in Tengchong, Yunnan Province, China. Geosci Front 3(3):273–288

Hedlund BP, Reysenbach A-L, Huang L, Ong JC, Liu Z, Dodsworth JA, Ahmed R, Williams AJ, Briggs BR, Liu Y, Hou W, Dong H (2015) Isolation of diverse members of the *Aquificales* from geothermal springs in Tengchong, China. Front Microbiol 6:157. https://doi.org/10.3389/fmicb.2015.00157

Hochstein MP, Regenauer-Lieb K (1998) Heat generation associated with collision of two plates: the Himalayan geothermal belt. J Volcanol Geotherm Res 83(1–2):75–92

Hou W, Wang S, Dong H, Jiang H, Briggs BR, Peacock JP, Huang Q, Huang L, Wu G, Zhi X, Li W, Dodsworth JA, Hedlund BP, Zhang C, Hartnett HE, Dijkstra P, Hungate BA (2013) A comprehensive census of microbial diversity in hot springs of Tengchong, Yunnan Province China using 16S rRNA gene pyrosequencing. PLoS One 8:e53350. https://doi.org/10.1371/journal.pone.0053350

Huang Q, Jiang H, Zhang C, Li W, Deng S, Yu B, Dong H (2010) Abundance of ammonia-oxidizing microorganisms in response to environmental variables of hot springs in Yunnan Province, China. Wei Sheng Wu Xue Bao 50(1):132–136

Huang Q, Dong CZ, Dong RM, Jiang H, Wang S, Wang G, Fang B, Ding X, Niu L, Li X, Zhang C, Dong H (2011) Archaeal and bacterial diversity in hot springs on the Tibetan Plateau, China. Extremophiles 15(5):549–563

Jian H, Bo C, Xin D, Wei H, Xiuling J, Yunlin W, Lianbing L (2010) Diversity of *Thermoacidophilic Sulfolobus* in Hot Springs in Tengchong of Yunnan, China. Chin J Appl Environ Biol 16(5):692–696

Jiang C, Liu Y, Liu Y, Guo X, Liu SJ (2009) Isolation and characterization of ferrous- and sulfur-oxidizing bacteria from Tengchong solfataric region, China. J Environ Sci (China) 21(9):1247–1252

Jiang H, Huang Q, Dong H, Wang P, Wang F, Li W, Zhang C (2010) RNA-based investigation of ammonia-oxidizing archaea in hot springs of Yunnan Province, China. Appl Environ Microbiol 76(13):4538–4541

Khan IU, Habib N, Hussain F, Xian WD, Amin A, Zhou EM, Ahmed I, Zhi XY, Li WJ (2017a) *Thermus caldifontis* sp. nov., a thermophilic bacterium isolated from a hot spring. Int J Syst Evol Microbiol 67(8):2868–2872

Khan IU, Hussain F, Tian Y, Habib N, Xian WD, Jiang Z, Amin A, Yuan CG, Zhou EM, Zhi XY, Li WJ (2017b) *Tibeticola sediminis* gen. nov., sp. nov., a thermophilic bacterium isolated from a hot spring. Int J Syst Evol Microbiol 67(5):1133–1139

Kikani BA, Sharma AK, Singh SP (2017) Metagenomic and culture-dependent analysis of the bacterial diversity of a hot spring reservoir as a function of the seasonal variation. Int J Environ Res 11:25. https://doi.org/10.1007/s41742-017-0003-9

Kumar B, Trivedi P, Mishra AK, Pandey A, Palni LM (2004) Microbial diversity of soil from two hot springs in Uttaranchal Himalaya. Microbiol Res 159:141–146

Lett MC, Muller D, Lievremont D, Silver S, Santini J (2012) Unified nomenclature for genes involved in prokaryotic aerobic arsenite oxidation. J Bacteriol 194(2):207–208

Li H, Yang Q, Li J, Gao H, Li P, Zhou H (2015) The impact of temperature on microbial diversity and AOA activity in the Tengchong Geothermal Field, China. Sci Rep 5:17056. https://doi.org/10.1038/srep17056

Liao Z (2018) Thermal springs and geothermal energy in the Qinghai-Tibetan Plateau and the surroundings, Springer hydrogeology. Springer, Singapore, pp 5–10

Lin LB, Chen CY, Peng Q, Ben KL, Zhou ZM (2002) *Thermus rehai* sp. nov., isolated from Rehai of Tengchong, Yunnan Province, China. J Basic Microbiol 42(5):337–344

Lin L, Hong W, Ji X, Han J, Huang L, Wei Y (2010) Isolation and characterization of an extremely long tail *Thermus* bacteriophage from Tengchong hot springs in China. J Basic Microbiol 50(5):452–456

Lin L, Han J, Ji X, Hong W, Huang L, Wei Y (2011) Isolation and characterization of a new bacteriophage MMP17 from *Meiothermus*. Extremophiles 15(2):253–258

Liu YY, Chen ZW, Jiang CY, Liu SJ (2007) Isolation and characterization of *Acidiphilium* strain teng-A and its metabolism of fe (III) during pure- and mixed cultivation. Wei Sheng Wu Xue Bao 47(2):350–354

Liu LJ, You XY, Guo X, Liu SJ, Jiang CY (2011a) *Metallosphaera cuprina* sp. nov., an acidothermophilic, metal-mobilizing archaeon. Int J Syst Evol Microbiol 61:2395–2400

Liu LJ, You XY, Zheng H, Wang S, Jiang CY, Liu SJ (2011b) Complete genome sequence of *Metallosphaera cuprina*, a metal sulfide-oxidizing archaeon from a hot spring. J Bacteriol 193(13):3387–3388

Liu L, Salam N, Jiao JY, Jiang HC, Zhou EM, Yin YR, Ming H, Li WJ (2016) Diversity of culturable thermophilic actinobacteria in hot springs in Tengchong, China and studies of their biosynthetic gene profiles. Microb Ecol 72(1):150–162

Lopez-Lopez O, Cerdan ME, Gonzalez-Siso MI (2013) Hot spring metagenomics. Life (Basel) 3(2):308–320

Ming H, Yin YR, Li S, Nie GX, Yu TT, Zhou EM, Liu L, Dong L, Li WJ (2014) *Thermus caliditerrae* sp. nov., a novel thermophilic species isolated from a geothermal area. Int J Syst Evol Microbiol 64:650–656

Ming H, Duan YY, Guo QQ, Yin YR, Zhou EM, Liu L, Li S, Nie GX, Li WJ (2015) *Meiothermus roseus* sp. nov., a thermophilic bacterium isolated from a geothermal area. Antonie Van Leeuwenhoek 108(4):897–905

Ming H, Duan YY, Yin YR, Meng XL, Li S, Zhou EM, Huang JR, Nie GX, Li WJ (2016) *Crenalkalicoccus roseus* gen. nov., sp. nov., a thermophilic bacterium isolated from alkaline hot springs. Int J Syst Evol Microbiol 66(6):2319–2326

Nobre MF, Truper HG, ds Costa MS (1996) Transfer of *Thermus ruber* (Loginova et al. 1984), *T. silvanus* (Tenreiro et al. 1995), and *T. chiliarophilus* (Tenreiro et al. 1995) to *Meiothermus* gen. nov. as *Meiothermus ruber* comb. nov., *Me. silvanus* comb. nov., and *Me. chilarophilus* comb. nov., respectively, and emendation of the genus *Thermus*. Int J Syst Bacteriol 46:604–606

Oremland RS, Stolz JF (2003) The ecology of arsenic. Science 300(5621):939–944

Pathak AP, Rathod MG (2014) Cultivable bacterial diversity of terrestrial thermal spring of Unkeshwar, India. J Biochem Technol 5:814–818

Peng TJ, Liu LJ, Liu C, Yang ZF, Liu SJ, Jiang CY (2015) *Metallosphaera tengchongensis* sp. nov., an acidothermophilic archaeon isolated from a hot spring. Int J Syst Evol Microbiol 65:537–542

Saiki T, Kimura R, Arima K (1972) Isolation and characterization of extremely thermophilic bacteria from hot springs. Agric Biol Chem 36(13):2357–2366

Satyanarayana T, Raghukumar C, Shivaji S (2005) Extremophilic microbes: diversity and perspectives. Curr Sci 89:78–90

Sen R, Maiti NK (2014) Genomic and functional diversity of bacteria isolated from hot springs in Odisha, India. Geomicrobiol J 31:541–550

Shang SM, Qian L, Zhang X, Li KZ, Chagan I (2013) *Themoanaerobacterium calidifontis* sp. nov., a novel anaerobic, thermophilic, ethanol-producing bacterium from hot springs in China. Arch Microbiol 195(6):439–445

Song ZQ, Zhi XY, Li WJ, Jiang H, Zhang CL, Dong H (2009) Actinobacterial diversity in hot springs in Tengchong (China), Kamchatka (Russia) and Nevada (USA). Geomicrobiol J 26:256–263

Song ZQ, Chen JQ, Jiang HC, Zhou EM, Tang SK, Zhi XY, Zhang LX, Zhang CLL, Li WJ (2010) Diversity of Crenarchaeota in terrestrial hot springs in Tengchong, China. Extremophiles 14(3):287–296

Song ZQ, Wang FP, Zhi XY, Chen JQ, Zhou EM, Liang F, Xiao X, Tang SK, Jiang HC, Zhang CL, Dong H, Li WJ (2013a) Bacterial and archaeal diversities in Yunnan and Tibetan hot springs, China. Environ Microbiol 15(4):1160–1175

Song ZQ, Wang L, Wang FP, Jiang HC, Chen JQ, Zhou EM, Liang F, Xiao X, Li WJ (2013b) Abundance and diversity of archaeal *accA* gene in hot springs in Yunnan Province, China. Extremophiles 17(5):871–879

Streit WR, Schmitz RA (2004) Metagenomics – the key to the uncultured microbes. Curr Opin Microbiol 7(5):492–498

Wang H, Zheng M, Huang X (2007) Cesium accumulation by bacterium *Thermus* sp. TibetanG7: hints for biomineralization of cesium-bearing geyserite in hot springs in Tibet, China. Chin Sci Bull 52:2680. https://doi.org/10.1007/s11434-007-0353-3

Wang S, Hou W, Dong H, Jiang H, Huang L, Wu G, Zhang C, Song Z, Zhang Y, Ren H, Zhang J, Zhang L (2013) Control of temperature on microbial community structure in hot springs of the Tibetan Plateau. PLoS One 8(5):e62901. https://doi.org/10.1371/journal.pone.0062901

Wang S, Dong H, Hou W, Jiang H, Huang Q, Briggs BR, Huang L (2014) Greater temporal changes of sediment microbial community than its waterborne counterpart in Tengchong hot springs, Yunnan Province, China. Sci Rep 4:7479. https://doi.org/10.1038/srep07479

Whitaker RJ, Grogan DW, Taylor JW (2003) Geographic barriers isolate endemic populations of hyperthermophilic archaea. Science 301(5635):976–978

Wu G, Jiang H, Dong H, Huang Q, Yang J, Webb L, Liu Q, Guo H, Yuan S, Li P (2015) Distribution of arsenite-oxidizing bacteria and its correlation with temperature in hot springs of the Tibetan-Yunnan geothermal zone in western China. Geomicrobiol J 32(6):482–493

Xian WD, Yin YR, Liu L, Yuan CG, Hussain F, Khan I, Habib N, Zhou EM, Li WJ (2016) *Brevibacillus sediminis* sp. nov., isolated from a hot spring. Int J Syst Evol Microbiol 66(2):548–553

Xiang XY, Dong XZ, Huang L (2003) *Sulfolobus tengchongensis* sp. nov., a novel thermoacidophilic archaeon isolated from a hot spring in Tengchong, China. Extremophiles 7(6):493–498

Xiang X, Chen L, Huang X, Luo Y, She Q, Huang L (2005) *Sulfolobus tengchongensis* spindle-shaped virus STSV1: virus-host interactions and genomic features. J Virol 79(14):8677–8686

Xu LH, Tiang YQ, Zhang YF, Zhao LX, Jiang CL (1998) *Streptomyces thermogriseus*, a new species of the genus *Streptomyces* from soil, lake and hot-spring. Int J Syst Evol Microbiol 48(4):1089–1093

Xue Y, Xu Y, Liu Y, Ma Y, Zhou P (2001) *Thermoanaerobacter tengcongensis* sp. nov., a novel anaerobic, saccharolytic, thermophilic bacterium isolated from a hot spring in Tengcong, China. Int J Syst Evol Microbiol 51(Pt 4):1335–1341

Xue Y, Zhang X, Zhou C, Zhao Y, Cowan DA, Heaphy S, Grant WD, Jones BE, Ventosa A, Ma Y (2015) *Caldalkalibacillus thermarum* gen. nov., sp. nov., a novel alkalithermophilic bacterium from a hot spring in China. Int J Syst Evol Microbiol 65:214–229

Yu TT, Ming H, Yao JC, Zhou EM, Park DJ, Hozzein WN, Kim CJ, Wadaan MA, Li WJ (2015) *Thermus amyloliquefaciens* sp. nov., isolated from a hot spring sediment sample. Int J Syst Evol Microbiol 65(8):2491–2495

Yuan CG, Chen X, Jiang Z, Chen W, Liu L, Xian WD, Asem MD, Zhi X-Y, Li WJ (2017) *Altererythrobacter lauratis* sp. nov, and *Altererythrobacter palmitatis* sp. nov., isolated from a Tibetan hot spring. Antonie Van Leeuwenhoek 110(8):1077–1086

Zhang CL, Ye Q, Huang Z, Li W, Chen J, Song Z, Zhao W, Bagwell C, Inskeep WP, Ross C, Gao L, Wiegel J, Romanek CS, Shock EL, Hedlund BP (2008) Global occurrence of archaeal *amoA* genes in terrestrial hot springs. Appl Environ Microbiol 74(20):6417–6426

Zou P, Wb Z, Lei T, Wang JK (2006) Study on bioleaching of primary chalcopyrite ore with *Thermoacidophilic* Archae. Acta Metall Sin Engl 19(5):341–346

Microbial Diversity of Terrestrial Geothermal Springs in Lesser Caucasus

4

Hovik Panosyan, Armine Margaryan, Liana Poghosyan, Ani Saghatelyan, Ekaterine Gabashvili, Ekaterine Jaiani, and Nils-Kåre Birkeland

Abstract

The geology of the Lesser Caucasus is complex, owing to accretion of terrains through plate-tectonic processes and to ongoing tectonic activity and volcanism. Numerous geothermal springs of different geotectonic origins and with different physicochemical properties are found on the territory of the Lesser Caucasus. Despite intensive microbiological studies on terrestrial geothermal springs in various regions of the globe, very little is known about microbial diversity of similar ecosystems in the Lesser Caucasus. Recently the phylogenetic diversity of the prokaryotic community thriving in some geothermal springs located on the

H. Panosyan (✉) · A. Saghatelyan
Department of Biochemistry, Microbiology and Biotechnology, Yerevan State University, Yerevan, Armenia
e-mail: hpanosyan@ysu.am

A. Margaryan
Department of Biochemistry, Microbiology and Biotechnology, Yerevan State University, Yerevan, Armenia

Research Institute of Biology, Yerevan State University, Yerevan, Armenia

L. Poghosyan
Department of Biological Sciences, University of Bergen, Bergen, Norway

Department of Microbiology, Radboud UniversityNijmegen, Nijmegen, The Netherlands

E. Gabashvili
Ilia State University, Tbilisi, Tbilisi, Georgia

George Eliava Institute of Bacteriophages, Microbiology and Virology, Tbilisi, Georgia

E. Jaiani
George Eliava Institute of Bacteriophages, Microbiology and Virology, Tbilisi, Georgia

N.-K. Birkeland
Department of Biological Sciences, University of Bergen, Bergen, Norway

D. Egamberdieva et al. (eds.), *Extremophiles in Eurasian Ecosystems: Ecology, Diversity, and Applications*, Microorganisms for Sustainability 8,
https://doi.org/10.1007/978-981-13-0329-6_4

territory of Armenia, Georgia, and Nagorno-Karabakh has been explored following both cultivation-based and culture-independent approaches. Despite previous efforts, a comprehensive census of the microbial communities in the Lesser Caucasus hot springs is still lacking. This chapter contains a review of the results of microbial diversity analyses of 11 geothermal springs of the Lesser Caucasus with special emphasis to its distribution, ecological significance, and biotechnological potential.

Keywords
Lesser Caucasus · Geothermal springs · Microbial diversity · Thermophiles · Culture-dependent and culture-independent techniques

4.1 Introduction

Natural geothermal springs, including terrestrial hot springs, are widely distributed in various regions of our planet and are primarily associated with tectonically active zones in areas where the Earth's crust is relatively thin. These habitats have attracted broad interest since they are analogs for primitive Earth (Stan-Lotter and Fendrihan 2012). Geothermal springs offer a new source of fascinating microorganisms with unique properties well adapted to these extreme environments (Hreggvidsson et al. 2012; Deepika and Satyanarayana 2013). The adaptation to these harsh habitats makes thermophiles and their thermostable proteins suitable for various industrial and biotechnological applications (Raddadi et al. 2015; DeCastro et al. 2016).

The scientific interest in the microbial diversity of these exotic niches has increased during the last decades. With time, the tools used for microbial exploration have improved. Initially, studies were incepted with culture-based approaches. In recent time, culture-independent techniques (16S rRNA gene-based clone library analysis, denaturing gradient gel electrophoresis (DGGE), pyrosequencing, metagenomics, and metatranscriptomics) are mostly being used (Bhaya et al. 2007; Liu et al. 2011; López-López et al. 2013; DeCastro et al. 2016). This has shifted the cultivation-based narrow view into a more detailed and holistic insight of hot spring microbial habitats in terms of diversity, adaptation, functions, and ecological significance. Using a combination of several approaches of traditional microbiology with state-of-the-art molecular biology techniques has substantially increased our understanding of the structural and functional diversity of the microbial communities. Such approaches has been extensively used to study microbiota of the geothermal springs located in Iceland (Krebs et al. 2014), Azores (Sahm et al. 2013), the United States (Meyer-Dombard et al. 2005; Bowen De León et al. 2013), Bulgaria (Stefanova et al. 2015), Russia (Kublanov et al. 2009), China (Hedlund et al., 2012; Hou et al. 2013), India (Singh and Subudhi 2016; Saxena et al. 2017; Poddar and Das 2017), Malaysia (Chan et al. 2015), Argentina (Urbieta et al. 2015), Turkay (Cihan et al. 2011), Italy (Maugeri et al. 2009), Thailand (Portillo et al. 2009), New Zeland (Hetzer et al. 2007), Tunisia (Sayeh et al. 2010), Marocco (Aanniz et al.

2015) Romania (Coman et al. 2013), Spain (López-López et al. 2015) and other parts of world.

Thermal springs located in the Lesser Caucasus still represent a challenge for exploring biodiversity and searching of undescribed biotechnological resource. The geology of the region where Armenia, Georgia, and Nagorno-Karabakh are situated is complex, owing to accretion of terrains through plate-tectonic processes and to ongoing tectonic activity and volcanism (Henneberger et al. 2000; Badalyan 2000). Numerous geothermal springs with different geochemical properties are found on the territory of Lesser Caucasus. Despite a wide distribution of hot springs throughout Lesser Caucasus with hints of intrinsic scientific interest, limited attention has been paid toward microbiological analysis of these hot springs. With the best of information available, it was noted that data of microbial communities of several hot springs distributed on the territory of Armenia and Nagorno-Karabakh were published to date (Panosyan 2010; Hedlund et al. 2013; Panosyan and Birkeland 2014; Panosyan 2017; Panosyan et al. 2017). Despite these previous efforts, a comprehensive census of the microbial communities in Lesser Caucasus hot springs is still lacking.

The primary objective of this chapter is to review the findings of microbiological studies of several geothermal springs in the Lesser Caucasus and to summarize investigations on relationships between thermophilic microbial communities and geochemical conditions of their habitats. The results of this study expand the current understanding of the microbiology of hot springs in Lesser Caucasus and provide a basis for comparison with other geothermal systems around the world.

4.2 Geographical Distribution and Physiochemical Profiling of Geothermal Springs

The Caucasus Mountains include the Greater Caucasus in the north and Lesser Caucasus in the south (Stokes 2011). The Lesser Caucasus Mountains are formed predominantly of the Paleogene rocks with a smaller portion of the Jurassic and Cretaceous rocks. The formation of the Caucasus began from the Late Triassic to the Late Jurassic during the Cimmerian orogeny at the active margin of the Tethys Ocean while the uplift of the Greater Caucasus is dated to the Miocene during the Alpine orogeny. The Caucasus Mountains formed largely as the result of a tectonic plate collision between the Arabian plate moving northwards with respect to the Eurasian plate. This collision caused the uplift and the Cenozoic volcanic activity in the Lesser Caucasus Mountains. This region is regularly subjected to strong earthquakes from this activity (Reilinger et al. 1997). While the Greater Caucasus Mountains have a mainly folded sedimentary structure, the Lesser Caucasus Mountains are largely of volcanic origin (Philip et al. 1989). The geology of the region is complex, owing to accretion of exotic terranes through plate-tectonic processes and to ongoing tectonic activity and volcanism which have taken place more or less continuously since Lower Pliocene or Upper Miocene time.

The distribution of natural geothermal springs, including terrestrial hot springs (with water temperature higher than 21.1 °C), in various regions of our planet are primarily associated with tectonically active zones in areas where the Earth's crust is relatively thin. On the territory of the Lesser Caucasus, where traces of recently active volcanic processes are still noticeable, many geothermal springs with different geotectonic origins and physicochemical properties are found (Mkrtchyan 1969, Kapanadze et al. 2010).

Although no high-temperature geothermal resources have been identified in Armenia, numerous low-temperature resource areas (cooler than 100 °C) are present. Geothermal springs distributed on the territory of Armenia have been catalogued and described, and hundreds of shallow wells have been drilled to investigate mineral water sources throughout the country (Mkrtchyan 1969).

Three main heat flow zones (northeastern, central, and southwestern) have been distinguished on the basis of heat flow and temperature gradients (Fig. 4.1). The central zone (Zone II), which coincides closely with the belt of Quaternary

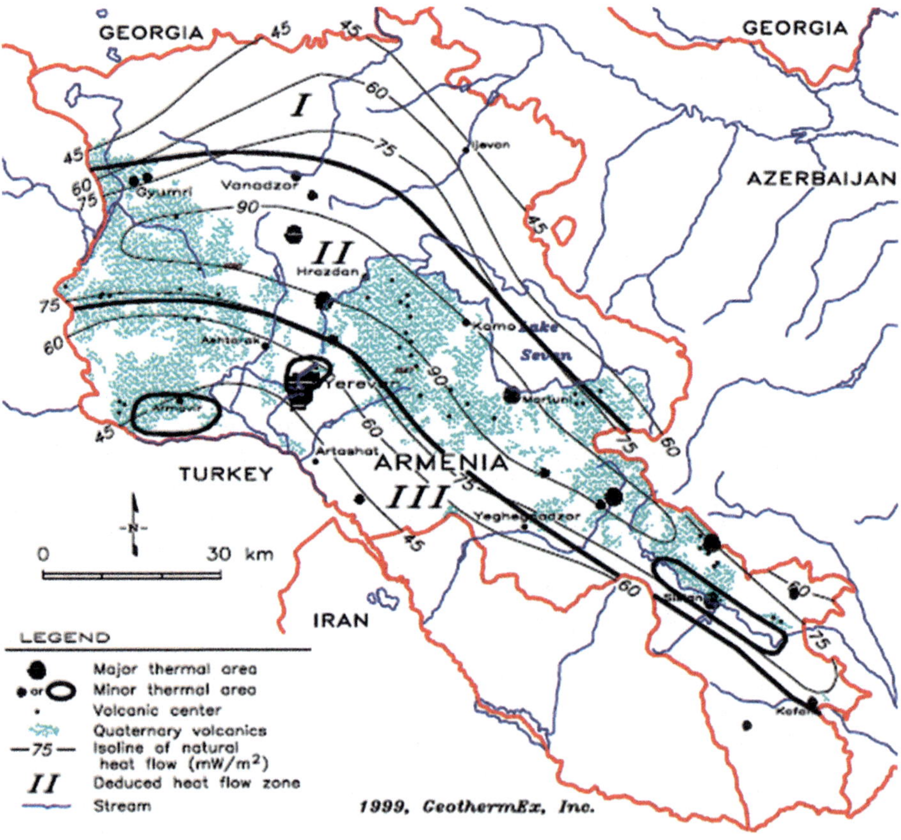

Fig. 4.1 Contour of heat flow with deduced heat flow zones in Armenia. (From Henneberger et al. 2000)

volcanoes, has highest heat flow (75 to more than 90 mW/m^2) and elevated temperature gradients (generally greater than 50 °C/km). The Zone I is considered to have no significant potential for geothermal resources. In Zone III there are scattered occurrences of thermal water, despite the overall low heat flow in this region (Karakhanian et al. 1997; Henneberger et al. 2000).

Nagorno-Karabakh is located in the southeastern part of the Lesser Caucasus. It is typically mountainous, embracing the eastern part of the Karabakh Plateau with the Artsakh valley, forming the great part of the Kura-Araks lowland. The Artsakh plateau like all Armenian plateaus is characterized by seismic activity. Volcanic rocks that appeared in ancient times are gaining ground: limestone and other sedimentary rocks from the Jurassic and Cretaceous period. Numerous geothermal springs at high elevations with different physicochemical properties are found also on the territory of Nagorno-Karabakh.

Georgia is located in the central and western parts of the Trans-Caucasus and lies between the Euro-Asiatic and Afro-Arabian plates. Apart from the Precambrian and Paleozoic formations that cover a smaller area, Mesozoic and Cenozoic rock assemblages mainly make up the geological structure of Georgia (Moores and Fairbridge 1998). Three major tectonic units can be distinguished according to the geologic development of Georgia: (1) the Greater Caucasus fold system, which represents a marginal sea in the geological past, (2) the Trans-Caucasus intermountain area which marks the northern part of the Trans-Caucasus island arc, and (3) the Lesser Caucasus fold system, the southern part of the ancient Trans-Caucasus island arc. The amount of thermal flow for the main parts of Georgia can be listed as follows:

1. The south flank of Caucasus Mountains, 100 mWm^2
2. Plate of Georgia:
   (a) For the west zone 40 mWm^2
   (b) For the east zone 30mWm^2
3. Adjara-Trialeti folded system:
   (a) Central part 90 mWm2
   (b) East zone 50 mWm2
4. Artvin-Bolnisi platform 60 mWm (Achmadova 1991)

The maximum heat flow is observed for the central zone of folded part of Georgia and the minimum for the plate, while the Adjara-Trialeti folded system is characterized by the middle range (Bunterbart et al. 2009).

Physical conditions, especially temperature, are regarded as a key factor for correlating microbial abundance and diversity of a spring (Everroad et al. 2012). Hot springs in the Lesser Caucasus could be grouped into three categories based on intrinsic temperature: warm springs (20–37 °C), moderately hot or mesothermal springs (37–50 °C), and hot springs (>50 °C). Using a cutoff temperature of 20 °C to distinguish thermal from nonthermal waters, several thermal areas are known to exist in Armenia (Mkrtchyan 1969). Hot springs at Uyts have the lowest temperature (25.8 °C). The highest temperature has been recorded for hot springs at Jermuk (>53 °C) and Karvachar (70 °C) (Fig. 4.2). The studies of some higher-temperature

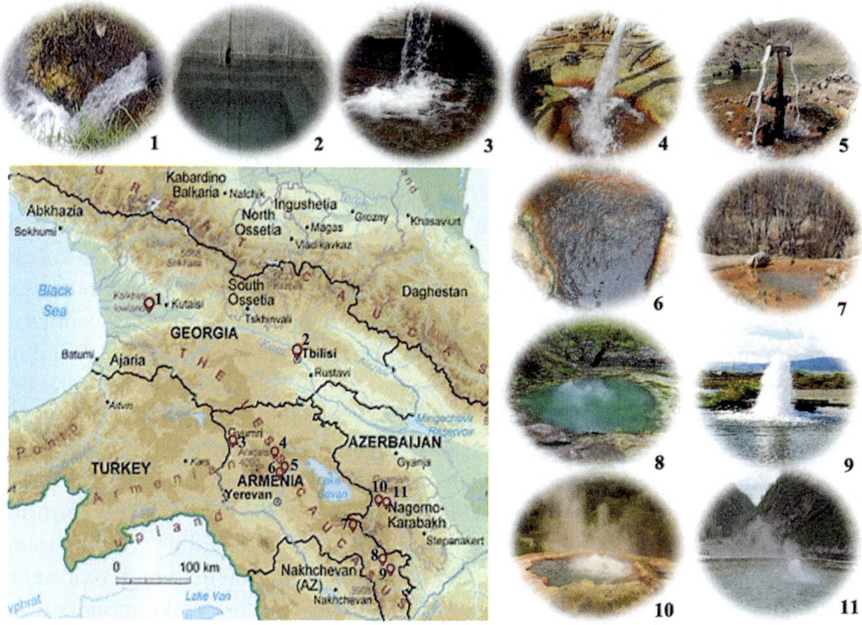

Fig. 4.2 Map of the locations of microbiologically explored terrestrial geothermal springs in the Lesser Caucasus. Closeup photographs of some geothermal springs. (1) Samtredia (2) Tbilisi sulfur spring (3) Akhurik (4) Hankavan (5) Bjni (6) Arzakan (7) Jermuk (8) Tatev (9) Uyts (10) Karvachar (11) Zuar. The source of the map is http://www.geocurrents.info/place/russia-ukraine-and-caucasus/where-is-the-caucasus

geothermal springs (for instance, Jermuk spring, located in the Karabakh Upland along Armenia's eastern border) using various geophysical surveys indicated that temperature at deeper levels (from 600 to 1000 m) can reach up to 99 °C (Karakhanian et al. 1997; Henneberger et al. 2000).

Geothermal springs found on the territory of Nagorno-Karabakh are also mainly classified as springs with moderate temperature. Two of Nagorno-Karabakh geothermal springs located in Karvachar (≥70 °C) and Zuar (42 °C) are characterized with higher water temperature (Fig. 4.2).

Up to 250 natural thermal springs and artificial wells are known in Georgia with water temperature ranging between 30–108 °C (Fig. 4.3) (Kapanadze et al. 2010). The lowest water temperature geothermal springs (30–35 °C) are distributed all over the territory of Georgia but are mainly found in Borjomi, Tsikhisjvari, Tskaltubo, and Saberio areas, while the highest water temperatures (78–108 °C) have been recorded for the waters from the artificial wells and boreholes in West Georgia, such as the Zugdidi-Tsaishi, Kvaloni, and Kindgi regions (Tsertsvadze et al. 1998).

All studied Armenian and Nagorno-Karabakhian hot springs are neutral, moderately alkaline, or alkaline in nature. Most of the spring samples have neutral pH (7–7.5), but hot springs at Tatev, Ajhurik, and Uyts have pH lower than 7. The hot

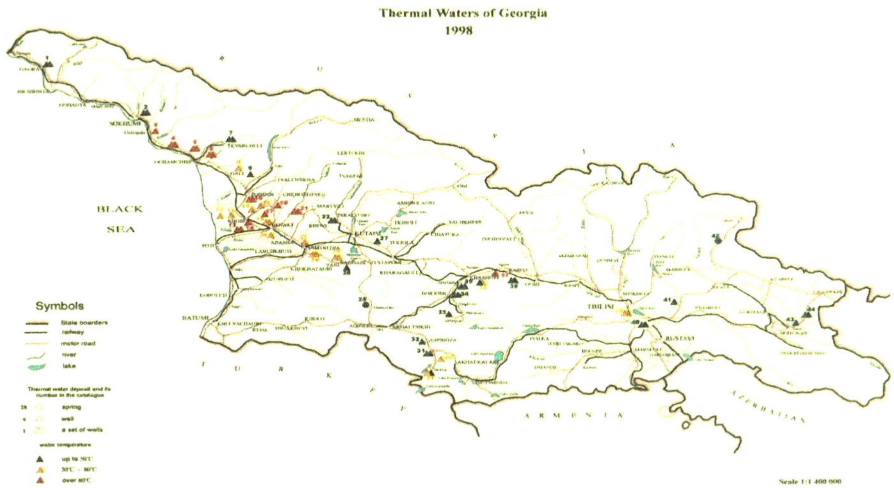

Fig. 4.3 Distribution of thermal waters in Georgia (Kapanadze et al. 2010)

springs in Georgia range from alkaline to acidic, but most of them are close to neutral or weak alkaline. The highly acidic hot spring in Georgia with pH 2.2 is located in Vani region, village Tsikhesulori, while the alkaline springs are found in Tbilisi area with pH 9.7 (Tsertsvadze et al. 1998).

Compared to physical analysis, limited attention has been paid to chemical profiling of hot spring water or sediment samples. Hot spring water usually has high concentrations of various elements owing to mineralization of dissolved solid elements from the adjacent areas. The composition of hot water is mainly determined by chemical interactions with reservoir rocks and rock-forming minerals along the ascent path, which may cause the spring water to be acidic or alkaline. All of the Armenian and Nagorno-Karabakhian thermal waters studied have mixed-cation mixed-anion compositions. Total dissolved solids contents tend to be less than about 0.5 mg/l but are occasionally higher. As is typically the case, the hotter and more saline samples tend to have higher ratios of (Na + K)/(Ca + Mg) and relatively high ratios of chloride to bicarbonate (Cl/HCO_3) or sulfate to bicarbonate (SO_4/HCO_3). The cooler waters tend to be higher in Ca + Mg and bicarbonate (Mkrtchyan 1969; Henneberger et al. 2000). For a few springs' major and minor elements, anions were analyzed by ionic coupled plasma optical emission spectrometry (ICP-OES; Thermo Iris), by mass spectrometry (ICP-MS; Thermo Element 2), and by ion chromatography (IC; Metrohm). Analyses of major and minor elements in the water sampled from the Arzakan geothermal spring revealed the following composition (in ppm): Na, 1183; Ca, 153; K, 108; Si, 47; Mg, 29; B, 15; Sr, 2.3; As, 1.6; Li, 1.3; Mn, 0.12; Fe, 0.72; Ba, 0.09; Cl, 297; and SO_4^2, 200. Nitrate was not detected (<2 ppm). For trace elements, the following concentrations were obtained (in ppb): Cr, 0.28; Co, 0.49; Cu, 0.82; and Zn, 6.73 (Panosyan and Birkeland 2014). The Georgian hot springs are characterized by diverse chemical composition, with mineralization ranging from 0.2 mg/L (Borjomi region) to 11.3 mg/L (Aspindza region). Similar to Armenian and Nagorno-Karabakhian region, the Georgian thermal waters also have mixed-cation and mixed-anion ratios mainly composed of hydrocarbonate, chloride, sulfate, sodium, potassium, magnesium, and calcium ions (Tsertsvadze et al. 1998). All studied springs are rich in heavy metals. Some of the springs contain gasses such as hydrogen sulfide, methane, nitrogen, and carbon dioxide (Mkrtchyan 1969; Tsertsvadze et al. 1998; Melikadze et al. 2010).

Most of the studies were focused on the hot springs at higher altitude and with high temperature. A majority of the hot springs found in the Lesser Caucasus are anthropogenically influenced and often used by tourists and local people for bath. Some of the geothermal springs are used for balneology (Mkrtchyan 1969; Melikadze et al. 2010).

The geographical locations, physicochemical profiling, and brief characteristic of main geothermal springs distributed on the territory of the Lesser Caucasus are summarized in Table 4.1.

4.3 Microbiological Analysis

Only a small fraction of the microorganisms found in a natural habitat can be cultivated under laboratory conditions and subsequently isolated. The knowledge of environmental microbial diversity has been largely aided by the development of culture-independent molecular phylogenetic techniques (Amann et al. 1995; DeLong and Pace 2001; Amann and Ludwig 2000; Zhou 2003; Bhaya et al. 2007; Liu et al. 2011; López-López et al. 2013; DeCastro et al. 2016). Using a combination of several approaches of traditional microbiology with state-of-the-art molecular biology techniques has substantially increased our understanding of the structural and functional diversity of microbial communities. Both culture-based and not culture-independent approaches have been used for addressing microbial diversity associated with geothermal springs. It has been reported that hot springs are inhabited by a variety of microbes belonging to the Bacteria and Archaea domains that tolerate environmental extremes and could have some yet undescribed biotechnological potential (Antranikian and Egorova 2007). Here we have summarized data of the phylogenetic diversity of the prokaryotic communities thriving in some of the geothermal springs in the Lesser Caucasus based on molecular- and culture-based methods (Tables 4.2 and 4.3).

4.3.1 Cultivation-Independent Studies

Up to date, two Armenian, two Georgian, and two thermal springs from Nagorno-Karabakh region have been analyzed using cultivation-independent approaches. Studies based on sequence analysis of 16S rRNA gene clone libraries from the mixed water and sediment sampled from the Arzakan (Armenia) geothermal spring have been done recently (Panosyan and Birkeland 2014). It was the first

Table 4.1 Geographical location, physiochemical profiling, and brief characteristic of main geothermal springs distributed on the territory of the Lesser Caucasus

Thermal mineral spring	Spring GPS location	Altitude, m, above sea level	pH	Conductivity, μS/cm	Temperature of water in the outlet, T, °C	Description
Armenia						
Akhurik	40°44′34.04″N 43°46′53.95″E	1490	6.5	2490	30	Near the village of Akhurik. This geothermal spring is a result of geological drillings. Geothermal water coming from a pipe. A shallow pool with a small continuous outflow from a 2-m man-made cement-fountain landscape. In composition, it belongs to the hydrocarbonate-sulfate sodium-magnesium type of spring. Slightly degassing. Sands at bottom. Multiple thick biomats (1–5 cm) of various colors (dark brown, red, dark green, and orange) are formed on the fountain, while a dark brown and green filamentous mat is present on the bank of the collecting pool. Tourist spot, believed to have medicinal values
Arzakan	40°27′36.10″N 44°36′17.76″E	1490	7.2	4378.3	44	Near the village of Arzakan. This geothermal spring is a result of geological drillings. Small pool source with a high flow rate. The hot spring belongs to the hydrocarbonate sodium class of mineral springs and possesses a high concentration of dissolved minerals (of which >20% is HCO^{3-} and > 20% is Na^{+}). Slightly degassing. Silicate sands at bottom. Biofilms with yellow, light brown, and light green colors are formed in the spring. Tourist spot, believed to have medicinal values
Bjni	40°45′94.44″N 44°64′86.11″E	1610	6.2–7.0	4138.3	30–37	Near the village of Bjni. This geothermal spring is a result of geological drillings. Geothermal water coming from a pipe. Small pool source with a low flow rate. In composition, it belongs to the chloride-hydrocarbonate sodium springs. Sands at bottom. Biofilms with yellow, light brown, and light green colors are formed in the spring

(continued)

Table 4.1 (continued)

Thermal mineral spring	Spring GPS location	Altitude, m, above sea level	pH	Conductivity, μS/cm	Temperature of water in the outlet, T, °C	Description
Hankavan	40°63′26.50″N 44°48′46. 00″E	1900	7.0–7.2	6722.9	42–44	Near the village of Hankavan located on the bank of Marmarik river. This geothermal spring is a result of geological drillings. Geothermal water coming from a pipe. Small pool source with a high flow rate. In composition, it belongs to the hydrocarbonate-chloride sodium springs. Vigorously degassing. Silicate sands at bottom. Biofilms with yellow, orange, light brown, and light green colors are formed in the spring tourist spot, believed to have medicinal values
Jermuk	39°96′63.90″N 45°68′52.80″E	2080	7.5	4340	>53	Near the town of Jermuk. Small pool with a high flow rate. In composition, it belongs to the carbon hydro-sulfate-sodium water sources. Sands at bottom. Biofilms with orange and light green are formed in the spring. Tourist spot, has medicinal value. Medicinal properties are similar to the springs in Karlovy Vary, Czech Republic
Tatev	39°23′76.00″ N 46°15′48.00″ E	960	6.0	1920	27.5	Hot spring located in Syunik region, near Satana's bridge (Satani Kamurj) on the bank of Vorotan River. In composition, it belongs to the carbon-bicarbonate calcium water sources. Many bubbling sources and no visible outflow. Roughly round-shaped pool with diameter ~3 m, and depth ~0.5 m. Clays and sands at the bottom. Biofilms with light green color are formed in the spring. Source is left in its natural form; no trace of human intervention is found. However, there are traces of ancient baths, which testify to the settlements dating BC
Uytc(Uz)	39°31′00″ N 46°03′09″ E	1600	6.23	2700	25.8	Near the village of Uz. This geothermal spring is a result of geological drillings. Big pool (5–6 m diameter, 20 cm depth) source with a high flow rate. In composition, it belongs to the hydrocarbonate-chloride-sulfate-sodium sources. Sands at bottom. Biofilms with dark green color are formed in the spring. Tourist spot, believed to have medicinal values

Nagorno-Karabakh						
Karvachar	40°17′41.00″N 46°27′50.00″ E	1584	7.3	4600	70	The spring is located on the bank of Tartar River, not far from Karvachar City (about 20 km). The hottest spring in Nagorno-Karabakh. Man-made round-shaped big pool (5 m diameter, 1 m depth) source with a high flow rate. Clear water and fine clays at the bottom. In composition, it belongs to the hydrocarbonate-sulfate sodium sources. Biofilms with yellow, orange, light brown, and light green colors are formed in the spring. Tourist spot, believed to have medicinal values
Zuar	40°02′47.60″N 46°14′09.30″E	1520	7.0	4300	42	The spring is located on the bank of Turon River. Round-shaped small pool (2 m diameter, 0.7 m depth) source with a high flow rate. Clays and sand at the bottom. In composition, it belongs to the hydrocarbonate-sulfate-sodium sources. Biofilms with yellow, orange, light brown, and light green colors are formed in the spring. Tourist spot, believed to have medicinal values
Georgia						
Samtredia borehole #1	42°10′12.04′′N 42°19′44.44′′E	24	7.15	5380	58	The spring is located in the town Samtredia, West Georgia. The spring resulted from geological drilling in 1969 to depth of 3045 m. In composition it belongs to chloride-hydrocarbonate-sulfate-calcium-sodium-potassium-magnesium sources. A yellow greenish biofilm is formed at the spring pool bottom. The water is used in greenhouses.
Tbilisi sulphur spring	41°41′18.87″N 44°48′52.30″E	405	7.8	737	37.7	The spring is located in Tbilisi and used for baths and balneology. In composition, it belongs to the sulfate-chloride-sodium type and contains hydrogen sulfide. The water has yellowish color

Table 4.2 Culture-independent studies of some geothermal spring microbiome in the Lesser Caucasus

Geothermal spring	Approach	Population proportion	Accession number	References
Armenia				
Arzakan	Shotgun pyrosequencing of V4 region on 454 GS FLX platform	Dominant bacterial phyla were cyanobacteria, in addition to Proteobacteria, Bacteroidetes, Chloroflexi, and Spirochaeta Dominant archaeal pyrotags which were affiliated with Euryarchaeota (*Methanosarcinales* and *Methanosaeta*) and Crenarchaeota (the yet-uncultivated group MCG)	SRR747863	Hedlund et al. (2013)
	Bacterial 16S rRNA gene library	Detected bacterial groups were Bacteroidetes (48%), Cyanobacteria (35%), Betaproteobacteria (22%), Gammaproteobacteria (13%), Epsilonproteobacteria (9%), Firmicutes (9%), and Alphaproteobacteria (8%)	JQ929026–JQ929037	Panosyan and Birkeland (2014)
	Archaeal 16S rRNA gene library	Archaeal population was presented by Euryarchaeota (methanogenic Archaea belonging to *Methanospirillum*, *Methanomethylovorans*, and *Methanoregula*), AOA Thaumarchaeota *Ca. Nitrososphaera gargensis*, and yet-uncultivated Crenarchaeota (MCG and DHVC1 groups)	KC682067–KC682083	Hedlund et al. (2013)
	DGGE	Dominant bacterial populations were related to Proteobacteria (affiliated with the Beta-, Epsilon-, and Gammaproteobacteria), Bacteroidetes, and Cyanobacteria	JX456536- JX456538	Panosyan and Birkeland (2014) and Panosyan (2017)

Jermuk	Shotgun pyrosequencing of V4 region on 454 GS FLX platform	Dominant bacterial pyrotags were affiliated with Proteobacteria and Bacteroidetes, and Synergistetes-dominant archaeal pyrotags were affiliated with Euryarchaeota (*Methanosarcinales*, *Methanosaeta*) and the yet-uncultivated Crenarchaeota groups MCG and DHVC1	SRR747864	Hedlund et al. (2013)
	Illumina HiSeq2500 paired-end sequencing	Dominant sequence reads were affiliated with Proteobacteria, Firmicutes, Bacteroidetes, candidate division WS6, and candidate phylum Ignavibacteria. Archaeal community (~1%) was dominated by Euryarchaeota, followed by Crenarchaeota, unclassified groups, and a minor fraction of Thaumarchaeota	–	Poghosyan (2015)
	Archaeal 16S rRNA gene library construction	Dominant archaeal populations were related to Euryarchaeota (methanogenic Archaea belonging to *Methanospirillum*, *Methanomethylovorans*, and *Methanosaeta*), AOA Thaumarchaeota *Ca. Nitrososphaera gargensis*, and yet-uncultivated Crenarchaeota (MCG group)	KC682084–KC682097	Hedlund et al. (2013)
	DGGE	Detected dominant groups were Epsilonproteobacteria, Bacteroidetes, Spirochaetes, Ignavibacteriae, and Firmicutes		Panosyan (2017)
Nagorno-Karabakh				
Karvachar	Bacterial 16S rRNA gene library construction	Dominant bacterial phyla were Proteobacteria (48.6%), Cyanobacteria (29.7%), Bacteroidetes (5.4%), Chloroflexi (5.4%), Verrucomicrobia (2.7%), and Planctomycetes (2.7%)	–	Saghatelyan and Panosyan (2015)
	DGGE	Detected bacterial groups were Bacteroidetes and Firmicutes	–	Panosyan (2017)
	Whole-metagenome shotgun sequencing, using Illumina HiSeq 4000 platform	Dominant sequence reads were affiliated with Actinobacteria, Alpha-, Beta-, Delta-, Epsilon-, and Gammaproteobacteria, Bacteroidetes/Chlorobi, Firmicutes, Chlamydiae, Cyanobacteria/Melainabacteria, Fusobacteria, and Synergistia	–	Unpublished data
Zuar	Bacterial 16S rRNA gene library construction	Dominant bacterial groups were Proteobacteria (42.3%), Firmicutes (19.2%), Bacteroidetes (15.4%), Cyanobacteria (3.8%), Tenericutes (3.8%), and yet-unclassified phylotypes (15.4%)	–	Saghatelyan et al. (2014)

(continued)

Table 4.2 (continued)

Geothermal spring	Approach	Population proportion	Accession number	References
Georgia				
Samtredia	Whole-metagenome shotgun sequencing, using Illumina HiSeq 2500 platform	Dominant bacterial sequence reads were affiliated with Firmicutes (33%), Gammaproteobacteria (32%), Actinobacteria (15.5%), Betaproteobacteria (9.1%), Alphaproteobacteria (2.9%), Chlamydia (1.6%), Bacteroidetes (1.5%).	–	Unpublished data
		Archaeal sequence reads were affiliated with Crenarchaeota (1.4%) and Euryarchaeota (0.2%)		
Tbilisi sulfur spring	Whole-genome shotgun sequencing using Illumina MiSeq platform	Bacterial sequence reads were affiliated with Firmicutes (20,6%), Gammaproteobacteria (46,4%), Actinobacteria (6,4%), Betaproteobacteria (16,4%), Alphaproteobacteria (5,7%), Chlamydia (1.7%), Bacteroidetes (1.5%), Deinococcus-Thermus (0.1%), delta/epsilon subdivisions (0.9%), Acidithiobacillia (0.2%), Cyanobacteria/Melainabacteria group (0.2%), and Synergistia (0.1%)	–	Unpublished data
		Archaeal sequence reads were affiliated with Euryarchaeota (0.2%)		
	DGGE	Detected dominant bacterial group was Betaproteobacteria (*Sulfurisoma*, *Thiobacillus*, *Oxalicibacterium faecigallinarum*)	–	Unpublished data
		Detected archaeal group was Euryarchaeota (*Methanosaeta harundinacea*)		

– data not available

Table 4.3 Summary of thermophilic bacteria isolated from geothermal springs of the Lesser Caucasus

Geothermal spring	Bacterial and Archaeal genera	Comments	References
Armenia			
Akhurik	*Bacillus* (*B. licheniformis*, *B. pumilus*, *B. murimartini*, *Bacillus* sp.), *Geobacillus* (*G. pallidus*), *Brevibacillus* (*B. borstelensis*), *Thermoactinomyces* (*Thermoactinomyces* sp.)	Used for studies of extracellular amylase, lipase, and protease. Lipase-producing *B. licheniformis* strain Akhourik 107 (accession number KY203975) consist genes encoding thermostable esterase GDSL (family II)	Panosyan (2010), Panosyan (2017), Shahinyan et al. (2017) and Shahinyan et al. (2015)
	Rhodobacter (*R. sulfidophilus*), *Thiospirillum* (*T. jenense*)	Used for studies of aspartase, aminoacylase, glucose isomerase, and inulinase activities, as well as sources of protein, carbohydrates, and vitamins	Paronyan (2002a)
	Methylocaldum-Methylococcus-Methyloparacoccus-Methylogaea *Methylocaldum* (*Methylocaldum* sp.)	*Methylocaldum* sp. strain AK-K6 (accession number KP272135) had a temperature range for growth of 8–35 °C (optimal 25–28 °C) and a pH range of 5.0–7.5 (optimal 6.4–7.0). 16S rRNA gene sequences showed that it was a new gammaproteobacterial methanotroph, which forms a separate clade in the family *Methylococcaceae*	Islam et al. (2015)
Arzakan	*Bacillus* (*B. licheniformis*, *B. simplex*), *Anoxybacillus* (*A. rupiensis*), *Geobacillus* (*G. toebii*, *G. thermodenitrificans*, *G. stearothermophilus*, *G. caldoxylosilyticus*), *Paenibacillus* (*Paenibacillus* sp.), *Sporosarcina* (*Sporosarcina* sp.), *Arcobacter* (*Arcobacter* sp.), *Methylocaldum* (*Methylocaldum* sp.)	Used for studies of extracellular amylase, lipase, and protease. The strain *A. rupiensis* Arzakan 2 produces thermostable amylases at optimal growth temperature 65 °C and pH 7. Two isolates, *G. thermodenitrificans* ArzA-6 (accession number JQ929020) and *G. toebii* ArzA-8 (accession number JQ929022), produce EPSs with high specific production yield (0.271 g g^{-1} dry cells and 0.222 g g^{-1} dry cells, respectively) at 65 °C and pH 7.0. Purified EPSs displayed a high molecular weight: 5×10^5 Da for *G. thermodenitrificans* ArzA-6 and 6×10^5 Da for *G. toebii* ArzA-8. Chemical composition and structure of the biopolymers, determined by GC-MS, HPAE-PAD, and NMR showed that both EPSs are heteropolymeric with mannose as major monomer unit	Hovhannisyan et al. (2016), Panosyan and Birkeland (2014), Panosyan (2010), Panosyan (2017) and Panosyan et al. (2014)

(continued)

Table 4.3 (continued)

Geothermal spring	Bacterial and Archaeal genera	Comments	References
	Rhodobacter (*R. sphaeroides*), *Rhodopseudomonas* (*R. palustris*), *Thiocapsa* (*T. roseopersicina*)	Used for studies of aspartase, aminoacylase, glucose isomerase, and inulinase activities	Paronyan (2002a)
	Acetoclastic and hydrogenotrophic methanogenic enrichment, *Methanoculleus* (*Methanoculleus* sp.)	Enrichment in defined media produced active and stable methanogenic cultures on acetate and H_2/CO_2 at 45 °C	Hedlund et al. (2013)
	Enrichment of nitrite-oxidizing bacteria (NOB)	Successful NOB enrichment incubated at 44 °C was obtained	Edwards et al. (2013)
Bjni	*Bacillus* (*B. licheniformis*, *B. aestuarii*), *Ureibacillus* (*U. thermosphaericus*), *Anoxybacillus* (*Anoxybacillus* sp.), *Geobacillus* (*G. toebii*)	Used for studies of extracellular amylase, lipase, and protease	Panosyan (2017)
	Rhodobacter (*R. sphaeroides*), *Rhodopseudomonas* (*R. palustris*), *Thiocapsa* (*T. roseopersicina*)	Used for studies of aspartase, aminoacylase, glucose isomerase, and inulinase activities, as well as sources of protein, carbohydrates, and vitamins	Paronyan (2002a)
Hankavan	*Bacillus* (*B. licheniformis*, *B. cirulans*), *Brevibacillus* (*B. thermoruber*), *Geobacillus* (*G. stearothermophilus*), *Anoxybacillus* (*Anoxybacillus* sp.)	Used for studies of extracellular amylase, lipase, and protease. The thermotolerant and metallotolerant bacilli *B. circulans* (3A) is able to grow at different concentrations (from 10 to 300 mg/l) of Cd^{2+}, Cu^{2+}, Zn^{2+}, and Ni^{2+}	Margaryan et al. (2010) and Panosyan (2017)
Jermuk	*Bacillus* (*B. licheniformis*, *B. amyloliquefaciens*), *Anoxybacillus* (*A. gonensis, A. kestanbolensis, Anoxybacillus* sp.), *Geobacillus* (*G. stearothermophilus, G. caldoxylosilyticus*)	Used for studies of extracellular amylase, lipase, and protease. *Anoxybacillus* sp. is a candidate of new species	Hovhannisyan et al. (2016) and Panosyan (2017); Poghosyan (2015)

	Desulfomicrobium (*D. thermophilum*) *Desulfovibrio* (*D. psychrotolerans*) *Treponema* (*Treponema* sp.)	One of the isolates is claimed to be a novel *Spirochaetes* species, *Treponema thermophilum* sp. nov., and two Deltaproteobacterial SRB strains possibly also representing novel species. The *T. thermophilum* genome was sequenced to near completion and the 3.2 Mb draft sequence supports the description of this isolate as a separate species	Poghosyan (2015)
	Rhodobacter (*R. Capsulatus, R. shpaeroides*) *Rhodopseudomonas* (*R. palustris*) *Thiospirillum* (*T. jenense*)	Used for studies of aspartase, aminoacylase, glucose isomerase, and inulinase activities, as well as sources of protein, carbohydrates, and vitamins. The strain *R. palustris* D-6 produces aspartase (with activity 33.05% per 100 mg dry biomass, at 37 °C, pH 6.0–9.0) The strain *R. capsulatus* D-4 produces L (+) lactic acid by 90% conversing sugars. The strain *R. shpaeroides* D-10 is a good producer of spheroiden and β-carotin	Paronyan (2002a), Paronyan (2002b) and Paronyan (2007)
	Methylotrophic, acetoclastic, and hydrogenotrophic methanogenic enrichments	Enrichments in defined media produced active and stable methanogenic cultures on methanol, acetate, and H_2/CO_2, at 45 and 55 °C	Hedlund et al. (2013)
	Enrichment of NOB, *Nitrospira* (*N. calida* and *N. moscoviensis*)	The enrichment had a wide temperature range (25–60 °C, with a temperature optimum of 45–50 °C) of nitrite oxidation yielding nitrite oxidation rates of 7.53 ± 1.20 to 23.0 ± 2.73 fmoles $cell^{-1}$ h^{-1}. The highest rates of NOB activity occurred with initial NO^{-2} concentrations of 0.5–0.75 mM; however, lower initial nitrite concentrations resulted in shorter lag times	Edwards et al. (2013)
Tatev	*Bacillus* (*B. licheniformis*), *Geobacillus* (*G. toebii, Geobacillus* sp.), *Anoxybacillus* (*Anoxybacillus* sp.), *Thermoactinomyces* (*T. vulgaris*)	Used for studies of extracellular amylase, lipases, and protease. The strains *Geobacillus* sp. Tatev N5 and Tatev N6 showed high lipases activity (70.3 U/ml) at 65 °C after 5 hours of incubation. Lipase-producing *Geobacillus* sp. Tatev 4 (KY203974) contains genes encoding thermostable lipases (family I). Lipase contains Zn^{2+} and Ca^{2+} as ligands	Vardanyan et al. (2015), Shahinyan et al. (2015), Panosyan (2017) and Shahinyan et al. (2017)

(continued)

Table 4.3 (continued)

Geothermal spring	Bacterial and Archaeal genera	Comments	References
Uyts	*Bacillus* (*B. licheniformis*), *Ureibacillus* (*U. terrenus*, *U. thermosphaericus*), *Anoxybacillus* (*Anoxybacillus* sp.), *Geobacillus* (*G. toebii*)	Used for studies of extracellular amylase, lipases or protease	Panosyan (2017)
Nagorno-Karabakh			
Karvachar	*Bacillus* (*B. Licheniformis*), *Anoxybacillus* (*Anoxybacillus* sp., *A. flavithermus*, *A. rupiensis*), *Geobacillus* (*G. toebii*), *Aeribacillus* (*A. pallidus*)	Used for studies of extracellular amylase, lipase, and protease. Lipase-producing *A. flavithermus* strain Karvachar QB2 contains genes encoding thermostable esterase GDSL (family II). Draft genome of amylase producer *Anoxybacillus* sp. strain K103 (accession number MQAD00000000) was sequenced using a HiSeq 400 Illumina genome sequencer and contains alpha-amylase and alpha-glucosidase genes. Sequence analysis supports the description of this isolate as a separate species	Shahinyan et al. (2015), Hovhannisyan et al. (2016), Hovhannisyan et al. (2017), Shahinyan et al. (2017) and Panosyan (2017)
	Thermus (*T. scotoductus*, *T. ruber*, *T. flavus*)	Draft genome of *T. scotoductus* K1 (accession number LJJR00000000) sequenced and assembled with PacBio RS technology and Celera Assembler. Used to study of DNA polymerase. *T. ruber* and *T. flavus* identified based only phenotypic characteristics	Achmadova (1991), Saghatelyan et al. (2015) and Saghatelyan et al. (2016)
	Enrichment of NOB	Successful NOB enrichment incubated at 56 °C was obtained	Edwards et al. (2013)
Zuar	*Anoxybacillus* (*A. rupiensis*), *Geobacillus* (*G. toebii*)	Used for studies of extracellular amylase, lipase, and protease	Hovhannisyan et al. (2016)

microbiological investigation on any hot spring in the Lesser Caucasus. The study indicated a predominance of Alphaproteobacteria (8%), Betaproteobacteria (22%), Gammaproteobacteria (13%), Epsilonproteobacteria (9%), Firmicutes (9%), Bacteroidetes (48%), and Cyanobacteria (35%). In addition, DGGE was employed to reveal the microbial profile of sediments of this hot spring. The authors reported an abundance of bacterial populations related to Proteobacteria (affiliated with the Beta-, Epsilon-, and Gammaproteobacteria), Bacteroidetes, and Cyanobacteria based on the DGGE profile, which was in good agreement with the clone library results. The sequence of dominating DGGE bands showed affiliation to *Rhodoferax* sp., a phototrophic, purple non-sulfur betaproteobacterium and to *Sulfurimonas* sp., a hydrogen-oxidizing chemolithoautotrophic bacterium isolated from a rearing tank with dissolved hydrogen (Panosyan and Birkeland 2014; Panosyan et al. 2017).

Samples from the Arzakan spring were screened also with advanced metagenomic approaches. Amplification of small-subunit rRNA genes using "universal" primers followed by pyrosequencing (pyrotags) on 454 GS FLX platform also revealed highly diverse microbial communities in Arzakan mat samples (Hedlund et al. 2013).The spring in Arzakan was colonized by a photosynthetic mat dominated by Cyanobacteria, in addition to Proteobacteria, Bacteroidetes, Chloroflexi, *Spirochaeta*, and a diversity of other Bacteria. It was shown that in Arzakan spring, relatively few (16%) of the total pyrotags could be assigned to known genera, underscoring the novelty of these ecosystem and the need for continued efforts to cultivate and describe microorganisms in geothermal systems.

The phylogenetic analysis of Bacteria identified the dominant phylotypes as members of Proteobacteria. The phylogeny for Proteobacteria revealed considerable diversity. While it is not possible to predict their metabolism from environmental sequences alone, the closest phylogenetic affiliations were to aerobic and anaerobic heterotrophs and methanotrophs (within the Proteobacteria lineage). It was established that the primary production of the Arzakan geothermal system supports by a complex microbial community composed of chemolithotrophs (hydrogen- and sulfide-oxidizing Epsilonproteobacteria and methanotrophic Gammaproteobacteria) and phototrophs (Cyanobacteria and purple non-sulfur anoxygenic phototrophic Betaproteobacteria). The most abundant Cyanobacteria OTUs were confidently assigned to the genera *Spirulina*, *Stanieria*, *Leptolyngbya*, and *Rivularia/Caldithrix*.

To study bacterial diversity of the hot spring in Jermuk (Armenia), 454 GS FLX pyrosequencing of V4–V8 variable regions of the small-subunit rRNA was applied. As reported, the most abundant phyla represented in the pyrotag dataset from Jermuk were the Proteobacteria, Bacteroidetes, and Synergistetes (Hedlund et al. 2013). Several abundant Proteobacteria OTUs were related to obligate or facultative chemolithoautotrophs capable of using sulfur compounds, Fe^{2+}, and/or H_2 as electron donors, including the genera *Thiobacillus*, *Sulfuricurvum*, *Sideroxydans*, and *Hydrogenophaga*, suggesting the importance of chemolithotrophy in primary productivity (Kampfer et al. 2005; Kellermann and Griebler 2009; Kodama and Watanabe 2004; Liu et al. 2012). The gross morphology of the mat was consistent with iron precipitation at the spring source as ferrous iron supplied from the subsurface is oxidized as the spring water becomes oxygenated. The Bacteroidetes were diverse, and many OTUs could not be assigned to known genera. An exception was an abundant OTU assigned to the genus *Lutibacter*, which contains chemoorganotrophs most commonly found in marine environments (Lee et al. 2006). Other Bacteroidetes and the Synergistetes in Jermuk are likely involved in heterotrophic processing of mat exudates and biomass.

DGGE analysis of the partial bacterial 16S rRNA gene PCR amplicons also was used to profile bacterial populations inhabiting the sediment and water fractions in the Jermuk geothermal spring. The sequence analysis of DGGE bands showed affiliation with Epsilonproteobacteria, Bacteroidetes, Spirochaetes, Ignavibacteriae, and Firmicutes. The sequences obtained from bands were related to anaerobic or facultatively anaerobic organotrophic or H_2-utilizing and thiosulfate-/sulfur-reducing bacteria. Heterotrophic microorganisms detected in the DGGE profile clustered among fermentative microorganisms, which are actively involved in C-cycle (Panosyan 2017).

Culture-independent technique with an emphasis on members of the Archaea was used to determine the composition and structure of microbial communities inhabiting microbial mats in the source pools of two geothermal springs, Arzakan and Jermuk. Based on an analysis of near full-length small-subunit rRNA genes amplified using Archaea-specific primers, it was shown that these springs are inhabited by a diversity of methanogens, including Methanomicrobiales and Methanosarcinales and relatives of *Methanomassiliicoccus luminyensis*, close relatives of the ammonia-oxidizing archaeon (AOA) "Candidatus *Nitrososphaera gargensis*," and the yet-uncultivated Miscellaneous Crenarchaeotal Group and Deep Hydrothermal Vent Crenarchaeota group 1 (Fig. 4.4) (Hedlund et al. 2013). Archaeal sequences were present at low abundance in both pyrotag datasets, with six archaeal pyrotags in three OTUs in Arzakan and nine pyrotags in six OTUs in Jermuk. The Methanosarcinales were represented in both pyrotag datasets, with *Methanomethylovorans* detected in Jermuk, and *Methanosaeta* and a sequence that could not be classified below the order level were detected in Arzakan. Close relatives of *Methanospirillum hungatei*, in the order *Methanomicrobiales*, were inferred to be abundant in both springs. In addition, two phylotypes in Arzakan were related to the genus *Methanoregula*, also in the *Methanomicrobiales*. Members of both genera use H_2/CO_2 and/or format as methanogenic substrates; however, their presence in the geothermal systems was somewhat surprising since they are not reported to grow above 37 °C. The other order of methanogens present in both springs was *Methanosarcinales*, represented by *Methanosaeta* and *Methanomethylovorans*. *Methanosaeta* was abundant in Jermuk and includes obligate acetoclastic species known to grow up to 60 °C (Liu and Whitman 2008).

Recently, Illumina HiSeq2500 paired-end sequencing of metagenomic DNA also was used to analyze water/sediment samples of the Jermuk hot spring. Taxonomic analyses of the metagenomic rRNA sequences revealed a prevalence of *Proteobacteria Firmicutes* and *Bacteroidetes*. However, many of the largest contigs represented uncharacterized or poorly characterized groups such as candidate division WS6 and candidate phylum Ignavibacteria. The archaeal community, constituting a minor fraction (~1%) of the community, was dominated by Euryarchaeota, followed by

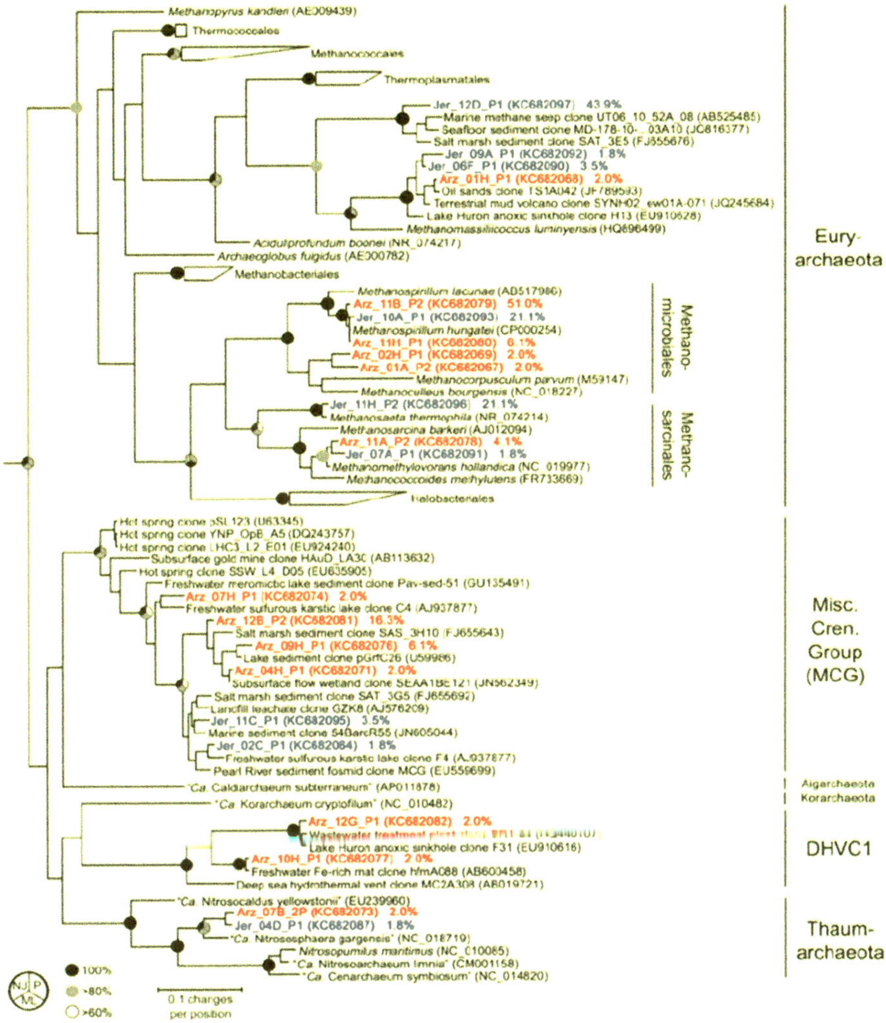

Fig. 4.4 Maximum-likelihood phylogeny depicting relationships between near-complete archaeal 16S rRNA genes recovered from Arzakan (red) and Jermuk (blue) and closely related sequences, including well-studied microbial isolates. Percent values for each OTU represent the percent abundance of the OTU in the clone library. Bootstrap support is indicated at major nodes for maximum-likelihood (ML; 100 replicates), parsimony (P; 1000 replicates), and distance (neighbor-joining, NJ; 1000 replicates) methods. Taxonomic designations for major phylogenetic groups are shown at the right (Hedlund et al. 2013)

Crenarchaeota, unclassified groups, and a minor fraction of Thaumarchaeota. The functional composition based on metagenomics sequence information indicated a dominance of heterotrophic types of metabolism (Poghosyan 2015).

For investigation of the bacterial composition of sediment and water samples from the Zuar geothermal spring (Nagorno-Karabakh), only a bacterial clone library based on 16S rRNA genes was constructed. It was shown that clones obtained from the Zuar geothermal spring originated from phyla Proteobacteria (42.3%), Firmicutes (19.2%), Bacteroidetes (15.4%), Cyanobacteria (3.8%), Tenericutes (3.8%), and yet-unclassified phylotypes (15.4% for Zuarr) (Saghatelyan et al. 2014).

According to the recent report of sequence analysis of clones obtained from bacterial 16S rRNA gene libraries, the presence of Proteobacteria (48.6%), Cyanobacteria (29.7%), Bacteroidetes (5.4%), Chloroflexi (5.4%), Verrucomicrobia (2.7%), and Planctomycetes (2.7%) in sediment and water samples in Karvachar (Nagorno-Karabakh) hot spring (Fig. 4.5) was indicated (Saghatelyan and Panosyan 2015). The dominating bacterial group was the phylum Proteobacteria. A few phylotypes belonging to the phylum Bacteroidetes were obtained. One of the dominating groups was Cyanobacteria, representatives of which dominate especially on top layer of microbial mats and are the most important primary producers in hot spring ecosystems (Roeselers et al. 2007).

Representatives of phylum Firmicutes were not detected in the clone library, while DGGE profiling of the same samples indicated presence of Firmicutes (genus *Geobacillus*) as a one of the major components in bacterial community of Karvachar geothermal spring (Panosyan 2017). This has been confirmed later by metagenome analysis of the Karvachar hot spring samples.

Based on recent data (unpublished data) obtained from the whole-genome shotgun sequencing of sediment samples of Karvachar, using an Illumina HiSeq 2500 platform, 580 bacterial sequences were aligned to reference genes (NCBI RefSeq), belonging to the following bacterial taxonomical groups: Actinobacteria; Alpha-, Beta-, Delta-, Epsilon-, and Gammaproteobacteria; Bacteroidetes/Chlorobi; Firmicutes; Chlamydiae; Cyanobacteria/Melainabacteria; *Fusobacteria*; and Synergistia. Among these groups, Proteobacteria (Alpha-, Beta-, and Gammaproteobacteria) and Firmicutes were the major components in the total bacterial sequence reads (Fig. 4.5). The sequences affiliated with Gammaproteobacteria were predominant (48.96% of Proteobacteria, 235 out of 480), and most of them were closely (98–100%) related to cultured Gammaproteobacteria. Representative of the groups of *Porphyrobacter*, *Paracoccus*, and *Oceanibaculum* was predominate Alphaproteobacteria found in study samples. The majority of sequences derived from spring were closely related (95–99% identity) to *Porphyrobacter cryptus*, a slightly thermophilic, aerobic, bacteriochlorophyll a-containing species isolated from a hot spring at Alcafache in Central Portugal (Rainey et al. 2003).

Betaproteobacterial-related sequences were the third major group of obtained bacterial sequences (20.6% of Proteobacteria, 99 out of 480). The majority of the obtained sequences showed 92–100% similarity to *Caldimonas taiwanensis*, an aerobic amylase-producing heterotrophic bacterium isolated from a hot spring located in Taiwan (Chen et al. 2005) and 94–99% of similarity to representatives of genus *Tepidimonas*, particularly to the species *T. taiwanensis*, *T. thermophilus*, and *T. fonticaldi*, isolated from hot springs in Taiwan and India (Chen et al. 2013; Poddar et al. 2014). Forty sequences (6.9%, 40 out of 580) were affiliated with Firmicutes. Around 10.3% (60 out of 580) of the total bacterial clone sequences were affiliated with some minor groups, such as Actinobacteria, Bacteroidetes/Chlorobi, Chlamydia,

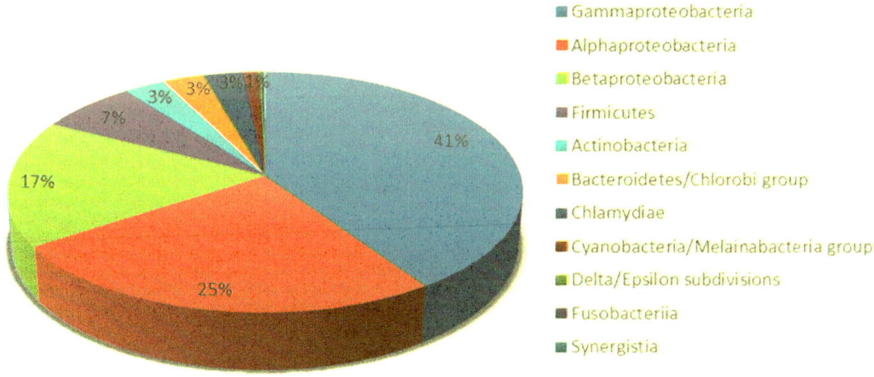

Fig. 4.5 Phylum level grouping of bacterial sequence read obtained from Karvachar geothermal spring

Cyanobacteria/Melainabacteria, Fusobacteria, and Synergistia. Most of these sequences were closely (98–99%) related to clones retrieved from water environments and different habitats (Anil Kumar et al. 2010; Yoon et al. 2009). Phototrophic bacteria belonging to genera *Neosynechococcus*, *Pseudanabaena*, and *Fischerella* represented the three most abundant and metabolically active primary producers of the analyzed community. Most Cyanobacteria detected were related to others previously reported in thermophilic environments (Portillo et al. 2009). Representatives of genus *Rhodobacter* (purple non-sulfur anoxygenic phototrophs) and other phototrophic microbes were found to share these environments with the cyanobacteria.

The sequence reads from the Samtredia geothermal spring (Georgia) water sample, obtained from the whole-genome shotgun sequencing on Illumina HiSeq 2500 platform, showed high similarity (>90%) to 938 bacterial and 15 archaeal reference sequences (Fig. 4.6). The majority of bacterial sequence reads were affiliated with the Firmicutes (33%) and Gammaproteobacteria (32%), followed by Actinobacteria (15.5%), Betaproteobacteria (9.1%), Alphaproteobacteria (2.9%), Chlamydia (1.6%), and Bacteroidetes (1.5%). Other groups of Prokaryotes (Aquificae, Deinococcus-Thermus, Deltaproteobacteria, Epsilonproteobacteria, Acidithiobacillia, Planctomycetes, Cyanobacteria/Melainabacteria group) comprised a minority, less than 1% of the communities. Archaeal sequence reads were affiliated with Crenarchaeota (1.4%) and Euryarchaeota (0.2%) (unpublished data).

The most dominant phylum, *Firmicutes*, was represented by genera *Streptococcus*, *Enterococcus*, *Clostridioides*, *Bacillus*, and *Listeria*. The majority of these bacteria can be recovered from a wide range of habitats. Firmicutes representatives considered as inhabitants of thermal waters include genera such as *Geobacillus*, *Thermoanaerobacter*, *Desulfotomaculum*, and *Desulfovirgula* have been revealed in the Samtredia hot spring. Similarly to other above described thermal waters, Proteobacteria were largely represented in the sequence reads. Four hundred and twenty two sequences (45%, 422 out of 938) were affiliated with Proteobacteria,

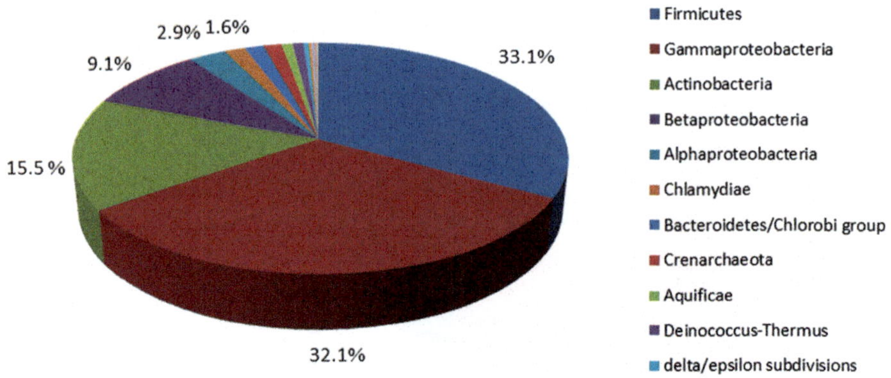

Fig. 4.6 Phylum level grouping of sequences obtained from Samtredia geothermal spring water sample

belonging to following subgroups: Alpha-, Beta-, Gamma-, and Deltaproteobacteria. The sequences affiliated with Gammaproteobacteria were predominant (72.5% of Proteobacteria, 306 out of 422). The dominant groups were *Escherichia*, *Acinetobacter*, *Pseudomonas*, *Salmonella*, and *Legionella.* Surprisingly most gammaproteobacterial sequences were *Escherichia*-related sequences. These microbes are not autochthons for hot springs and could be considered as contaminants.

Betaproteobacterial-related sequences were the second major group of obtained proteobacterial sequences (20.6% of Proteobacteria, 87 out of 422). The genera of *Caldimonas* and *Tepidiphilus*, representing the hot spring microbiota, were one of the minor groups of Betaproteobacteria found in the study samples. Alphaproteobacterial-related sequences comprised 6.6% of Proteobacteria and were represented mainly by nonindigenous bacteria. Actinobacteria accounted for a significant portion of bacteria, composing 15.8% (148 out of 938) of total bacterial populations dominated by *Mycobacteria*, while the Deinococcus-Thermus group was mainly represented by thermophilic bacteria belonging to the genus *Thermus*. Aquificales accounted for 0.8% of the reads, affiliated to facultatively anaerobic, hydrogen- or sulfur-/thiosulfate-oxidizing, thermophilic bacteria belonging to genus *Sulfurihydrogenibium.* Less than 5% of the total bacterial sequences were aligned with some other minor groups, such as Acidithiobacillia, Bacteroidetes/Chlorobi, Chlamydia, Cyanobacteria/Melainabacteria, and Planctomycetes. Most of these sequences were closely related to clones retrieved from water and soil environments.

The microbial diversity of the Tbilisi sulfur spring (Georgia) was analyzed using whole-genome shotgun sequencing using Illumina MiSeq platform (unpublished data). The sequences obtained from metagenomic DNA showed high similarity (>90%) to 1090 RefSeq database reference sequences, revealing 240 species. The thermal water was dominated by Gammaproteobacteria (46.4% of total reads) followed by Firmicutes (20.6%), Betaproteobacteria (16.4%), Actinobacteria (6.1%), Alphaproteobacteria (5.7%), Chlamydiae (1.7%), and Bacteroidetes (1.5%).

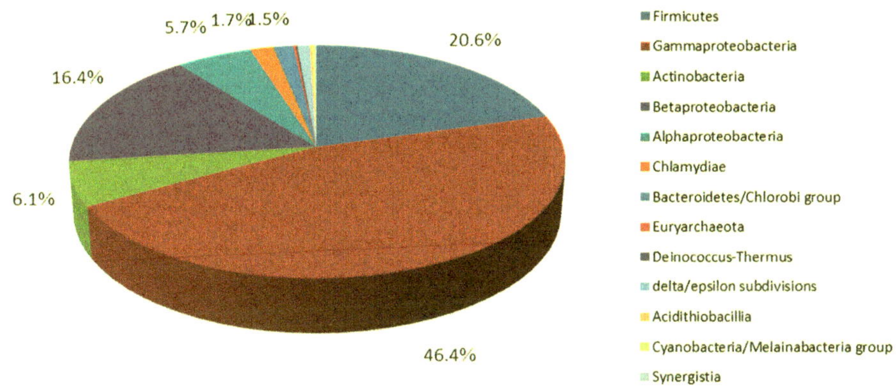

Fig. 4.7 Phylum level grouping of sequences obtained from Tbilisi sulfur spring water sample

Deinococcus-Thermus, delta/epsilon subdivisions, Acidithiobacillus Cyanobacteria/Melainabacteria group, and Synergistia comprised a minority of the prokaryotic populations accounting for less than 1% of total reads for each group (Fig. 4.7). Archaeal sequence reads were also in minority, belonging to the Euryarchaeota and comprising 0.2% of total reads.

Gammaproteobacteria were represented by 35 bacterial genera, dominated by *Escherichia* (30%), *Pseudomonas* (24%), *Xanthomonas* (11%), *Legionella* (5%), *Salmonella* (3.5%), and *Acinetobacter* (2.8%). Some of these bacteria are found in diverse habitats and may also cause diseases in humans. Interestingly, *Pseudomonas*, *Legionella*, and *Acinetobacter* were reported in a variety of geothermal springs (Lin et al. 2007; Petursdottir et al. 2009; Saxena et al. 2017). The sulfur spring also harbored *Silanimonas lenta* (3,5%), belonging to moderately thermophilic alkaliphilic bacteria isolated from a hot spring in Korea (Lee et al. 2005), purple sulfur bacteria *Ectothiorhodospira*, and thermophilic bacterium *Thermomonas hydrothermalis* isolated from a hot spring in Central Portugal (Alves et al. 2003).

The second most abundant group of bacteria was Firmicutes, including representatives belonging to the *Bacillus* and related genera such *Geobacillus* and *Tepidibacillus*.

Betaproteobacteria were dominated by *Neisseria* (29%), presumably allochthonous bacteria. The other two most prevalent betabacteria inhabiting the studied spring were amylase-producing *Caldimonas taiwanensis* (22%) and alkaline-protease-producing *Tepidimonas taiwanensis* (14%), thermophilic bacteria reported in geothermal springs in Taiwan (Chen et al. 2005; Chen et al. 2006) and, as described above, have been found in the Karvachar hot spring as well.

Actinobacteria were represented by 12 genera dominated by *Streptomyces* and *Mycobacteria* that may inhabit thermal spring environments. *Streptomyces* spp. are known to produce various enzymes and biological active compounds, including antimicrobials, and can be readily isolated from geothermal environments (Al-Dhabi et al. 2016). Mycobacteria, with potential to cause diseases in humans, have been also found and isolated from the sulfur hot springs (Lee et al. 2015).

Alphaproteobacteria of the sulfur spring comprised 18 genera dominated by *Mesorhizobium* (23% of total Alphaproteobacteria reads) and *Thalassobius* (16%) species. Though these bacteria were not described in the hot springs, they have been found in diverse environments such as marine waters and soils (Arahal et al. 2005; Yuan et al. 2016), indicating possibility of their presence in thermal waters as well.

The Bacteroidetes/Chlorobi group was represented by 12 genera with 25% and 12% of sequence reads aligned to *Bacteroides* and *Pedobacter* reference genes, respectively. *Bacteroides* spp. have not been described in geothermal waters and can be considered as a contaminant, while the presence of *Pedobacter* has been reported in an alkaline hot spring in Thermopolis (Buckingham et al. 2013).

The delta/epsilon subdivision comprised a minority of the sulfur spring microbial population represented only with five genera, including the sulfur-reducing microaerophilic bacterium *Sulfurospirillum* that could be considered as natural inhabitant of this geothermal spring. The Deinococcus-Thermus group was also in minority, represented by *Meiothermus taiwanensis*, aerobic, thermophilic, non-sporulating, filamentous bacteria reported in a hot spring in Taiwan (Chen et al. 2002).

The sulfur spring was inhabited by two methanogenic Euryarchaeota species, *Methanolacinia paynteri* and *Methanosarcina mazei*. The optimum growth conditions for *Methanolacinia paynteri* are pH 6.6–7.2, temperature 40 °C, and the sulfide may serve as the sulfur source (Zellner et al. 1989), thus presence of this archaeon in the sulfur spring is not surprising. *Methanosarcina* spp. can survive in a variety of habitats, including extreme environments and may use different metabolic pathways to produce methane (Assis das Graças et al. 2013).

In addition to whole metagenomic DNA sequencing, the microbial diversity of the sulfur spring was also analyzed using a PCR/DGGE approach. The majority of DGGE bands were affiliated with Betaproteobacteria involved in sulfur cycle, such as species belonging to the genera *Sulfurisoma*, *Thiobacillus*, and oxalotrophic bacterium *Oxalicibacterium faecigallinarum*.

The study has also revealed the presence of the methanogen *Methanosaeta harundinacea*, belonging to Euryarchaeota, confirming that the methanogenic Euryarchaeota dominate archaeal populations of the sulfur spring.

4.3.2 Cultivation-Dependent Studies

Cultivable approaches have been used for analysis of microbial diversity associated with hot springs. Several studies have been performed on the description of novel genera, species and strains, characterization of different bio-resources, and whole-genome analysis of a few isolates from geothermal springs in the Lesser Caucasus. Many thermostable enzymes, including lipase, protease, amylase, DNA polymerase, aspartase, aminoacylase, glucose isomerase and inulinase, producers of EPS, protein and vitamins, enrichments of nitrite-oxidizing bacteria (NOB), and methylotrophic, acetoclastic, and hydrogenotrophic methanogens with potential biotechnological applications have been reported by several authors (Table 4.3).

Overall, all isolates of bacteria and Archaea from the Lesser Caucasus belong to more than 40 distinct species of 21 different genera, namely, *Bacillus*, *Geobacillus*, *Anoxybacillus*, *Paenibacillus*, *Brevibacillus*, *Aeribacillus*, *Ureibacillus*, *Thermoactinomyces*, *Sporosarcina*, *Thermus*, *Rhodobacter*, *Thiospirillum*, *Thiocapsa*, *Rhodopseudomonas*, *Methylocaldum*, *Desulfomicrobium*, *Desulfovibrio*, *Treponema*, *Arcobacter*, *Nitrospira*, and *Methanoculleus*. The members of phylum Firmicutes were most dominant among the identified bacteria isolated from all thermal springs. Culture-dependent studies indicate that *Bacillus* and related genera were ubiquitous and predominant in harsh environments of high temperatures. Representatives of the genera *Geobacillus* and *Anoxybacillus* are the most highly distributed obligate thermophiles in the Lesser Caucasus hot springs. All isolates from the hot springs that belonged to the genus *Bacillus* were thermotolerant microorganisms among which *B. licheniformis* appeared as the dominating species. All studied springs demonstrated significantly lower content of species belonging to genera *Brevibacillus*, *Ureibacillus*, *Paenibacillus*, *Thermoactinomyces*, and *Sporosarcina*.

Bacteria belonging to the genera *Bacillus* and *Thermus* were mostly reported as aerobic, heterotrophic thermophiles and found in thermal systems with neutral to alkaline pH (Spanevello and Patel 2004). Although *Thermus* spp. may be predominant heterotrophs in many hot springs (Hjorleifsdottir et al. 2001), they were isolated only from the Karvachar hot spring.

Several strains representing potentially novel species were reported from the Akhurik, Jermuk, and Karvachar geothermal springs. Two novel strains belonging to genera *Anoxybacillus* and *Treponema* were reported from the hot spring at Jermuk. A novel species belonging to genus *Anoxybacillus* and a new strain belonging to *Thermus scotoductus* were reported from the Karvachar spring (Saghatelyan et al. 2015; Hovhannisyan et al. 2017). 16S rRNA gene sequences of a methanotrophic isolate from Akhurik geothermal spring showed that it was a new gammaproteobacterial methanotroph, forming a separate clade in the *Methylococcaceae* family. It fell into a cluster with thermotolerant and mesophilic methanotrophs, comprising the genera *Methylocaldum*-*Methylococcus*-*Methyloparacoccus*-*Methylogaea*. The genes *pmoA*, *mxaF*, *cbbL*, and *nifH* were detected, but no *mmoX* gene was found. The strain probably represents a novel methanotrophic genus (Islam et al. 2015).

Whole-genome analysis of the hot spring isolates was a major thrust area of investigation. Whole-genome shotgun sequencing of novel species isolated from hot springs at Jermuk (*Treponema thermophilum* sp. nov) and Karvachar (*Anoxybacillus* sp. strain K103) was performed (Poghosyan 2015; Hovhannisyan et al. 2017). Similarly, the whole-genome sequence of *Thermus scotoductus* K1 was reported following its isolation from the Karvachar spring (Saghatelyan et al. 2015).

Attention was also paid to the bioprospecting of geothermal spring's microbes with an intention of using these resources for commercial applications. In total, 135 thermophilic and thermotolerant bacilli strains were isolated under aerobic conditions at 55–65 °C and identified based on 16S rRNA gene sequence analysis as representatives of genera *Bacillus*, *Geobacillus*, *Anoxybacillus*, *Paenibacillus*, *Brevibacillus*, *Aeribacillus*, *Ureibacillus*, *Thermoactinomyces*, and *Sporosarcina*. These thermophilic bacilli were tested for hydrolytic enzyme production capacities, and biotechnologically valuable enzyme producers were selected (Panosyan 2017). The majority of the studies focused on hydrolytic enzymes like lipase (Vardanyan et al. 2015; Shahinyan et al. 2017), amylase (Hovhannisyan et al. 2016), and protease (Panosyan 2017).

Some phototrophic bacteria isolated from Armenian hot springs were good producers of enzymes such as aspartase, aminoacylase, glucose isomerase, and inulinase, as well as sources of protein, carbohydrates, and vitamins (Paronyan 2003). Two isolates belonging to the genus *Geobacillus* are able to produce heteropolymeric EPSs with high molecular weight (Panosyan et al. 2014).

Prospective microbes from hot springs offer a major advantage of preserving those strains for future studies and exploring them in due course for potential biotechnological applications in medical, industrial, and agricultural processes.

4.4 Correlation Between Geophysiology and Microbiology of the Hot Springs in the Lesser Caucasus

Understanding the microbial community structure in hot springs with different ecologies is important to elucidate community functions and their importance for the maintenance of hot spring ecosystems.

In general, microbial diversity was inversely correlated with temperature, and temperature has been shown to be a key factor in controlling the microbial diversity in hot springs (Wang et al. 2013). Thermophilic or hyperthermophilic Bacteria are commonly present in high-temperature hot springs (>75 °C) (Hou et al. 2013). When temperature is suitable for photosynthesis (<75 °C), moderately thermophilic and mesophilic phototrophic Bacteria are important members in terrestrial thermal springs, such as Cyanobacteria, Chloroflexi, and phototrophic representatives of Proteobacteria (Cox et al. 2011). In addition to Bacteria, members of the archaeal phyla Crenarchaeota, Euryarchaeota, and Thaumarchaeota are also commonly detected in geothermal systems (Ochsenreiter et al. 2003; Zhang et al. 2008).

The comparison of microbial species abundance and diversity in the Lesser Caucasus hot springs with those available internationally displays similar patterns. It was shown earlier that there is a negative correlation between spring temperature and diversity of microbes (Wang et al. 2013; Poddar and Das 2017).

Prokaryotic diversity was found to be low at high-temperature springs in contrast to low-temperature springs. Temperature has also been shown to drive phylum diversity in hot springs. Most of the studied hot springs in the Lesser Caucasus have a temperature below 50 °C and harbor bacterial species pertaining to phyla Firmicutes, Proteobacteria, Bacteroidetes, and Cyanobacteria, although with varying abundance between springs. Springs with higher temperatures also contained thermophiles belonging to Actinobacteria, Deinococcus-Thermus, and Aquificae. Representatives of the phylum Firmicutes were most versatile in the investigated hot springs and could populate hot springs with a wide range of temperatures. These observations are in accordance with many global studies indicating that thermophilic bacteria belonging to phyla Aquificae, Deinococcus-Thermus, and Firmicutes were abundant in the hot springs with high temperatures, whereas mesophilic bacterial members of Cyanobacteria, Chloroflexi, and Proteobacteria mostly occupy mesothermal hot springs (Wang et al. 2013). Cyanobacteria are the most commonly reported microbial group in these types of environments and are considered to be the major primary producers in these habitats (Castenholtz 1973). It was shown earlier that moderate-temperature geothermal systems cool enough to permit phototrophy at the source with neutral or alkaline pH are often colonized by visible microbial growth that forms laminated mats or streamers dominated by phototrophic bacteria (Klatt et al. 2011). Relatively low-temperature (>75 °C) and neutral pH in all studied springs can support growth of phototrophic bacteria due to obvious light effect in the outlet of the spring.

A comparison of the optimum growth temperature of the closest cultivated relatives of the microorganisms detected in the clone libraries, DGGE profiles, or pyrotags suggested that most of the microorganisms, including microorganisms representing some of the most dominant groups, are likely able to grow at reservoir temperature and, therefore, should not be regarded as contaminants. The bacterial metagenomic DNA sequences also affiliated with taxa that are not described in the literature as being associated with geothermal environments. This can be explained by the presence of contamination from surrounding soils. Although most of the retrieved sequences were most similar to environmental sequences representing uncultured bacteria from various habitats, some of them were phylogenetically associated with environmental clones obtained from similar habitats.

Archaea appeared to be a minority in the prokaryotic community. High-temperature environments were previously generally believed to be the realm of Archaea (Li et al. 2015; Urbieta et al. 2015; Chan et al. 2017). However, recent studies applying molecular methods have revealed that bacteria rather are the predominant prokaryotic communities in such environments (Badhai et al. 2015; López-López et al. 2015). The factors that allow bacteria to dominate in high-temperature habitats are not well understood.

All reported Lesser Caucasus springs have circumneutral pH and, therefore, harbor a microbial community different from acidic hot springs environments (Purcell et al. 2007; Poddar and Das 2017). Acidic springs have been reported to contain chemolithotrophic acidophiles belonging to genera *Acidithiobacillus*, *Sulfobacillus*, *Hydrogenobaculum*, *Acidobacteria*, *Acidimicrobium*, etc. that participate in Fe and sulfur oxidation in those environments (Burgess et al. 2012; Urbieta et al. 2015; Skirnisdottir et al. 2000). Acidophiles were hardly detected in Lesser Caucasian hot springs. Bacterial species isolated from the studied hot springs exhibited optimal growth at neutral pH and could not grow at low pH conditions. It was shown by many investigators that Firmicutes and Proteobacteria are the phyla consistently present in circumneutral hot springs. Results obtained from Lesser Caucasus geothermal springs also are in line with observations of microbial assemblages distributed in hot springs with pH $\geq$ 7 globally (Nakagawa and Fukui

2002; Wang et al. 2013). In general, Archaea are not dominant in circumneutral hot springs, which is in agreement with several recent reports with similar pH ranges (Wang et al. 2013; Merkel et al. 2017).

Environmental conditions and the nutritional status in a natural habitat may drive the development of a particular microbial group or population. The set of abiotic factors allow natural selection of a few species that can dominate and multiply in the ecologically relevant niche. Limited carbon and nitrogen sources and high temperature of the springs located in the Lesser Caucasus allowed also the development of a unique population dominated by a large number of bacilli including *Geobacillus* and *Anoxybacillus* spp.

Besides temperature and pH, the limiting factor for microbial diversity and biomass could be a combination of abiotic factors including dissolved gasses (H_2, CO_2, H_2S, CH_4) and high mineralization. The geothermal systems of the Lesser Caucasus are known to contain high concentrations of minerals, and thus, the mineralization may also have a strong influence on the community composition. Recent studies have also highlighted that other factors, such as biogeography and geological history, can be important in determining the thermophilic diversity of geothermal springs (Whitaker et al. 2003; Takacs-Vesbach et al. 2008).

4.5 Conclusion

Investigations of the geothermal springs' microbiome are important for understanding the microbe-mediated biogeochemical cycles and ecosystem functioning as well as exploring the biotechnological potency of thermophilic isolates. This is the first comprehensive census of the microbial communities thriving in 11 geothermal springs of the Lesser Caucasus. Firmicutes, Proteobacteria, Bacteroidetes, and Cyanobacteria were the signature phyla in all 11 hot springs that along with the presence of site-specific taxa contributed to the uniqueness of each spring. Archaea appeared to be a minority in the prokaryotic community composing less than 1% of all microbial population. Overall, microbial diversity and richness were negatively affected by increasing temperature. Other influential factors shaping the microbiota of the studied Lesser Caucasus circumneutral geothermal springs appear to be pH and mineralization. Biogeography and geological history should not be ignored in microbial ecology studies, as all abiotic factors collectively contribute to the dynamics of the microbial populations. Many new thermophilic microbes mainly belonging to the *Bacillus* and related genera have been isolated, identified, and evaluated taking into account their biotechnological potency.

The present work, therefore, extends the previous sphere of information regarding the thermophilic bacterial diversity of thermal springs in the Lesser Caucasus.

Acknowledgments This work was supported by grants from the EURASIA Program of the Norwegian Center for International Cooperation in Education (CPEA-2011/10081, CPEA-LT-2016/10095) and partially supported by the RA MES State Committee of Science, in the frames of the research project №15 T-1F399, Armenian National Science and Education Fund based in New York, USA, to HP (ANSEF-NS-microbio 2493, 3362 and 4676) and Project CRDF/GRDF/DTRA A60920.

References

Aanniz TM, Ouadghiri M, Melloul M, Swings J, Elfahime E, Ibijbijen J, Ismaili M, Amar M (2015) Thermophilic bacteria in Moroccan hot springs, salt marshes and desert soils. Braz J Microbiol 46(2):443–453

Achmadova FR (1991) Rasprostranenie termofilnikh bakterii vidov Bacillus и Thermus v gorjachikh istochnikov Azerbadjanskoj SSR (Distribution of bacteria belonging to the Bacillus and Thermus in hot springs of Azerbadjan SSR). Dissertation, Timiriazev KA Agriculture Academy, Moscow (in Russian)

Al-Dhabi NA, Esmail GA, Duraipandiyan V, Valan Arasu M, Salem-Bekhit MM (2016) Isolation, identification and screening of antimicrobial thermophilic *Streptomyces* sp. Al-Dhabi-1 isolated from Tharban hot spring, Saudi Arabia. Extremophiles 20:79–90

Alves MP, Rainey FA, Nobre MF, da Costa MS (2003) *Thermomonas hydrothermalis* sp. nov., a new slightly thermophilic gamma-proteobacterium isolated from a hot spring in Central Portugal. Syst Appl Microbiol 26:70–75

Amann R, Ludwig W (2000) Ribosomal RNA-targeted nucleic acid probes for studies in microbial ecology. FEMS Microbiol Rev 24:555–565

Amann RI, Ludwig W, Schleifer KH (1995) Phylogenetic identification and in-situ detection of individual microbial-cells without cultivation. Microbiol Rev 59:143–169

Anil Kumar P, Srinivas TN, Madhu S, Manorama R, Shivaji S (2010) *Indibacter alkaliphilus* gen. nov., sp. nov., an alkaliphilic bacterium isolated from a haloalkaline lake. Int J Syst Evol Microbiol 60(4):721–726

Antranikian G, Egorova K (2007) Extremophiles, a unique resource of biocatalysts for industry. In: Gerday C, Glansdorff N (eds) Physiology and biochemistry of extremophiles. ASM Press, Washington, DC, pp 361–406

Arahal DR, Macian MC, Garay E, Pujalte MJ (2005) *Thalassobius mediterraneus* gen. nov., sp. nov., and reclassification of *Ruegeria gelatinovorans* as *Thalassobius gelatinovorus* comb. nov. Int J Syst Evol Microbiol 55:2371–2376

Assis das Graças D, Thiago Jucá Ramos R, Vieira Araújo AC, Zahlouth R, Ribeiro Carneiro A, Souza Lopes T et al (2013) Complete genome of a *Methanosarcina mazei* strain isolated from sediment samples from an Amazonian flooded area. Genome Announc 1:e00271–13

Badalyan M (2000) Geothermal features of Armenia: a country update. In: Proceedings world geothermal congress, Kyushu-Tohoku, Japan, 28 May–10 June, vol 2000, pp 71–75

Badhai J, Ghosh TS, Das SK (2015) Taxonomic and functional characteristics of microbial communities and their correlation with physicochemical properties of four geothermal springs in Odisha, India. Front Microbiol 6:1166. https://doi.org/10.3389/fmicb.2015.01166

Bhaya D, Grossman AR, Steunou A-S, Khuri N et al (2007) Population level functional diversity in a microbial community revealed by comparative genomic and metagenomic analyses. ISME J 1:703–713

Bowen De León K, Gerlach R, Peyton BM, Fields MW (2013) Archaeal and bacterial communities in three alkaline hot springs in Heart Lake Geyser Basin, Yellowstone National Park. Front Microbiol 4:330. https://doi.org/10.3389/fmicb.2013.00330

Buckingham AR, Law RM, Holley GA, Yeomans NS and Smaglik SM (2013) An intergrated study of a hot springs microbial community in Thermopolis, WY. In: The Geological Society of America abstracts with programs of the 125th anniversary annual meeting&expo, Denver, Colorado, USA, 27–30 October 2013. vol 45(7), p 582

Bunterbart G, Kapanadze N, Kobzev G, Melikadze G (2009) Re-assessment of the geothermal potential of Tbilisi region: the hydrodynamic digital model project. In: Environment and resources, Association of Academics of Science in Asia Workshop, Izmir, 25–27 September, vol 2009, pp 154–161

Burgess AE, Unrine JM, Mills GL, Romanek CS, Wiegel J (2012) Comparative geochemical and microbiological characterization of two thermal ponds in the Uzon Caldera, Kamchatka, Russia. Microb Ecol 63:471–489

Castenholtz RW (1973) Ecology of blue-green algae in hot springs. In: Carr NG, Whitton BA (eds) The biology of blue-green algae. University of California Press, Los Angeles, pp 379–414

Chan CS, Chan KG, Tay YL, Chua YH, Goh KM (2015) Diversity of thermophiles in a Malaysian hot spring determined using 16S rRNA and shotgun metagenome sequencing. Front Microbiol 6:177. https://doi.org/10.3389/fmicb.2015.00177

Chan CS, Chan KG, Ee R, Hong KW, Urbieta MS, Donati ER, Shamsir MS, Goh KM (2017) Effects of physiochemical factors on prokaryotic biodiversity in Malaysian Circumneutral Hot Springs. Front Microbiol 8:1252. https://doi.org/10.3389/fmicb.2017.01252

Chen MY, Lin GH, Lin YT, Tsay SS (2002) *Meiothermus taiwanensis* sp. nov., a novel filamentous, thermophilic species isolated in Taiwan. Int J Syst Evol Microbiol 52:1647–1654

Chen WM, Chan JS, Chiu CH, Chang SC, Chen WC, Jiang CM (2005) *Caldimonas taiwanensis* sp. nov., α-amylase producing bacterium isolated from a hot spring. Syst Appl Microbiol 28:415–420

Chen TL, Chou YJ, Che WM, Arun B, Young CC (2006) *Tepidimonas taiwanensis* sp. nov., a novel alkaline-proteaseproducing bacterium isolated from a hot spring. Extremophiles 10(1):35–40

Chen WM, Huang HW, Chang JS, Han YL, Guo TR, Sheu SY (2013) *Tepidimonas fonticaldi* sp. nov., a slightly thermophilic betaproteobacterium isolated from a hot spring. Int J Syst Evol Microbiol 63(5):1810–1816

Cihan AC, Ozcan B, Tekin N, Cokmus C (2011) Phylogenetic diversity of isolates belonging to genera *Geobacillus* and *Aeribacillus* isolated from different geothermal regions of Turkey. World J Microbiol Biotechnol 27:2683–2696

Coman C, Drugă B, Hegedus A, Sicora C, Dragoş N (2013) Archaeal and bacterial diversity in two hot spring microbial mats from a geothermal region in Romania. Extremophiles 17(3):523–534. https://doi.org/10.1007/s00792-013-0537-5

Cox A, Shock EL, Havig JR (2011) The transition to microbial photosynthesis in hot spring ecosystems. Chem Geo 280:344–351

DeCastro ME, Rodríguez-Belmonte E, González-Siso MI (2016) Metagenomics of thermophiles with a focus on discovery of novel thermozymes. Front Microbiol 7:1521. https://doi.org/10.3389/fmicb.2016.01521

Deepika M, Satyanarayana T (2013) Diversity of hot environments and thermophilic microbes. In: Satyanarayana T, Littlechild J, Kawarabayasi Y (eds) Thermophilic microbes in environmental and industrial biotechnology biotechnology of thermophiles. Springer, Dordrecht/Heidelberg/New York/London, pp 3–60

DeLong EE, Pace NR (2001) Environmental diversity of bacteria and archaea. Syst Biol 50:470–478

Edwards TA, Calica NA, Huang DA, Manoharan N, Hou W, Huang L, Panosyan H, Dong H, Hedlund BP (2013) Cultivation and characterization of thermophilic Nitrospira species from geothermal springs in the U.S. Great Basin, China, and Armenia. FEMS Microbiol Ecol 85(2):283–292

Everroad RC, Otaki H, Matsuura K, Haruta S (2012) Diversification of bacterial community composition along a temperature gradient at a thermal spring. Microbes Environ 27:374–381

Hedlund BP, Cole JK, Williams AJ, Hou W, Zhou E, Li WJ, Dong H (2012) A review of the microbiology of the Rehai geothermal field in Tengchong, Yunnan Province, China. Geosci Front 3(3):273–288

Hedlund BP, Dodsworth JA, Cole JK, Panosyan HH (2013) An integrated study reveals diverse methanogens, Thaumarchaeota, and yet-uncultivated archaeal lineages in Armenian hot springs. Antonie Van Leeuwenhoek 104(1):71–82

Henneberger R, Cooksley D, Hallberg J (2000) Geothermal resources of Armenia: in: proceedings world geothermal congress, Kyushu-Tohoku, Japan, 28 May–10 June 2000, pp 1217–1222

Hetzer A, Morgan HW, McDonald IR, Daughney CJ (2007) Microbial life in champagne pool, a geothermal spring in Waiotapu, New Zealand. Extremophiles 11(4):605–614

Hjorleifsdottir S, Skirnisdottir S, Hreggvidsson G, Holst O, Kristjansson J (2001) Species composition of cultivated and noncultivatedbacteria from short filaments in an Icelandic hot spring at 88 °C. Microb Ecol 42:117–125

Hou W, Wang S, Dong H, Jiang H, Briggs BR, Peacock JP, Huang Q, Huang L, Wu G, Zhi X, Li W, Dodsworth JA, Hedlund BP, Zhang C, Hartnett HE, Dijkstra P, Hungate BA (2013) A comprehensive census of microbial diversity in hot springs of Tengchong, Yunnan province China using 16S rRNA gene pyrosequencing. PLoS One 8(1):e53350

Hovhannisyan P, Turabyan A, Panosyan H, Trchounian A (2016) Thermostable amylase production bacilli isolated from Armenian geothermal springs. Biol J Armenia 68:6–15

Hovhannisyan P, Panosyan H, Trchounian A, Birkeland NK (2017) Amplification and cloning of an alpha-amylase gene from *Anoxybacillus flavithermus* K103 isolated from an Armenian geothermal spring. In: Abstracts of the 7th congress of European microbiologists (the FEMS 2017), Valencia, Spain, 9–13 July, 2017

Hreggvidsson GO, Petursdottir SK, Bjornsdottir SH, Fridjonsson OH (2012) Microbial speciation in the geothermal ecosystem. In: Stan H, Fendrihan LS (eds) Adaption of microbial life to environmental extremes: novel research results and application. Springer, Wien, pp 37–68

Islam T, Larsen Ø, Torsvik V, Øvreås L, Panosyan H, Murrell C, Birkeland NK, Bodrossy L (2015) Novel methanotrophs of the family Methylococcaceae from different geographical regions and habitats. Microorganisms 3:484–499

Kampfer P, Schulze R, Jackel U, Malik KA, Amann R, Spring S (2005) *Hydrogenophaga defluvii* sp. nov. and *Hydrogenophaga atypica* sp. nov., isolated from activated sludge. Int J Syst Evol Microbiol 55:341–344

Kapanadze N, Melikadze GI, Janelidze P (2010) Estimation of Zugdidi and Tbilisi thermal water deposits. In: Weller A, Melikadze G, Kapanadze N (eds) Exploration and exploitation of groundwater and thermal water systems in Georgia. Tbilisi, 2010

Karakhanian AS, Trifonov VG, Azizbekian OG, Hondkarian DG (1997) Relationship of late Quaternary tectonics and volcanism in the Khanarassar active fault zone, the Armenian Upland. Terra Nova 16:131–134

Kellermann C, Griebler C (2009) Thiobacillus thiophilus sp. nov., a chemolithoautotrophic, thiosulfate-oxidizing bacterium isolated from contaminated aquifer sediments. Int J Syst Evol Microbiol 59:583–588

Klatt CG, Wood JM, Rusch DB, Bateson MM, Hamamura N, Heidelberg JF, Grossman AR, Bhaya D, Cohan FM, Kühl M, Bryant DA, Ward DM (2011) Community ecology of hot spring cyanobacterial mats: predominant populations and their functional potential. ISME J 5:1262–1278

Kodama Y, Watanabe K (2004) *Sulfuricurvum kujiense* gen. nov., sp. nov., a facultatively anaerobic, chemolithoautotrophic, sulfur-oxidizing bacterium isolated from an underground crude-oil storage cavity. Int J Syst Evol Microbiol 54:2297–2300

Krebs JE, Vaishampayan P, Probst AJ, Tom LM, Marteinsson VT, Andersen GL, Venkateswaran K (2014) Microbial community structures of novel Icelandic hot spring systems revealed by PhyloChip G3 analysis. Astrobiology 14(3):229–240

Kublanov IV, Perevalova AA, Slobodkina GB, Lebedinsky AV, Bidzhieva SK, Kolganova TV, Kaliberda EN, Rumsh LD, Haertle T, Bonch-Osmolovskaya EA (2009) Biodiversity of thermophilic prokaryotes with hydrolytic activities in hot springs of Uzon caldera, Kamchatka (Russia). Appl Environ Microbiol 75(1):286–291

Lee EM, Jeon CO, Choi I, Chang KS, Kim CJ (2005) *Silanimonas lenta* gen. nov., sp. nov., a slightly thermophilic and alkaliphilic gammaproteobacterium isolated from a hot spring. Int J Syst Evol Microbiol 55:385–389

Lee SY, Lee MH, Oh TK, Yoon JH (2006) *Lutibacter aestuarii* sp. nov., isolated from a tidal flat sediment, and emended description of the genus *Lutibacter* Choi and Cho. Int J Syst Evol Microbiol 62(2):420–424

Lee MR, Sheng WH, Hung CC, Yu CJ, Lee LN, Hsueh PR (2015) *Mycobacterium abscessus* complex infections in humans. Emerg Infect Dis 21(9):1638–1646

Li H, Yang Q, Li J, Gao H, Li P, Zhou H (2015) The impact of temperature on microbial diversity and AOA activity in the Tengchong geothermal field, China. Sci Rep 5:17056. https://doi.org/10.1038/srep17056

Lin YE, Lu WM, Huang HI, Huang WK (2007) Environmental survey of *Legionella pneumophila* in hot springs in Taiwan. J Toxicol Environ Health A 70(1):84–87

Liu Y, Whitman WB (2008) Metabolic, phylogenetic, and ecological diversity of the methanogenic archaea. Ann N Y Acad Sci 1125:171–189

Liu Z, Klatt CG, Wood JM, Rusch DB, Ludwig M et al (2011) Metatranscriptomic analyses of chlorophototrophs of a hotspring microbial mat. ISME J 5:1279–1290

Liu J, Wang Z, Belchik SM, Edwards MJ, Liu C, Kennedy DW, Merkley ED, Lipton MD, Butt JN, Richardson DJ, Zachara JM, Fredrickson JK, Rosso KM, Shi L (2012) Identification and characterization of MtoA: a decaheme c-type cytochrome of the neutrophilic Fe(II)-oxidizing bacterium *Sideroxydans lithotrophicus* ES-1. Front Microbiol 3:37. https://doi.org/10.3389/fmicb.2012.00037

López-López O, Cerdán ME, González-Siso MI (2013) Hot spring metagenomics. Life 3(2):308–320

López-López O, Knapik K, Cerdán ME, González-Siso MI (2015) Metagenomics of an alkaline hot spring in Galicia (Spain): microbial diversity analysis and screening for novel lipolytic enzymes. Front Microbiol 6:1291. https://doi.org/10.3389/fmicb.2015.01291

Margaryan A, Panosyan H, Popov YG (2010) Isolation and characterization of new metallotolerant bacilli strains. Biotechnol Biotechnol Equip 24(2):450–454

Maugeri TL, Lentini V, Gugliandolo C, Italiano F, Cousin S, Stackebrandt E (2009) Bacterial and archaeal populations at two shallow hydrothermal vents off Panarea Island (Eolian Islands, Italy). Extremophiles 13(1):199–212

Melikadze G, Tsertsvadze L, Tsertsvadze N, Vardigoreli O, Barabadze T (2010) Country update from Georgia. In: Proceedings of world geothermal congress, Bali, Indonesia, 25–29 April 2010

Merkel AY, Pimenov NV, Rusanov II, Slobodkin AI, Slobodkina GB, Tarnovetckii IY et al (2017) Microbial diversity and autotrophic activity in Kamchatka hot springs. Extremophiles 21(2):307–317

Meyer-Dombard DR, Shock EL, Amend JP (2005) Archaeal and bacterial communities in geochemically diverse hot springs of Yellowstone National Park, USA. Geobiology 3:211–227

Mkrtchyan S (ed) (1969) Geology of Armenian SSR. Publishing house of AS of ASSR, Yerevan (in Russian)

Moores EM, Fairbridge RW (eds) (1998) Encyclopedia of European and Asian Regional Geology. Springer Netherlands

Nakagawa T, Fukui M (2002) Phylogenetic characterization of microbial mats and streamers from a Japanese alkaline hot spring with a thermal gradient. J Gen Appl Microbiol 48:211–222

Ochsenreiter T, Selezi D, Quaiser A, Bonch-Osmolovskaya L, Schleper C (2003) Diversity and abundance of Crenarchaeota in terrestrial habitats studied by 16S RNA surveys and real time PCR. Environ Microbiol 5:787–797

Panosyan H (2010) Phylogenetic diversity based on 16S rRNA gene sequence analysis of aerobic thermophilic endospore-forming bacteria isolated from geothermal springs in Armenia. Biol J Armenia 62(4):73–80

Panosyan HH (2017) Thermophilic bacilli isolated from Armenian geothermal springs and their potential for production of hydrolytic enzymes. Int J Biotech Bioeng 3(8):239–244

Panosyan H, Birkeland NK (2014) Microbial diversity in an Armenian geothermal spring assessed by molecular and culture-based methods. J Basic Microbiol 54(11):1240–1250

Panosyan H, Anzelmo G, Nicolaus B (2014) Production and characterization of exopolysaccharadies synthesized by geobacilli isolated from an Armenian geothermal spring. FEBS J 281(1):667

Panosyan HH, Margaryan AA, Trchounian AH (2017) Denaturing gradient gel electrophoresis (DGGE) profiles of the partial 16S rRNA genes defined bacterial population inhabiting in Armenian geothermal springs. Biol J Armenia 68(3):102–109

Paronyan A (2002a) Ecology and biodiversity of phototrophic bacteraia in various ecosystems of Armenia. Biol J Armenia 54(1–2):91–98 (in Russion)

Paronyan AK (2002b) Consumption of organic carbon sources and biosynthesis of lactic acid by the photosynthetic bacterium *Rhodobacter* sp. D-4. Appl Biochem Microbiol 38(1):53–58

Paronyan AKh (2003) Ekologia, biologicheskie osobennosti fototrofnikh bakterii Armenii i perspektiwi ikh ispolzovanija. (Ecology, biological pecularities of phototrophic bacteria of Armenia and prospects its application). Disertation, Institute of Microbiology NAS of RA

Paronyan A (2007) Ecophisiological charachteristics of phototrophic bacteria *Rhodopseudomonas palustris* isolated from mineral geothermal Jermuk. Biol J Armenia 59(1–2):73–77 (in Russion)

Petursdottir SK, Bjornsdottir SH, Hreggvidsson GO, Hjorleifsdottir S, Jakob KK (2009) Analysis of the unique geothermal microbial ecosystemof the Blue Lagoon. FEMS Microbiol Ecol 70(3):425–432

Philip H, Cisternas A, Gvishiani A, Gorshukov A (1989) The Caucasus: an actual example of the initial stages of continental collision. Tectonophysics 161(1–2):1–21

Poddar A, Das SK (2017) Microbiological studies of hot springs in India: a review. Arch Microbiol 200:1. https://doi.org/10.1007/s00203-017-1429-3

Poddar A, Lepcha RT, Das SK (2014) Taxonomic study of the genus *Tepidiphilus*: transfer of Petrobacter succinatimandens to the genus *Tepidiphilus* as *Tepidiphilus succinatimandens* comb. nov., emended description of the genus *Tepidiphilus* and description of *Tepidiphilus thermophilus* sp. nov., isolated from a terrestrial hot spring. Int J Syst Evol Microbiol 64(1):228–235

Poghosyan L (2015) Prokaryotic diversity in an Armenian geothermal spring using metagenomics, anaerobic cultivation and genome sequencing. Master thesis, University of Bergen

Portillo MC, Sririn V, Kanoksilapatham W, Gonzalez JM (2009) Differential microbial communities in hot spring mats from Western Thailand. Extremophiles 13(2):321–331

Purcell D, Sompong U, Lau CY, Barraclough TG et al (2007) The effects of temperature, pH and sulphide on community structure of hyperthermophilic streamers in hot springs of northern Thailand. FEMS Microbiol Ecol 60(3):456–466

Raddadi N, Cherif A, Daffonchio D, Neifar M, Fava F (2015) Biotechnological applications of extremophiles, extremozymes and extremolytes. Appl Microbiol Biotechnol 99(19):7907–7913

Rainey FA, Silva J, Nobre MF, Silva MT, da Costa MS (2003) *Porphyrobacter cryptus* sp. nov., a novel slightly thermophilic, aerobic, bacteriochlorophyll a-containing species. Int J Syst Evol Microbiol 53(1):35–41

Reilinger RE, McClusky SC, Oral MB, King RW, Toksoz MN, Barka AA, Kinik I, Lenk O, Sanli I (1997) Global Positioning System measurements of present-day crustal movements in the Arabia-Africa-Eurasia plate collision zone. J Geophys Res 102(B5):9983–9999

Roeselers G, Norris T, Castenholz R, Rysgaard S, Glud R, Kuhl M, Muyzer G (2007) Diversity of phototrophic bacteria in microbial mats from Arctic hot springs (Greenland). Environ Microbiol 9(1):26–38

Saghatelyan A, Panosyan H (2015) Study of bacterial diversity of Karvachar geothermal spring using culture-independent methods. In: Collection of scientific articles of YSU SSS: materials of the scientific session dedicated to the 95th anniversary of YSU. Yerevan State University, Yerevan, pp 13–18

Saghatelyan A, Panosyan H, Trchounian A, Birkeland NK (2014) Investigation of microbial diversity of geothermal springs in Nagorno-Karabakh based on metagenomic and culture-based approaches. In: Abstracts book of international scientific workshop trends in microbiology and microbial biotechnology, YSU, Yerevan, Armenia, 5–8 October 2014

Saghatelyan A, Poghosyan L, Panosyan H, Birkeland NK (2015) Draft genome sequence of Thermus scotoductus strain K1, isolated from a geothermal spring in Karvachar, Nagorno Karabakh. Genome Announc 3(6):e01346-15. https://doi.org/10.1128/genomeA.01346-15

Saghatelyan A, Panosyan H, Trchounian A, Birkeland NK (2016) Amplification and cloning of DNA polymerase 1 (pol1) of Thermus scotoductus K1 isolated from an Armenian geothermal spring. FEBS J 283(1):165–166

Sahm K, John P, Nacke H, Wemheuer B, Grote R, Daniel R et al (2013) High abundance of heterotrophic prokaryotes in hydrothermal springs of the Azores as revealed by a network of 16S rRNA gene-based methods. Extremophiles 17(4):649–662

Saxena R, Dhakan DB, Mittal P, Waiker P, Chowdhury A, Ghatak A, Sharma VK (2017) Metagenomic analysis of hot springs in Central India reveals hydrocarbon degrading thermophiles and pathways essential for survival in extreme environments. Front Microbiol 7:2123. https://doi.org/10.3389/fmicb.2016.02123

Sayeh R, Birrien JL, Alain K, Barbier G, Hamdi M, Prieur D (2010) Microbial diversity in Tunisian geothermal springs as detected by molecular and culture-based approaches. Extremophiles 14(6):501–514

Shahinyan GS, Margaryan AA, Panosyan HH, Trchounian AH (2015) Isolation and characterization of lipase-producing thermophilic bacilli from geothermal springs in Armenia and Nagorno-Karabakh. Biol J Armenia 67(2):6–15

Shahinyan G, Margaryan AA, Panosyan HH, Trchounian AH (2017) Identification and sequence analysis of novel lipase encoding novel thermophilic bacilli isolated from Armenian geothermal springs. BMC Microbiol 17:103. https://doi.org/10.1186/s12866-017-1016-4

Singh A, Subudhi E (2016) Profiling of microbial community of Odisha hot spring based on metagenomic sequencing. Genomics Data 7:187–188

Skirnisdottir S, Hreggvidsson GO, Hjorleifsdottir S, Marteinsson VT, Petursdottir SK, Holst O, Kristjansson JK (2000) Influence of sulfide and temperature on species composition and community structure of hot spring microbial mats. Appl Environ Microbiol 66:2835–2841

Spanevello MD, Patel BK (2004) The phylogenetic diversity of Thermus and Meiothermus from microbial mats of an Australian subsurface aquifer runoff channel. FEMS Microbiol Ecol 50:63–73

Stan-Lotter H, Fendrihan S (2012) Adaption of microbial life to environmental extremes. Springer, Wien

Stefanova K, Tomova I, Tomova A, Radchenkova N, Atanassov I, Kambourova M (2015) Archaeal and bacterial diversity in two hot springs from geothermal regions in Bulgaria as demonstrated by 16S rRNA and GH-57 genes. Int Microbiol 18:217–223

Stokes CR (2011) Caucasus Mountains. In: Singh VP, Singh P, Haritashya UK (eds) Encyclopedia of snow, ice and glaciers. Springer, Dordrecht, pp 127–128

Takacs-Vesbach C, Mitchell K, Jakson-Weaver O, Reysenbach AL (2008) Volcanic calderas delineate biogeographi provinces among Yellowstone thermophiles. Environ Microbiol 10(7):1681–1689

Tsertsvadze N, Buachidze G, Vardigoreli O, Vashakidze B, Inaishvili T, Kotrikadze N, Tsertsvadze L (1998) Thermal water of Georgia. Georgian Geothermal Association, Tbilisi

Urbieta MS, Gonzalez-Toril E, Bazan AA, Giaveno MA, Donati E (2015) Comparison of the microbial communities of hot springs waters and the microbial biofilms in the acidic geothermal area of Copahue (Neuquen, Argentina). Extremophiles 19(2):437–450

Vardanyan G, Margaryan A, Panosyan H (2015) Isolation and characterization of lipase-produvting bacilli from Tatev geothermal spring (Armenia). In: Collection of scientific articles of YSU SSS: materials of the scientific session dedicated to the 95th anniversary of YSU. Yerevan State University, Yerevan, pp 33–36

Wang S, Hou W, Dong H, Jiang H, Huang L, Wu G et al (2013) Control of temperature on microbial community structure in hot springs of the Tibetan Plateau. PLoS One 8:e62901. https://doi.org/10.1371/journal.pone.0062901

Whitaker RJ, Grogan DW, Taylor JW (2003) Geographical barriers isolate endemic population of hyperthermophilic archaea. Science 301:976–978

Yoon HS, Aslam Z, Song GC, Kim SW, Jeon CO, Chon TS, Chung YR) (2009) Flavobacterium sasangense sp. nov., isolated from a wastewater stream polluted with heavy metals. Int J Syst Evol Microbiol 59(P5):1162–1166

Yuan CG, Jiang Z, Xiao M, Zhou EM, Kim CJ, Hozzein WN, Park DJ, Zhi XY, Li WJ (2016) *Mesorhizobium sediminum* sp. nov., isolated from deep-sea sediment. Int J Syst Evol Microbiol 66:4797–4802

Zellner G, Messner P, Kneifel H, Tindall BJ, Winter J, Stackebrandt E (1989) Methanolacinia gen. nov., incorporating Methanomicrobium paynteri as Methanolacinia paynteri comb. nov. J Gen Appl Microbiol 35:185–202

Zhang CL, Ye Q, Huang Z, Li W, Chen J et al (2008) Global occurrence of archaeal amoA genes in terrestrial hot springs. Appl Environ Microbiol 74:6417–6426

Zhou JH (2003) Microarrays for bacterial detection and microbial community analysis. Curr Opin Microbiol 6:288–294

Geobacillus and *Anoxybacillus* spp. from Terrestrial Geothermal Springs Worldwide: Diversity and Biotechnological Applications

5

Armine Margaryan, Grigor Shahinyan, Pargev Hovhannisyan, Hovik Panosyan, Nils-Kåre Birkeland, and Armen Trchounian

Abstract

A large number of thermophilic representatives of the *Geobacillus* and *Anoxybacillus* genera have been isolated from geographically distant and physicochemically different environments, including high-, moderate-, and low-temperature habitats. However, terrestrial hot springs are the main habitats for *Geobacillus* and *Anoxybacillus* species. The members of these genera possess a variety of thermo-adaptive features that enable them to thrive at elevated temperatures. Due to their ability to withstand harsh environmental conditions, geobacilli and anoxybacilli are a valuable source for provision of thermostable enzymes, such as amylases, lipases, proteases, etc., and other components. Thermostable enzymes obtained from thermophilic bacilli have found a plethora of commercial applications due to their sturdiness and toughness in withstanding the heat generated in various biotechnological and industrial processes. This

A. Margaryan · A. Trchounian
Department of Biochemistry, Microbiology and Biotechnology, Yerevan State University, Yerevan, Armenia

Research Institute of Biology, Yerevan State University, Yerevan, Armenia

G. Shahinyan
Research Institute of Biology, Yerevan State University, Yerevan, Armenia

P. Hovhannisyan · H. Panosyan (✉)
Department of Biochemistry, Microbiology and Biotechnology, Yerevan State University, Yerevan, Armenia
e-mail: hpanosyan@ysu.am

N.-K. Birkeland
Department of Biological Sciences, University of Bergen, Bergen, Norway

D. Egamberdieva et al. (eds.), *Extremophiles in Eurasian Ecosystems: Ecology, Diversity, and Applications*, Microorganisms for Sustainability 8,
https://doi.org/10.1007/978-981-13-0329-6_5

chapter contains a review of studies of geobacilli and anoxybacilli from terrestrial geothermal springs worldwide with special emphasis on their distribution and diversity, ecological significance, adaptive mechanisms, enzymes, and biotechnological potential.

Keywords
Terrestrial geothermal springs · *Geobacillus* · *Anoxybacillus* · Thermostability · Thermostable enzymes · Extremophiles

5.1 Introduction

Terrestrial hot springs are manifestations of geological activity and represent aquatic microcosms that are formed by the emergence of geothermally heated groundwater from the Earth's crust (Mehta and Satyanarayana 2013b). Terrestrial hydrothermal springs represent extreme environments and have been found worldwide, like those in Yellowstone National Park, which harbor the closest relatives to the original organisms that lived on our planet. Finding these features on Mars (or any other planet) could have big implications for the question of extraterrestrial life (Van Kranendonk et al. 2017). Hence, the microbiota thriving in geothermal hot springs have been the subject of extensive research. Among the diversity of microbes harboring the hot springs in different parts of the world, members of the *Geobacillus* and *Anoxybacillus* genera were frequently isolated and extensively studied during the last decades. The members of *Geobacillus* and *Anoxybacillus* genera are thermophilic bacilli, which have adapted to grow optimally at temperatures ranging from 35 to 75 °C (Bergey et al. 2009). The ability of these microorganisms to grow at high temperatures has made them suitable objects for studying and understanding the thermostability mechanisms for the microbial adaptations to harsh conditions.

Thermophilic bacilli constitute valuable sources for various biotechnological products (Antranikian 2007; Satyanarayana et al. 2012). The members of the *Geobacillus* and *Anoxybacillus* genera have shown tremendous potential in biotechnology because of their ability to produce unique thermostable enzymes and proteins with high industrial and economical values (Antranikian 2007; Gurumurthy and Neelagund 2012). The recent interest in biotechnology, coupled with the discovery of novel thermophilic bacilli, has prompted studies on the utilization of thermophiles and their enzymes, such as amylase (Gurumurthy and Neelagund 2010, 2012, Rekadwad 2015, Acer et al. 2016), lipase (Balan et al. 2012; Mahadevan and Neelagund 2014; Ozdemir et al. 2015; Ay et al. 2011), protease (Hawumba et al. 2002; Zhu et al. 2007; Nakamichi et al. 2010), xylanase (Sunna et al. 1997; Kacagan et al. 2008; Ellis and Magnuson 2012; Inan et al. 2013), and cellulase (Ibrahim and El-diwany 2007; Padilha et al. 2015).

Enzymes from these microorganisms are in great demand as they are not usually denatured at high temperatures but are rather more active. These enzymes are also more resistant to chemical reagents and extreme pH values in comparison with their mesophilic homologues (Synowiecki 2010; Pinzon-Martinez et al. 2010). Moreover, their thermostability is associated with higher biochemical reaction rates, lower viscosity, and less risk of contamination (Turner et al. 2007). All these factors have stimulated a renewed interest in the exploration of extracellular enzymatic activities of thermostable organisms.

The objective of this chapter is to review the findings of the diversity, thermostability mechanisms, and biotechnological applications of microbes belonging to genera *Geobacillus* and *Anoxybacillus* from different terrestrial geothermal springs worldwide.

5.2 Taxonomy and Species Diversity

The genera *Geobacillus* and *Anoxybacillus* of the phylum Firmicutes comprise a group of Gram-positive, endospore-forming, rod-shaped, chemoorganotrophic thermophilic bacteria, including obligate aerobes, denitrifiers, and facultative anaerobes that can grow over a temperature range of 35–75 °C. Their catabolic versatility, particularly in the degradation of starch, xylene, cellulose, and lipids, and rapid growth rates have raised their profile as organisms with high potential for industrial and biotechnological applications.

5.2.1 The Genus *Geobacillus*

The members of the genus *Geobacillus* were originally classified in the genus *Bacillus*, as thermophilic variants of *Bacillus* spp. For many years *Bacillus stearothermophilus* (Donk 1920) was the only obligate thermophilic species of the genus *Bacillus* with a validly published name. After 1980, additional thermophilic species were proposed based on phenotypic analyses of novel isolates. Subsequent 16S rRNA gene sequencing indicated that *B. stearothermophilus*, *Bacillus kaustophilus*, and *Bacillus thermoglucosidasius* formed a phylogenetic lineage that was distinct from other *Bacillus* spp. (Ash et al. 1991). The continued development of genetic tools to facilitate both fundamental investigations and metabolic engineering and accumulating evidence for clustering of many of the thermophiles in a separate subgroup (group 5) supported by 16S rRNA analysis led to their reclassification as a separate genus (Nazina et al. 2001). Nazina et al. (2001) proposed that the six species of that lineage, namely, *Bacillus stearothermophilus*, *B. kaustophilus*, *B. thermoglucosidasius*, *B. thermocatenulatus*, *B. thermoleovorans*, and *B. thermodenitrificans*, should be placed in a new genus, *Geobacillus*, with *G. stearothermophilus* as the type species and along with two novel species, *G. subterraneus* and *G. uzenensis*. *B. pallidus* (Scholz et al. 1987), *Saccharococcus caldoxylosilyticus* (Ahmad et al. 2000), and *B. vulcani* (Caccamo et al. 2000) were also transferred to the genus *Geobacillus* (Fortina et al. 2001; Banat et al. 2004; Nazina et al. 2004). Subsequently, six additional species, *G. toebii* (Sung et al. 2002), *G. gargensis* (Nazina et al. 2004), *G. debilis* (Banat et al. 2004), *G. lituanicus* (Kuisiene et al.

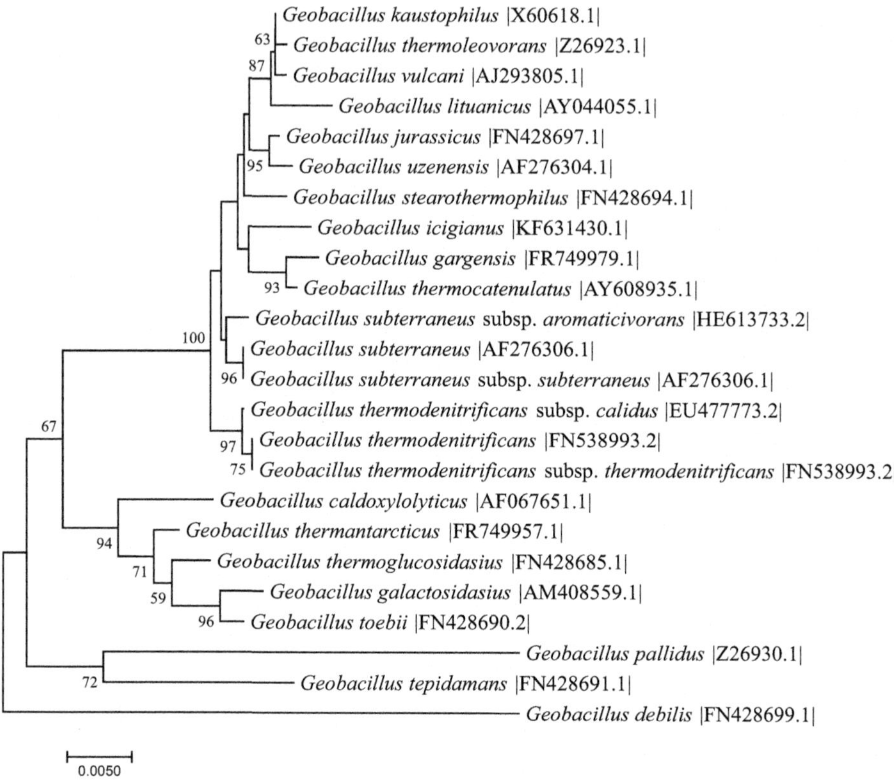

Fig. 5.1 Evolutionary relationships of species of the genus *Geobacillus*. The evolutionary history was inferred using the neighbor-joining method. The optimal tree with the sum of branch length = 0.17390653 is shown. The percentage of replicate trees in which the associated taxa clustered together in the bootstrap test (1000 replicates) is shown next to the branches. Significant bootstrapping values (>59%) are shown on the nodes. The analysis involved 24 16S rRNA nucleotide sequences of the type strains of *Geobacillus* species, obtained from NCBI GenBank (accession numbers shown between bars). All positions containing gaps and missing data were eliminated. There were a total of 1218 positions in the final dataset. Evolutionary analyses were conducted in MEGA7

2004), *G. tepidamans* (Coorevits et al. 2012), and *G. jurassicus* (Nazina et al. 2005), have been described. Currently the genus *Geobacillus* includes 20 species and 4 subspecies (Bergey et al. 2009). Miñana-Galbis et al. (2010) proposed the further transfer of *Geobacillus pallidus* to the new genus *Aeribacillus*.

The evolutionary tree showing the phylogenetic relationships of *Geobacillus* species is presented in Fig. 5.1.

The majority of geobacilli strains grow in the temperature range 35–75 °C, with the optimum at 55–65 °C. Vegetative cells are rod-shaped and occur either singly or in short chains and are motile by means of peritrichous flagella, or they are nonmotile. The cell-wall structure is Gram-positive, but the Gram-stain reaction may vary. Endospores are ellipsoidal or cylindrical and located terminally or subterminally in slightly swollen or non-swollen sporangia. Colony morphology and size are variable; pigments may be produced on certain media. They are aerobic or facultatively anaerobic anaerobic. Oxygen is the terminal electron acceptor, replaceable in some species by nitrate. They are neutrophilic. Growth occurs at pH 6.0–8.5, with optimal growth at pH 6.2–7.5. Growth factors, vitamins, NaCl, and KCl are not required by most species. Most species can utilize n-alkanes as carbon and energy sources. Most species produce acid but not gas from fructose, glucose, maltose, mannose, and sucrose. Catalase and oxidase reaction varies. Most species produce extracellular thermostable hydrolytic enzymes that have high potential of use in industry. The major cellular fatty acids are C15:0 iso, C16:0 iso, and C17:0 iso, which make up more than 60% of the total. The main menaquinone type is MK-7. The lowest level of 16S rRNA gene sequence similarity between all *Geobacillus* species is around 93%, which indicates that at least some species need to be reclassified at the genus level (Bergey et al. 2009). The average genome size for *Geobacillus* spp. ranges from 3.5 to 3.9 Mbp. The smallest genome was found in *G. kaustophilus* and the largest in *G. thermoglucosidasius*. This might reflect the additional coding requirements associated with anaerobic growth, additional CRISPR regions, as well as genes of unassigned function found between transposable elements in the genome of *G. thermoglucosidasius*. Despite the small genome, it was shown that the highest number of IS/transposable elements was present in the *G. kaustophilus* (Hussein et al. 2015).

Geobacillus species are widely distributed in nature, and being catabolically diverse, they are readily isolated from active communities growing in compost, hot springs, and deep geothermal sites, including oil wells and deep sediments. However, it has long been known that *Geobacillus* spp. can be isolated from a wide range of moderate- and low-temperature environments including temperate soils and have also been isolated from low-temperature environments such as the Bolivian Andes, deep seawater, and even the Mariana Trench (Hussein et al. 2015).

5.2.2 The Genus *Anoxybacillus*

Genus *Anoxybacillus* has only been described recently by Pikuta et al. (2000, 2003). Since then, the number of *Anoxybacillus* species has rapidly increased and now contains 22 validly described species and 2 subspecies. The following species of the genus have been reported up to date: *Anoxybacillus pushchinoensis*, *A. flavithermus* (Pikuta et al. 2000), *A. gonensis* (Belduz et al. 2003), *A. contaminans* (De Clerck et al. 2004), *A. voinovskiensis* (Yumoto et al. 2004), *A. kestanbolensis*, *A. ayderensis* (Dulger et al. 2004), *A. kamchatkensis* (Kevbrin et al. 2005), *A. amylolyticus* (Poli et al. 2006), *A. rupiensis* (Derekova et al. 2007), *A. bogrovensis* (Atanassova et al. 2008), *A. kamchatkensis* subsp. *asaccharedens* (Gul-Guven et al. 2008), *A. thermarum* (Poli et al. 2009), *A. eryuanensis*, *A. tengchongensis* (Zhang et al. 2011), *A. salavatliensis* (Cihan et al. 2011), *A. mongoliensis* (Namsaraev et al. 2010), *A. flavithermus* subsp. *flavithermus*, *A. flavithermus* subsp. *yunnanensis* (Dai et al. 2011), *A. caldiproteolyticus* (Coorevits et al. 2012), *A. tepidamans* (Coorevits et al. 2012),

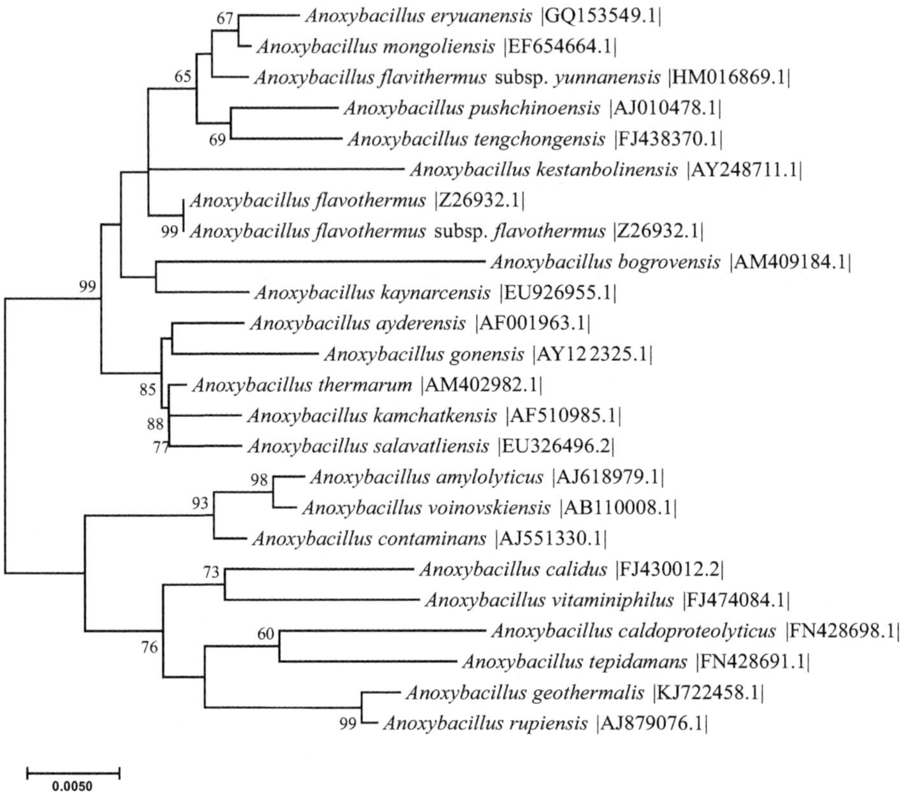

Fig. 5.2 Evolutionary relationships of species of the genus *Anoxybacillus*. The evolutionary history was inferred using the neighbor-joining method. The optimal tree with the sum of branch length = 0.18764567 is shown. The percentage of replicate trees in which the associated taxa clustered together in the bootstrap test (1000 replicates) is shown next to the branches. Significant bootstrapping values (>60%) are shown on the nodes. The analysis involved 24 16S rDNA nucleotide sequences of the type strains of *Anoxybacillus* species, obtained from NCBI GenBank. All positions containing gaps and missing data were eliminated. There were a total of 1031 positions in the final dataset. Evolutionary analyses were conducted in MEGA7

A. kaynarcensis (Inan et al. 2013), *A. vitaminiphilus* (Zhang et al. 2013), *A. calidus* (Cihan et al. 2014), and *A. geothermalis* (Filippidou et al. 2016). *A. kaynarcensis* and *A. kamchatkensis* subsp. *asaccharedens* are still not included in the validation list.

Most of the species belonging to genus *Anoxybacillus* were found to be a homogeneous phylogenetic group of thermophilic bacilli with high 16S rRNA gene sequence similarity values. A tree showing the phylogenetic relationships of species of the genus *Anoxybacillus* is shown in Fig. 5.2.

As it can be deduced from the genus name ("anoxybacillus" means small rod living without oxygen), the members of the genus *Anoxybacillus* are aerotolerant anaerobes or facultative anaerobes (Pikuta et al. 2000). *A. pushchinoensis*, the type strain of this genus, was first described as obligate anaerobe (Pikuta et al. 2000) but was later described as an aerotolerant anaerobe (Pikuta et al. 2003).

The majority of *Anoxybacillus* species are moderate thermophiles (grow in the temperature range 30–75 °C, with the optimum at 50–62 °C). Vegetative cells are rod-shaped or straight or slightly curved, sometimes with angular division and Y-shaped cells, often in pairs or short chains, with rounded ends. The cells are motile or nonmotile. Endospores are round, oval, or cylindrical and have a terminal location. Colony morphology and size are variable. Most of the species produce cellular carotenoid like pigments, which yields yellow colonies. They are catalase-variable. Many members of the genus are alkaliphilic, but most of the species can grow at neutral pH. Only *A. amylolyticus* grows optimally at slightly acidic conditions (pH 5.6). *Anoxybacillus* species are chemoorganotrophic, with a fermentative or aerobic respiration metabolism. They can use oxygen or nitrate as electron acceptors, and in the absence of electron acceptors, they perform fermentation by the Embden-Meyerhof-Parnas pathway.

Many species produce a variety of thermostable enzymes, such as amylase (Poli et al. 2006; Baltas et al. 2016), glucosidase (Cihan et al. 2011), esterase (Shahinyan et al. 2017; Chis et al. 2013), proteinase (Matpan Bekler et al. 2015; Nakamichi et al. 2010), and xylanase (Inan et al. 2013; Ellis and Magnuso 2012; Kacagan et al. 2008).

Most species of the genus have been isolated from hot springs. They have been found also in geothermal soils, manure, hydrothermal vents, etc. (Bergey et al. 2009).

5.3 Distribution of *Geobacillus* and *Anoxybacillus* in Terrestrial Hot Springs

Since Thomas Brock made the remarkable discovery in 1966 that microorganisms were growing in the boiling hot springs of Yellowstone National Park, the search for thermophiles in terrestrial hot springs has increased. Terrestrial hot springs are created by the emergence of geothermally heated groundwater from the Earth's crust (Mehta and Satyanarayana 2013a, b). Thermophilic microbes have been discovered in geothermal springs all over the world, including areas in Asia, America, Kamchatka, Iceland, New Zealand, Italy, China, etc.

Thermophilic representatives of the *Geobacillus* and *Anoxybacillus* genera have been recovered from a variety of environments, including high-, moderate-, and low-temperature environments. However, terrestrial hot springs are the main habitats for *Geobacillus* and *Anoxybacillus* species (Fig. 5.3). An overview of the various *Geobacillus* and *Anoxybacillus* species isolated from terrestrial hot springs is given in Tables 5.1 and 5.2.

Fig. 5.3 Geographical distribution of sites from where *Geobacillus* and *Anoxybacillus* species have been isolated. Green and purple circles represent *Geobacillus* and *Anoxybacillus* isolates, respectively. Each circle denotes a published report that describes one or several strains

5.4 Adaptations of Growth at High Temperatures

Thermophilic bacilli, growing at high temperatures, have developed different mechanisms to survive in these extreme environments. Understanding the adaptations that enable thermophilic organisms to survive at extreme temperatures is a challenge that has interested researchers since 1897, and a vast amount of literature exists regarding this issue (England et al. 2003). The main mechanistic determinants of thermoadaptation in bacilli are adaptation of membrane phospholipid composition, synthesis of heat shock proteins (HSPs), and enzyme adaptation to give molecular stability as well as structural flexibility. The high GC content in the genome of the thermophiles also contributes to their thermoadaptation (Chakravorty and Petra 2013).

5.4.1 Adaptation of Membrane Phospholipid Composition at High Temperatures

It has been shown that the lipids isolated from a psychrophilic (*Psychrobacter* sp.), mesophilic (*Escherichia coli*), and thermophilic (*G. stearothermophilus*) bacteria are different depending on the bacterial growth temperature. With increasing growth temperature, bacteria reduce the number of unsaturated bonds or increase the degree of branching in their lipid acyl chains (van de Vossenberg et al. 1995). Thus, the lipid of the cytoplasmic membrane of *Psychrobacter* sp. (optimal growth temperature 21–29 °C) is mainly represented by monounsaturated (93%) and short-chain lipids. The monounsaturated and short-chain lipids compose 32% of the cytoplasmic membrane lipids in *E. coli* (optimal growth temperature 37–42 °C), whereas the

Table 5.1 *Geobacillus* species recovered from hot springs worldwide

Species	Source	Place	Temp. range (°C)	Opt. temp. (°C)	pH range	Opt. pH	Comments	References
G. gargensis	Garga hot spring, in the valley of the river Barguzin	Baikal region, Russia	45–70	60–65	5.5–8.5	6.5–7.0	–	Nazina et al. (2004)
G. stearothermophilus			ND	ND	ND	ND	The draft genome sequence is available (GenBank AN:JQCS00000000 and JPYV00000000)	Rozanov et al. (2014)
	Hot springs in Yellowstone National Park	USA	ND	55–65	ND	ND	DNA polymerase has been characterized	Stenesh and McGowan (1977) and Zeigler (2001)
	Hot spring in Chiang Mai	Thailand	ND	65	ND	ND	Produces an extracellular superoxide dismutase (SOD)	Sookkheo et al. (2002)
	Hammam Pharaon hot spring	Egypt	ND	70	ND	ND	Producing thermostable cellulases	Ibrahim and El-diwany (2007)
	Hot spring areas in Bulgaria	Bulgaria	50–82	55	5.5–8.5	7.0	Producing thermostable gellan lyase	Derekova et al. (2006)
	Dikili-Bergama Kaynarca hot spring in Izmir	Turkey	40–70	55	5.5–8.5	7.0	High xylanase and arabinofuranosidase activities	Canakci et al. (2007)
	Savusavu hot spring	Fiji	ND	ND	5–8	ND	High cadmium ion adsorption potential	Narayan et al. (2008)
	Dalupirip hot spring	Benguet, Philippines	ND	60	ND	7.0	Xylan-degrading ability	Daupan and Rivera (2015)

(continued)

Table 5.1 (continued)

Species	Source	Place	Temp. range (°C)	Opt. temp. (°C)	pH range	Opt. pH	Comments	References
	Soldhar hot spring	Garhwal Himalaya, India	40–90	65–70	4–11	6–8	–	Sharma et al. (2009)
	Arzakan hot geothermal mineral spring	Kotayk province, Armenia	37–70	55	6.5–8.5	ND	–	Panosyan (2010), Panosyan and Birkeland (2014), Hovhannisyan et al. (2016), and Panosyan (2017)
	Jermuk hot spring	Vayots Dzor province, Armenia	ND	ND	ND	ND	Active amylase producer	Vardanyan et al. (2015) and Hovhannisyan et al. (2016)
G. thermodenitrificans	Hammam Pharaon hot spring	Egypt	ND	70	ND	ND	Producing thermostable cellulases	Ibrahim and El-diwany (2007)
	Alangullu, Omerbeyli, and Camkoy Camur hot springs	Aydin, Turkey	40–75	55	5.5–8.5	7.0	Showed high xylanase and arabinofuranosidase activities	Canakci et al. (2007)
	Tattapani hot spring	Northwest Himalayas, India	60–80	60	ND	7.0	Produced a thermotolerant cellulose GenBank AN:KP842609	Priya et al. (2016)

	Arzakan hot geothermal mineral spring	Kotayk province, Armenia	37–70	55	6.5–8.5	ND	–	Panosyan (2010), Panosyan and Birkeland (2014), Hovhannisyan et al. (2016), and Panosyan (2017)
G. thermoparaffinivorans	Badekbek hot spring	Benguet, Philippines	ND	60	ND	7.0	Xylan-degrading ability	Daupan and Rivera (2015)
G. thermoglucosidasius	Obsidian hot spring, Yellowstone National Park	Montana, USA	55–75	65	5.8–8.0	7.5	Complete genome is reported (GenBank AN:CP002835)	Brumm et al. (2015)
	North Shuna hot spring	Jordan	40–80	60	6–9	6–8	Exhibited high hydrolytic activities	Obeidat et al. (2012)
G. thermoleovorans	Ulu Slim hot spring	Malaysia	ND	ND	ND	ND	Complete genome is reported	Muhd Sakaff et al. (2012)
	Hot spring of the Waimangu Volcanic Valley	New Zealand	ND	70	ND	7.0	Producing thermostable alpha amylase	Uma Maheswar Rao and Satyanarayana (2007)
	Aguas Calientes geothermal spring	Amazon rainforest, Peru	50–70	ND	ND	7.4	Characterized with high cellulase activity	Cortez et al. (2016)
	Hot spring in Kobe	Japan	50–75	60	3.0–11	7.0	Produce thermoactive xylanases	Sunna et al. (1997)
	Badekbek hot spring	Benguet, Philippines	ND	60	ND	7.0	Showed xylan-degrading ability	Daupan and Rivera (2015)

(continued)

Table 5.1 (continued)

Species	Source	Place	Temp. range (°C)	Opt. temp. (°C)	pH range	Opt. pH	Comments	References
G. thermocatenulatus	El Tatio geyser field and Liquiñe hot springs	Chile	ND	60	ND	ND	Phytase activity in crude protein extracts	Jorquera et al. (2018)
G. icigianus	Hot spring of the Valley of Geysers	Kamchatka, Russia	50–75	60–65	5–9	6.5–7.0	The draft genome sequence is available (GenBank AN: JPYA00000000)	Bryanskaya et al. (2015)
G. subterraneus subsp. *aromaticivorans*	Guclukonak hot spring	Sirnak, Turkey	30–65	60	5.5–10	9.0	The isolate hydrolyses lipase, ONPG, phosphatase, urease, oxidase, gelatin, and starch	Poli et al. (2012)
G. caldoxylosilyticus	Fosso Bianco hot springs	Bagni di Filippo, Mount Amiata, Tuscany, Italy	45–70	55–60	ND	7.0	Able to grow in the presence of Hg(II) (>1 mM)	Chatziefthimiou et al. (2007)
	Selayang hot spring	Malaysia	45–85	55	5.0–9.0	6.5	Able to reduce toxic chromium (VI) to non-harmful of chromium (III)	Che Ibrahim and Wan Ahmad (2017)
	Jermuk hot spring	Vayots Dzor province, Armenia	ND	ND	ND	ND	Active amylase producer	Vardanyan et al. (2015) and Hovhannisyan et al. (2016)

G. kaustophilus	Aguas Calientes geothermal spring	Amazon rainforest, Peru	50–70	ND	ND	7.4	High cellulase activity	Cortez et al. (2016)
	Dalupirip hot spring	Benguet, Philippines	ND	60	ND	7.0	Xylan-degrading ability	Daupan and Rivera (2015)
	Soldhar hot spring	Garhwal Himalaya, India	40–90	65–70	4–11	6–8	–	Sharma et al. (2009)
G. pallidus	Tanjung Sakti hot spring	South Sumatera, Indonesia	ND	ND	ND	ND	–	Yohandini et al. (2015)
G. pallidus	Hammamt Al-Burbita, Afra hot springs, Ma'in hot springs	Jordan	ND	ND	ND	ND	–	Al-Batayneh et al. (2011)
G. toebii	Arzakan hot geothermal mineral spring	Kotayk Province, Armenia	37–70	55	6.5–8.5	ND	–	Panosyan (2010), Panosyan and Birkeland (2014), Hovhannisyan et al. (2016), and Panosyan (2017)
Geobacillus spp.	Hammamat Afra, Jordan Himma, Zara-Bani Hamida, Ma'in-Roman hot spring	Jordan	40–85	65	4–11	6–8	High hydrolytic activities	Obeidat et al. (2012)
	Soldhar hot spring	Garhwal Himalaya, India	40–90	65–70	4–11	6–8	–	Sharma et al. (2009)

(continued)

Table 5.1 (continued)

Species	Source	Place	Temp. range (°C)	Opt. temp. (°C)	pH range	Opt. pH	Comments	References
	Irde geothermal spring	Konkan region, Southern India	40–95	60	ND	8.0	Active thermostable lipase and amylase producer	Mahadevan and Neelagund (2010, 2014)
	Tattapani hot spring	Azad Kashmir, Pakistan	45–75	65	5.5–8.5	7.0	Produced significant amount of industrially important enzymes, i.e., extracellular α-amylase, CMCase, FPase, xylanase, protease and lipase, and intracellular CMCase and FPase	Zahoor et al. (2016)
	Tapovan hot spring	Chamoli, Uttarakhand, India	40–90	70	6–8.5	7–8	Active α-amylase producer. High amylase activity at 80 °C and pH 8.0	Jugran et al. (2015), Arya et al. (2015)
	Tengchong hot spring	China	ND	60–65	ND	ND	α-amylase producer	Wang et al. (2011b)
	Double Hot Springs, Nevada	Nevada, USA	ND	ND	ND	ND	Draft genome is reported (GenBank AN: SAMN0017395)	De Maayer et al. (2014)
	Larijan hot spring	Iran	40–80	65	ND	6.8	Producing alpha amylase with 52 kDa molecular mass	Mollania et al. (2010)
	Hot spring in Rosario de la Frontera	Salta, Argentina	ND	ND	ND	ND	Whole-genome shotgun project is reported (GenBank AN: LDNZ00000000)	Ortiz et al. (2015)

	Alangullu, Omerbeyli, and Camkoy Camur hot springs	Aydin, Turkey	35–65	55	5.5–9	7.0	Showed high xylanase and arabinofuranosidase activities	Canakci et al. (2007)
	Buranga hot springs	Western Uganda, Africa	37–72	60–62	5–10	7.5–8.5	Producing thermostable protease	Hawumba et al. (2002)
	Arzakan hot geothermal mineral spring	Kotayk province, Armenia	37–70	55	6.5–8.5	ND	–	Panosyan (2010), Panosyan and Birkeland (2014), Hovhannisyan et al. (2016), and Panosyan (2017)
	Tatev geothermal spring	Syunik province, Armenia	45–70	65	6–9	7.0	Lipase encoding genes have been reported	Shahinyan et al. (2017)

ND not determined, – data not available

Table 5.2 *Anoxybacillus* species recovered from hot springs worldwide

Species	Source	Place	Temp. range (°C)	Opt. temp. (°C)	pH range	Opt. pH	Comments	References
A. rupiensis	Unnamed hot spring	Rupi basin, Bulgaria	35–67	55	5.5–8.5	6.0–6.5	Strictly aerobic, producing amylolytic enzymes	Derekova et al. (2007)
	Arzakan hot geothermal mineral spring	Kotayk province, Armenia	ND	ND	ND	ND	Active amylase producer	Hovhannisyan et al. (2016), Vardanyan et al. (2015), Panosyan (2017)
	Karvachar and Zuar geothermal springs	Nagorno-Karabakh						
	Tanjung Sakti hot spring	South Sumatera, Indonesia	ND	ND	ND	ND	–	Yohandini et al. (2015)
A. tepidamans	Geothermal heated soil	Yellowstone National Park, USA	39–67	55	6.0–9.0	7.0	Covered with an oblique S-layer lattice, composed of identical S-layer glycoprotein protomers	Schäffer et al. (2004)
A. voinovskiensis	Voinovskie hot spring	Kamchatka, Russia	30–64	54	7.0–8.0	7.0–8.0	–	Yumoto et al. (2004)
A. kaynarcensis	Kaynarca hot spring	Izmir, Turkey	35–70	60	6.0–10.0	7.0	Active alkaline xylanase producer	Inan et al. (2013)
A. flavithermus	Tanjung Sakti hot spring	South Sumatera, Indonesia	ND	ND	ND	ND	–	Yohandini et al. (2015)
	Unnamed hot spring	Northern Island of New Zealand	30–72	60–65	5.5–9.0	7.0	Produces a carotenoid	Heinen et al. (1982)
	Savusavu hot spring	Fiji	ND	ND	5–8	ND	Showed gelatinase activity	Narayan et al. (2008)
	Al-Ain Alhara thermal hot spring	Gazan, Saudi Arabia	ND	55	ND	7.0	The draft genome sequence is reported (GenBank AN: APCD00000000)	Khalil et al. (2015)

	Mickey Hot Springs area of the Alvord Basin hydrothermal system	Oregon, USA	ND	65	ND	9.0	Extracellular xylanase enzymes producer	Ellis and Magnuson (2012)
	Dalupirip hot spring	Benguet, Philippines	ND	60	ND	7.0	Degrading xylan	Daupan and Rivera (2015)
	Hot spring in Tăşnad	Satu Mare county, Romania	40–70	50–60	6.5–8.5	7.5	Thermostable esterase/lipase with molecular weight of 28.03 kDa was characterized	Chis et al. (2013)
	The New Lorne bore thermal basin	Great Artesian Basin, Australia	50–65	ND	ND	ND	Can use Fe(III) as electron acceptor and yeast extract as carbon source	Ogg et al. (2013)
	Karvachar geothermal spring	Nagorno-Karabakh	40–70	60	7.0–11.0	9.0–10.0	GDSL family lipase encoding genes have been reported. Active thermostable alpha amylase producer	Shahinyan et al. (2017), Panosyan et al. (2014), Panosyan (2017)
	Omer hot spring	Afyonkarahisar, Turkey	25–85	55	5.0–10.0	6.0	Thermostable α-amylase was characterized, and some of its industrial applications were examined	Agüloğlu Fincan et al. (2014), Ozdemir et al. (2015)
	Hammam Pharaon hot spring	Egypt	ND	70	ND	ND	Producing thermostable cellulases	Ibrahim and El-diwany (2007)
	Fosso Bianca hot springs	Bagni di Filippo, Mount Amiata, Tuscany, Italy	45–65	55–60	ND	7.0	Fosso Bianca hot springs are naturally enriched with mercury of geological origin. Isolates able to grow in the presence of Hg(II) (40 μM–1 mM)	Chatziefthimiou et al. (2007)
A. contaminans			45–65	55–60	ND	7.0		

(continued)

Table 5.2 (continued)

Species	Source	Place	Temp. range (°C)	Opt. temp. (°C)	pH range	Opt. pH	Comments	References
A. flavithermus subsp. *yunnanensis*	Unnamed hot springs	Yunnan, China	30–66	60	5.5–10.0	7.0–7.5	Produce thermostable β-glucosidase. Optimal enzyme activity at 60 °C and pH 7.0	Dai et al. (2011), Liu et al. (2017)
A. bogrovensis	Geothermal spring	Sofia, Bulgaria	40–69	65	6.0–10.0	8.0	Hydrolysis of starch and gelatin	Atanassova et al. (2008)
A. eryuanensis	Eryuan hot spring	Yunnan, China	35–75	55	7.0–11.0	8.0	–	Zhang et al. (2011)
A. mongoliensis	Tsenkher hot spring	Central Mongolia	35–75	60	5.0–10.8	8.0	Produced thermostable alkaline subtilisin like serine proteinase	Namsaraev et al. (2010)
A. tengchongensis	Tengchong hot spring	Yunnan, China	30–75	50	7.0–11.0	8.5	Hydrolysis of starch and gelatin	Zhang et al. (2011)
A. gonensis	Gonen hot springs	Balikesir, Turkey	40–70	55–60	6.0–10.0	7.5–8.0	Produces xylose isomerase, carboxylesterase, and fructose-1,6-bisphosphate aldolase	Belduz et al. (2013), Lim et al. (2015)
A. ayderensis	Ayder hot springs	Rize, Turkey	30–70	50	6.0–11.0	7.5–8.5	Genes encoding various glycoside hydrolases are reported in the genome of the strain	Dulger et al. (2004), Belduz et al. (2015)
A. kestanbolensis	Kestanbol hot spring	Canakkale, Turkey	40–70	50–55	6.0–10.5	7.5–8.5	–	

A. kamchatkensis	Hot spring in the Valley of Geysers	Kamchatka, Russia	33–67	60	5.7–9.9	6.8–8.5	–	Kevbrin et al. (2005)
	Hot spring	Indonesia	ND	ND	ND	ND	This whole-genome shotgun project has been deposited (DDBJ/ EMBL/ GenBank AN: ALJT00000000)	Lee et al. (2012)
	Los Baños hot spring	Mexico	37–60	50–70	6.0–11.0	6.5–7.5	Produces thermostable lipases, proteases, and amylases	Pinzón-Martínez et al. (2010)
	Kuala Woh hot spring	Malaysia	25–60	50–60	5–10	ND	Thermostable lipase producer	Olusesan et al. (2009)
	Dongda hot spring	Xi'an, Shaanxi province, China	ND	ND	ND	ND	Producing Cyclo(Gly-l-Pro) 20, which are immunomodulators of carp	Wang et al. (2011a)
A. kamchatkensis subsp. *asaccharedens*	Talisdere hot spring	Batman, Turkey	35–65	55	5.5–9.5	7.5	Not consuming sugars and carbohydrates as carbon source	Gul-Guven et al. (2008)
A. thermarum	Euganean hot spring	Abano Terme, Padova, Italy	55–67	65	6.0–7.5	7.2	Draft genome sequence is reported	Poli et al. (2009)
A. vitaminiphilus	Puge hot spring	Puge county, Sichuan province, China	38–66	57–60	6.0–9.3	7.0–7.5	A distinctive characteristic of this isolate was its extreme reliance on vitamin mixture or yeast extract for growth	Zhang et al. (2013)
A. suryakundensis	Surya Kund hot spring	Jharkhand, India	40–60	55	5.5–11.5	7.5	Draft genome sequence is reported. Hydrolysis of gelatin, starch, esculin, and DNA	Deep et al. (2016)

(continued)

Table 5.2 (continued)

Species	Source	Place	Temp. range (°C)	Opt. temp. (°C)	pH range	Opt. pH	Comments	References
Anoxybacillus spp.	Dargeçit hot spring	Turkey	30–65	60	5.5–10.0	7.0–7.5	Exhibited high alpha amylase activity (2668.4–3627 U/mL)	Acer et al. (2015)
	Seferihisar Karakoc hot spring	Turkey	ND	55	ND	7.4	Producing thermostable carboxylesterase	Ay et al. (2011)
	North Shuna hot spring	Jordan	ND	ND	ND	ND	–	Al-Batayneh et al. (2011)
	Sungai Klah and Dusun Tua hot springs	Malaysia	30–65	55	6.0–10.0	7.0	Strains SK3-4 and DT3-1 were able to degrade pullulan and to produce maltotriose and glucose, respectively, as their main end products	Chai et al. (2012)
	Hot spring in Beppu city	Oita, Japan	40–70	60	5.5–10.0	7.5–8.0	Producing thermostable protease	Nakamichi et al. (2010)

ND not determined, *AN* accession number, – data not available cytoplasmic membrane lipids of *G. stearothermophilus* (optimal growth temperature 60–70 °C) are composed of saturated and branched acyl chains (80% of total lipids) (van de Vossenberg et al. 1995).

The thermophilic bacilli differ from mesophilic ones also in the fatty acid and polar headgroup compositions of their phospholipids. Hence, the major cellular fatty acid components of *Geobacillus* species following incubation at 55 °C are iso-$C_{15:0}$ (20–40%, mean 29%), iso-$C_{16:0}$ (6–39%, mean 25%), and iso-$C_{17:0}$ (7–37%, mean 19.5%), which account for 60–80% of the total (Nazina et al. 2001). The high levels of is-$C_{15:0}$ and iso-$C_{17:0}$ are also found in *Anoxybacillus* species (Dulger et al. 2004). It has been shown that the major fatty acid patterns in *G. toebii*, *G. subterraneus* subsp. *aromaticivorans*, *G. icigianus*, *A. bogrovensis*, and *A. suryakundensis* are iso-$C_{15:0}$>iso-$C_{17:0}$>anteiso-$C_{17:0}$ (Atanassova et al. 2008; Poli et al. 2012; Deep et al. 2016; Cihan et al. 2014; Bryanskaya et al. 2015). The membrane fatty acid patterns in *Bacillus* species are mainly represented by iso-$C_{15:0}$ and anteiso-$C_{15:0}$ in contrast with *Geobacillus* and *Anoxybacillus* species growing at higher temperatures (>50 °C) (Table 5.3). The acyl chains such as iso-$C_{17:0}$ have higher melting point than other acyl chains, which explains its synthesis at maximum growth temperatures, whereas iso-$C_{15:0}$ is predominating at minimum growth temperature.

Llarch et al. (1997) showed that any potential distinctions between the rather variable fatty acid profiles of *Geobacillus* species and *Bacillus* species are largely lost when strains of each group are incubated at the same temperature, clearly underpinning their role in thermoadaptation.

More detailed studies of the effect of temperature on the membrane composition of *G. stearothermophilus* showed that ratio of phosphatidylglycerol (PG) and cardiolipin (CL), which comprise about 90% of the membrane phospholipids, is changed at different growth temperatures. The PG content increases at the expense of the CL content at the high temperatures. The acyl-chain composition of all the membrane lipids also changes; the longer, saturated-linear, and iso-fatty acids with relatively high melting points increase in abundance, and anteiso-fatty acids and unsaturated components with lower melting points decrease (Tolner et al. 1997). Nicolaus et al. (1995), reclassifying some of the *Bacillus* species, showed that the strains tentatively identified as *Bacillus* showed increased phosphoglycolipid contents with increased growth temperature, at the expense of phosphoaminolipid and phospholipids. As a result, the organism is able to maintain nearly constant membrane fluidity across its whole growth temperature range; this has been termed homeoviscous adaptation (a homeostatic process that regulates the viscosity of membrane lipids). An alternative theory, homeophasic adaptation (adaptation of the cell membrane lipid composition), considers that maintenance of the liquid-crystalline phase is more important than an absolute value of membrane fluidity in bacteria (Tolner et al. 1997).

Table 5.3 The major cellular fatty acid composition of *Geobacillus* and *Anoxybacillus* species

Species	Temperature growth range (optimum) in °C	Major cellular fatty acids (>10%)	References
G. gargensis	45–70 (60–65)	Iso-$C_{15:0}$>iso-$C_{16:0}$>iso-$C_{17:0}$	Nazina et al. (2004)
G. icigianus	50–75 (60–65)	Iso-$C_{15:0}$>iso-$C_{17:0}$>anteiso-$C_{17:0}$	Bryanskaya et al. (2015)
G. subterraneus subsp. *aromaticivorans*	30–65 (60)	Iso-$C_{15:0}$>iso-$C_{17:0}$>anteiso-$C_{17:0}$	Poli et al. (2012)
G. stearothermophilus	37–65 (ND)	Iso-$C_{15:0}$>iso-$C_{17:0}$>anteiso-$C_{17:0}$	Nazina et al. (2001)
G. thermoglucosidasius	37–68 (ND)	Iso-$C_{17:0}$>iso-$C_{15:0}$>anteiso-$C_{17:0}$>iso-$C_{16:0}$	
G. uzenensis	45–65 (55–60)	Iso-$C_{17:0}$>iso-$C_{15:0}$>anteiso-$C_{17:0}$>iso-$C_{16:0}$	
G. toebii	45–70 (60)	Iso-$C_{15:0}$>iso-$C_{17:0}$	Cihan et al. (2014)
A. kaynarcensis	35–70 (60)	Iso-$C_{15:0}$>iso-$C_{17:0}$	Inan et al. (2013)
A. bogrovensis	40–69 (65)	Iso-$C_{15:0}$>iso-$C_{17:0}$>anteiso-$C_{17:0}$	Atanassova et al. (2008)
A. suryakundensis	40–60 (55)	Iso-$C_{16:0}$>iso-$C_{15:0}$>anteiso-$C_{17:0}$>iso-$C_{17:0}$	Deep et al. (2016)
A. pushchinensis	37–66 (62)	Iso-$C_{15:0}$> $C_{16:0}$> $C_{18:0}$	Pikuta et al. (2000)
A. rupiensis	35–67 (55)	Iso-$C_{15:0}$>iso-$C_{17:0}$	Cihan et al. (2014)
A. flavithermus	30–72 (60)	Iso-$C_{15:0}$>iso-$C_{17:0}$>$C_{16:0}$	
A. kamchatkensis	38–67 (60)	$C_{16:}$0>iso-$C_{16:0}$>anteiso-$C_{17:0}$	
A. calidus	35–70 (55)	Iso-$C_{15:0}$>iso-$C_{17:0}$>iso-$C_{16:0}$	

5.4.2 Heat Shock Proteins

Although a wide variety of survival strategies are deployed when cells are exposed to environmental challenges such as heat stress, synthesis of the effector proteins generally referred to as heat shock proteins (HSPs) is increased. HSPs are diverse in structure and function and are usually classified based on their subunit molecular weights. Classes that occur in microorganisms and in the majority of thermophiles include Hsp100, Hsp90, Hsp70, Hsp60, and the small HspS (Trent 1996). Most of these proteins function as molecular chaperones, catalyzing the refolding of denatured proteins, assisting the folding of newly synthesized proteins, as well as assisting in protein translocation across membranes and assembly/disassembly of protein complexes (Chang et al. 2008).

The 70-kDa heat shock proteins (Hsp70s) are highly conserved and are ATP dependent. Together with J-domain ATPase-activating proteins or nucleotide exchange factors, Hsp70s bind and release their substrates in ATP-driven cycles (Chang et al. 2008; Goh et al. 2014). Hsp70s chaperone family proteins, DnaK (GkDnaK) from *G. kaustophilus* (Chang et al. 2008), *G. thermoleovorans* (Graham et al. 2005), *G. thermoglucosidasius* (Brumm et al. 2015), *Geobacillus* sp. (Shih and Pan 2011), and *Anoxybacillus* sp. (Goh et al. 2014), have been identified and characterized.

Besides Hsp70s, the low-molecular-weight Hsp20 and Hsp33 proteins from *G. thermoglucosidasius* strain C56-YS93 have been described (Brumm et al. 2015). Proteome analysis of *G. thermoleovorans* strain T80 revealed the presence of sigma factors, such as σ^A, which initiates transcription of the heat shock operons controlled by the HRCA-CIRCE complex. This operon encodes some of the proteins involved in heat shock response, such as GroEL (Hsp60), GroES (Hsp10), and peptidyl-prolyl cis-trans isomerase (Graham et al. 2005).

The heat shock protein Hsp70 (DnaK) in *Anoxybacillus* works not only in the presence of ATP but also in cooperation with Hsp40 (DnaJ, J-protein). The genes encoding Hsp70 and Hsp40 proteins are located near each other. Other proteins related to temperature adaptations such as GroEL (Hsp60) and its co-chaperonin GroES (Hsp10), a few small Hsp20 molecular chaperones, Hsp33, and ClpC (Hsp100) and its related Clp-protease were identified in the genomes of many species of *Anoxybacillus* (Goh et al. 2014).

5.4.3 Protein and Enzyme Adaptation

Thermophilic bacilli, under constant threat of temperature-induced damage, maintain the stability and functionality of their proteins and enzymes by changing the ratio of charged to uncharged amino acids, increased ionic interactions and hydrogen bonding, metal coordination and the compactness of their proteins, and the preference of certain amino acids (Scandurra et al. 1998).

Lobry and Chessel (2003) reported that larger amounts of Ala, Gly, Ser, Asp, and Glu and smaller amounts of Cys in the transmembrane proteins of thermophiles have significant roles for their protein thermostability. Change in amino acids from Lys to Arg, Ser to Ala, Gly to Ala, Ser to Thr, and Val to Ile has been observed in comparison with mesophilic versus thermophilic organisms (Scandurra et al. 1998; Wang et al. 2015). For example, in *G. stearothermophilus*, it was reported that Gly is preferred over Ile and Ala over Tyr (Trivedi et al. 2006). Schneider et al. (2002) studied sequence differences between predicted transmembrane helices in the genomes of thermophilic and mesophilic membrane proteins. They observed a striking depletion of Cys residues in thermophiles and an increase in Gly, Ser, and Ala pair motifs, suggesting a preference for the packing of small residues. The integral membrane proteins of thermophiles have lower amounts of Glu, Lys, and Asp residues, as a mode of adaptation to increased temperature (Lobry and Chessel 2003).

Pertaining to secondary and three-dimensional structure, thermostable proteins have high levels of α-helical and β-sheet content (Chakravorty and Patra 2013). The thermostable lipases from *Geobacillus* and *Anoxybacillus* contain terminal α-helices and a central β-sheet (Arpigny and Jaeger 1999; Shahinyan et al. 2017) possibly contributing to its thermostability.

Sawle and Ghosh (2011) suggested that entropic stabilization may be largely responsible for the high melting temperature in hyperstable proteins and hints at residual structure or compactness of the denatured state in thermophiles. They showed that the gain in enthalpy upon folding is lower in thermophiles than in mesophiles, whereas the loss in entropy upon folding is higher in mesophiles than in thermophiles. The thermostable proteins have a slow unfolding rate, which helps to retain their near-native structures (Sawle and Ghosh 2011).

The thermostability of some enzymes is due to the presence of an extra repeat N-terminal domain (NTD) in the enzyme. For example, a novel thermostable SOD from *G. thermodenitrificans* NG80-2 exhibits maximum activity at 70 °C and high thermostability over a broad range of temperatures (20–80 °C). Unlike other reported SODs, this enzyme contains an extra repeat-containing NTD of 244 residues adjacent to the conserved functional SODA domain. It has been showed that the deletion of the NTD dramatically decreased its optimum active temperature (OAT) to 30 °C and also impaired its thermostability. Conversely, appending the NTD to a mesophilic counterpart from *B. subtilis* led to a moderately thermophilic enzyme (OAT changed from 30 to 55 °C) with improved heat resistance. The NTD also contributes to the stress resistance of host proteins without altering their metal ion specificity or oligomerization form except for a slight effect on their pH profile (Wang et al. 2014).

Metals such as zinc and calcium are often found in enzymes where they can stabilize a loop structure or hold secondary structures. The zinc ions, involved in the Zn-binding domain of thermoalkalophilic lipases from *G. thermocatenulatus* stabilized the structural arrangements of around 70 amino acids and the concerted movement of two lids, the 6- and 7-helices, during enzyme activation (Carrasco-Lopez et al. 2009). Ca ions restrict the conformational flexibility of certain helices and loops and bring about the stabilization of His residues through hydrogen bonding and thus lead to lipase thermostability (Sharma et al. 2013). Alpha-amylases and proteases isolated from various *Geobacillus* and *Anoxybacillus* spp. have been shown to contain Ca ions, which is enhancing the stability and activity of the enzymes at high temperatures (Eijsink et al. 2011; Chai et al. 2016).

5.4.4 Other Mechanisms for Thermostability

Large-scale genomic comparisons between thermophiles and mesophiles have shown that the genomes of thermophilic organisms have a higher guanine and cytosine (GC) content than mesophiles (Takami et al. 2004; Wang et al. 2015). It was hypothesized that a high GC content contributes to the thermostability of the genome and correlated with the optimum growth temperature of bacteria (Musto et al. 2005; Musto et al. 2006). Additionally, tRNAs and rRNAs, the translational machinery of some thermophilic organisms, were reported to have high GC contents as well (Higgs and Ran 2008; Satapathy et al. 2010).

It has been observed that higher tRNA diversity usually occurs in thermophiles in comparison with non-thermophiles. Among psychrophiles, the total number of tRNA was found to be more than twofold higher than in the non-psychrophiles. The fact that growth temperature correlates with diversity and total amount of cellular tRNA (Satapathy et al. 2010) extends the list of molecular features undergoing adaptation due to growth temperature and supports the view that growth temperature acts as a strong selecting factor at the molecular level during evolution.

Small RNA (sRNA) has been shown to play important gene regulatory roles in the prokaryotes and can be involved in the adaptation at high temperatures. The sRNAs from *Geobacillus thermoleovorans* CCB_US3_UF5 strain, growing at 60–70 °C, were reverse transcribed to cDNA and sequenced. Sequencing data identified 83 putative sRNAs classified as antisense, intergenic region, untranslated region, or noncoding. Out of this total, 44 sRNA candidates were specific to growth at elevated temperature, suggesting that regulatory sRNA may play an important role in high-temperature adaptation in thermophilic bacteria (Tan and Alam 2010).

5.5 Biotechnological Potential of *Geobacillus* and *Anoxybacillus* Species

Members of the genera *Geobacillus* and *Anoxybacillus* can be used both in whole-cell applications and in biofuel and chemical production through engineered cells. One of the main advantages of using bacteria from these taxa is faster rate of growth, decreased contamination, and easier maintenance (Bertoldo and Antranikian 2002; Antranikian 2007). *Geobacillus* and *Anoxybacillus* species can be used as cell factories for multiple products, from gold nanoparticles using *Geobacillus* sp. strain ID17 (using NADH-dependent enzymes which convert Au^{3+} to elemental gold) (Correa-Llantén et al. 2013) to provision of thermostable enzymes (Zahoor et al. 2016; Sharma et al. 2013).

The members of these genera have a strong potential for application in bioremediation, especially with regard to degradation of aromatic compounds and removal of heavy metals. For example, 50 mg/L dried cells of *Geobacillus thermantarcticus* remove Cd^{2+}, Cu^{2+}, Co^{2+}, and Mn^{2+} up to 85.4%, 46.3%, 43.6%, and 65.1%, respectively, whereas *Anoxybacillus amylolyticus* removes 74.1%, 39.8%, 35.1%, and 36.6%, respectively, and *Anoxybacillus amylolyticus* removes the mentioned metal ions up to 74.1%, 39.8%, 35.1%, and 36.6%, respectively (Özdemir et al. 2013).

The ability of *Geobacillus* strains to metabolize aromatic compounds has been described by Feitkenhauer et al. (2003). They studied the kinetics of phenol degradation in continuous culture at 65 °C using *G. thermoleovorans*. Al-Jailawi et al. (2016) suggested that *A. rupiensis* strain Ir3 could be used as alternative to hydrodenitrogenation (HDN) for nitroaromatic compounds elimination (biotreatment) of crude oil and its derivatives. The quantitative analysis (HPLC) indicated that this bacterium showed as much as 99.62% consumption of carbazole, 99.4% of ρ-nitrophenol, 97.73% of nitrobenzene, and 98.89% of naphthalene (Al-Jailawi et al. 2016).

Geobacillus and *Anoxybacillus* species demonstrate great versatility for adaptation and catalytic metabolism in a wide variety of environmental niches and are valuable sources of various thermostable enzymes. Thermophilic bacilli are of special interest as a source of novel thermostable enzymes and possess properties suitable for biotechnological and commercial use. There is, indeed, a considerable demand for a new generation of stable enzymes that are able to withstand severe conditions in industrial processes by replacing or supplementing traditional chemical processes. Their ability to conduct various reactions to higher process rates because of increase in substrate diffusion coefficient and reduced viscosity at higher temperatures makes them a preferred choice over mesophilic sources (Niehaus et al. 1999; Sharma et al. 2013).

Geobacillus and *Anoxybacillus* isolated from different hot springs show high potential as biocatalysts suitable for industrial biotechnology applications. The ability of these bacteria to produce a variety of extracellular enzymes, such as amylases, lipases, xylanases, proteases, esterases, and ureases, has ranked them among the most important enzyme producers (Bruins et al. 2001; Satyanarayana et al. 2012).

5.5.1 Amylases

Amylases are among the most important industrial enzymes and are of significance for their specific use in the starch conversion processes, having approximately 25% of the world enzyme market (Reddy et al. 2003). Amylolytic enzymes act on starch and related oligo- and polysaccharides, catalyzing the hydrolysis of internal α-1,4-glycosidic linkages in starch into low-molecular-weight products, such as glucose, maltose, and maltotriose units (Antranikian 2007).

A number of studies on starch-hydrolyzing enzymes based on the DNA sequence, structural analysis, and catalytic mechanism have led to the concept of one enzyme family: the alpha amylase. The amylolytic and related enzymes have been classified as glycoside hydrolases. They have been categorized as exoenzyme, endoenzyme, de-branching, and cyclodextrin-producing enzymes. The application of these enzymes has been established in a number of industrial processes such as food, fermentation, textiles, and paper industries (Antranikian 2007).

Amylolytic enzymes have been produced by a wide range of microorganisms. Heat-adapted amylases derived from the genus *Geobacillus* and *Anoxybacillus* have a big potential for commercial applications. α-Amylases from the members of genera *Geobacillus* and *Anoxybacillus* from the terrestrial hot springs are characterized by a high thermophilicity and stability (reacting between 30 and 120 °C) and activity within a wide range of pH values (from 5.5 to 13) (Table 5.4).

The α-amylases from thermophilic bacilli were purified and characterized with 42–97 kDa molecular weight (Table 5.4). Gurumurthy and Neelagund (2012) completed the molecular characterization of an extremely thermostable α-amylase

Table 5.4 Characteristics of thermostable α-amylases from *Geobacillus* and *Anoxybacillus* species

Microorganism	Source	Molecular weight (kDa)	pH optimal/stability	Temperature optimal/stability	Key observation	References
G. stearothermophilus	Mae'en hot springs in Jordan	–	7.0/5.5–13	55/25–75	Ca, Mn, Mg, and Cu slightly improve the enzymatic activity	Zakaria Al-Qodah (2006)
G. thermoleovorans	Hot spring of the Waimangu Volcanic Valley, New Zealand	–	7.0/ND	70/ND	Enzyme titer increased significantly in cane-molasses medium (60 U ml/1) as compared to that in the synthetic medium (26 U ml/1)	Maheswar Rao and Satyanarayana (2007)
G. thermoleovorans	Unkeshwar hot spring sediment, Nanded, India	42	7.5/5.5–9.0	68/65–90	The Km and Vmax values were 2.702 mg/ml and 7692.3 mmol, respectively. Ca, Cu, and Co ions increased activity	Rekadwad (2015)
Geobacillus sp.	Uttarakhand hot spring, Himalayan region, India	97	6.5/5.0–10	60/40–120	The values of *K*m and *V*max were 36 mg/ml and 222 μmol/mg/min, respectively	Dheeran et al. (2010)
Geobacillus sp.	Irde geothermal spring, Karnataka, India	43	8.0/5.0–11	90/45–95	Enzyme revealed about 55% α-helix, 5% β-strand, and 40% of unordered structure	Gurumurthy and Neelagund (2010, 2012)
Geobacillus sp.	Tengchong hot spring, China	67	5.6/ND	70/ND	Fe^{3+}, Cu^{2+}, EDTA inhibiting an enzyme activity	Wang et al. (2011b)
A. beppuensis	Tulsi Shyam hot spring reservoir, Gujarat, India	43	7.0/ND	80/50–90	*K*m and *V*max were 0.5 mg/ml and 3571.42 μmol/ml/m. Enzyme Ca ion independent and resistant to chemical	Kikani and Singh (2012)
A. thermarum	Hot mineral spring in Erzurum, Turkey	50	ND/5.5–10.5	70/20–90	The specific activity is 1203.7 U/mg. Enzyme activated by Ca, Cu, Ba, Co, and Zn ions	Baltas et al. (2016)

(continued)

Table 5.4 (continued)

Microorganism	Source	Molecular weight (kDa)	pH optimal/stability	Temperature optimal/stability	Key observation	References
Anoxybacillus sp.	Dargeçit hot spring, Turkey	85	7.0/ND	60/ND	*K*m and *V*max values were 0.102 μmol and 0.929 μmol/min, respectively	Acer et al. (2016)
Anoxybacillus sp.	Diyadin hot spring in Agri, Turkey	–	8.0/ND	60/ND	The maximum α-amylase production was secreted in the presence of 2% (w/v) soluble starch and casamino acid (14,310.6 U/mL)	Matpan Bekler and Güven (2014)
Anoxybacillus sp.	Hot spring in Malaysia	50	8.0/6.0–9.0	60/ND	The high immobilized enzyme activity retention (100%) and activity recovery (93%) achieved using ReliZyme HFA403/M	Kahar et al. (2016)
A. flavithermus	Omer hot springs, Afyonkarahisar, Turkey	60	7.0/6.0–8.0	55/35–70	Enzyme hydrolyzed soluble starch at 55 °C with Km: 0.005 mM and Vmax: 3.5 μmol/min	Fincan et al. (2013)
A. flavithermus	Karvachar hot spring, Nagorno-Karabakh	75	ND/5.5–8.5	ND/40–100	Amylase production started in early log phase and reached a maximum in late exponential phase with an activity of 205 U/ml	Panosyan et al. (2014)

ND not determined, *AN* accession number; – data not available produced by a *Geobacillus* sp. for industrial applications. This α-amylase is considered as a novel enzyme due to its optimum activity at a very high temperature (90 °C) and at an alkaline condition (pH 8.0).

5.5.2 Lipases/Esterases

One of the important groups of biotechnologically relevant enzymes are lipases (EC 3.1.1.3 – triacylglycerol hydrolases), which have found large applications in food, diary, detergent, and pharmaceutical industries (Sharma et al. 2013; Gudiukaite et al. 2017). Lipases catalyze the hydrolysis of ester bonds of triacylglycerol at the interface between an insoluble substrate and water. In nonaqueous media these reactions are reversed due to a hydrophobic domain (lid), covering the active site of the lipase. The three-dimensional structures of lipases have a structural similarity with the α-/β-hydrolase family which contain terminal α-helices and a central β-sheet including the active Ser placed in a loop termed the catalytic elbow. Most α-/β--hydrolases contain a consensus sequence, Gly-X-Ser-X-Gly, around the active site serine, with a catalytic triad (Ser-Asp-His) (Ollis et al. 1992; Arpigny and Jaeger 1999; Gudiukaite et al. 2014).

Lipase-coding genes and activities have been reported in a wide range of microorganisms. However, lipases derived from thermophiles have privileges compared to the mesophilic lipases due to their unique attributes (Lotti and Alberghina 2007). Among the huge diversity of thermophilic bacteria, mainly bacilli have been reported as active thermostable lipase producers (Leow et al. 2004; Antranikian 2007; Sharma et al. 2013, Yang et al. 2013). A number of thermophilic bacilli species belonging to the genera *Geobacillus* and *Anoxybacillus* have been isolated from different geothermal springs and reported as thermostable lipase producers (Table 5.5).

The purified lipase from *Anoxybacillus* sp. isolated from the hot springs in Tăşnad (Romania) and Seferihisar Karakoc (Turkey) has a molecular weight of 25–26 kDa characterized by extremely high thermostability (25–90 °C) with optimum activity at 60–65 °C (Ay et al. 2011; Chis et al. 2013). Additional lipolytic enzymes from thermophilic bacilli were purified and characterized and possess molecular weights between 25 and 47 kDa (Ay et al. 2011; Balan et al. 2012; Chis et al. 2013; Mahadevan and Neelagund 2014).

The lipases purified from *G. thermodenitrificans* and *Geobacillus* sp. are characterized with 30–45 kDa molecular weight and act at the temperatures from 60 to 85 °C (Balan et al. 2012; Mahadevan and Neelagund 2014).

The lipases are stable at a wide range of pH values (5.0–11.0). The lipases from geobacilli mostly act at neutral pH, while the lipases from anoxybacilli are slightly alkaliphilic (Table 5.4).

Table 5.5 Thermostable lipase/esterase from *Geobacillus* and *Anoxybacillus* species and their characteristics

Microorganism	Source	Molecular weight (kDa)	pH optimal/stability	Temperature optimal/stability	Key observation	References
Geobacillus sp.	Tattapani hot springs, Himachal Pradesh, India	–	8.5/ND	60/ND	The *Km* and *Vmax* values of enzyme were 14 mM and 17.86μmol/ml/min, respectively. Enzyme activity increased in the presence of metal ions	Mehta et al. (2012)
G. thermodenitrificans	Hot spring in Labok, Kelantan, Malaysia	30	7.0/6.0–8.0	65/60–70	Enzyme showed elevated activity when pretreated with BaCl2, CaCl2, and KCl with 112%, 108%, and 106%, respectively. Lipase hydrolyzed tripalmitin (C16) and olive oil with optimal activity (100%) compared to other substrates	Balan et al. (2012), Yang et al. (2013)
Geobacillus sp.	Irde geothermal springs, Karnataka, India	47	8.0/8.0–12	70/60–85	The enzyme activity was promoted in the presence of Ca and Mg ions. The secondary structure of purified lipase contains 36% α-helix and 64% β-sheet	Mahadevan and Neelagund (2014)
Geobacillus sp.	Tatev hot spring, Armenia	–	6.0/4.0–8.0	65/45–75	The lipases belong to true lipases from family I and characterized with the presence of aspartic residues involved in Ca^{2+}-binding site	Shahinyan (2015), Shahinyan et al. (2017)
A. flavithermus	Hot spring in Tăşnad, Romania	25	6.5–8.0/3.0–9.0	60–65/25–80	Est/Lip is highly enantioselective, with preference for the (S)-enantiomer of substrates	Chis et al. (2013)
A. gonensis	Gonen hot springs, Turkey	–	7.5/5.5–9.5	60/50–70	Vmax and Km for the esterase activity of crude enzyme in the presence of p-nitrophenyl butyrate were 50 U/L and 0.125 mM, respectively. Ca^{2+} ion is cofactor	Colak et al. (2005)

Anoxybacillus sp.	Kuala Woh hot spring, Peninsular Malaysia	–	–	–	Lipase activity 0.56–2.62 U/ml	Olusesan et al. (2009)
Anoxybacillus sp.	Seferihisar Karakoc hot spring, Turkey	26	8.0/5.0–10.0	60/25–90	The enzyme exhibited a high level of activity with p-nitrophenyl butyrate with apparent Km, Vmax, and Kcat values of 0.348 ± 0.030 mM, 3725.8 U/mg, and 1500 ± 54.50/s, respectively	Ay et al. (2011)
Anoxybacillus sp.	Omer hot spring, Afyonkarahisar, Turkey	–	7.0/6.0–11	80/50–90	Co and Mg ions activated the enzyme by 188% and 149%, respectively	Ozdemir et al. (2015)
Anoxybacillus sp.	Karvachar hot spring, Nagorno-Karabakh	–	10.0/7.0–11.0	65/55–75	The lipases are close to the esterases and in primary structure display a Gly-Asp-Ser-(Leu) (GDSL) motif containing the active-site serine residue	Shahinyan (2015), Shahinyan et al. (2017)

ND not determined, – data not available

5.5.3 Proteases

Proteases are cleaving proteins into short peptides or free amino acids and are mainly divided into two major groups depending on their site of action: exopeptidases and endopeptidases (Sookkheo et al. 2002). Exopeptidases cleave the peptide bond proximal to the amino or carboxy termini of the substrate, whereas endopeptidases cleave the peptide bonds distant from the termini of the substrate. They can also be further divided into four groups based on the functional group present at the active site. These are serine, aspartic, cysteine, and metalloproteases (Rao et al. 1998).

Protease enzymes constitute one of the most important groups of industrial enzymes which are extensively used in the food, pharmaceutical, protein hydrolysis, detergent, cheese-making, brewing, photographic, baking, meat, and leather industries and inclusions in animal and human food as digestive aids (Seifzadeh et al. 2008; Synowiecki 2010).

There is a good correlation between growth temperature of the organism and the stability of its extracellular proteases. Thermophilic bacteria from hot springs are often good sources of thermostable proteases. *Geobacillus* and *Anoxybacillus* strains producing thermostable proteases are listed in Table 5.6.

The studied thermostable proteases of geobacilli and anoxybacilli isolated from the hot springs mostly exhibited optimum activity at a slightly alkaline pH (7–8) and are stable at the wide range of pH values (6–10). The thermostable alkaline proteases have been found to be the most appropriate enzyme in detergent industry, as the enzymes used in detergent formulations should have high activity and stability over a broad range of pH and temperature (Rao et al. 1998; Seifzadeh et al. 2008; Matpan Bekler et al. 2015).

5.5.4 Xylanases

Xylanase (EC 3.2.1.8) degrades the linear polysaccharide β-1,4-xylan into xylose, thus breaking down hemicellulose, one of the major components of plant cell walls. Biodegradation of xylan requires action of several enzymes, among which xylanase plays a key role. In the nature, the xylanase degrades the plant matter into usable nutrients and plays a major role in microbial thriving on plant sources. Microbial xylanases have large application in industry including the food, feed, fuel, textile, detergents, paper, and pulp industries and, also, in waste treatment (Kumar et al. 2013).

A number of thermophilic bacilli isolated from different terrestrial hot springs in Bulgaria (Derekova et al. 2008), Turkey (Kacagan et al. 2008; Inan et al. 2011, 2013), Japan (Sunna et al. 1997), India (Sharma et al. 2007), Pakistan (Zahoor et al. 2016), and the USA (Ellis and Magnuson 2012) were described as active xylanase producers. The purified and characterized xylanases of the species of *Geobacillus* and *Anoxybacillus* from the hot springs are listed in Table 5.7.

Table 5.6 Thermostable proteases from *Geobacillus* and *Anoxybacillus* species and their characteristics

Microorganism	Source	Molecular weight (kDa)	pH optimal/stability	Temperature optimal/stability	Key observation	References
Geobacillus sp.	Buranga hot spring, Uganda	97, 72, 50, 27, 22, 17, and 12	6.5/5.0–9.0	70/37–80	Proteases with 97 and 72 kDa molecular weight are heat-activated proteases, while the 12-kDa proteases are responsible for the observed protease activity at low temperature	Hawumba et al. (2002)
Geobacillus sp.	Tengchong hot spring, China	59.2	7.5/6.0–9.0	85/45–100	The activity of the protease is activated by Ca^{2+} and Mg^{2+} but inhibited partially by Ba^{2+}, Zn^{2+}, Pb^{2+}, Co^{2+}, Mn^{2+}, and Cu^{2+}	Zhu et al. (2007)
G. stearothermophilus	Hot spring in Chiang Mai, Thailand	36, 53, and 71	8.5, 7.5, and 7.0/6.0–10.0	70, 85, and 90/50–100	Classified as Zn^{2+} metalloproteases. The cleavage specificities of proteases S, N, and B on a 30-residue synthetic peptide from pro-BPN subtilisin were Tyr-Ile, Phe-Lys, and Gly-Phe, respectively	Sookkheo et al. (2002)
Anoxybacillus sp.	Köprü hot spring, Turkey	106	9.0/ND	50/50–60	The enzyme activity was increased in the presence of Ca^{2+}, Cu^{2+}, Tween 80, and Triton X-100	Matpan Bekler et al. (2015)
Anoxybacillus sp.	Hot spring Beppu city, Oita, Japan	57	6.0–8.0/5.5–10.0	70/50–90	The solubilization of sewage sludge by thermophilic protease is secreted	Nakamichi et al. (2010)

ND not determined

Table 5.7 Thermostable xylanases from *Geobacillus* and *Anoxybacillus* species and their characteristics

Microorganism	Source	Molecular weight (kDa)	pH optimal/ stability	Temperature optimal/ stability	Key observation	References
G. thermoleovorans	Hot spring in Kobe (Japan)	40 and 69	7.0/6.0–9.0	70–80/30–90	The crude xylanase complex is composed of two active bands	Sunna et al. (1997)
	Uttaranchal hot spring (India)	–	8.5/6.0–11.0	80/50–100	The treatment of the pulp with the xylanase (50 U g/1 dry pulp) was showed	Sharma et al. (2007)
A. gonensis *A. voinovskiensis* *A. kestanbolensis*	Dikili-Bergama Kaynarca hot spring, Camkoy Camur hot spring, Omerbeyli hot spring, Alangullu hot spring (Turkey)	–	9.0/ND	65–75/40–90	The alkaline active xylanases allow the direct enzymatic treatment of the alkaline pulp and avoid the cost incurring and time-consuming steps of pH readjustment	Inan et al. (2011)
A. kaynarcensis	Kaynarca hot Spring (Turkey)	100–150	7.0–9.0/ND	65/ND	The presence of three xylanases in the cell supported by the zymogram of SDS-PAGE	Inan et al. (2013)
A. flavithermus	Alvord Basin hydrothermal system in Oregon (USA)	≈250	6.0–8.0/5.0–9.0	65/65–85	Pentameric protein complex consisting of protein subunits ranging from 25 to 75 kDa is suggested	Ellis and Magnuso (2012)
A. pushchinoensis	Diyadin hot springs (Turkey)	≈83	6.5/6.5–11.0	55/50–60	Vmax and Km determined at optimum temperature and were found to be 59.88 U/mg protein and 0.909 mg/ml, respectively	Kacagan et al. (2008)

ND not determined, – data not available

Thermophilic and alkaline active xylanases from *Anoxybacillus* species drive higher interest than other ones. The use of alkaline active xylanases allows direct enzymatic treatment of the alkaline pulp and avoids the cost of incurring and time-consuming steps of pH readjustment. Due to better solubility of xylan under alkaline conditions, alkaline active xylanase may also find other potential applications in addition to pulp bleaching (Inan et al. 2011, 2013).

5.5.5 Cellulases

Cellulose is the most abundant organic compound on Earth and has been extensively used as a substrate for the production of single-cell proteins, biofuels, and various other chemicals through microbial enzymatic degradation. The conversion of cellulosic biomass to fermentable sugars requires different types of cellulases, namely, β-1,4 endoglucanase (EC 3.4.1.4), β-1,4 exoglucanase (EC 3.2.1.91), and β-1,4 glucosidase (EC 3.2.1.21) (Sharma et al. 2015). Cellulose-degrading enzymes have various applications in starch processing, grain alcohol fermentation, deinking, drainage improvement, malting, and brewing. Thermostable cellulase is extensively used in the bio-stoning of denim fabrics and production of environment-friendly washing powders. In wine production cellulases are applied to obtain better fruit skin degradation, improved color extraction, easier must clarification, and better extraction (Kuhad et al. 2011).

Extracellular cellulases-producing anoxybacilli and geobacilli were mainly isolated from hot springs in Turkey (Cihan et al. 2014) and India (Sharma et al. 2015; Priya et al. 2016). The cellulases produced by *G. kaustophilus* PW11, *G. toebii* PW12, *G. thermoleovorans* PW13, *G. toebii* PS4, and *G. thermodenitrificans* IP_WH1strains isolated from Tattapani hot spring (India) were thermostable and exhibited activity even at 100 °C. Among the metal ions tested, Mn^{2+}, Co^{2+}, and Fe^{2+} significantly enhanced the cellulase activity, while Hg^{2+} (1 mM) strongly inhibited enzyme activity (Sharma et al. 2015; Priya et al. 2016). The activity of cellulase produced from *G. thermodenitrificans* IP_WH1 was higher (0.94 IU/ml at 60 °C) (Priya et al. 2016) than the activity of other thermophilic cellulases reported in the literature, such as *Bacillus* sp. with 0.14–0.37 IU/ml (Padilha et al. 2015) and *Bacillus* sp. SMIA-2 with 0.29 IU/ml (Ladeira et al. 2015) at 50 °C and pH 7.0.

Anoxybacillus gonensis isolated from Agri Diyadin hot spring (Turkey) produces a cellulase with approximately 40 kDa molecular weight, with highest activity at 50 °C and with an unusual broad optimum pH range (3–10) (Genc et al. 2015).

Cellulase-producing bacteria, such as *A. flavithermus* EHP1, *G. stearothermophilus* EHP2, and *G. thermodenitrificans* EHP3, have been isolated from Egyptian hot spring. The crude *A. flavithermus* EHP1 enzyme was produced at the end of the stationary phase and exhibited highest activity at 75 °C and pH 7.5 (Ibrahim and El-diwany 2007).

5.5.6 Exopolysaccharides

Exopolysaccharides (EPSs) are high-molecular-weight polymers composed of sugar residues. Bacteria produce diverse and multifunctional polysaccharides including intracellular, structural, and extracellular polysaccharides (exopolysaccharides). EPSs generally consist of polymers of monosaccharides and some non-carbohydrate substituents (such as acetate, pyruvate, succinate, and phosphate). EPSs play an important role for microbial cells, as they can form a protective layer for the cell against harsh external environments, serve as carbon and energy sources during starvation, mediate cell-cell interactions, facilitate the adherence of the cell to surface, and induce microbial aggregation or biofilm formation (Nwodo et al. 2012). Nichols et al. (2005), Junge et al. (2004), and Tourney and Ngwenya (2014) suggest also functions which include cryoprotection for growth at low temperatures, high-salinity tolerance with reference to sea ice microbial communities, and heavy metal precipitation on the cell surface.

The various properties of microbial EPS have found large application in the industry. EPS like xanthan and gellan are already utilized in the food industry as gelling agents and thickeners for salad dressings, desserts, sauces, syrups, and ice cream (Kornmann et al. 2003). New areas for the application of microbial polysaccharides include improving the efficiency of liquid herbicides and insecticides; stabilization of emulsified pharmaceutical and cosmetic creams (Moonmangmee et al. 2002; Sutherland 1999), as thickeners and stabilizers in shampoos, toothpaste, and makeup; and solidifier of microbiological and plant tissue culture media. In recent years there has been an increasing interest in their biological activities, like antitumor, antiviral, immunostimulatory (Arena et al. 2006; Weiner et al. 1995), and anti-inflammatory effects (De Stefano et al. 2007).

EPSs from thermophilic bacteria offer numerous applications in various fields of industry, as the thermophiles provide more suitable processes for polymer production with decreased viscosity at high temperature. Extremophiles offer a great diversity in chemical and physical properties of their EPS compared to anywhere else in the biosphere (Guezennec 2002). Additionally, EPSs synthesized by thermophilic bacteria are likely to keep their structural properties at high temperature, which is a desired feature of the polymer solution (Radchenkova et al. 2013).

EPSs from geobacilli and anoxybacilli isolated from different geothermal springs are promising for their use in the industry. One gram of EPS from *Anoxybacillus* sp. R4-33 isolated from a hot spring in China absorbed 1.9783 mg Zn(II) and 1.4095 mg Cd(II) at pH 6.0 (Zhao et al. 2014). This EPS was a heteropolysaccharide, composed of D-mannose and D-glucose as its principal monosaccharide components in the relative proportions 1:0.45 (Table 5.8). Production of thermostable EPS was also reported for *G. tepidamans* (Kambourova et al. 2009), *G. thermodenitrificans* (Panosyan 2017, Panosyan et al. 2014), *G. toebii*, and *A. kestanbolensis* (Radchenkova et al. 2013; Panosyan 2017).

Table 5.8 Thermostable EPS from *Geobacillus* and *Anoxybacillus* species and their characteristics

Microorganism	Isolation source	Carbon Source	EPS yield (mg l^{-1})	EPS molecular weight (kDa), chemical composition (relative ratio)	References
G. toebii	Rupi hot spring, Bulgaria	Sucrose	50	ND	Radchenkova et al. (2013)
G. tepidamans V264	Velingrad hot spring, Bulgaria	Maltose	111.4	1000, Glc/Gal/Fuc/Fru (1/0.07/0.04/0.02)	Kambourova et al. (2009) and Coorevits et al. (2011)
G. thermodenitrificans ArzA-6	Arzakan geothermal spring, Armenia	Fructose/glucose	76	500, Man/Gal/Ara/Fru/Glc (1/0.13/0.1/0.06/0.05)	Panosyan (2017) and Panosyan et al. (2014)
G. toebii ArzA-8			80	600, Man/Gal/Glc/Ara (1/0.5/0.2/0.05)	
A. kestanbolensis 415	Mizinka hot spring, Bulgaria	Sucrose	25.3	ND	Radchenkova et al. (2013)
Anoxybacillus sp. *R4-33*	Radioactive radon hot spring, China	Glucose	1083	EPSII, 1000, Man/Glc (1/0.45)	Zhao et al. (2014)

ND not determined

5.6 Conclusion

Isolation and study of thermophilic bacilli from terrestrial geothermal springs are important for understanding of the diversity of thermophilic microbes and exploring their biotechnological potency. Many new thermophilic microbes belonging to the genera *Anoxybacillus* and *Geobacillus* have been isolated from different terrestrial geothermal springs worldwide, identified, and evaluated taking into account their biotechnological potency. *Anoxybacillus* is a relatively new genus compared to the well-studied *Geobacillus*. Most of the reported data has revealed that the members of both genera produce interesting enzymes that are thermostable and tolerant to alkaline conditions. Some of the studied enzymes were discovered through partnerships with industry. The interest in heat-adapted industrial enzymes is expected to increase. The present work, therefore, extends the previous sphere of information regarding the thermophilic bacilli diversity of terrestrial geothermal springs worldwide and their biotechnological applications and potency.

Acknowledgments This work was supported by grants from the EURASIA Programme of the Norwegian Center for International Cooperation in Education (CPEA-2011/10081, CPEA-LT-2016/10095) and partially supported by the RA MES State Committee of Science, in the frames of the research project no. 15T-1F399, and the Armenian National Science and Education Fund based in New York, USA, to HP (ANSEF-NS-microbio 2493, 3362, and 4676).

References

Acer Ö, Bekler FM, Pirinççioğlu H, Güven RG, Güven K (2016) Purification and Characterization of Thermostable and Detergent-Stable α-Amylase from *Anoxybacillus* sp. AH1. Food Technol Biotechnol 54(1):70–77

Acer Ö, Pırınççiğlu H, Bekler Fatma M, Gül-Güven R (2015) *Anoxybacillus* sp. AH1, an α-amylase-producing thermophilic bacterium isolated from Dargeçit hot spring. Biologia Sect Cell Mol Biol 70(7):853–862

Agüloğlu Fincan S, Enez B, Özdemir S, Matpan Bekler F (2014) Purification and characterization of thermostable α-amylase from thermophilic *Anoxybacillus flavithermus*. Carbohydr Polym 102(Supplement C):144–150

Ahmad S, Scopes RK, Rees GN, Patel BKC (2000) *Saccharococcus caldoxylosilyticus* sp. nov., an obligately thermophilic, xylose-utilizing, endospore-forming bacterium. Int J Syst Evol Microbiol 50:517–523

Al-Batayneh KM, Jacob JH, Hussein EI (2011) Isolation and molecular identification of new thermophilic bacterial strains of *Geobacillus pallidus* and *Anoxibacillus flavithermus*. Int J Integr Biol 11(1):39–43

Al-Jailawi MH, Mahdi MS, Fadhil AMA (2016) Investigation the Consumption of Aromatic Hydrocarbons by *Anoxybacillus rupeinsis* Ir3 (JQ912241) using FTIR and HPLC Analyses. Br Biotechnol J 10(2):1–12 Article no. BBJ.18203

Al-Qodah Z (2006) Production and characterization of thermostable alpha-amylase by thermophilic *Geobacillus stearothermophilus*. Biotechnol J 1(7-8):850–857

Antranikian G (2007) Industrial relevance of thermophiles and their enzymes. In: Thermophiles. CRC Press, pp 113–160

Arena A, Maugeri TL, Pavone B, Iannello D, Gugliandolo C, Bisignano G (2006) Antiviral and immunoregulatory effect of a novel exopolysaccharide from a marine thermotolerant *Bacillus licheniformis*. Int Immunopharmacol 6(1):8–13

Arpigny JL, Jaeger KE (1999) Bacterial lipolytic enzymes: classification and properties. Biochem J 343(Pt 1):177–183

Arya M, Joshi GK, Gupta AK, Kumar A, Raturi A (2015) Isolation and characterization of thermophilic bacterial strains from Soldhar (Tapovan) hot spring in Central Himalayan Region, India. Ann Microbiol 65(3):1457–1464. https://doi.org/10.1007/s13213-014-0984-y

Ash C, Farrow JAE, Wallbanks S, Collins MD (1991) Phylogenetic heterogeneity of the genus *Bacillus* revealed by comparative analysis of small subunit ribosomal RNA sequences. Lett Appl Microbiol 13:202–206

Atanassova M, Derekova A, Mandeva R, Sjoholm C, Kambourova M (2008) *Anoxybacillus bogrovensis* sp. nov., a novel thermophilic bacterium isolated from a hot spring in Dolni Bogrov, Bulgaria. Int J Syst Evol Microbiol 58(Pt 10):2359–2362

Ay F, Karaoglu H, Inan K, Canakci S, Belduz AO (2011) Cloning, purification and characterization of a thermostable carboxylesterase from *Anoxybacillus* sp. PDF1. Protein Expr Purif 80(1):74–79

Balan A, Ibrahim D, Abdul Rahim R, Ahmad Rashid FA (2012) Purification and characterization of a thermostable lipase from *Geobacillus thermodenitrificans* IBRL-nra. Enzyme Res 2012:987523. https://doi.org/10.1155/2012/987523

Baltas N, Dincer B, Ekinci AP, Kolayli S, Adiguzel A (2016) Purification and characterization of extracellular α-amylase from a thermophilic *Anoxybacillus thermarum* A4 strain, Braz. arch. biol. technol. Eng Technol Tech 59:e16160346. https://doi.org/10.1590/1678-4324-2016160346

Banat IM, Marchant R, Rahman TJ (2004) *Geobacillus debilis* sp. nov., a novel obligately thermophilic bacterium isolated from a cool soil environment, and reassignment of *Bacillus pallidus* to *Geobacillus pallidus* comb. nov. Int J Syst Evol Microbiol 54:2197–2201

Belduz AO, Canakci S, Chan K-G, Kahar UM, Chan CS, Yaakop AS, Goh KM (2015) Genome sequence of *Anoxybacillus ayderensis* AB04(T) isolated from the Ayder hot spring in Turkey. Stand Genomic Sci 10:70. https://doi.org/10.1186/s40793-015-0065-2

Belduz AO, Dulger S, Demirbag Z (2003) *Anoxybacillus gonensis* sp. nov., a moderately thermophilic, xylose-utilizing, endospore-forming bacterium. Int J Syst Evol Microbiol 53(Pt 5):1315–1320

Bergey DH, Whitman WB, De Vos P, Garrity GM and Jones D (2009) Bergey's manual of systematic bacteriology. 3, New York: Springer.: https://doi.org/10.1002/9781118960608.gbm00529

Bertoldo C, Antranikian G (2002) Starch-hydrolyzing enzymes from thermophilic archaea and bacteria. Curr Opin Chem Biol 6(2):151–160. https://doi.org/10.1016/S1367-5931(02)00311-3

Bruins ME, Janssen AE, Boom RM (2001) Thermozymes and their applications: a review of recent literature and patents. Appl Biochem Biotechnol 90(2):155–186

Brumm P, Land ML, Hauser LJ, Jeffries CD, Chang Y-J, Mead DA (2015) Complete Genome Sequence of *Geobacillus* strain Y4.1MC1, a Novel CO-Utilizing *Geobacillus thermoglucosidasius* Strain Isolated from Bath Hot Spring in Yellowstone National Park. BioEnergy Res 8(3):1039–1045

Bryanskaya AV, Rozanov AS, Slynko NM, Shekhovtsov SV, Peltek SE (2015) *Geobacillus icigianus* sp. nov., a thermophilic bacterium isolated from a hot spring. Int J Syst Evol Microbiol 65(Pt 3):864–869

Caccamo D, Gugliandolo C, Stackebrandt E, Maugeri TL (2000) *Bacillus vulcani* sp. nov., a novel thermophilic species isolated from a shallow marine hydrothermal vent. Int J Syst Evol Microbiol 50:2009–2012

Canakci S, Inan K, Kacagan M, Belduz AO (2007) Evaluation of arabinofuranosidase and xylanase activities of *Geobacillus* spp. isolated from some hot springs in Turkey. J Microbiol Biotechnol 17(8):1262–1270

Carrasco-Lopez C, Godoy C, de Las Rivas B, Fernandez-Lorente G, Palomo JM, Guisan JM, Fernandez-Lafuente R, Martinez-Ripoll M and Hermoso JA (2009) Activation of bacterial thermoalkalophilic lipases is spurred by dramatic structural rearrangements, J Biol Chem 7. 4365-4372

Chai KP, Othman NF, Teh AH, Ho KL, Chan KG, Shamsir MS, Goh KM, Ng CL (2016) Crystal structure of *Anoxybacillus* alpha-amylase provides insights into maltose binding of a new glycosyl hydrolase subclass. Sci Rep 6:23126. https://doi.org/10.1038/srep23126

Chai YY, Kahar UM, Md Salleh M, Md Illias R, Goh KM (2012) Isolation and characterization of pullulan-degrading *Anoxybacillus* species isolated from Malaysian hot springs. Environ Technol 33(10-12):1231–1128

Chakravorty D, Patra S (2013) Attaining Extremophiles and Extremolytes: Methodologies and Limitations. In: Singh OV (ed) Extremophiles: sustainable resources and biotechnological implications, 1st edn. Wiley, Hoboken, pp 29–74

Chang YW, Sun YJ, Wang C, Hsiao CD (2008) Crystal structures of the 70-kDa heat shock proteins in domain disjoining conformation. J Biol Chem 283(22):15502–15511

Chatziefthimiou AD, Crespo-Medina M, Wang Y, Vetriani C, Barkay T (2007) The isolation and initial characterization of mercury resistant chemolithotrophic thermophilic bacteria from mercury rich geothermal springs. Extremophiles 11(3):469–479. https://doi.org/10.1007/s00792-007-0065-2

Che Ibrahim MA, Ahmad WA (2017) Growth optimization of a thermophilic strain *Geobacillus caldoxylosilyticus* utm6 isolated from selayang hot spring. eProc Chem 2(1):119–123

Chis L, Hriscu M, Bica A, Tosa M, Nagy G, Rona G, Vertessy BG, Dan Irimie F (2013) Molecular cloning and characterization of a thermostable esterase/lipase produced by a novel *Anoxybacillus flavithermus* strain. J Gen Appl Microbiol 59(2):119–134

Cihan AC, Cokmus C, Koc M, Ozcan B (2014) *Anoxybacillus calidus* sp. nov., a thermophilic bacterium isolated from soil near a thermal power plant. Int J Syst Evol Microbiol 64(Pt 1):211–219

Cihan AC, Ozcan B, Cokmus C (2011) *Anoxybacillus salavatliensis* sp. nov., an alpha-glucosidase producing, thermophilic bacterium isolated from Salavatli, Turkey. J Basic Microbiol 51(2):136–146

Colak A, Sisik D, Saglam N, Guner S, Canakci S, Belduz AO (2005) Characterization of a thermoalkalophilic esterase from a novel thermophilic bacterium, *Anoxybacillus gonensis* G2. Bioresour Technol 96(5):625–631

Coorevits A, Dinsdale AE, Halket G, Lebbe L, De Vos P, Van Landschoot A, Logan NA (2012) Taxonomic revision of the genus *Geobacillus*: emendation of *Geobacillus, G. stearothermophilus, G. jurassicus, G. toebii, G. thermodenitrificans* and *G. thermoglucosidans* (nom. corrig., formerly 'thermoglucosidasius'); transfer of *Bacillus thermantarcticus* to the genus as *G. thermantarcticus* comb. nov.; proposal of *Caldibacillus debilis* gen. nov., comb. nov.; transfer of *G. tepidamans* to *Anoxybacillus* as *A. tepidamans* comb. nov.; and proposal of *Anoxybacillus caldiproteolyticus* sp. nov. Int J Syst Evol Microbiol 62(Pt 7):1470–1485

Correa-Llantén DN, Muñoz-Ibacache SA, Castro ME, Muñoz PA, Blamey JM (2013) Gold nanoparticles synthesized by *Geobacillus* sp. strain ID17 a thermophilic bacterium isolated from Deception Island, Antarctica. BMC Microb Cell Fact 12(75):1–6. https://doi.org/10.1186/1475-2859-12-75

Cortez Y, Paul S, Villena G, Gutiérrez-Correa M (2016) Isolation and characterization of cellulase producing bacterial strains from an Amazonian geothermal spring in Peru. Br Microbiol Res J 15(5):1–8

Dai J, Liu Y, Lei Y, Gao Y, Han F, Xiao Y, Peng H (2011) A new subspecies of *Anoxybacillus flavithermus* ssp. *yunnanensis* ssp. nov. with very high ethanol tolerance. FEMS Microbiol Lett 320(1):72–78

Daupan SMMV, Rivera WL (2015) Isolation and characterization of culturable thermophilic bacteria from hot springs in Benguet, Philippines. Int J Philipp Sci Technol 8(2):14–18

De Clerck E, Rodríguez-Díaz M, Vanhoutte T, Heyrman J, Logan NA, De Vos P (2004) *Anoxybacillus contaminans* sp. nov., and *Bacillus gelatini* sp. nov., isolated from contaminated gelatin batches. Int J Syst Evol Microbiol 54:941–946

De Maayer P, Brumm PJ, Mead DA, Cowan DA (2014) Comparative analysis of the *Geobacillus hemicellulose* utilization locus reveals a highly variable target for improved hemicellulolysis. BMC Genomics 15(1):836. https://doi.org/10.1186/1471-2164-15-836

Deep K, Poddar A, Whitman WB, Das SK (2016) Draft Genome Sequence of *Anoxybacillus suryakundensis* Strain JS1T (DSM 27374T) Isolated from a Hot Spring in Jharkhand, India. Genome Announc 4(4):pii: e00824-16. https://doi.org/10.1128/genomeA.00824-16

Derekova A, Mandeva R, Kambourova M (2008) Phylogenetic diversity of thermophilic carbohydrate degrading bacilli from Bulgarian hot springs. World J Microbiol Biotechnol 24(9):1697–1702

Derekova A, Sjoholm C, Mandeva R, Kambourova M (2007) *Anoxybacillus rupiensis* sp. Nov., a novel thermophilic bacterium isolated from Rupi basin (Bulgaria). Extremophiles 11(4):577–583

Derekova A, Sjoholm C, Mandeva R, Michailova L, Kambourova M (2006) Biosynthesis of a thermostable gellan lyase by newly isolated and characterized strain of *Geobacillus stearothermophilus* 98. Extremophiles 10(4):321–326

De Stefano D, Tommonaro G, Simeon V, Poli A, Nicolaus B, Carnuccio R (2007) A polysaccharide from tomato (*Lycopersicon esculentum*) peels affects NF-jB activation in LPS-stimulated J774 macrophages. J Nat Prod 70:1636–1639

Dheeran P, Kumar S, Jaiswal YK, Adhikari DK (2010) Characterization of hyperthermostable alpha-amylase from *Geobacillus* sp. IIPTN. Appl Microbiol Biotechnol 86(6):1857–1866

Donk PJ (1920) A Highly Resistant Thermophilic Organism. Journal of Bacteriology 5(4):373–374

Dulger S, Demirbag Z, Belduz AO (2004) *Anoxybacillus ayderensis* sp. nov. and *Anoxybacillus kestanbolensis* sp. nov. Int J Syst Evol Microbiol 54(Pt 5):1499–1503

Eijsink VG, Matthews BW, Vriend G (2011) The role of calcium ions in the stability and instability of a thermolysin-like protease. Protein Sci 20(8):1346–1355

Ellis JT, Magnuson TS (2012) Thermostable and Alkalistable Xylanases Produced by the Thermophilic Bacterium *Anoxybacillus flavithermus* TWXYL3. ISRN Microbiology 12:ID 517524. https://doi.org/10.5402/2012/517524

England JL, Shakhnovich BE, Shakhnovich EI (2003) Natural selection of more designable folds: a mechanism for thermophilic adaptation. Proc Natl Acad Sci USA 100(15):8727–8731

Feitkenhauer H, Muller R, Markl H (2003) Degradation of polycyclic aromatic hydrocarbons and long chain alkanes at 60–70 degrees C by *Thermus* and *Bacillus* spp [corrected]. Biodegradation 14(6):367–372

Filippidou S, Jaussi M, Junier T, Wunderlin T, Jeanneret N, Palmieri F, Palmieri I, Roussel-Delif L, Vieth-Hillebrand A, Vetter A, Chain PS, Regenspurg S, Junier P (2016) *Anoxybacillus geothermalis* sp. nov., a facultatively anaerobic, endospore-forming bacterium isolated from mineral deposits in a geothermal station. Int J Syst Evol Microbiol 66(8):2944–2951

Fortina MG, Mora D, Schumann P, Parini C, Manachini PL, Stackebrandt E (2001) Reclassification of *Saccharococcus caldoxylosilyticus* as *Geobacillus caldoxylosilyticus* (Ahmad et al. 2000) comb. nov. Int J Syst Evol Microbiol 51:2063–2071

Genc B, Nadaroglu H, Adiguzel A, Baltaci O (2015) Purification and characterization of an extracellular cellulase from *Anoxybacillus gonensis* O9 isolated from geothermal area in Turkey. J Environ Biol 36(6):1319–1324

Goh KM, Gan HM, Chan K-G, Chan GF, Shahar S, Chong CS, Kahar UM, Chai KP (2014) Analysis of *Anoxybacillus* genomes from the aspects of lifestyle adaptations, prophage diversity, and carbohydrate metabolism. PLOS One 9(3):e90549

Graham R, Ternan N, McMullan G (2005) Proteomic investigation of *Geobacillus thermoleovorans* T80. In: Proceedings of international symposium on extremophiles and their applications, pp 97–101

Gudiukaite R, Sadauskas M, Gegeckas A, Malunavicius V, Citavicius D (2017) Construction of a novel lipolytic fusion biocatalyst GDEst-lip for industrial application. J Ind Microbiol Biotechnol 44(6):799–815

Gudiukaite R, Gegeckas A, Kazlauskas CD (2014) Influence of N- and C-terminal regions on activity, expression, characteristics and structure of lipase from *Geobacillus* sp. 95. Extremophiles 18:131–145

Guezennec J (2002) Deep-sea hydrothermal vents: A new source of innovative bacterial exopolysaccharides of biotechnological interest. J Ind Microbiol Biotechnol 29:204–208

Gul-Guven R, Guven K, Poli A, Nicolaus B (2008) *Anoxybacillus kamchatkensis* subsp. *asaccharedens* subsp. nov., a thermophilic bacterium isolated from a hot spring in Batman. J Gen Appl Microbiol 54(6):327–334

Gurumurthy DM, Neelagund SE (2010) *Geobacillus* sp. Iso 5, a novel amylase-producing thermophile from thermal springs in Konkan region of Southern India. J Earth Sci 21(1):319–322

Gurumurthy DM, Neelagund SE (2012) Molecular characterization of industrially viable extreme thermostable novel α-amylase of *Geobacillus* sp. Iso5 isolated from geothermal spring. J Pure Appl Microbiol 6(4):1–15

Hawumba JF, Theron J, Brozel VS (2002) Thermophilic protease-producing *Geobacillus* from Buranga hot springs in Western Uganda. Curr Microbiol 45(2):144–150

Heinen W, Lauwers AM, Mulders JW (1982) *Bacillus flavothermus*, a newly isolated facultative thermophile. Antonie Van Leeuwenhoek 48(3):265–272

Higgs PG, Ran W (2008) Coevolution of codon usage and tRNA genes leads to alternative stable states of biased codon usage. Mol Biol Evol 25(11):2279–2291

Hovhannisyan P, Panosyan H, Turabyan A, Trchounian A (2016) Termostable amylase producing bacilli isolated from Armenian geothermal springs. Biol J Armenia 68(Special Issue):68–73

Hussein AH, Lisowska BK, Leak DJ (2015) The genus *Geobacillus* and their biotechnological potential. Adv Appl Microbiol 92:1–48

Ibrahim ASS, El-diwany AI (2007) Isolation and identification of new cellulases producing thermophilic bacteria from an Egyptian hot spring and some properties of the crude enzyme. Aust J Basic Appl Sci 1(4):473–478

Inan K, Belduz AO, Canakci S (2013) *Anoxybacillus kaynarcensis* sp. nov., a moderately thermophilic, xylanase producing bacterium. J Basic Microbiol 53(5):410–419

Inan K, Canakci S, Belduz A (2011) Isolation and characterization of xylanolytic new strains of *Anoxybacillus* from some hot springs in Turkey. Turk J Biol 35:529–542

Jorquera MA, Gabler S, Inostroza NG, Acuna JJ, Campos MA, Menezes-Blackburn D, Greiner R (2018) Screening and characterization of phytases from bacteria isolated from Chilean hydrothermal environments. Microb Ecol 75(2):387–399

Jugran J, Rawat N, Joshi G (2015) Amylase production by *Geobacillus* sp. GJA1 isolated from a hot spring in Uttarakhand. ENVIS Himalayan Ecol 23:120–125

Junge K, Eicken H, Deming JW (2004) Bacterial Activity at −2 to −20 degrees C in Arctic wintertime sea ice. Appl Environ Microbiol 70(1):550–557

Kacagan M, Canakci S, Sandalli C, Inan K, Colak DN, Belduz AO (2008) Characterization of a xylanase from a thermophilic strain of *Anoxybacillus pushchinoensis* A8. Biologia 63(5):599–606

Kahar UM, Sani MH, Chan KG, Goh KM (2016) Immobilization of alpha-Amylase from *Anoxybacillus* sp. SK3-4 on ReliZyme and Immobead Supports. Molecules 21(9):1196. https://doi.org/10.3390/molecules21091196

Kambourova M, Mandeva R, Dimova D, Poli A, Nicolaus B, Tommonaro G (2009) Production and characterization of a microbial glucan, synthesized by *Geobacillus tepidamans* V264 isolated from Bulgarian hot spring. Carbohydrate Polymers 77(2):338–343

Kevbrin VV, Zengler K, Lysenko AM, Wiegel J (2005) *Anoxybacillus kamchatkensis* sp. nov., a novel thermophilic facultative aerobic bacterium with a broad pH optimum from the Geyser valley, Kamchatka. Extremophiles 9(5):391–398

Kikani BA, Singh SP (2012) The stability and thermodynamic parameters of a very thermostable and calcium-independent α-amylase from a newly isolated bacterium, *Anoxybacillus beppuensis* TSSC-1. Process Biochem 47(12):1791–1798

Khalil A, Sivakumar N, Qarawi S (2015) Genome Sequence of *Anoxybacillus flavithermus* Strain AK1, a Thermophile Isolated from a Hot Spring in Saudi Arabia. Genome Announc 3(3):e00604-15

Kornmann H, Duboc P, Marison I, von Stockar U (2003) Influence of nutritional factors on the nature, yield, and composition of exopolysaccharides produced by *Gluconacetobacter xylinus* I-2281. Appl Environ Microbiol 69(10):6091–6098

Kuisiene N, Raugalas J, Chitavichius D (2004) *Geobacillus lituanicus* sp. nov. Int J Syst Evol Microbiol 54:1991–1995

Kuhad RC, Gupta R, Singh A (2011) Microbial Cellulases and Their Industrial Applications. Enzyme Res 2011(280696). https://doi.org/10.4061/2011/280696

Kumar V, Verma D, Archana A, Satyanarayana T (2013) Thermostable Bacterial Xylanases. In: Satyanarayana T, Littlechild J, Kawarabayasi Y (eds) Thermophilic microbes in environmental and industrial biotechnology: biotechnology of thermophiles. Springer, Dordrecht, pp 813–857

Ladeira SA, Cruz E, Delatorre AB, Barbosa JB, Martins MLL (2015) Cellulase production by thermophilic *Bacillus* sp. SMIA-2 and its detergent compatibility. Electron J Biotechnol 18(2):110–115. https://doi.org/10.1016/j.ejbt.2014.12.008

Lee SJ, Lee YJ, Ryu N, Park S, Jeong H, Kim BC, Lee DW, Lee HS (2012) Draft genome sequence of the thermophilic bacterium *Anoxybacillus kamchatkensis* G10. J Bacteriol 194(23):6684–6685

Leow TC, Rahman RN, Basri M, Salleh AB (2004) High level expression of thermostable lipase from *Geobacillus* sp. strain T1. Biosci Biotechnol Biochem 68(1):96–103

Lim YL, Chan KG, Ee R, Belduz AO, Canakci S, Kahar UM, Yaakop AS, Goh KM (2015) Complete genome of the potential thermozyme producer *Anoxybacillus gonensis* G2(T) isolated from the Gonen hot springs in Turkey. J Biotechnol 212:65–66. https://doi.org/10.1016/j.jbiotec.2015.08.007

Liu Y, Li R, Wang J, Zhang X, Jia R, Gao Y, Peng H (2017) Increased enzymatic hydrolysis of sugarcane bagasse by a novel glucose- and xylose-stimulated β-glucosidase from *Anoxybacillus flavithermus* subsp. *yunnanensis* E13(T). BMC Biochem 18:art4, 1–8. https://doi.org/10.1186/s12858-017-0079-z

Llarch A, Logan NA, Castellvi J, Prieto MJ, Guinea J (1997) Isolation and characterization of thermophilic *Bacillus* spp. from geothermal environments on Deception Island, South Shetland Archipelago. Microb Ecol 34(1):58–65

Lobry JR, Chessel D (2003) Internal correspondence analysis of codon and amino-acid usage in thermophilic bacteria. J Appl Genet 44(2):235–261

Lotti M, Alberghina L (2007) Lipases: molecular structure and function. In: Polaina J, MacCabe AP (eds) Industrial enzymes: structure, function and applications, vol 16. Springer, Dordrecht, pp 264–282

Mahadevan GD, Neelagund SE (2014) Thermostable lipase from *Geobacillus* sp. Iso5: bioseparation, characterization and native structural studies. J Basic Microbiol 54(5):386–396. https://doi.org/10.1002/jobm.201200656

Mahadevan GD, Neelagund SE (2010) *Geobacillus* sp. Iso 5, a novel amylase producing thermophile from konkan region thermal springs in southern India. J Earth Sci 21:318–322

Matpan Bekler F, Acer O, Guven K (2015) Production and purification of novel thermostable alkaline protease from *Anoxybacillus* sp. KP1. Cell Mol Biol (Noisy-le-grand) 61(4):113–120

Matpan Bekler F, Güven K (2014) Isolation and production of thermostable α-amylase from thermophilic *Anoxybacillus* sp. KP1 from Diyadin hot spring in Ağri, Turkey. Biologia 69(4):419–427

Mehta D, Satyanarayana T (2013a) Biochemical and molecular characterization of recombinant acidic and thermostable raw-starch hydrolysing α-amylase from an extreme thermophile *Geobacillus thermoleovorans*. J Mol Catal B: Enzym 85–86(0):229–238

Mehta D, Satyanarayana T (2013b) Diversity of hot environments and thermophilic microbes'. In: Satyanarayana T, Littlechild J, Kawarabayasi Y (eds) Thermophilic microbes in environmental and industrial biotechnology: biotechnology of thermophiles. Springer, Dordrecht, pp 3–60

Mehta A, Kumar R, Gupta R (2012) Isolation of lipase producing thermophilic bacteria: optimization of production and reaction conditions for lipase from *Geobacillus* sp. Acta Microbiol Immunol Hung 59(4):435–450

Miñana-Galbis D, DL P´ n, JG L´n, Manresa A, OliartRos RM (2010) Reclassification of *Geobacillus pallidus* (Scholz et al. 1988) Banat et al. 2004 as *Aeribacillus pallidus* gen. nov., comb. nov. Int J Syst Evol Microbiol 60:1600–1604

Mollania N, Khajeh K, Hosseinkhani S, Dabirmanesh B (2010) Purification and characterization of a thermostable phytate resistant alpha-amylase from *Geobacillus* sp. LH8. Int J Biol Macromol 46(1):27–36

Moonmangmee S, Toyama H, Adachi O, Teerakool G, Lotong N, Matsushita K (2002) Purification and characterization of a novel polysaccharide involved in the pellicle produced by a thermotolerant *Acetobacter* strain. Biosci Biotechnol Biochem 66:777–783

Muhd Sakaff MK, Abdul Rahman AY, Saito JA, Hou S, Alam M (2012) Complete genome sequence of the thermophilic bacterium *Geobacillus thermoleovorans* CCB_US3_UF5. J Bacteriol 194(5):1239

Musto H, Naya H, Zavala A, Romero H, Alvarez-Valin F, Bernardi G (2005) The correlation between genomic G+C and optimal growth temperature of prokaryotes is robust: a reply to Marashi and Ghalanbor. Biochem Biophys Res Commun 330(2):357–360

Musto H, Naya H, Zavala A, Romero H, Alvarez-Valin F, Bernardi G (2006) Genomic GC level, optimal growth temperature, and genome size in prokaryotes. Biochem Biophys Res Commun 347(1):1–3

Nakamichi T, Nakashima T, Fujisaki H, Nobue T, Muramatsu T, Takahashi Y, Ishibashi Y (2010) Characteristics of *Anoxybacillus* sp. MU3 isolated from a hot spring and its application to the hyper thermal solubilization of sewage sludge. Environ Eng Sci 27(12):993–999

Namsaraev ZB, Babasanova OB, Dunaevsky YE, Akimov VN, Barkhutova DD, Gorlenko VM, Namsaraev BB (2010) *Anoxybacillus mongoliensis* sp. nov., a novel thermophilic proteinase producing bacterium isolated from alkaline hot spring, central Mongolia. Mikrobiologiia 79(4):516–523

Narayan VV, Hatha MA, Morgan HW, Rao D (2008) Isolation and characterization of aerobic thermophilic bacteria from the savusavu hot springs in fiji. Microbes Environ 23(4):350–352

Nazina TN, Lebedeva EV, Poltaraus AB, Tourova TP, Grigoryan AA, Sokolova D, Lysenko AM, Osipov GA (2004) *Geobacillus gargensis* sp. nov., a novel thermophile from a hot spring, and the reclassification of *Bacillus vulcani* as *Geobacillus vulcani* comb. nov. Int J Syst Evol Microbiol 54(Pt 6):2019–2024

Nazina TN, Sokolova D, Grigoryan AA, Shestakova NM, Mikhailova EM, Poltaraus AB, Tourova TP, Lysenko AM, Osipov GA, Belyaev SS (2005) *Geobacillus jurassicus* sp. nov., a new thermophilic bacterium isolated from a high-temperature petroleum reservoir, and the validation of the *Geobacillus* species. Syst Appl Microbiol 28(1):43–53

Nazina TN, Tourova TP, Poltaraus AB, Novikova EV, Grigoryan AA, Ivanova AE, Lysenko AM, Petrunyaka VV, Osipov GA, Belyaev SS, Ivanov MV (2001) Taxonomic study of aerobic thermophilic bacilli: descriptions of *Geobacillus subterraneus* gen. nov., sp. nov. and *Geobacillus uzenensis* sp. nov. from petroleum reservoirs and transfer of *Bacillus stearothermophilus, Bacillus thermocatenulatus, Bacillus thermoleovorans, Bacillus kaustophilus, Bacillus thermodenitrificans* to *Geobacillus* as the new combinations *G. stearothermophilus,* G. th. Int J Syst Evol Microbiol 51(Pt 2):433–446

Niehaus F, Bertoldo C, Kahler M, Antranikian G (1999) Extremophiles as a source of novel enzymes for industrial application. Appl Microbiol Biotechnol 51(6):711–729

Nicolaus B, Manca MC, Lama L, Esposito E, Gambacorta A (1995) Effects of growth temperature on the polar lipid pattern and fatty acid composition of seven thermophilic isolates from the Antarctic continent. Syst Appl Microbiol 18(1):32–36

Nichols CM, Lardiere SG, Bowman JP, Nichols PD, Gibson JAE, Guezennec J (2005) Chemical characterization of exopolysaccharides from Antarctic marine bacteria. Microb Ecol 49(4):578–589

Nwodo UU, Green E, Okoh AI (2012) Bacterial exopolysaccharides: functionality and prospects. Int J Mol Sci 13(11):14002–14015

Obeidat M, Khyami-Horani H, Al-Zoubi A, Otri I (2012) Isolation, characterization, and hydrolytic activities of *Geobacillus* species from Jordanian hot springs. Afr J Biotechnol 11(25):6763–6768

Ogg CD, Spanevello MD, Patel BKC (2013) Exploring the ecology of thermophiles from Australia's Great Artesian basin during the genomic era. In: Satyanarayana T, Littlechild J, Kawarabayasi Y (eds) Thermophilic microbes in environmental and industrial biotechnology: biotechnology of thermophiles. Springer, Dordrecht, pp 61–97

Ollis DL, Cheah E, Cygler M, Dijkstra B, Frolow F, Franken SM, Harel M, Remington SJ, Silman I, Schrag J et al (1992) The alpha/beta hydrolase fold. Protein Eng 5(3):197–211

Olusesan AT, Azura LK, Abubakar F, Hamid NSA, Radu S, Saari N (2009) Phenotypic and molecular identification of a novel thermophilic *Anoxybacillus* species: a lipase-producing bacterium isolated from a Malaysian hotspring. World J Microbiol Biotechnol 25(11):1981–1988

Ortiz EM, Berretta MF, Navas LE, Benintende GB, Amadio AF, Zandomeni RO (2015) Draft genome sequence of *Geobacillus* sp. Isolate T6, a thermophilic bacterium collected from a thermal spring in Argentina. Genome Announc 3(4):e00743-15

Ozdemir S, Okumus V, Ulutas MS, Dundar A, Akarsubası AT, Dumonted S (2015) Isolation of a novel thermophilic *Anoxybacillus flavithermus* SO- 13, production, characterization and industrial applications of its thermostable α-amylase. J Bioprocess Biotech 5(7):1–9. https://doi.org/10.4172/2155-9821.1000237

Özdemir S, Kılınç E, Poli A, Nicolaus B (2013) Biosorption of heavy metals (Cd^{2+}, Cu^{2+}, Co^{2+}, and Mn^{2+}) by thermophilic bacteria, *Geobacillus thermantarcticus* and *Anoxybacillus amylolyticus*: equilibrium and kinetic studies. Bioremed J 17(2):86–96

Padilha I, Carvalho L, Dias P, Grisi T, Silva F, Santos S, Araujo D (2015) Production and characterization of thermophilic carboxymethyl cellulase synthesized by Bacillus sp. growing on sugarcane bagasse in submerged fermentation. Braz J Chem Eng 32(1):35–42. https://doi.org/10.1590/0104-6632.20150321s00003298

Panosyan HH (2017) Thermophilic bacilli isolated from Armenian geothermal springs and their potential for production of hydrolytic enzymes. Int J Biotech Bioeng 3(8):239–244

Panosyan H (2010) Phylogenetic diversity based on 16S rRNA gene sequence analysis of aerobic thermophilic endospore-forming bacteria isolated from geothermal springs in Armenia. Biol J Armenia 4(65):73–80

Panosyan H, Birkeland NK (2014) Microbial diversity in an Armenian geothermal spring assessed by molecular and culture-based methods. J Basic Microbiol 54(11):1240–1250

Panosyan H, Anzelmo G, Nicolaus B (2014) Production and characterization of exopolysaccharadies synthesized by geobacilli isolated from an Armenian geothermal spring. FEBS J 281(1):667

Pikuta E, Cleland D, Tang J (2003) Aerobic growth of *Anoxybacillus pushchinoensis* K1(T): emended descriptions of *A. pushchinoensis* and the genus *Anoxybacillus*. Int J Syst Evol Microbiol 53(Pt 5):1561–1562

Pikuta E, Lysenko A, Chuvilskaya N, Mendrock U, Hippe H, Suzina N, Nikitin D, Osipov G, Laurinavichius K (2000) *Anoxybacillus pushchinensis* gen. nov., sp. nov., a novel anaerobic, alkaliphilic, moderately thermophilic bacterium from manure, and description of *Anoxybacillus flavitherms* comb. nov. Int J Syst Evol Microbiol 50(Pt 6):2109–2117

Pinzon-Martinez DL, Rodriguez-Gomez C, Minana-Galbis D, Carrillo-Chavez JA, Valerio-Alfaro G, Oliart-Ros R (2010) Thermophilic bacteria from Mexican thermal environments: isolation and potential applications. Environ Technol 31(8-9):957–966

Poli A, Guven K, Romano I, Pirinccioglu H, Guven RG, Euzeby JP, Matpan F, Acer O, Orlando P, Nicolaus B (2012) *Geobacillus subterraneus* subsp. *aromaticivorans* subsp. nov., a novel thermophilic and alkaliphilic bacterium isolated from a hot spring in Sirnak, Turkey. J Gen Appl Microbiol 58(6):437–446

Poli A, Romano I, Cordella P, Orlando P, Nicolaus B, Ceschi Berrini C (2009) *Anoxybacillus thermarum* sp. nov., a novel thermophilic bacterium isolated from thermal mud in Euganean hot springs, Abano Terme, Italy. Extremophiles 13(6):867–874

Poli A, Romano I, Caliendo G, Nicolaus G, Orlando P, Falco A, Lama L, Gambacorta A, Nicolaus B (2006) *Geobacillus toebii* subsp. *decanicus* subsp. nov., a hydrocarbon-degrading, heavy metal resistant bacterium from hot compost. J Gen Appl Microbiol 52:223–234

Priya I, Dhar MK, Bajaj BK, Koul S, Vakhlu J (2016) Cellulolytic activity of thermophilic bacilli isolated from Tattapani hot spring sediment in North West Himalayas. Indian J Microbiol 56(2):228–231. https://doi.org/10.1007/s12088-016-0578-4

Radchenkova N, Vassilev S, Panchev I, Anzelmo G, Tomova I, Nicolaus B, Kuncheva M, Petrov K, Kambourova M (2013) Production and properties of two novel exopolysaccharides synthesized by a thermophilic bacterium *Aeribacillus pallidus* 418. Appl Biochem Biotechnol 171(1):31–43

Rao MB, Tanksale AM, Ghatge MS, Deshpande VV (1998) Molecular and biotechnological aspects of microbial proteases. Microbiol Mol Biol Rev 62(3):597–635

Reddy NS, Nimmagadda A, Sambasiva Rao KRS (2003) An overview of the microbial α-amylase family. Afr J Biotechnol 2:645–648

Rekadwad BN (2015) Characterization of amylase from industrially important thermophilic microorganism: *Geobacillus thermoleovorans* strain Rekadwadsis. Int J Life Sci Biotechnol Pharma Res 4(1):26–30

Rozanov AS, Logacheva MD, Peltek SE (2014) Draft genome sequences of *Geobacillus stearothermophilus* strains 22 and 53, isolated from the garga hot spring in the Barguzin River Valley of the Russian Federation. Genome Announc 2(6):e01205-14

Satapathy SS, Dutta M, Ray SK (2010) Higher tRNA diversity in thermophilic bacteria: a possible adaptation to growth at high temperature. Microbiol Res 165(8):609–616

Satyanarayana T, Sharma A, Mehta D, Puri AK, Kumar V, Nisha M, Joshi S (2012) Biotechnological applications of biocatalysts from the firmicutes *Bacillus* and *Geobacillus* species. In:

Satyanarayana T, Johri BN (eds) Microorganisms in sustainable agriculture and biotechnology. Springer, Dordrecht, pp 343–379

Sawle L, Ghosh K (2011) How do thermophilic proteins and proteomes withstand high temperature? Biophys J 101(1):217–227

Seifzadeh S, Hassan Sajedi R, Sariri R (2008) Isolation and characterization of thermophilic alkaline proteases resistant to sodium dodecyl sulfate and ethylene diamine tetraacetic acid from *Bacillus* sp. GUS1. Iran J Biotechnol 6(4):214–221

Scandurra R, Consalvi V, Chiaraluce R, Politi L, Engel PC (1998) Protein thermostability in extremophiles. Biochimie 80(11):933–941

Schäffer C, Franck WL, Scheberl A, Kosma P, McDermott TR, Messner P (2004) Classification of isolates from locations in Austria and Yellowstone National Park as *Geobacillus tepidamans* sp. nov. Int J Syst Evol Microbiol 54(Pt 6):2361–2238

Schneider D, Liu Y, Gerstein M, Engelman DM (2002) Thermostability of membrane protein helix-helix interaction elucidated by statistical analysis. FEBS Lett 532:231–236

Scholz T, Demharter W, Hensel R, Kandler O (1987) *Bacillus pallidus* sp. nov., a new thermophilic species from sewage. Syst Appl Microbiol 9:91–96

Shahinyan G (2015) Temperature and pH effect on lipase production and enzymatic activity of thermophilic bacilli isolated from Armenian geothermal springs. Sci Med J 10(1):9–15

Shahinyan G, Margaryan A, Panosyan H, Trchounian A (2017) Identification and sequence analyses of novel lipase encoding novel thermophillic bacilli isolated from Armenian geothermal springs. BMC Microbiol 17(103). https://doi.org/10.1186/s12866-017-1016-4

Sharma A, Pandey A, Shouche YS, Kumar B, Kulkarni G (2009) Characterization and identification of *Geobacillus* spp. isolated from Soldhar hot spring site of Garhwal Himalaya, India. J Basic Microbiol 49(2):187–194

Sharma R, Thakur V, Sharma M and Birkeland N-K (2013) Biocatalysis through thermostable lipases: adding flavor to chemistry, In Satyanarayana, T., Littlechild, J. & Kawarabayasi, Y. (eds.) Thermophilic microbes in environmental and industrial biotechnology: DordrechtSpringer : 905-927

Sharma P, Gupta S, Sourirajan A, Dev K (2015) Characterization of extracellular thermophillic cellulase from thermophilic *Geobacillus* sp. isolated from Tattapani Hot spring of Himachal Pradesh, India. Int JAdv Biotechnol Res (IJBR) 6(3):433–442

Sharma A, Adhikari S, Satyanarayana T (2007) Alkali-thermostable and cellulase-free xylanase production by an extreme thermophile *Geobacillus thermoleovorans*. World J Microbiol Biotechnol 23(4):483–490

Shih TW, Pan TM (2011) Stress responses of thermophilic *Geobacillus* sp. NTU 03 caused by heat and heat-induced stress. Microbiol Res 166(5):346–359

Sookkheo B, Sinchaikul S, Thannan H, Thongprasong O, Phutrakul S, Chen ST (2002) Proteomic analysis of a thermostable superoxide dismutase from *Bacillus stearothermophilus* TLS33. Proteomics 2(9):1311–1315

Stenesh J, McGowan GR (1977) DNA polymerase from mesophilic and thermophilic bacteria. I. Purification and properties of DNA polymerase from *Bacillus licheniformis* and *Bacillus stearothermophilus*. Biochim Biophys Acta 272:156–166

Sung MH, Kim H, Bae JW, Rhee SK, Jeon CO, Kim K, Kim JJ, Hong SP, Lee SG, Yoon JH, Park YH, Baek DH (2002) *Geobacillus toebii* sp. nov., a novel thermophilic bacterium isolated from hay compost. Int J Syst Evol Microbiol 52(Pt 6):2251–2255

Sunna A, Prowe SG, Stoffregen T, Antranikian G (1997) Characterization of the xylanases from the new isolated thermophilic xylan-degrading *Bacillus thermoleovorans* strain K-3d and *Bacillus flavothermus* strain LB3A. FEMS Microbiol Lett 148(2):209–216

Sutherland IW (1999) Polysaccharases for microbial exopolysaccharides. Carbohydr Polym 38(4):319–328

Synowiecki J (2010) Some applications of thermophiles and their enzymes for protein processing. Afr J Biotechnol 9:7020–7025

Takami H, Takaki Y, Chee GJ, Nishi S, Shimamura S, Suzuki H, Matsui S, Uchiyama I (2004) Thermoadaptation trait revealed by the genome sequence of thermophilic *Geobacillus kaustophilus*. Nucleic Acids Res 21:6292–6303

Tan H Alam M (2010) sRNAs are involved in high temperature adaptation of Geobacillus thermoleovorans. EMBO Symposium Non-coding genome, At Heidelberg, Germany

Tolner B, Poolman B, Konings WN (1997) Adaptation of microorganisms and their transport systems to high temperatures. Comp Biochem Physiol 118A:423–428

Tourney J, Ngwenya BT (2014) The role of bacterial extracellular polymeric substances in geomicrobiology. Chem Geol 386(Supplement C):115–132

Trent JD (1996) A review of acquired thermotolerance, heat-shock proteins, and molecular chaperones in archaea. FEMS Microbiol Rev 18(2-3):249–258

Trivedi S, Gehlot HS, Rao SR (2006) Protein thermostability in Archaea and Eubacteria. Genet Mol Res 5(4):816–827

Turner P, Mamo G, Karlsson EN (2007) Potential and utilization of thermophiles and thermostable enzymes in biorefining. Microbial Cell Fact 6(1):9. https://doi.org/10.1186/1475-2859-6-9

Uma Maheswar Rao JL, Satyanarayana T (2007) Improving production of hyperthermostable and high maltose-forming alpha-amylase by an extreme thermophile *Geobacillus thermoleovorans* using response surface methodology and its applications. Bioresour Technol 98(2):345–352

van de Vossenberg JL, Ubbink-Kok T, Elferink MG, Driessen AJ, Konings WN (1995) Ion permeability of the cytoplasmic membrane limits the maximum growth temperature of bacteria and archaea. Mol Microbiol 18(5):925–932

van Kranendonk MJ, Djokic T, Campbell KA, Deamer D, Damer B, Walter MR, Steller L, Ota T, Nakamura E, Tadbiri S (2017) An origin of life at terrestrial hot springs: support from early earth and implications for the search for life on mars. Astrobiology Science Conference 2017 (LPI Contrib. No. 1965)

Vardanyan G, Margaryan A, Panosyan H (2015) Isolation and characterization of lipase-producing bacilli from Tatev geothermal spring. Collection of Scientific Articles of YSU SSS 1.1(4): 33–36

Wang GX, Wang Y, Wu ZF, Jiang HF, Dong RQ, Li FY, Liu XL (2011a) Immunomodulatory effects of secondary metabolites from thermophilic *Anoxybacillus kamchatkensis* XA-1 on carp, Cyprinus carpio. Fish Shellfish Immunol 30(6):1331–1338

Wang Q, Cen Z, Zhao J (2015) The survival mechanisms of thermophiles at high temperatures: an angle of omics. Physiology (Bethesda) 30(2):97–106

Wang X, Gao R, Liao C, Meng Y, Tang X, Li J, Xu B, Huang Z (2011b) Isolation of *Geobacillus* sp. A27 and characterization of its α-amylase. Sci Technol Food Ind 32:151–155 (in Chinese with English abstract)

Wang W, Ma T, Zhang B, Yao N, Li M, Li G, Ma Z, Cheng J (2014) A novel mechanism of protein thermostability: a unique N-terminal domain confers heat resistance to Fe/Mn-SODs. Sci Rep 4(7284):1–9. https://doi.org/10.1038/srep07284

Weiner R, Langille S, Quintero E (1995) Structure, function and immunochemistry of bacterial exopolysaccharides. J Ind Microbiol 15:339–346

Yang Z, Zhang Y, Shen T, Xie Y, Mao Y, Ji C (2013) Cloning, expression and biochemical characterization of a novel, moderately thermostable GDSL family esterase from *Geobacillus thermodenitrificans* T2. J Biosci Bioeng 115(2):133–137

Yohandini H, Julinar, Muharni (2015) Isolation and phylogenetic analysis of thermophile community within Tanjung Sakti Hot Spring, South Sumatera, Indonesia. HAYATI J Biosci 22(3):143–148

Yumoto I, Hirota K, Kawahara T, Nodasaka Y, Okuyama H, Matsuyama H, Yokota Y, Nakajima K, Hoshino T (2004) *Anoxybacillus voinovskiensis* sp. nov., a moderately thermophilic bacterium from a hot spring in Kamchatka. Int J Syst Evol Microbiol 54(Pt 4):1239–1242

Zahoor S, Javed M, Babarl M (2016) Characterization of a novel hydrolytic enzyme producing thermophilic bacterium isolated from the Hot Spring of Azad Kashmir-Pakistan. Braz Arch Biol Technol 56:e16150622. https://doi.org/10.1590/1678-4324-2016150662

Zeigler R (2001) *Bacillus* genetic stock center catalog of strains (7th edn), Volume 3: The genus *Geobacillus*: 11

Zhang XQ, Zhang ZL, Wu N, Zhu XF, Wu M (2013) *Anoxybacillus vitaminiphilus* sp. nov., a strictly aerobic and moderately thermophilic bacterium isolated from a hot spring. Int J Syst Evol Microbiol 63(Pt 11):4064–4071

Zhang CM, Huang XW, Pan WZ, Zhang J, Wei KB, Klenk HP, Tang SK, Li WJ, Zhang KQ (2011) *Anoxybacillus tengchongensis* sp. nov. and *Anoxybacillus eryuanensis* sp. nov., facultatively anaerobic, alkalitolerant bacteria from hot springs. Int J Syst Evol Microbiol 61(Pt 1):118–122

Zhao S, Cao F, Zhang H, Zhang L, Zhang F, Liang X (2014) Structural characterization and biosorption of exopolysaccharides from *Anoxybacillus* sp. R4-33 isolated from radioactive radon hot spring. Appl Biochem Biotechnol 172(5):2732–2746

Zhu W, Cha D, Cheng G, Peng Q, Shen P (2007) Purification and characterization of a thermostable protease from a newly isolated *Geobacillus* sp. YMTC 1049. Enzyme Microb Technol 40:1592–1597

Thermophiles and Their Exploration for Thermostable Enzyme Production

6

Nikoleta Boteva and Margarita Kambourova

Abstract

Currently no more than 1% of the total microbial species that exists in nature is known, the deal of known microorganisms in extremophilic niches being even much less. Thermophiles are a type of extremophiles which study is related to clarifying a number of fundamental issues such as origin of life and molecular mechanisms of thermostability, revealing the vast potential of their enzymes for biotechnological use. Microbial biodiversity in Eurasian hot springs is still badly known, and molecular analyses revealed a presence of significant part of unknown groups comparable with those of well-studied Yellowstone and Iceland springs. Intensive studies on ecology, physiology, and molecular biology of extremophiles provide valuable insight into the life processes at each level, as well as the potential for numerous industrial applications. Thermophilic molecules suggest many advantages in their exploration as biocatalysts; however known enzymes still are not able to satisfy evolving new needs and requirements of biotechnological processes, among which their stability under industrial conditions is of particular importance. Nowadays, different approaches are used to find the desired enzyme activities including direct screening in big microbial collections, metagenome screening, and shotgun sequencing, the last two based on analysis of the coding regions of the known enzymes. Direct screening confirms unambiguously the real existence and several functional characteristics of an enzyme activity. Metagenome is accepted as a huge reservoir of taxonomic and functional genes and therefore is seeing as a feasible possibility to introduce more and diverse enzymes with better performance meeting the global demand for new catalysts. The discovery of new microbial species as well as the

N. Boteva · M. Kambourova (✉)
Institute of Microbiology, Bulgarian Academy of Sciences, Sofia, Bulgaria
e-mail: margikam@microbio.bas.bg

D. Egamberdieva et al. (eds.), *Extremophiles in Eurasian Ecosystems: Ecology, Diversity, and Applications*, Microorganisms for Sustainability 8,
https://doi.org/10.1007/978-981-13-0329-6_6

sequencing of new genomes and metagenomes allows access to new enzymes with new application capabilities.

Keywords
Thermophiles · Biodiversity · Thermophilic enzymes · Metagenomic identification of enzymes

6.1 Introduction

Microorganisms are an integral part of the history of life on Earth, and the diversity of microbial world has been created for billions of years of evolution. Because they are the oldest form of life, prokaryotic microorganisms have been the subject of twice longer evolution resulted into a vast variety of organisms with the widest range of known metabolic pathways (Stetter 2001).

It is generally accepted that the degree of microbial diversity is not adequately characterized and that there is a huge gap between the knowledge of this diversity and its actual importance for environmental processes and economic development (Vibha and Neelam 2012). Perception of the low degree of their knowledge was developed with development of research approaches, and a real danger from the loss of species, sometimes even before being identified, is increasingly clear with expansion of man's influence on natural ecosystems. Culturable microorganisms (those microorganisms for which cultivation parameters in the laboratory were established) are no more than 1% of the total microbial diversity that exists in nature (Stewart 2012). So far, about 10,000 microbial species have been recognized (Mora et al. 2011). Despite the common belief that prokaryotes are characterized by the highest taxonomic diversity (Oren 2009), this number is considerably smaller than the number of species described for eukaryotes (e.g., insects only are 300,000 species). Even if we assume that microbial species are only one million, and that a thousand new species are described per year, it will take 1000 years to reach a complete knowledge of the diversity of the microbial world. That is why scientists assume that we are still far from the actual knowledge of microbial diversity; moreover according to common belief, the presumed species are much more than a million and that the new microorganisms described are a little less than 1000 (Amann 2000). The number of major phyla has increased from 12 identifiable lineages in 1987, to 30 in 2014, or over 50 including candidate phyla. The total real number has been estimated to exceed 1000 bacterial phyla (Yarza et al. 2014).

Particularly large is the share of unknown microorganisms in the extreme niches as the more extremophilic is the niche the more difficult is to reproduce growth conditions in the laboratory. Extremophiles not only endure but are also functionally active in some of the harshest living conditions found on Earth. The term "extreme" was introduced by MacElroy (1974). "Extreme" is a relative term to mean conditions too raw for human existence like temperature from −12 °C to> 100 °C, pH from 0 to 13, hydrostatic pressure up to 1400 atmospheres, and salt

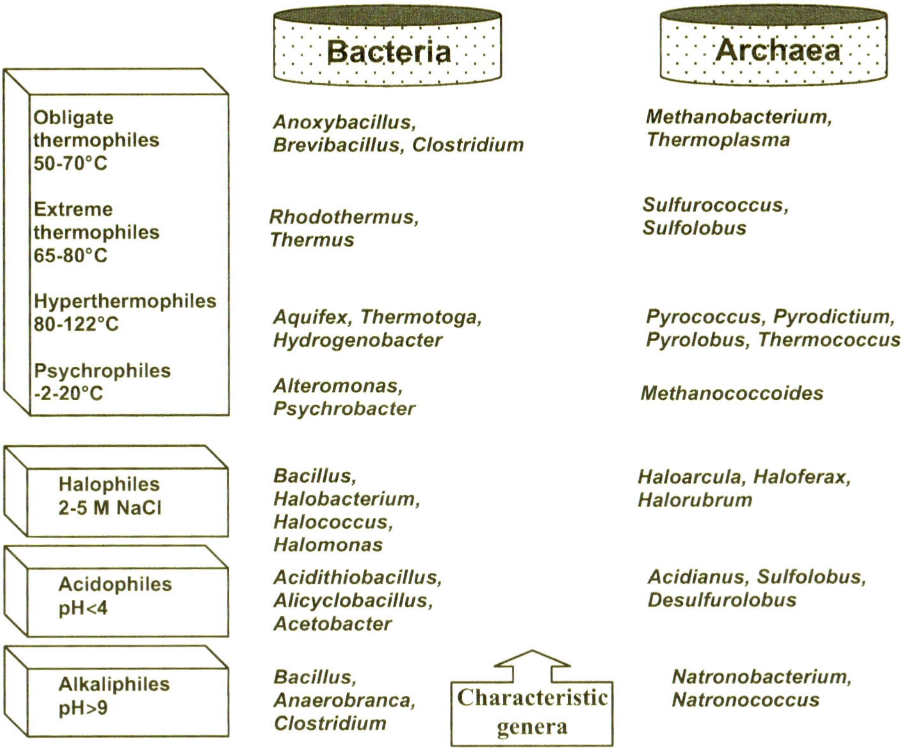

Fig. 6.1 Examples of the some common genera of extremophiles in the specific type of niches

concentration equal to that of saturated solutions (Satyanarayana et al. 2005). In the last decades, extremophiles have been found in such inhospitable niches as the active volcanoes, the bottom of the world's ocean, the salt marshes. Unlike most organisms, extremophiles develop optimally when one or more of the environmental parameters have extreme values such as high or low temperature, high or low pH, high salinity, high pressure, reduced water content and nutrients, and high radiation (Rothschild and Mancinelli 2001). Examples of typical extremophilic species are shown in Fig. 6.1. Intensive studies on ecology, physiology, and molecular biology of extremophiles provide valuable insight into the life processes at each level, as well as the potential for numerous industrial applications. These studies reveal the physiological differences of the inhabitants of the extreme niches in comparison with the inhabitants of the other niches, as well as the unusual molecular and regulatory mechanisms of life in them. Demonstration of the scientific interest to extremophilic microorganisms, their unusual properties, and the enormous possibilities for exploiting their biosynthetic capacity are 2-year International Congresses on Extremophiles last one performed in 2016 in Kyoto, Japan (11th), next one (12th) – in 2018 in Naples, Italy. In addition to the International Congresses on Extremophiles,

International Meetings on Thermophiles are also organized every 2 years, in 2017 being in Mpumalanga, South Africa.

Thermophiles have been the subject of an increased scientific interest for last several decades. They are a type of extremophiles, growing at relatively high temperatures, between 45 °C and 122 °C (Takai et al. 2008). Their research is related to clarifying a number of fundamental issues such as origin of life and molecular mechanisms of thermostability, revealing the vast potential of their enzymes for biotechnological use.

6.2 Origin of Thermophilic Microorganisms in Time and Space

The question of life origin is directly related with thermophiles. Despite the considerable interest of scientists in this issue, there is still no single opinion on whether life originated on Earth or was transferred from another planet. The second hypothesis for independent evolution of macromolecules on Earth is still officially accepted, and most scientists accept common ancestor of life, probiont, although the theory of panspermia (the emergence of a probiont on one planet and then transfer to others) more easily explains one-time appearance of life. Evidence of a single occurrence are as follows: the almost universal genetic code; the general principles of metabolic processes; and the existence of orthologous proteins in the three kingdoms – bacteria, archaea, and eukaryotes.

With regard to the temperature at which the probiont has developed, both "hot" and "cold" versions exist, arguing about its thermophilic or mesophilic nature. According to the hot theory shared by the majority of scientists, the last common ancestor of the three domains was a hyperthermophilic organism that only had the ability to survive under the harsh conditions of ancient Earth like no free oxygen, 10 times higher atmospheric pressure than today, absence of ozone layer. Akanuma et al. (2013) have estimated that the probiont probably was a (hyper)thermophile that lived at 75 °C or higher temperature. In favor of the hypothesis that thermophiles are direct descendants of the common ancestor is the fact that members of the lowest and shortest branches of life (genera *Thermotoga*, *Thermodesulfobacterium*, *Aquifex*, *Hydrogenobacter*) have the highest growth temperatures (Stetter 1994). Thermophilic genera are also positioned in the basal archaeal branch in the tree of life (*Pyrolobus*, *Pyrodictium*, *Methanopyrus*). These slowly evolving microorganisms, as could be assumed by their 16S rRNA, are the most primitive, still existing microorganisms and therefore closest to the common ancestor. These microorganisms are predominantly anaerobic chemotrophs, which need water, mineral elements, and heat for growth, i.e., from conditions typical for ancient Earth. Microbial fossils of 4.2 billion years detected in the rocks suggest morphological similarity with today's prokaryotes, including thermophiles (O'Neil et al. 2008). According to the less supported cold version, the thermophiles have adapted to survive at high temperatures as a result of a secondary adaptation developed after the origin of life (Islas et al. 2003). The supporters of this theory assume that on the basis of geochemical and fossil remains, as well as biochemical features of macromolecules, it can be concluded that the mesophilic microorganisms have existed long before thermophilic. Brochier and Philippe (2002) reanalyzed bacterial phylogeny based on rRNA gene and assumed that the mesophilic order *Planctomycetales* is closest to the life tree root.

6.3 Mechanisms for High-Temperature Life

It is still unclear the upper limit for growth of microorganisms and all specific factors that determine this limit, but it is commonly accepted that these are the factors that dictate the stability of the macromolecules in the cell. Development of omics technologies (genomics, transcriptomics, proteomics, and phenomics) promoted understanding of the mechanisms of temperature adaptation (Fondi et al. 2016).

The increased stability of the DNA helix at high temperatures is associated with the presence of a powerful reparative system, super-coiling, increased G + C content, presence of polyamines and histones, and increased salt content in the microenvironment (Daniel and Cowan 2000). High-frequency recombination allows recovery of a viable genome even from highly damaged cells by combining elements from several different chromosomes. Thermal resistance is determined by features in the organization of the genome, such as positive superspiralization and accelerated genetic repair of single-stranded DNA damage. Among protein genes in hyperthermophiles, the reverse gyrase associated with positive superspiralization of DNA is the only specific for hyperthermophiles and absent in mesophiles (Forterre 2002). In most cases, the adaptation to thermophilicity is associated with significant modifications of genomes involving the use of codons with increased G + C content which affects the stability of the double helix (Lynn et al. 2002). A characteristic feature of the thermophilic RNAs is that they are unusually short and have a tendency to reduce single-stranded regions, which is important for their stability. *Pyrococcus furiosus* cells growing at 100 °C contain three times more dimethyl and trimethyl guanosine than those growing at 70 °C (Imanaka 2011), suggesting that the modified nucleosides contribute to stabilizing the mRNA.

The increased G + C content in DNA in some thermophiles results in enhancing of amino acids encoded by codons rich in G + C like alanine, arginine, and proline. The amount of histidine, asparagine, glutamine, cysteine, methionine, and threonine, which are thermolabile amino acids, is also reduced. The theoretical models indeed show that the proportion of some amino acids and protein stability are proportional over a wide temperature range (Vendittis et al. 2008). A statistical analysis comparing the amino acid composition of mesophilic and thermophilic proteins revealed a tendency for glycine replacement with alanine, thereby reducing the conformational flexibility of the molecule (Taylor and Vaisman 2010). Two different evolutionary strategies probably have developed in prokaryotes, depending on the evolutionary history of the organism – whether the thermophilic character appeared early in the evolutionary process or it was obtained later during the adaptation to thermal niches. Proteins from organisms evolved at high temperatures are significantly more compact and more hydrophobic than their mesophilic analogs. Genomic analyses indicate smaller encoding genes are typical for thermophilic genomes (Islas et al. 2003). Organisms that have colonized hot habitats relatively late have evolved by the key mechanism of replacing amino acids with those that allow the formation of hydrogen bonds and electrostatic interactions within the subunit or between individual protein subunits, such a way influencing positively the interaction with chaperones (Jollivet et al. 2012). Thermophilic proteins are smaller and more rigid than mesophilic. The more rigid structure of the thermophilic enzymes is also responsible for their lower enzymatic activity than the mesophilic ones (Daniel and Cowan 2000). It is commonly accepted that the influence of temperature on enzyme activity constitutes of two effects – increased temperature gives increased activity and at the same time results in a loss of activity as a result of enzyme denaturation (the classical model). According to the model of Daniel and collaborators, the active site is more flexible than the whole protein, and loss of its activity occurs before denaturation (Bergquist et al. 2014). An important mechanism of thermostability is the presence of heat shock proteins (HSP) (Sharma et al. 2009). The higher the temperature of the habitat, the greater variety of constitutive HSP families is found.

Particular feature of thermophilic membranes is a large amount of saturated fatty acids with unbranched chains into lipid composition. Their high melting temperature makes the membrane more resistant to the temperature. In some cases, specific adaptations increase membrane thermostability like glycerol-ether lipids in thermophilic archaea and hyperthermophilic bacterium *Thermotoga maritima* (Koga and Morii 2005).

Specific low-molecular organisms, called solutes, which do not affect cell metabolism even in molar concentrations are found in thermophile cells under temperature or osmotic shock (Borges et al. 2010). They are sugars, polyols, amino acids, and derivatives thereof such as mannosylglycerate, mannosyl glyceramide, diaminoinositol phosphate, mannosyl diaminoinositol phosphate, etc. The mechanism of their action is connected with their participation in water retention by creating common structures with water molecules.

6.4 Ecology and Phylogeny of Thermophilic Microorganisms

Natural biotopes of thermophiles are volcanic and geothermal niches, mainly located on ground surface or underwater tectonic fractures, where earth plates collide or move away from one another. Deepwater wells are subdivided into so-called black and white smokers. The black smoker fluid is acidic, rich in metal ions and reduced substances (CH_4 and H_2S) and has temperature of 300–400 °C. White smoker fluid is relatively cooler (250–300 °C), white color is connected with a lack of metal compounds in it, and the fluid is relatively slower.

Microorganisms can also be found in underground hot habitats, mainly in the fields of hardened sediments and rock cracks. The oil fields resulting from the

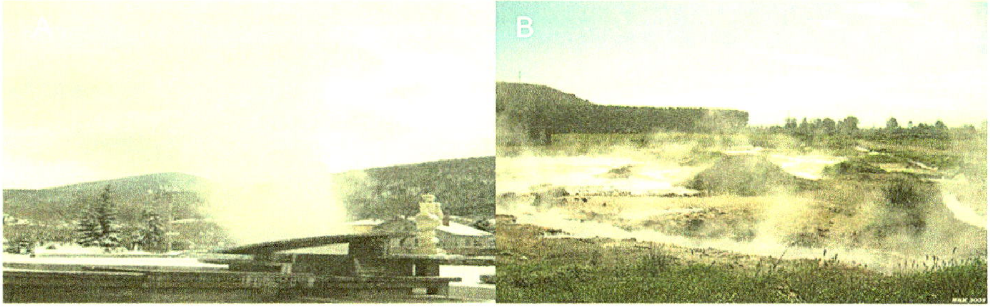

Fig. 6.2 Bulgarian hot springs: A, Geyser, Sapareva Banya has the highest temperature in continental Europe – 103 °C at the point of drilling; B, Rupi Basin

transformation of organic matter into hydrocarbons are located 1.5–4 km below the ground or seabed, at a pressure between 15 and 40 MPa and temperatures of 60–130 °C. Almost all underground habitats are characterized by anoxic conditions.

High-temperature continental springs are located in active volcanic zones 2–5 km below the surface where the magma chamber serves as a heat source. High-temperature springs are usually in the form of fumarole. The high-temperature springs are acidic as H_2S is oxidized to sulfuric acid on the surface, thus decreasing pH to 2–2.5. Solfatara (fumaroles that emit sulfur) fields are located in Iceland, Yellowstone National Park (USA), New Zealand, Kamchatka, Japan, and Italy. High-temperature springs include geysers whose water has a neutral or slightly alkaline pH. Globally, there are about 1000 geysers, half of which are in Yellowstone (USA). Low-temperature continental springs are located outside the active volcanic zones in geographically more stable regions in which extinguished or deep lava flows and dead magma chambers serve as sources of heat. The water temperature at a depth of 500–3000 m is up to 150 °C, temperature and flow of spring water are constant, and water in these springs is neutral to alkaline. Continental hot springs are highly distributed on the territory of Europe, like Italy and Portugal, and Asia, like Japan, China, and Armenia. Many continental hot springs are concentrated on the territory of Balkan Peninsula. There are about 140 natural habitats with water temperature 40–103 °C and a pH of 6.0–9.5 only on the territory of Bulgaria (Fig. 6.2). Diverse properties of water derived from their different geotectonic origin are a prerequisite for a considerable diversity of microorganisms living there.

Thermophilic bacteria refer to more than 50 genera distributed in 14 phyla representing almost half of the validly recognized phyla (Boone et al. 2001; Hreggvidsson and Kristjansson 2003; http://www.bacterio.net/-classifphyla.html). Phyla *Aquificae*, *Thermotogae*, *Thermodesulfobacteria*, *Thermomicrobia*, and *Deinococcus-Thermus* are presented only by thermophiles. Other phyla contain both thermophilic and mesophilic bacteria. Highest Topt have the bacteria from the families Thermotogaceae и Aquificae, above 85 °C, which together with a number of thermophilic archaea are referred as hyperthermophiles. The most common pH range for growth of thermophilic bacteria is between 5 and 9. Biodiversity in

Eurasian hot springs is still badly known, and molecular analyses revealed a presence of significant part of unknown groups comparable with those of well-studied Yellowstone and Iceland springs (Table 6.1). Study on diversity of thermophilic bacteria inhabiting a hot spring located in Rupi Basin (RB), Southwest Bulgaria, revealed a high phylogenetic richness in it (genotypic diversity is 0.37). One third of the sequence types showed less than 97% similarity to the closest neighbor and referred as new sequences. Four of them were distantly related to validly described bacteria (showed ≤90% similarity) suggesting new taxons on at least genus level (Tomova et al. 2010). Similarly, most of the sequences retrieved from Arzakan hot spring, Armenia, were most closely related to uncultivated microorganisms and shared less than 96% similarity with their closest matches in GenBank, indicating that this spring harbors a unique community of novel microbial species or genera (Panosyan and Birkeland 2014). As can be seen from Table 6.1, *Proteobacteria* and *Aquificales* are usually among the dominant phyla, and at the same type, different specific groups were observed in dependence of geographic location and spring temperature. Intensive investigations on bacterial diversity in environmental samples using 16S rRNA gene suggest that most the high taxa will be discovered by the end of the current decade (Yarza et al. 2014).

Culturable thermophilic archaea refer to two among the three archaeal phyla, Proteoarchaeota (predominantly Crenarchaeota) and Euryarchaeota. Crenarchaeota consists entirely of thermophiles, represented by a single recognized class (Thermoprotei) and one "*Candidatus* Nitrosocaldus yellowstonii" (de la Torre et al. 2008). Culture-independent molecular phylogenetic analyses revealed high Archaea diversity in Eurasian terrestrial hot springs. Investigation of the structure of the microbial community in a hot spring Varvara, Bulgaria (Fig. 6.3) showed high proportion of OTUs representing uncultivated archaeal phylogroups, the abundance of novel phylotype sequences (almost a quarter of the sequenced 16S rDNAs), the presence of high proportions of Crenarchaeota phylotypes unrelated to cultivated organisms (four OTUs formed a new archaeal subgroup without close described sequences or culturable relatives), and the presence of a sequence only distantly related to "Korarchaeota" phylum. "Korarchaeota" sequences showed 90% similarity to the closest neighbor forming unique branch in this phylum (Ivanova et al. 2011).

Analysis of near full-length archaeal rRNA genes retrieved from Arzakan and Jermuk hot springs, Armenia, showed that both springs are inhabited by a diversity of methanogens, including Methanomicrobiales, Methanosarcinales, relatives of *Methanomassiliicoccus luminyensis*, close relatives of the ammonia-oxidizing archaeon (AOA) "*Candidatus Nitrososphaera gargensis*," and the yet-uncultivated miscellaneous Crenarchaeotal group and Deep Hydrothermal Vent Crenarchaeota group 1 (Hedlund et al. 2013).

Table 6.1 Comparison of the bacterial diversity in different Eurasian hot springs, Obsidian Pool-Yellowstone, and Hengill's spring, Iceland

Hot spring	t°C	pH	Main ions in water (mg l^{-1})	No of clones	No of phylotypes presented	Dominant bacterial divisions[a]	Bacterial groups specific for the spring	New sequences[b]	Reference
Arzakan hot spring, Armenia	44	7.0–7.2	>20% $HCO3^-$, >20% is Na^+	23	12	*Proteobacteria* (52%) *Cyanobacteria* (35%)	*Cyanobacteria*	4	Panosyan and Birkeland (2014)
Rupi Basin, Bulgaria	79	8.6	Anions: Cl^-, SO_4^{2-}, HCO_3^-, HS^-; cations: Na, K, Ca.	120	32	*Hydrogenobacter* (29.2%) *β-Proteobacteria* (27.5%) *Thermus* (18.3%)	*Spirochaetes*	14	Tomova et al. (2010)
Agnikunda sediment, India	66–69	9.1–9.3	Fe^{3+} and Fe^{2+}	100	14	*γ-Proteobacteria* (40%) *Cyanobacteria* (32%)	No specific groups	12	Ghosh et al. (2003)
Hengill's spring, Iceland	67	6.7	Sulfide	171	14	*Aquificales* (68.4%) *Thermodesulfobacterium* group (18.1%)	No specific groups	9	Skirnisdottir et al. (2000)
Obsidian pool, Yellowstone (3 springs), USA	75–93	No data	Reduced iron, sulfide	312 (3 libraries)	58	*Aquificales* (26%) *δ-Proteobacteria* (12.8%)	*Acidobacterium* *Verrucomicrobium* *Dictyoglomus* group	12	Hugenholtz et al. (1998)
Hot spring, Japan (SM)	52–72	7.2–8.0	Sulfide	25	25	*β-Proteobacteria* (40%) *Aquificales* (40%)	*α-Proteobacteria*	No data	Yamamoto et al. (1998)

[a]divisions presented with >10%

[b]sequences related with less than 97% to the closest phylogenetic neighbors are accepted as new

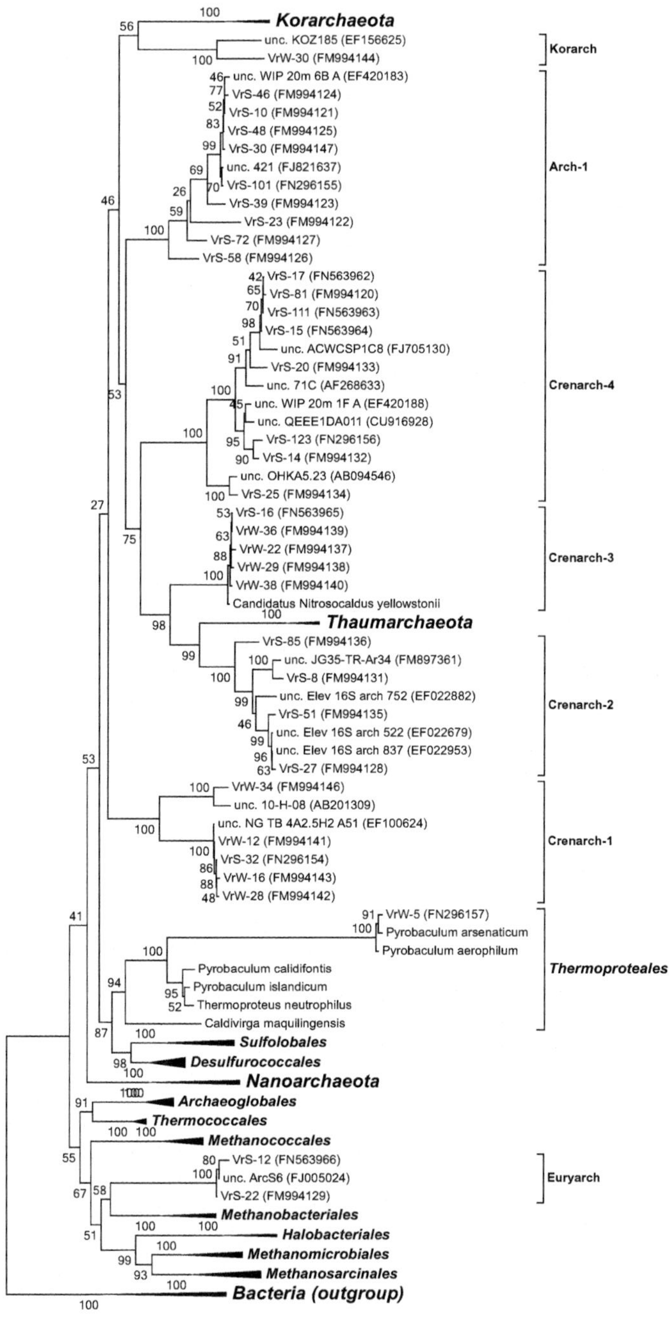

Fig. 6.3 Phylogeny of the archaeal phylotypes identified in Varvara hot spring. A neighbor-joining phylogenetic tree was constructed from: archaeal 16S rRNA sequences from Varvara hot spring; sequences of their closest relatives and sequences of culturable and characterized archaea (Ivanova et al. 2011)

6.5 Discovery of Novel Thermophilic Enzymes with Biotechnological Potential

Still most of the industrial enzymes used are derived from mesophilic organisms whose exploitation is limited by their low stability at high temperatures and salinity and extreme pH values. Nowadays about 20 enzymes are produced in large industrial scale (Li et al. 2012). Despite the fact that to date more than 3000 different biocatalysts have been identified (Kumar et al. 2011), mainly by culturable bacteria, they are still far from being able to respond to the ever-evolving new needs and requirements, among which their stability under industrial conditions is of particular importance. Reaching to the desired activity depends of the technical capability to explore the huge microbial diversity (Boehmwald et al. 2016). The largest enzyme companies like Diversa, Genencor International Inc., and Novozymes invest in search for new extremozymes (Gomes and Steiner 2004).

One of the main stimuli for intensive studies on thermophilic microorganisms is the potential biotechnological application of their enzymes. Thermophiles are characterized by high growth rates, which accelerate the fermentation process several times compared to those using mesophilic producers; the risk of microbial contamination in the biotechnological processes is significantly lower; diffusion rate and mass turnover, respectively, are higher; solubility of poorly soluble components in particular polymeric substrates is improved. The rigidity of thermophilic molecules determines their ability to be active and stable under the extreme conditions in which their producers live, allowing their use in industrial conditions at extreme temperatures, pH, salinity, organic solvents, and detergents. Particularly important in commercial preparations is longer process of "aging" the enzyme, which allows long storage at room temperature. Nowadays, different approaches are used to find the desired enzyme activities including direct screening in big microbial collections, metagenome screening, and shotgun sequencing, the last two based on analysis of the coding regions of the known enzymes (Boehmwald et al. 2016). The discovery of new microbial species and the sequencing of new genomes and metagenomes allow access to new enzymes with new application capabilities. As a result of the development of genome technology, new extremozymes offering new biocatalytic processes in biotechnology and pharmaceutical industries, green technologies, cosmetics, and food additives will be found in near future. Particular hopes are placed on archaeal thermophilic enzymes, whose potential is still practically unused.

6.5.1 Direct Screening in Microbial Collections

Still the best approach to discover new enzymes is growing of a microbial culture. Most of the papers are devoted to the biocatalytic activity of recombinant strains, despite of their disadvantages like genetic instability and interference with other biosynthetic pathways (Mühling et al. 2013). Direct screening for a desired activity has several advantages in comparison with other approaches. It confirms unambiguously its real existence and functional characteristics like substrate specificity, pH and temperature optima, and stability. Despite that classical screening is time and labor consuming, still many new enzymes or enzymes with improved properties have been reported exploring this approach in different biotechnologically important enzyme groups like glycosyl hydrolases, proteases, and lipases.

Thermostability of glycosyl hydrolases is particularly valuable due to the insolubility of carbohydrates at temperatures below 50–60 °C. Among glycosyl hydrolases probably the need in effective degradation of the huge plant biomass composed mainly by cellulose and hemicellulose is sharpest and still scarcely explored. The effective transformation of lignocellulosic biomass to cheap fermentable sugars could be used for different purposes like further conversion into ethanol or production of dietary fibers. Most studied cellulases and xylanases are optimally active at mesophilic temperatures (40–60 °C) and neutral or low acidic pH (Kulkarni 2003). Main disadvantages in use of thermophilic cellulases and xylanases are their low yield and low specific activity (Karmakar and Ray 2011). The lack of effective biocatalysts suitable for use in harsh industrial environments determines the growing scientific interest toward high-yield synthesis of enzymes that function at temperatures approaching and exceeding 80 °C that could overcome the need in the preliminary treatment of these substrates. Last reports concern predominantly anaerobic thermophiles as their higher growth temperature suggest more thermophilic enzymes like *Caldicellulosiruptor saccharolyticus* (VanFossen et al. 2011) and *C. bescii* (Su et al. 2012). The half-life of inactivation of xylanase from *Dictyoglomus thermophilum* in 80 °C and 500 MPa was over 30 h (Li et al. 2015).

So far, more than 40 proteases, active at temperatures between 50 and 95 °C, have been isolated and described, the most thermophilic produced by Archaea (Białkowska et al. 2016). Although the most intensive search for thermostable proteases was two to three decades ago, still interesting enzymes have been described in the last several years. Several workers have reported protease activity from thermophilic *Bacillus* species, subtilisin-like protease from *Bacillus* sp. MLA64 having the highest temperature optimum of 95 °C and $t_{1/2}$ 25 min at 110 °C (Lagzian and Asoodeh 2012), and alkaline serine protease from *Geobacillus stearothermophilus* B-1172 has $t_{1/2}$ 60 min at 100 °C (Iqbal et al. 2015). The enzyme from *Coprothermobacter proteolyticus* expressed in *E. coli* has a temperature optimum of 85 °C (Toplak et al. 2013).

The natural substrates of the lipases are practically insoluble in water at room temperature, which determines the interest in thermostable enzymes. Although prevalent mesophilic lipases are still available in commercially available preparations, the instability of well-characterized mesophilic enzymes in extreme conditions provokes the industrial demand for thermostable lipases (Tirawongsaroj et al. 2008). Over the years unrelenting interest toward thermostable lipases is observed especially to the group of thermophilic bacilli. Lipolytic enzymes from recently isolated thermophilic aerobes *Geobacillus* sp. EPT9, *Thermus thermophilus* HB27, and *Acidicaldus* USBA-GBX-499 have a temperature optimum 55–65 °C (Fuciños et al. 2014; Lopez- Lopez et al. 2014; Zhu et al. 2015). Usually thermophilic anaerobic lipases suggest higher temperature optimum of 70–78 °C (Cai et al. 2011; Tao et al. 2013); however the lipases from *Geobacillus stearothermophilus* MC 7 and

Thermus scotoductus SA-01 showed a temperature optimum of 80 °C (Kambourova et al. 2003; du Plessis et al. 2010).

Although several novel enzymes were described in last years by direct functional discovery of the enzyme, the practical disadvantages of this approach determined the development of high-throughput screening (HTS) by which simultaneous detection of the desired activity in thousand samples became possible. However as many factors could influence enzyme activity in the crude cell extract, this approach has a limited application (Boehmwald et al. 2016). This approach was used for screening of coenzyme preference change of thermophilic 6-phosphogluconate dehydrogenase (Huang et al. 2016) and for screening and characterization of xylose-utilizing, ethanol-tolerant thermophilic bacteria for bioethanol production (Qi et al. 2011).

6.5.2 Metagenomic Approach for Discovering of Novel Enzymes

Metagenomics as a cultivation-independent approach provides the opportunity to study and explore the uncultivated fraction of the microorganisms in nature. The method is based on a direct extraction of total DNA from environmental habitats, subsequent sequencing, assembly, and analyzing using computational tools (Sharon and Banfield 2013) or digesting and cloning into suitable vectors, expression, and functional screening. Metagenome is accepted as a huge reservoir of taxonomic and functional genes coming from a vast number of genomes in an environmental sample and therefore is seen as a feasible possibility to introduce more and diverse enzymes with better performance meeting the global demand for new catalysts. The accessibility and wide usage of metagenomic sequencing in recent years resulted in generating of lots of sequencing reads and available data and contribute its developing as a powerful tool for screening and discovery of novel enzymes (Lämmle et al. 2007). Enzyme discovery could be based on searching for enzyme homology with known enzymes by sequence comparison; amplification of genes by using of specific primers designed according to conserved regions in the gene; and metagenomic libraries built and screened using DNA cloned directly from environmental metagenomes and functional metagenomics (Lopez-Lopez et al. 2014; Lam et al. 2015). Genomic sequencing technology and functional genomics have been successful in amassing large amount of genomic data on extremophiles, including information about genes at the DNA sequence level (Lee et al. 2008). The rapid development of sequencing techniques after the introduction of next-generation DNA sequencing technologies (NGS) has emerged and substantially influenced the progress of the genomic era. Nowadays NGS allows large-scale analysis of microbial communities including comparative community metagenomics, metatranscriptomics, and metaproteomics (Lämmle et al. 2007).

There are lots of studies addressed to understanding the taxonomical composition of different high-temperature environments. In parallel the functional- and sequence-based metagenomic approaches (Fig. 6.4) along with metatranscriptomics and metaproteomics provide information for community functional activity (DeCastro et al. 2016).

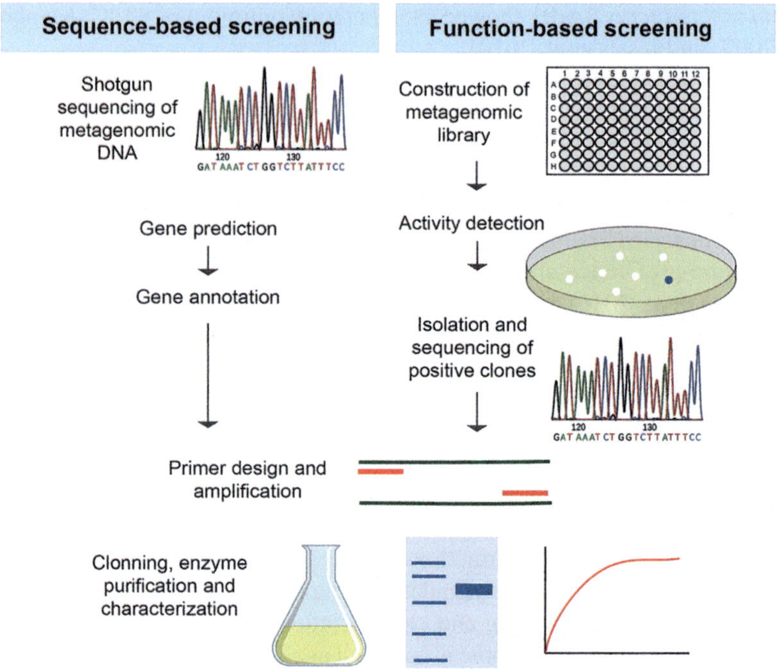

Fig. 6.4 Strategies used for screening metagenomes in search of new thermozymes. (According to DeCastro et al. 2016)

Most hope in discovery of truly new enzymes is assigned on the searching for open reading framework (ORF) in metagenomes from hot environments; however in this case, the information for the possible enzyme activity, specificity, and properties could be ambiguous (Chan et al. 2015). If ORF of the novel enzymes differs significantly from known enzymes, it could be difficult to reveal their real biotechnological impact. According to Galperin and Koonin (2010), about 30–40% of genes in newly sequenced genomes remained unknown. Proteomic studies on extremophiles can provide information about the number of proteins induced under specific conditions (Burg et al. 2011). Therefore, improving screening methods using proteomic technique is essential for identifying novel proteins in extremophiles. Additionally to functional screening, the novel enzymes could be searched by shotgun sequencing and the search for the target enzyme in metagenomic libraries. In this case the novel enzyme searching is based on the sequence homology with known enzymes.

Several hot spring metagenomic studies based on sequence similarity search reported for significant amount of sequences with unknown function assumed as potential novel bio-products (Mangrola et al. 2015; Mehetre et al. 2016). Solid and constantly developing new and diverse bioinformational tools supported a correct annotation against number of databases as the KEGG orthology database (Kanehisa et al. 2016), SEED annotation system (Overbeek et al. 2014), and Pfam database (Finn et al. 2016). A disadvantage of sequence-based enzyme screening is the impossibility to discover completely new enzymes as the methods rely on sequence homology to already described enzymes. Directed evolution and protein engineering are often further required to enhance the desired enzyme performance. However the insertion of synthetic genes as an alternative to amplification of native sequences provides several advantages as codon optimization (Te'o et al. 2000), vector engineering, gene design, and ORF engineering (Gustafsson et al. 2012). The tools of modern synthetic biology are increasingly applied in the research and industrial scale.

One functional metagenomic approach for novel enzyme screening is direct cloning of DNA extracted from the environment including several steps: metagenomic DNA extraction, purification and cloning into expression vector, transformation into suitable host, and expression and further observation for functional activities (Lämmle et al. 2007). Construction of cosmid- or fosmid-based metagenomic libraries containing high-molecular DNA inserts is a laborious, time-consuming, and expensive process, despite the fact that several biotechnologically perspective enzymes were identified by functional metagenomics (Carvalho 2017). The method suffers from several disadvantages as codon usage bias and improper protein folding, not or poorly supported by the host machinery posttranslational modifications (Lämmle et al. 2007). The gap between the expression level and produced clones from metagenomic libraries imposes the implementation of different strategies to overcome the disadvantages. Additionally, the problem with protein expression level could arise if an appropriate host and expression system have not been developed. It is known that in *E. coli*-based expression, systems up to 40% of the genes are successfully expressed and active (Gabor et al. 2004). For thermophilic enzyme expression, other hosts like bacilli and *Sulfolobus* are often explored. Bidirectional plasmid vector was used for direct expression cloning and screening for different enzyme activities from compost (Lämmle et al. 2007). The functional screening detected the following activities: protease, phosphatase, and lipolytic activities. The provided data about the active enzyme frequency among the clones proved the usage of duo-orientated vector as advantageous for direct metagenomic gene cloning and expression. Providing enzyme in sufficient amounts at reasonable costs depends also of the clone number that should be tested and the available equipment. Metagenomics gives insides on the theoretical functional capacity of the microbial community, but further understanding of proteins activities and their biological role is essential. Naturally proteomic studies have developed as a new functional genomics tool for novel extremophilic enzymes discovery (Yun et al. 2016).

The processes of transcription and translation of the heterologous sequences into functional proteins are still poorly understood. The number of factors is recognized to affect these processes as codon usage, toxicity of the expressed proteins and foreign DNA, posttranslational modification of the protein and secretion, transcriptional signals (promoters) recognition, mRNA stability, ribosomal binding, and translation initiation. These limitations can be overcome by integrated approach comprising all varying factors (Fig. 6.5).

There are lots of published reports describing different enzymes derived from hot spring metagenomic libraries. The carbon, sulfur, and nitrogen metabolic diversity

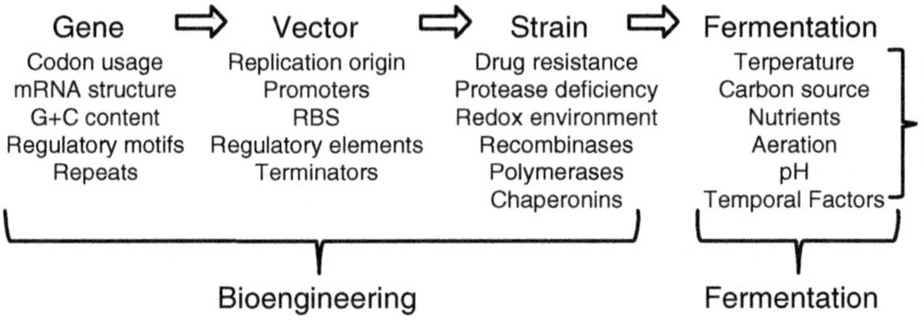

Fig. 6.5 Recombinant enzyme expression is influenced by numerous factors. (According to Gustafsson et al. 2012)

of the community from Malaysian hot spring were revealed by shot gun metagenome sequencing (Chan et al. 2015). The cloned and overexpressed thermostable Fe-superoxide dismutase from hot spring metagenomic library in *E. coli* had broad pH range from 4.0 to 11.0 and activity at 80 °C and retained 50% activity after heat treatment at 95 °C for 2 h (He et al. 2007).

Bidirectional plasmid vector was used for direct expression cloning and screening for different enzyme activities from compost (Lämmle et al. 2007). The functional screening detected the following activities: protease, phosphatase, and lipolytic activities. The provided data about the active enzyme frequency among the clones proved the usage of duo-orientated vector as advantageous for direct metagenomic gene cloning and expression.

Recently, novel genes coding lipolytic and proteolytic enzymes were identified by mining a thermal spring volcanic metagenome from Kamchatka peninsula (Wemheuer et al. 2013). Three new lipolytic and one proteolytic enzymes were detected in small-insert metagenomic libraries, successfully cloned, overexpressed in *E. coli,* and characterized by multiple displacement amplification for library construction. The described lipolytic enzymes showed maximal activities at 85 °C, 90 °C, and 65 °C, respectively, with no significant effect of EDTA, KCl, or NaCl on enzyme activity. Cloned highly thermostable xylanase from *Thermotoga thermarum* with optimal activity at 95 °C retained almost 100% activity after incubation at 85 °C for 2 h at pH 7.0 (Shi et al. 2013).

Another approach for introduction of improved enzyme features as thermal stability is protein rational design engineering including disulfide bond insertion, optimization of protein surface charge, and the free energy of unfolding (Yang et al. 2015). Several bioinformatic tools are used to compare sequences of mesophilic and thermophilic homologous enzymes in order to understand the genetic bases of high-temperature stability. Understanding the adaptive evolutionary response under elevating temperatures is an important prerequisite for thermal stability enhance and enzyme rational design achievement (Sammond et al. 2016). An example for improved expression of proofreading DNA polymerase (taqIIRM) from *Thermus aquaticus* using a synthetic gene "one amino acid–one codon" method was described to be more applicable for industrial production compared to the recombinant wild gene (Zylicz-Stachula et al. 2014).

References

Akanuma S, Nakajima Y, Yokobori S, Kimura M, Nemoto N, Mase T, Miyazono K, Tanokura M, Yamagishi A (2013) Experimental evidence for the thermophilicity of ancestral life. Proc Natl Acad Sci U S A 110(27):11067–11072

Amann R (2000) Who is out there? Microbial aspects of diversity. Syst Appl Microbiol 23:1–8

Bergquist PL, Morgan HW, Saul D (2014) Selected enzymes from extreme thermophiles with applications in biotechnology. Curr Biotechnol 3(1):45–59

Białkowska A, Gromek E, Florczak T, Krysiak J, Szulczewska K, Turkiewicz M (2016) Extremophilic proteases: developments of their special functions, potential resources and biotechnological applications. In: Rampelotto PH (ed) Biotechnology of extremophiles. Springer, Cham, pp 399–444

Boehmwald F, Muñoz P, Flores P, Blamey JM (2016) Functional screening for thc discovery of new extremophilic enzymes. In: Rampelotto PH (ed) Biotechnology of extremophiles. Springer, Cham, pp 321–350

Boone DR, Castenholz RW, Garrity GM (2001) The Archaea and the deeply branching and phototrophic Bacteria. In: Bergey's manual of systematic bacteriology, vol 1, 2nd edn. Springer, New York, pp 169–620

Borges N, Matsumi R, Imanaka T, Atomi H, Santos H (2010) *Thermococcus kodakaraensis* mutants deficient in di-myo-inositol phosphate use aspartate to cope with heat stress. J Bacteriol 192(1):191–197

Brochier C, Philippe H (2002) Phylogeny: a non-hyperthermophilic ancestor for Bacteria. Nature 417:244

Burg D, Ng C, Ting L, Cavicchioli R (2011) Proteomics of extremophiles. Environ Microbiol 13(8):1934–1955

Cai J, Xie Y, Song BK, Wang Y, Zhang Z, Feng Y (2011) *Fervidobacterium changbaicum* Lip1: identification, cloning, and characterization of the thermophilic lipase as a new member of bacterial lipase family V. Appl Microbiol Biotechnol 89(5):1463–1473

Carvalho CC (2017) Whole cell biocatalysts: essential workers from nature to the industry. Microb Biotechnol 10(2):250–263

Chan CS, Chan KG, Tay YL, Chua YH, Goh KM (2015) Diversity of thermophiles in a Malaysian hot spring determined using 16S rRNA and shotgun metagenome sequencing. Front Microbiol 6:Article 177

Daniel RM, Cowan DA (2000) Biomolecular stability and life at high temperatures. Cell Mol Life Sci 57(2):250–264

De la Torre JR, Walker CB, Ingalls AE, Könneke M, Stahll DA (2008) Cultivation of a thermophilic ammonia oxidizing archaeon synthesizing crenarchaeol. Environ Microbiol 10(3):810–818

DeCastro ME, Rodríguez-Belmonte E, González-Siso MI (2016) Metagenomics of thermophiles with a focus on discovery of novel thermozymes. Front Microbiol 7:Article 1521

du Plessis EM, Berger E, Stark T, Louw ME, Visser D (2010) Characterization of a novel thermostable esterase from *Thermus scotoductus* SA-01: evidence of a new family of lipolytic esterases. Curr Microbiol 60(4):248–253

Finn RD, Coggill P, Eberhardt RY, Eddy SR, Mistry J, Mitchell AL, Potter SC, Punta M, Qureshi M, Sangrador-Vegas A, Salazar GA (2016) The Pfam protein families database: towards a more sustainable future. Nucleic Acids Res 44(D1):D279–D285

Fondi M, Bosi E, Giudice AL, Fani R (2016) A systems biology view on bacterial response to temperature shift. In: Rampelotto PH (ed) Biotechnology of extremophiles. Springer, Cham, pp 597–618

Forterre PA (2002) Hot story from comparative genomics: reverse gyrase is the only hyperthermophile-specific protein. Trends Genet 18(5):236–237

Fuciños P, Atanes E, Lopez-lopez O, Solaroli M, Cerdan ME, Gonzalez-Siso MI, Pastrana L, Rua ML (2014) Cloning, expression, purification and characterization of an oligomeric His-tagged thermophilic esterase from *Thermus thermophilus* HB27. Process Biochem 49(6):927–935

Gabor EM, Alkema WB, Janssen DB (2004) Quantifying the accessibility of the metagenome by random expression cloning techniques. Appl Environ Microbiol 6(9):879–886

Galperin MY, Koonin EV (2010) From complete genome sequence to "complete" understanding? Trends Biotechnol 28(8):398–406

Ghosh D, Bal B, Kashyap VK, Pal S (2003) Molecular phylogenetic exploration of bacterial diversity in a Bakreshwar (India) hot spring and culture of Shewanella-related thermophiles. Appl Environ Microbiol 69(7):4332–4336

Gomes J, Steiner W (2004) The biocatalytic potential of extremophiles and extremozymes. Food Technol Biotechnol 42(4):223–235

Gustafsson C, Minshull J, Govindarajan S, Ness J, Villalobos A, Welch M (2012) Engineering genes for predictable protein expression. Protein Expr Purif 83(1):37–46

He YZ, Fan KQ, Jia CJ, Wang ZJ, Pan WB, Huang L, Yang KQ, Dong ZY (2007) Characterization of a hyperthermostable Fe-superoxide dismutase from hot spring. Appl Microbiol Biotechnol 75(2):367–376

Hedlund BP, Dodsworth JA, Cole JK, Panosyan HH (2013) An integrated study reveals diverse methanogens, Thaumarchaeota, and yet-uncultivated archaeal lineages in Armenian hot springs. Antonie Leeuwenhoek 104(1):71–82

Hreggvidsson GO, Kristjansson JK (2003) Thermophily. In: Gerday C, Glansdorff N (eds) Extremophiles – encyclopedia of life support systems (EOLSS). EOLSS Publishers, Oxford

Huang R, Chen H, Zhong C, Kim JE, Zhang YHP (2016) High-throughput screening of coenzyme preference change of thermophilic 6-phosphogluconate dehydrogenase from $NADP^+$ to NAD^+. Sci Rep 6:32644

Hugenholtz P, Pitulle C, Hershberger KL, Pace NR (1998) Novel division level bacterial diversity in a Yellowstone hot spring. J Bacteriol 180(2):366–376

Imanaka T (2011) Molecular bases of thermophily in hyperthermophiles. Proc Jpn Acad Ser B Phys Biol Sci 87(9):587–602

Iqbal I, Aftab MN, Afzal M, Ur-Rehman A, Aftab S, Zafat A, Ud-Din Z, Khuharo AR, Iqbal J, Ul-Haq I (2015) Purification and characterization of cloned alkaline protease gene of *Geobacillus stearothermophilus*. J Basic Microbiol 55(2):160–171

Islas S, Velasco AM, Becerra A, Delaye L, Lazcano A (2003) Hyperthermophily and the origin and earliest evolution of life. Int Microbiol 6(2):87–94

Ivanova I, Atanassov I, Lyutskanova D, Stoilova-Disheva M, Dimitrova D, Tomova I, Derekova A, Radeva G, Buchvarova V, Kambourova M (2011) High archaea diversity in Varvara hot spring, Bulgaria. J Basic Microbiol 51(2):163–172

Jollivet D, Mary J, Gagniere N, Tanguy A, Fontanillas E, Boutet I, Hourdez S, Segurens B, Weissenbach J, Poch O, Lecompte O (2012) Proteome adaptation to high temperatures in the ectothermic hydrothermal vent Pompeii worm. PLoS One 7(2):e31150

Kambourova M, Kirilova N, Mandeva R, Derekova A (2003) Purification and properties of thermostable lipase from a thermophilic *Bacillus stearothermophilus* MC 7. J Mol Catal B Enzym 22(5–6):307–313

Kanehisa M, Sato Y, Kawashima M, Furumichi M, Tanabe M (2016) KEGG as a reference resource for gene and protein annotation. Nucleic Acids Res 44(D1):D457–D462

Karmakar M, Ray RR (2011) Current trends in research and application of microbial cellulases. Res J Microbiol 6(1):41–53

Koga Y, Morii H (2005) Recent advances in structural research on ether lipids from archaea including comparative and physiological aspects. Biosci Biotechnol Biochem 69(11):2019–2034

Kulkarni SV (2003) Role of actinomycetes in environment. In: Kumar A et al (eds) Environment, pollution and management. A.R.H. Publishing Corporation, New Delhi, pp 531–542

Kumar L, Awasthi G, Singh B (2011) Extremophiles: a novel source of industrially important enzymes. Biotechnology 10(2):121–135

Lagzian M, Asoodeh A (2012) An extremely thermotolerant, alkaliphilic subtilisin-like protease from hyperthermophilic *Bacillus* sp. MLA64. Int J Biol Macromol 51(5):960–967

Lam KN, Cheng J, Engel K, Neufeld JD, Charles TC (2015) Current and future resources for functional metagenomics. Front Microbiol 6:Article 1196

Lämmle K, Zipper H, Breuer M, Hauer B, Buta C, Brunner H, Rupp S (2007) Identification of novel enzymes with different hydrolytic activities by metagenome expression cloning. J Biotechnol 127(4):575–592

Lee HS, Kang SG, Bae SS, Lim JK, Cho Y, Kim YJ, Jeon JH, Cha SS, Kwon KK, Kim HT, Park CJ, Lee HW, Kim SI, Chun J, Colwell RR, Kim SJ, Lee JH (2008) The complete genome sequence of *Thermococcus onnurineus* NA1 reveals a mixed heterotrophic and carboxydotrophic metabolism. J Bacteriol 190(22):7491–7499

Li S, Yang X, Yang S, Zhu M, Wang X (2012) Technology prospecting on enzymes: applications, marketing and engineering. Comput Struct Biotechnol J 2:e201209017

Li H, Voutilainen S, Ojamo H, Turunen O (2015) Stability and activity of *Dictyoglomus thermophilum* GH11 xylanase and its disulphide mutant at high pressure and temperature. Enzym Microb Technol 70:66–71

López-López O, Cerdán ME, González Siso MI (2014) New extremophilic lipases and esterases from metagenomics. Curr Protein Peptide Sci 15(5):445–455

Lynn DJ, Singer GAC, Hickey DA (2002) Synonymous codon usage is subject to selection in thermophilic bacteria. Nucleic Acids Res 30(19):4272–4277

MacElroy RD (1974) Some comments on the evolution of extremophiles. Biosystems 6:74–75

Mangrola A, Dudhagara P, Koringa P, Joshi CG, Parmar M, Patel R (2015) Deciphering the microbiota of Tuwa hot spring, India using shotgun metagenomic sequencing approach. Genomics Data 4:153–155

Mehetre GT, Paranjpe AS, Dastager SG, Dharne MS (2016) Complete metagenome sequencing based bacterial diversity and functional insights from basaltic hot spring of Unkeshwar, Maharashtra, India. Genomics Data 7:140–143

Mora C, Tittensor DP, Adl S, Simpson AGB, Worm B (2011) How many species are there on Earth and in the ocean? PLoS Biol 9(8):e1001127

Mühling M, Joint I, Willetts AJ (2013) The biodiscovery potential of marine bacteria: an investigation of phylogeny and function. Microb Biotechnol 6(4):361–370

O'Neil J, Carlson RW, Francis E, Stevenson RK (2008) Neodymium-142 evidence for Hadean mafic crust. Science 321(5897):1828–1831

Oren A (2009) Systematics of Archaea and Bacteria. In: Minelli A, Contrafatto G (eds) Biological science: fundamentals and systematics, vol 2. Eolss Publishers Co Ltd., Oxford, pp 244–269

Overbeek R, Olson R, Pusch GD, Olsen GJ, Davis JJ, Disz T, Edwards RA, Gerdes S, Parrello B, Shukla M, Vonstein V (2014) The SEED and the rapid annotation of microbial genomes using subsystems technology (RAST). Nucleic Acids Res 42(D1):D206–D214

Panosyan H, Birkeland NK (2014) Microbial diversity in an Armenian geothermal spring assessed by molecular and culture-based methods. J Basic Microbiol 54(11):1240–1250

Qi X, Zhang Y, Tu R, Lin Y, Li X, Wang Q (2011) High-throughput screening and characterization of xylose-utilizing, ethanol-tolerant thermophilic bacteria for bioethanol production. J Appl Microbiol 110(6):1584–1591

Rothschild LJ, Mancinelli RL (2001) Life in extreme environments. Nature 409:1092–1101

Sammond DW, Kastelowitz N, Michael E, Himmel ME, Yin H, Crowley MF, Yannick J, Bomble YJ (2016) Comparing residue clusters from thermophilic and mesophilic enzymes reveals adaptive mechanisms. PLoS One 11(1):e0145848

Satyanarayana T, Raghukumar C, Shivaji S (2005) Extremophilic microbes: diversity and perspectives. Curr Sci 89(1):78–90

Sharma SK, Christen P, Goloubinoff P (2009) Disaggregating chaperones: an unfolding story. Curr Protein Pept Sci 10(5):432–446

Sharon I, Banfield J (2013) Genomes from metagenomics. Science 342(6162):1057–1058

Shi H, Zhang Y, Li X, Huang Y, Wang L, Wang Y, Ding H, Wang F (2013) A novel highly thermostable xylanase stimulated by Ca^{2+} from *Thermotoga thermarum*: cloning, expression and characterization. Biotechnol Biofuels 6(1):26

Skirnisdottir S, Hreggvidsson GO, Hjorleifsdottir S, Marteinsson VT, Petursdottir SK, Holst O, Kristjansson JK (2000) Influence of sulfide and temperature on species composition and community structure of hot spring microbial mats. Appl Environ Microbiol 66(7):2835–2841

Stetter KO (1994) The lesson of Archaebacteria. In: Bengtson S (ed) Early life on Earth, Nobel symposium no. 84. Columbia University Press, New York, pp 143–151

Stetter KO (2001) Hyperthermophilic microorganisms. In: Horneck G, Baumstark-Khan C (eds) Astrobiology – the quest for the conditions of life. Springer, Berlin, pp 169–184

Stewart EJ (2012) Growing unculturable bacteria. J Bacteriol 194(16):4151–4160

Su X, Mackie RI, Cann IK (2012) Biochemical and mutational analyses of a multidomain cellulase/mannanase from *Caldicellulosiruptor bescii*. Appl Environ Microbiol 78(7):2230–2240

Takai K, Nakamura K, Toki T, Tsunogai U, Miyazaki M, Miyazaki J, Hirayama H, Nakagawa S, Nunoura T, Horikoshi K (2008) Cell proliferation at 122°C and isotopically heavy CH4 production by a hyperthermophilic methanogen under high-pressure cultivation. Proc Natl Acad Sci U S A 105(31):10949–10954

Tao W, Shengxue F, Duobin M, Xuan Y, Congcong D, Xihua W (2013) Characterization of a new thermophilic and acid tolerant esterase from *Thermotoga maritima* capable of hydrolytic resolution of racemic ketoprofen ethyl ester. J Mol Catal B Enzym 85–86:23–30

Taylor TJ, Vaisman II (2010) Discrimination of thermophilic and mesophilic proteins. BMC Struct Biol 10(Suppl 1):S5

Te'o VS, Cziferszky AE, Bergquist PL, Nevalainen KH (2000) Codon optimization of xylanase gene xynB from the thermophilic bacterium *Dictyoglomus thermophilum* for expression in the filamentous fungus *Trichoderma reesei*. FEMS Microbiol Lett 190(1):13–19

Tirawongsaroj P, Sriprang R, Harnpicharnchai P, Thongarama T, Champreda V, Tanapongpipat S, Pootanakit K, Eurwilaichitr L (2008) Novel thermophilic and thermostable lipolytic enzymes from a Thailand hot spring metagenomic library. J Biotechnol 133(1):42–49

Tomova I, Stoilova-Disheva M, Lyutskanova D, Pascual J, Petrov P, Kambourova M (2010) Phylogenetic analysis of the bacterial community in a geothermal spring, Rupi basin, Bulgaria. World J Microbiol Biotechnol 26(11):2019–2028

Toplak A, Wu B, Fusetti F, Quaedfl ieg PJ, Janssen DB (2013) Proteolysin, a novel highly thermostable and cosolvent-compatible protease from the thermophilic bacterium Coprothermobacter proteolyticus. Appl Environ Microbiol 79(18):5625–5632

VanFossen AL, Ozdemir I, Zelin SL, Kelly RM (2011) Glycoside hydrolase inventory drives plant polysaccharide deconstruction by the extremely thermophilic bacterium *Caldicellulosiruptor saccharolyticus*. Biotechnol Bioeng 108(7):1559–1569

Vendittis E, Castellano I, Cotugno R, Ruocco MR, Raimo G, Masullo M (2008) Adaptation of model proteins from cold to hot environments involves continuous and small adjustments of average parameters related to amino acid composition. J Theor Biol 250(1):156–171

Vibha B, Neelam G (2012) Importance of exploration of microbial biodiversity. Int Res J Biol Sci 1(3):78–83

Wemheuer B, Taube R, Akyol A, Wemheuer F, Daniel R (2013) Microbial diversity and biochemical potential encoded by thermal spring metagenomes derived from the Kamchatka peninsula. Archaea 2013:Article ID 136714

Yamamoto H, Hiraishi A, Kato K, Chiura HX, Maki Y, Shimizu A (1998) Phylogenetic evidence for the existence of novel thermophilic bacteria in hot spring sulfur-turf microbial mats in Japan. Appl Environ Microbiol 64(5):1680–1687

Yang H, Liu L, Li J, Chen J, Du G (2015) Rational design to improve protein thermostability: recent advances and prospects. Chem Bio Eng Rev 2(2):87–94

Yarza P, Yilmaz P, Pruesse E, Glöckner FO, Ludwig W, Schleifer KH, Whitman WB, Euzéby J, Amann R, Rosselló-Móra R (2014) Uniting the classification of cultured and uncultured bacteria and archaea using 16S rRNA gene sequences. Nat Rev Microbiol 12(9):635–645

Yun SH, Choi CW, Lee SY, Park EC, Kim SI (2016) A proteomics approach for the identification of novel proteins in extremophiles. In: Rampelotto PH (ed) Biotechnology of extremophiles. Springer, Cham, pp 303–319

Zhu Y, Li H, Ni H, Xiao A, Li L, Cai H (2015) Molecular cloning and characterization of a thermostable lipase from deep-sea thermophile *Geobacillus* sp. EPT9. World J Microbiol Biotechnol 31(2):295–306

Zylicz-Stachula A, Zolnierkiewicz O, Sliwinska K, Jezewska-Frackowiak J, Skowron P (2014) Modified 'one amino acid-one codon' engineering of high GC content TaqII-coding gene from thermophilic *Thermus aquaticus* results in radical expression increase. Microb Cell Factories 13:7

Thermophilic Chemolithotrophic Bacteria in Mining Sites

7

Narine S. Vardanyan and Arevik K. Vardanyan

Abstract

Microorganisms which inhabit extremely acidic environments are increasingly attracting the attention of researchers because of their peculiar physiology. These extremophiles play a huge role in geochemical processes in mining sites and environmental pollution by heavy metals. They also have important applications in biotechnology of metals. The study of biodiversity and relevant biogeochemical processes is of great interest for improving metal leaching technologies and developing countermeasures for the formation of acid mine drainage (AMD). Due to the insufficient data of ecology of chemolithotrophic bacteria inhabit natural and technogenic biotopes of sulfide ores in Armenia, studies of biodiversity and dissemination of these bacteria in the copper, copper-molybdenum, gold-bearing, and polymetallic ore deposits of Armenia were performed. Using enrichment media and isolation techniques, new and original strains of sulfur- and/or iron-oxidizing bacteria (SIOB) were isolated and studied. Based on physiological and biochemical peculiarities as well as molecular biological studies, the isolated strains were identify as *Acidithiobacillus ferrooxidans*, *Leptospirillum ferrooxidans*, *L. ferriphilum*, and *Sulfobacillus thermosulfidooxidans* subsp. asporogenes. In this paper we have made an attempt to summarize the data obtained concerning dissemination of moderate thermophilic and thermotolerant SIOB and their biological properties as well as abilities to oxidize the most abundant minerals, pyrite and chalcopyrite. Their role in geochemical processes occurring in mining sites as well as bioleaching of the most abundant minerals pyrite, chalcopyrite, and refractory gold-bearing ores were evaluated.

N. S. Vardanyan (✉) · A. K. Vardanyan
Laboratory of Geomicrobiology of Scientific and Production Center "Armbiotechnology" of National Academy of Sciences of Armenia, Yerevan, Armenia
e-mail: nvard@sci.am

D. Egamberdieva et al. (eds.), *Extremophiles in Eurasian Ecosystems: Ecology, Diversity, and Applications*, Microorganisms for Sustainability 8,
https://doi.org/10.1007/978-981-13-0329-6_7

Keywords
Extremely acidic environments · Moderate thermophiles · Iron- and sulfur-oxidizing bacteria · Biooxidation · Bioleaching · Metal recovery

7.1 Introduction

Microorganisms which inhabit extremely acidic environments (pH <3) are increasingly attracting the attention of researchers because of their peculiar physiology. These extremophiles play a huge role in geochemical processes in mining sites and environmental pollution by heavy metals. They also have important applications in biotechnology of metals. The study of biodiversity and relevant biogeochemical processes is of great interest for improving metal leaching technologies and developing countermeasures for the formation of acid mine drainage (AMD).

Due to the insufficient data of ecology of chemolithotrophic bacteria inhabit natural and technogenic biotopes of sulfide ores in Armenia, studies of biodiversity and dissemination of these bacteria in the copper, copper-molybdenum, gold-bearing, and polymetallic ore deposits of Armenia were performed.

By using enrichment media and isolation techniques, it has been shown that natural and technogenic sulfide ores are mainly represented by the following genera of iron-oxidizing bacteria: *Acidithiobacillus*, *Leptospirillum*, and *Sulfobacillus*.

In this paper we have made an attempt to summarize the data obtained concerning dissemination of moderate thermophilic and thermotolerant sulfur- and/or iron-oxidizing bacteria and their biological properties as well as abilities to oxidize sulfide ores and minerals.

7.2 Extremely Acidic Environments

Extremely acidic environments may be formed by naturally occurring geochemical processes in sulfide ores. However, the majority of extremely acidic sites worldwide is connected with human activity, particularly metal mining. Many commercially important metals occur in forms of sulfides (copper, zinc, etc.). Mining increases the surface area of sulfide ores exposed to air and water and, thus, increases metal sulfide oxidation and acid generation processes. Pyrite (FeS_2) is the most abundant sulfide mineral. Exposure of pyrite surfaces to oxygen and water results in the formation of sulfuric acid:

$$FeS_2 + 3.5O_2 + H_2O \rightarrow Fe^{2+} + 2SO_4{}^{2-} + 2H^+ \tag{7.1}$$

Investigations have established that ferric iron is more effective oxidant for sulfide minerals than oxygen (indirect oxidation). Depending on the type of mineral, two different ways of indirect oxidation (by Fe (III)) of minerals are distinguished (Schippers et al. 1996, 1999; Schippers and Sand 1999; Sand et al. 1995). Metal sulfides, valence bonds of which are obtained exclusively from metal orbitals, are oxidized with Fe (III) and cannot be subjected to proton attack (FeS_2, MoS_2, and WS_2). The dissolution of these minerals according to the work of Steudel (Steudel 1996) proceeds through the formation of thiosulphate:

$$FeS_2 + 6Fe^{3+} + 3H_2O \rightarrow S_2O_3^{2-} + 7Fe^{2+} + 6H^+ \quad (7.2)$$

In acidic solutions containing Fe (III), thiosulfate via a variety of polythionates is finally oxidized to sulfate (Sand et al. 1995):

$$S_2O_3^{2-} + 8Fe^{3+} + 5H_2O \rightarrow 2SO_4^{2-} + 8Fe^{2+} + 10H^+ \quad (7.3)$$

Other sulfides, in the formation of the valence bonds of which the orbitals of metals and sulfur participate, are soluble in acid and are attacked by both protons and Fe (III) ions (ZnS, CdS, NiS, CoS, CuS, and CuS_2). Dissolution of these sulfides proceeds by a different mechanism – through the formation of polysulfides.

$$MS + Fe^{3+} + H^+ \rightarrow M^{2+} + 0.5H_2S_n + Fe^{2+} \quad (7.4)$$

$$0.5H_2S_n + Fe^{3+} \rightarrow 0.125S_8 + Fe^{2+} + H^+ \quad (7.5)$$

Consequently oxidation of pyrite and other sulfide minerals occurs by series of intermediate reduced inorganic sulfur compounds (RISCs) (Schippers et al. 1996; Schippers and Sand 1999). RISCs and ferrous iron are potential energy sources for some acidophilic chemolithotrophic prokaryotes, SIOB (Eq. 7.6):

$$Fe^{2+} + 0.5O_2 + H^+ \xrightarrow{At.\,\mathit{ferrooxidans}} Fe^{3+} + H_2O \quad (7.6)$$

$$0.125S_8 + 1.5O_2 + H_2O \xrightarrow{At.\,\mathit{thiooxidans},\,At.\,\mathit{caldus}} SO_4^{2-} + 2H^+ \quad (7.7)$$

In general metal sulfide oxidation reactions are energy producing or highly exothermic reactions. The generation of elevated temperatures in bioleaching environments depends on climate, ambient temperature, mineralogy, etc. Besides extremely acidic environments are characterized by high concentrations of heavy metals originated from the oxidation of sulfide minerals.

7.3 Biodiversity of Microorganisms

Despite the extreme acidity, heat, and high concentration of various metals, extremely acidic environments are populated by a wide range of microorganisms. Molecular biological methods such as fluorescent in situ hybridization (FISH), CARD-FISH, polymerase chain reaction (PCR) combined with cloning, and denaturing gradient gel electrophoresis (DGGE), widely used in recent years to study biodiversity of microorganisms, allow to detect potentially all microorganisms involved in geochemical processes in natural and technogenic biotopes of sulfide ores (Hedrich and Johnson 2013; Schippers 2007; Johnson and Hallberg 2003;

Johnson et al. 2009; Zhang et al. 2010), while the cultivation method can detect only culturable constituents of microbial communities.

According to a number of researchers, natural bioleaching niches are characterized by a wide variety of microorganisms. More than 33 species of bacteria belonging to 14 genera and three domains were detected in these niches (Baker and Banfield 2003; Bond et al. 2000; Edwards et al. 1999; Johnson 1998; Johnson and Hallberg 2003; Rawlings 2002; Rawlings and Johnson 2007; Sand et al. 2007). More frequent microorganisms detected by cultivation and molecular biological techniques in mine dumps or heaps belong to bacteria. Moreover, they represent not only different genera or families but even phyla. Metal leaching bacteria belong to phyla *Proteobacteria* (*Acidithiobacillus*, *Acidiphilium*, *Acidiferrobacter*, *Acidisphaera*), *Nitrospirae* (*Leptospirillum*), *Firmicutes* (*Alicyclobacillus*, *Sulfobacillus*), and *Actinobacteria* (*Ferrimicrobium*, *Acidimicrobium*). All phyla have both mesophilic and moderate thermophilic forms (Coram and Rawlings 2002; Norris et al. 2000).

Acidithiobacillus ferrooxidans is the first and the most well-studied representative of Proteobacteria. This microorganism exhibits a wide range of metabolic activities and can oxidize RISCs, iron, and sulfide minerals (Kelly and Wood 2000). Some strains may also oxidize molecular hydrogen (H_2) and acetic acid.

The *Firmicutes* is represented by moderate thermophilic gram-positive spore-forming bacteria of the genus *Sulfobacillus*. All species are mixotrophs: the growth on Fe (II), S^o, and RISCs is possible only in the presence of yeast extract (Golovacheva and Karavaiko 1978; Norris and Kelly 1978; Norris et al. 1996; Melamud et al. 2003; Johnson et al. 2008). Some species grow on S^o under anaerobic conditions using Fe (III) as the final acceptor of electrons (Bridge and Johnson 1998).

Phylum *Nitrospirae* includes bacteria that oxidize only Fe (II). The representatives of the genus *Leptospirillum* are gram-negative motile vibrios or spirilla. Leptospirilla are strict aerobes, obligate chemolithotrophs, and grow only by the oxidation of Fe^{2+}. *Leptospirillum* includes both mesophilic, thermoresistant, and thermophilic forms (Battaglia et al. 1994; Golovacheva et al. 1992; Hippe 2000; Markosyan 1972; Sand et al. 1992).

Acidimicrobium ferrooxidans and *Ferrimicrobium ferrooxidans* representing phylum *Actinobacteria* are gram-positive rod-shaped, aerobic, thermoresistant, or moderately thermophilic bacteria that grow autotrophically by the oxidation of Fe (II) and heterotrophically using the yeast extract (Clark and Norris 1996).

The Archaea are mainly thermophiles. Archaea mainly belong to the class of Sulfolobales that represent extreme thermophilic sulfur and iron oxidizers (*Acidianus*, *Sulfolobus*, *Metallosphaera*, *Sulfurococcus*). Class Thermoplasmatales are presented in two species: *Ferroplasma acidiphilum* (Golyshina et al. 2000) and *Ferroplasma acidarmanus* (Edwards et al. 2000). Eukarya detected in these environments include algae, fungi, yeasts, and protozoa.

The exceptional diversity of ecogeographical conditions of Armenia and the richness of nonferrous metals represent a great and valuable potential for the investigation of biodiversity of acidophilic CB in mining sites, as well as for the isolation of new highly efficient strains and their communities.

The studies showed that in the ore deposits of Armenia during their exploitation, powerful oxidation zones and acidic mine drainage were formed leading to dissolution of heavy metals and contamination of aquatic ecosystems, soil, and other econiches.

Dissemination of acidophilic CB in sulfide ores of different mineralization (copper, copper-molybdenum, gold-bearing, and polymetallic ore) in Armenia has been studied using enrichment media and isolation techniques. It has been revealed that natural and technogenic biotopes of sulfide ores are represented by the following genera of IOB: *Acidithiobacillus*, *Leptospirillum*, and *Sulfobacillus* that can operate in the temperature range of 10–50 °C. It is noteworthy that the microflora of copper and copper-molybdenum ore deposits are mainly represented by *At. ferrooxidans* and *S. thermosulfidooxidans*, whereas in polymetallic ores rich in pyrite, *Leptospirillum* spp. bacteria dominate, which can be explained by their physiological properties, in particular, high resistance to low pH values and high concentrations of Fe^{3+} (Rawlings 1995; Rawlings et al. 1999; Dew et al. 1997; Olson et al. 2003). The fact that the activity of *At. ferrooxidans* and *S. thermosulfidooxidans* is mainly associated with copper deposits is also confirmed by other researchers (Pizarro et al. 1996; Dopson and Lindstrom 2004). Besides, *Acidithiobacillus* spp. bacteria dominate in the population of the majority of ore and acid drainage water samples. *Sulfobacillus* spp. bacteria were mainly found in ore dumps in quantity of 2–3 orders lower in comparison with *Acidithiobacillus* and *Leptospirillum* spp. bacteria. About half of IOB in ore dump samples of polymetallic ore deposits were *Leptospirillum* spp. bacteria. This indicates their important role in the bioleaching processes taking place in sulfide ore deposits. The obligatory constituent part of stable communities of IOB is sulfur-oxidizing bacteria and acidophilic heterotrophs (Vardanyan et al. 2015a, b). As a result of ecological studies carried out, new and efficient original strains of acidophilic chemolithotrophic bacteria belonging to genera *Acidithiobacillus* as well as *Leptospirillum* and *Sulfobacillus* have been isolated from sulfide ores of different mineralization (copper, copper-molybdenum polymetallic, gold-bearing ores) in Armenia. The main characteristics of the isolated bacteria are given below (Table 7.1).

7.4 Moderate Thermophilic Bacteria

Recently particular attention of researchers was attracted to moderately thermophilic and thermotolerant microorganisms as promising for improving the efficiency of bioleaching of complex concentrates and biooxidation of refractory gold-bearing ores and concentrates.

Moderate thermophiles include representatives of the genus *Sulfobacillus*, which are able to exist at elevated temperatures and low pH values and also oxidize elemental sulfur, ferrous iron, and sulfide minerals, actively participating in the cycle of chemical elements in nature and metal biotechnology (Pivovarova and Golovacheva 1985; Karavaiko et al. 1988).

Table 7.1 Characteristics of CB isolated from sulfide ores of different mineralization in Armenia

Isolated bacterial strains	Source of energy	Source of isolation	Temp., °C	Cell morphology
S. thermosulfidooxidans subsp. asporogenes str. 41	Fe^{2+}, S^{o}, FeS_2	Armanis polymetallic ore	30–55	Rods
S. thermosulfidooxidans subsp. asporogenes str. 69	Fe^{2+}, S^{o}, FeS_2	Drmbon concentrate (gold ore)	37–55	Rods
S. thermosulfidooxidans str. 86	Fe^{2+}, S^{o}, FeS_2	Tandzut (Polymet) ore	37–60	Rods
L. ferrooxidans str. ZC	Fe^{2+}, FeS_2	Zinc concentrate bioleaching pulp	37	Curved rods, spirilla
Leptospirillum sp. str. 64	Fe^{2+}, FeS_2	Akhtala (copper) ore	37	Curved rods, spirilla
Leptospirillum sp. str. 72	Fe^{2+}, FeS_2	Alaverdi (copper) ore		Curved rods, spirilla
L. ferrooxidans str. Teg	Fe^{2+}, S^{o}, FeS_2	Teghout (Cu-Mo) ore	37	Curved rods
Acidithiobacillus tandzuti str. 5	S^{o}	Tandzut (Polymet) ore	37	Straight rods
Acidithiobacillus sp. str. 13Zn	Fe^{2+}, S^{o}, FeS_2	Zinc concentrate	30–35	Straight rods
Acidithiobacillus ferrooxidans str.18	Fe^{2+}, S^{o}, FeS_2	Tandzut (Polymet) ore dump	30	Straight rods
At. ferrooxidans str. 61	Fe^{2+}, S^{o}, FeS_2	Tandzut (Polymet) ore	30	Straight rods
At. ferrooxidans str. Dr	Fe^{2+}, FeS_2	Drmbon (copper) ore	37	Straight rods
At. ferrooxidans str. Tz	Fe^{2+}, S^{o}, FeS_2	Tandzut (Polymet) ore	30	Straight rods

The second genus of moderately thermophilic gram-positive, iron-oxidizing bacteria at present contains a single species *Acidimicrobium ferrooxidans* (Clark and Norris 1996). It was found in a commercial copper leaching dump but has not been extensively studied so far.

It was found that sulfobacilli, together with representatives of the genus *Alicyclobacillus*, form a single phylogenetic group and represent an independent branch within the subdivision of gram-positive bacteria, which separated from the bacilli at relatively early stages of evolution (Fig. 7.1) (Karavaiko et al. 1990; Turova et al. 1995).

The general origin of the genera *Sulfobacillus* and *Alicyclobacillus* is indicated by the presence of differential phenotypic characters. Thus, representatives of both species are obligate acidophiles. Cells of sulfobacilli contain specific lipids, which contain ω-cyclohexanoic acids, which are a distinctive feature of the genus *Alicyclobacillus* (Tsaplina et al. 1994; Turova et al. 1995). However, bacteria of the genus *Alicyclobacillus* are less acidophilic and unable to oxidize inorganic substrates (Wisotzkey et al. 1992) (Table 7.2).

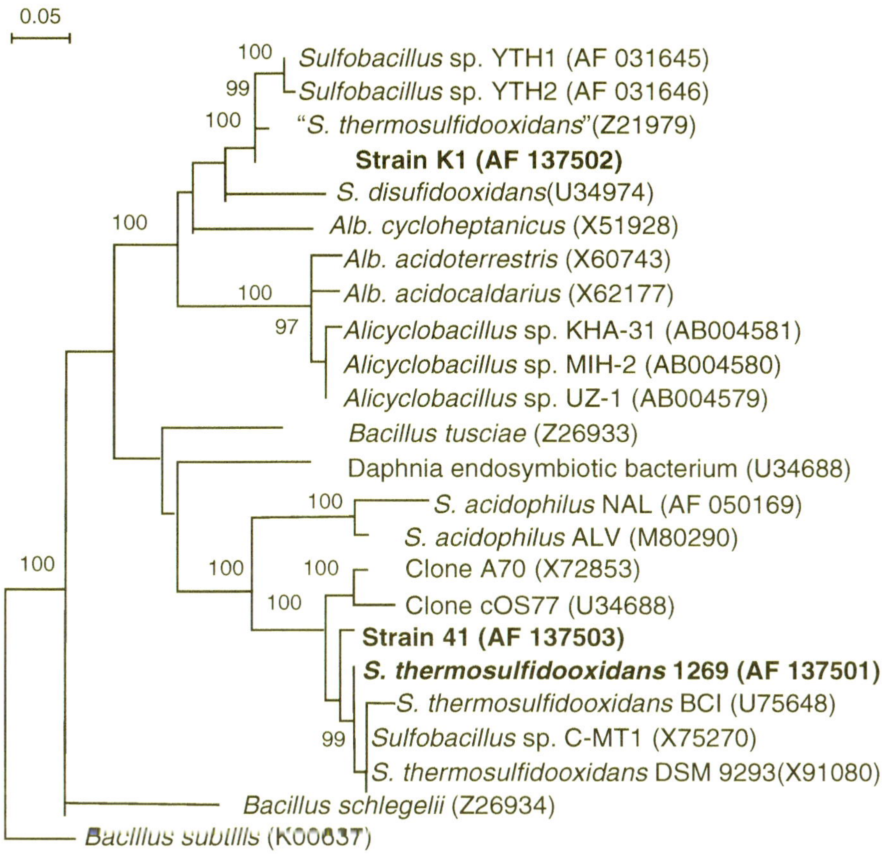

Fig. 7.1 Phylogenetic tree of genera *Sulfobacillus* and *Alicyclobacillus*. The scale corresponds to 5 nucleotide replacement for every 100 nucleotides. The figures show the statistical reliability of the order of branching, determined using the "bootstrap" analysis of 100 alternative trees; values less than 95% are not indicated (Karavaiko et al. 2000)

At present, this genus is represented by five species of moderately thermophilic and mesophilic bacteria: *S. thermosulfidooxidans* VKM V-1269 = DSM 9293 (Golovacheva and Karavaiko 1978), BC1 (Norris et al. 1996), *S. acidophilus* (NAL and ALV) (Norris et al. 1996), *S. disulfidooxidans* SD –11 (Dufresne et al. 1996), S. sibiricus N1 (VKM V-2280) (Melamud et al. 2003), and *S. benefaciens* (Johnson et al. 2008).

A gram-positive thermoacidophilic non-spore-forming bacterium *S. thermosulfidooxidans* subsp. asporogenes str. 41 has been isolated from acid mine drainage water of Armanis polymetallic ore deposit in Armenia (Vartanyan et al. 1988). By its physiological properties, the bacterium is similar to type strain *S. thermosulfidooxidans* VKM V-1269 (Table 7.2). The G + C content in DNA is 45.5 mol %. The strain shows high level of DNA-DNA hybridization (81%) with *S. thermosulfidooxidans* VKM V-1269. The bacterium differs from the type strain by a

Table 7.2 Morphophysiological features of bacteria from the genera *Sulfobacillus* and *Alicyclobacillus* (Karavaiko et al. 2006)

	Alicyclobacillus						
	S. thermosulfidooxidans VKM V-1269 (DSM 9293)	*S. thermosulfidooxidans* sp. asporogenes 41	*S. disulfidooxidans* SD-11	*S. acidophilus* ALV	*S. sibiricus N1* (VKM V-2280	A. *acidocaldarius* (DSM-446)	*A. acidoterrestris* (DSM-3922)
Cell size, (μm)	0.6–0.8 × 1.0–3.0	0.5–0.9 × 2.0–4.0	0.6–1.0 × 1.0–6.0	0.5–0.8 × 3.0–5.0	0.75–1.1 × 1.0–3.0	0.7–0.8 × 2.0–3.0	0.6–0.8 × 2.9–4.3
pH range, optimum	1.5–5.5	2.0–4.5	1.5–2.5	2.0	2.0–3.5	2.6–6.0	2.2–5.8
	1.7–2.4	1.6–1.8			2.2–2.5		
t°C (range and optimum)	20–60	30–60	4–40	45–50	17–60	45–70	35–55
	50–55	50	35		55		42–53
G + C composition (mol %)	47.2–47.5	46.5	48.2	55–57	48.2	60.3	52.2
Size of genome (Da)	3.7×10^9	3.0×10^8	3.0×10^9	4.0×10^9	Nd	Nd	Nd
Inorganic substrates	Fe^{2+}, S°, sulfide minerals					–	–

smaller size of genome (3.0×10^9Da vs. 3.7×10^9Da) and by the fact that it does not form spores. Therefore, the bacterium was considered as a new subspecies of *S. thermosulfidooxidans* – *S. thermosulfidooxidans* subsp. asporogenes str. 41 (Vartanyan et al. 1988).

Previously, bioleaching of metals and acid mine drainage (AMD) production are more often found associated with bacteria from genus *Acidithiobacillus* (former *Thiobacillus*): *At. ferrooxidans* and *At. thiooxidans*. Later, studies have shown that *Leptospirillum* is dominating microorganisms in bioleaching niches at temperatures higher than 30 °C and lower pH (<1.0) (Coram and Rawlings 2002; Rawlings 1995). Thus, at present, different strains of *Leptospirillum* have been isolated from natural and technological systems, which represent a great interest both in fundamental and practical aspects due to their unique capability of obtaining energy for their activity only from oxidation of Fe (II) (Coram and Rawlings 2002; Okibe et al. 2003; Rawlings 1995; Sand et al. 1992; Vardanyan and Akopyan 2003; Zhang et al. 2010).

The first representative of the genus *Leptospirillum* – *L .ferrooxidans* – was isolated and described in Armenia (Markosyan 1972). Subsequently, leptospirilla were found in uranium deposits, heaps of coal, acidic mine waters of copper deposits (Norris 1983, Harrison and Norris 1985; Johnson 1995; Sand et al. 1992), as well as laboratory systems for continuous leaching of pyrite and cobalt-iron pyrite (Battaglia et al. 1994; Helle and Onken 1988). The isolated strains of leptospirilla were assigned to *L. ferrooxidans* or *Leptospirillum*-like bacteria (Goebel and Stackebrandt 1994). Later studies revealed significant differences between individual strains, indicating the heterogeneity of the genus *Leptospirillum* (Harrison 1986; Hallmann et al. 1993). However, only recently molecular genetic studies particularly differences in G + C content in DNA carried out served as the basis for separating isolated strains of leptospirilla into two groups, I and II (Bond et al. 2000) (Table 7.3).

The genus *Leptospirillum* was included in the phylum *Nitrospirae* as a phylogenetically separate cluster and has been divided into three groups – I, II, and III – on the basis of 16S rRNA gene phylogeny (Bond et al. 2000) (Fig. 7.2).

L. ferrooxidans is the representative of group I, and *L. ferriphilum* is the representative of group II. Representatives of group III were identified in the biofilm analyzed by community genomics. *L. ferrodiazotrophum* was proposed for a nitrogen-fixing representative of the group III rRNA sequence cluster of strains (Tyson et al. 2005). No cultured representatives of group III have been described up to now (Bond et al. 2000; Tyson et al. 2005).

In zones of spontaneous heating of sulfide ore dumps, moderately thermophilic leptospirilla were found which were described by the content of G + C in DNA and by the degree of homology of DNA (26.7%) as a new species of *Leptospirillum thermoferrooxidans* sp. nov. (Karavaiko et al. 1980; Karavaiko and Golovacheva 1986; Golovacheva et al. 1992). However, *L. thermoferrooxidans* was not preserved, and data on 16SrRNA are absent, which makes it impossible to compare this bacterium with the abovementioned groups of leptospirilla.

Table 7.3 Genetic characteristics of leptospirilla, isolated from different sources (Coram and Rawlings 2002)

Strains	Groups	Subgroups	Habitat	References	G+C, %	Growth at 45 °C
P_3 a	I	1.1	Coal mine, North Wales (UK)	Sand et al. (1992)	51.9	–
DSM 2705	I	1.1	Copper mine deposit, Armenia	Markosyan (1972)	51.7	–
ATCC 49879	I	1.1	Romania	Sand et al. (1992)	51.7	–
Sy	I	1.2	Sygun Cu mine North Wales, UK	Johnson (1995)	48.8	–
Parys	I	1.2	Parys Mountain, Anglesey Cu mine, Wales	Johnson (1995)	51.5	–
BCT2	I	1.2	Birch Coppice Warwickshire, UK	Johnson (1995)	51.0	–
Chil-L f2	I	1.2	Cu mine, Chile	Johnson	51.2	–
Warwick	II		Warwick, UK	Norris (1983)	54.9	–
ATCC 49880	II		Romania	Sand et al. (1992)	57.8	+
ATCC 49881	II		Peru	Sand et al. (1992)	56.6	+
BU-1	II		Gramatikovo, South-East Bulgaria	Harrison (1986)	55.4	–

7.4.1 Morphology and Ultrastructure

In the logarithmic phase of growth on media with S°, Fe^{2+}, and sulfide minerals, moderately thermophilic sulfobacilli are rods with rounded ends and 0.5–1.0 x 1.0–6.0 μm in size. They occur as single cells, pairs, or in the form of short chains. Cells are nonmotile, devoid of flagella. Microcapsules are found around the cell. With the growth of the bacterium on pyrite, there is a tendency to form pseudococci and filamentous forms (Golovacheva 1979; Vartanyan et al. 1988). Reproduction occurs by binary division (Fig. 7.3d).

In the central part of the cell of str. 41, well-defined electron-transparent zones of a nucleoid with DNA strands 2 nm thick were found (Fig. 7.3b, d). In the cytoplasm, a significant number of small electronically dense granules – polyribosome – were observed. Large polyphosphate granules were found at the poles of the cells. On ultrathin sections it is seen that sulfobacilli have a one-component cell wall typical of gram-positive bacteria. Outside, it is covered with an electronically transparent S-layer, connected with the murein layer underneath it. The electronically dense murein layer closely adjoins the outer layer of the cytoplasmic membrane, due to which the latter looks asymmetric and the outer layer appears thickened. S-layer in *S. thermosulfidooxidans*, together with the polysaccharide substance of the capsule lying outside, forms a complex glycocalyx that plays an important role in the

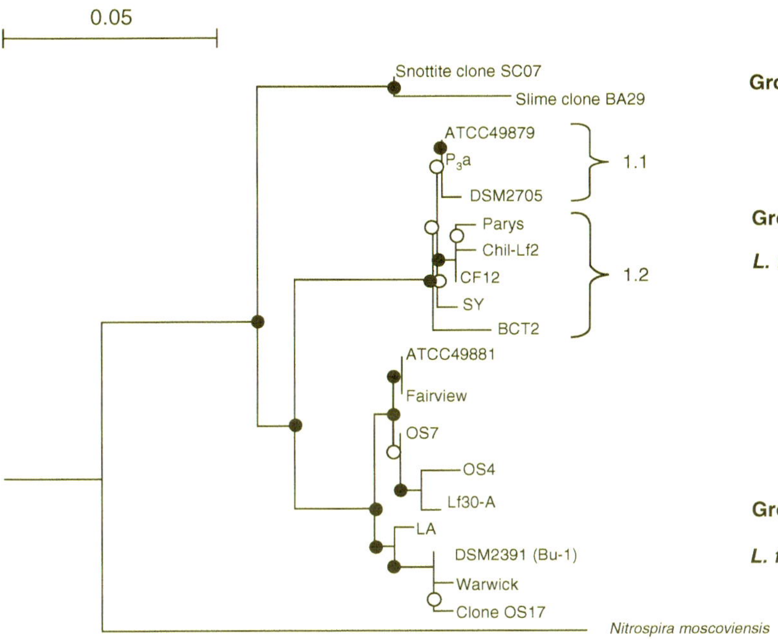

Fig. 7.2 Dendrogram of leptospirilla, composed on the basis of the analysis of nucleotide sequence of 16S rRNA. Branches showing values above 75% in the "bootstrap" analysis are indicated by black circles, the rest (50–75%) – hollow circles. Two subgroups of hybridization of leptospirilla (1.1 and 1.2) are distinguished by large brackets, which contain three copies of the rrn gene (group I) (Coram and Rawlings 2002)

interactions of microorganisms with the environment (Severina et al. 1995; Senyushkin et al. 1997; Karavaiko et al. 2006).

In comparison with the bacilli, the S-layer of *S. thermosulfidooxidans* is characterized by a higher content of hydrophobic amino acids. Another important feature of the S-layer of S. thermosulfidooxidans is the predominance of acidic amino acids.

It is known that the most adhesive part is the carbohydrate part of the layer. In carbohydrates of S-layer, *S. thermosulfidooxidans* VKM B-1269 dominates mannose. Glucosamine was also detected (Severina et al. 1995).

Intracytoplasmic membrane structures are very diverse. Along with simple loop-shaped invagination, complex four-contour and tubular-vesicular membrane structures are observed (Pivovarova and Golovacheva 1985). The functions of intracytoplasmic membrane structures have not been fully studied. It is assumed that they are the "depots" of whole membrane blocks and lipids, which are used by bacteria for rapid growth and when restoring the surface parts of the cytoplasmic membrane. According to another hypothesis, invert membranes regulate the transport of certain substances, in particular hydrophobic S^{o}, into the cell. According to

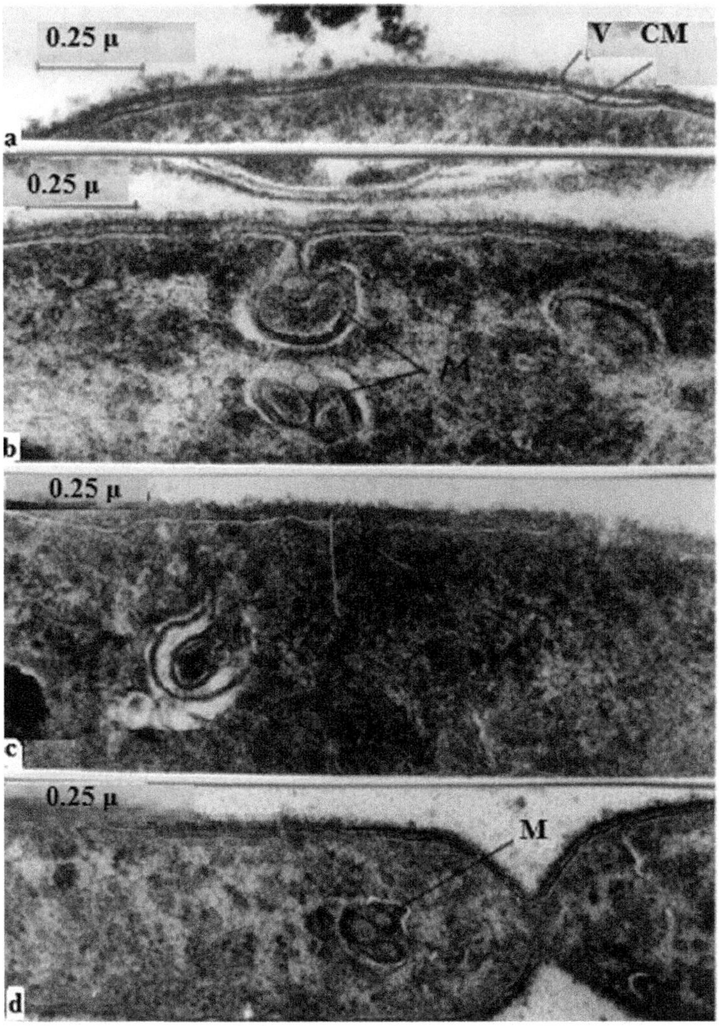

Fig. 7.3 The structure of the cell wall and the various types of mesosomes in *S. thermosulfidooxidans* subsp. asporogenes str. 41
V vascular structures, *CM* cytoplasmic membrane, *M* mesosomes

the third hypothesis, these structures play a role in the cell cycle and participate in the formation of septa during cell division (Suzina et al. 1999).

All representatives of sulfobacilli are capable of forming endospores (Golovacheva 1976; Golovacheva and Karavayko 1978; Kovalenko and Malakhova 1983; Karavaiko et al. 2006). Moreover, spore formation is especially intense when growing on copper-zinc-pyrite ores and sulfide minerals such as pyrite, chalcopyrite, arsenopyrite, and antimonite. Less intensively, it proceeds with the growth of bacteria on sulfur and ferrous iron. In the latter case, bacteria practically do not manage to form spores due to the rapid depletion of the energy substrate (Golovacheva and Karavayko 1978). For a type strain of *S. thermosulfidooxidans* VKM B-1269, spherical spores are characteristic, and for *S. thermosulfidooxidans* subsp. thermotolerans – oval, slightly inflating sporangium (Kovalenko and Malakhova 1983). The location of spores in sporangia is terminal or subterminal (Golovacheva 1979). The spores of *S. thermosulfidooxidans* have a structure similar to that of thermophilic and mesophilic bacilli. It is a protoplast, enclosed in a powerful cortex, which has multilayer covers and an exosporium (Golovacheva 1979).

Electron microscopic observations in dynamics, as well as the special staining procedure, did not reveal spores in *S. thermosulfidooxidans* subsp. asporogenes str. 41, isolated from Armanis polymetallic ore deposits in Armenia (Vartanyan et al. 1988). Investigations indicated that the genome size in str. 41 was 3.0×10^8 Da, which is 20% less than the genome of *S. thermosulfidooxidans* VKM B-1269 (3.7×10^9 Da). It is assumed that as a result of the deletion part of the genome of the isolated bacteria, responsible for sporulation has been lost.

Representatives of the genus *Leptospirillum* are characterized by well-defined polymorphism. The 4-day cultures are vibrios with diameters of 0.9–1.1 µm and length of 0.3–0.6 µm. Spiral-shaped cells with a number of turns up to 4 (Fig. 7.4) (Pivovaarova et al. 1981; Golovacheva et al. 1992; Sand et al. 1992; Vardanyan and Akopyan 2003) dominate in a 1–2 week culture. Under adverse conditions, in particular at pH 1.5, the cells are transformed into pseudococci with a diameter of 400–800 nm. They are also characterized by the formation of cocci at the terminal ends of the spirilla (Pivovaarova et al. 1981). They have one polar flagellum 18–22 nm in diameter. *L. ferrooxidans* is multiplied by cell division. In general, there is uniform division, but uneven cell division is also possible by separating one of the spiral turns (Pivovaarova et al. 1981).

Outside, the cells of the leptospirilla are covered with a mucous layer 100–450 nm thick, which, due to a unique chemical structure, forms complexes with Fe^{3+} ions. It has been established that on the third day of cultivation, the *L. ferrooxidans* cell is surrounded by an iron cover, which gradually thickens and leads to the cessation of cell development (Pivovaarova et al. 1981; Golovacheva et al. 1992) (Fig. 7.5).

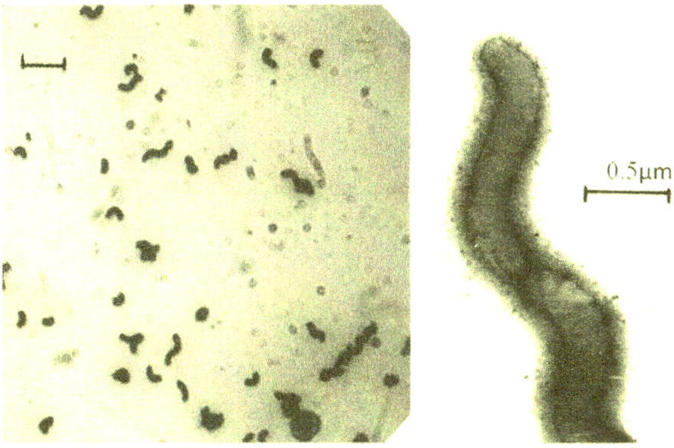

Fig. 7.4 General view of *Leptospirillum* sp. str. 72. (**a**) Light microscopy, (**b**) electron microscopy

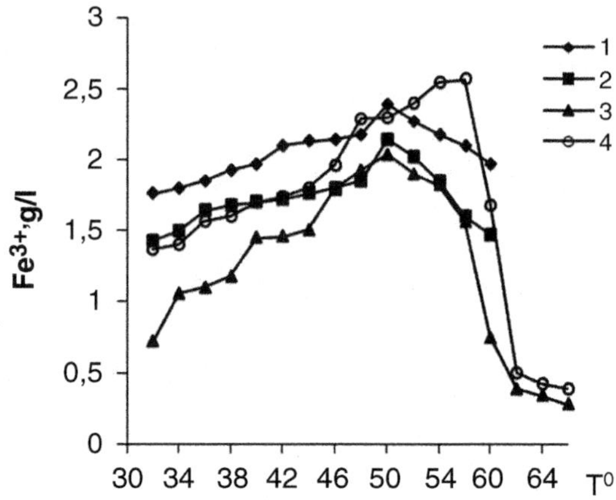

Fig. 7.5 Oxidation of Fe^{2+} strains 41 (1), 42 (2), 69 (3), and 86 (4) at different temperatures

The cell wall of *L. ferrooxidans* has a typical structure for gram-negative cells. The outer layer is represented by a cell membrane consisting of 2 electronically dense layers 0.6–1.0 and 0.35–0.6 nm thick and an electron-transparent layer between them 0.7–0.8 nm thick. A rigid murein layer is found only in its individual parts, which, in the opinion of the authors, may be due to the insufficient course of the fixation process. Between the cell wall and plasma membrane, there is a small electron-dense layer 0.8–1.55 nm thick (Pivovaarova et al. 1981).

Intracytoplasmic structures were not found in leptospirilla. In the central part of the cell, there is an electronically transparent layer with DNA strands. In the cytoplasm there are polyribosomes in large numbers. The accumulation of hydroxybutyrate, inherent for the Spirilaceae family, was not detected.

7.4.2 Physiology

On the basis of phenotypic properties, sulfobacilli are divided into two groups: moderate thermophiles (str. 1269, 41, BC1, N1, NAL, ALV) and mesophiles (strains K1 and SD-11). Representatives of the first group oxidize Fe^{2+}, S^0, and sulfide minerals in the presence of 0.2 g/l of yeast extract or other organic substances in the medium. Strains of the second group oxidize inorganic substrates partially and grow at higher concentrations of organic substances (1.0–2.5 g/l) (Bogdanova et al. 2006; Kovalenko and Malakhova 1983; Dufresne et al. 1996). The latter group of bacteria is characterized by larger cell sizes.

In autotrophic conditions, sulfobacilli can grow due to inorganic sources of energy, such as tetrathionate ($S_4O_6^{2-}$), thiosulfate ($S_2O_3^{2-}$), and elemental sulfur (S^o). At the atmospheric concentration of carbon dioxide, the autotrophic growth of sulfobacilli is weak and ceases after several passages. The specific growth rate is in the range of 0.018–0.026 hr./1 (Krasilnikova et al. 1998). The autotrophic growth of bacteria on Fe^{2+} is also limited, but the addition of tetrathionate or thiosulphate to the medium leads to an increase in their growth rate.

One aspect of the metabolic diversity of sulfobacilli is associated with their ability to grow anaerobically. Studies of recent years have shown that IOB are rather facultative anaerobes than strict aerobes. When growing in conditions of limited amounts of oxygen, five strains belonging to the species *S. thermosulfidooxidans* (strain TH1), *S. acidophilus* (ALV, YTF1), and *Acidimicrobium ferrooxidans* (TH3) showed the ability to reduce Fe^{3+} (Bridge and Johnson 1998). It was found that in some strains the reduction of Fe^{3+} is associated with mixotrophic or heterotrophic growth, when glycerol acts as electron donor. And in *S. acidophilus* ALV and THWX and *S. thermosulfidooxidans* TH1, the reduction of Fe^{3+} was accompanied by oxidation of tetrathionate:

$$S_4O_6^{2-} + 10H_2O + 14Fe^{3+} \rightarrow 4H_2SO_4 + 12H^+ + 14^{2+}$$

Of special interest is the fact that the *S. acidophilus* str. YTF1 is capable of reducing the minerals containing Fe^{3+} (iron hydroxide, jarosite, goethite) when growing under limited aeration using glycerol as a source of carbon and energy:

$$Fe(III) \xrightarrow{\text{solid phase}} Fe^{3+} \xrightarrow{\text{soluble}} Fe^{2+}$$

This ability of moderately thermophilic bacteria can have wide practical application for the removal of jarosite from other minerals accumulating in the reactors of bacterial leaching of metals, providing further efficient oxidation of ores (Bridge and Johnson 1998).

From dumps and water flows of Armanis and Akhtala polymetallic ores and Shamlug copper ore, three strains of acidophilic moderately thermophilic SIOB were isolated: str. 41, str. 69, and str. 86. The studies showed that the optimal growth temperature for strains 41 and 69 is 50 °C and for str. 86 is 55 °C. The lower temperature limit is 30 °C. Oxidation of Fe^{2+} with isolated strains was observed up to 60 °C (Fig. 7.4). Optimal pH values for the growth of bacteria on Fe^{2+} lie in the range of 1.6–1.8. At pH values above 2.5, a sharp decrease in the Fe^{2+} oxidation activity is observed. The growth of bacteria on medium with elemental sulfur occurs in the pH range 2.0–4.5; the optimal pH values are 2.3–2.5.

The optimum growth temperature for *L. ferrooxidans* is 28 °C (Markosyan 1972). However, in *Leptospirillum*-like bacteria, according to a number of authors, it varied from 28 to 35 °C (Norris 1983; Harrison and Norris 1985; Sand et al. 1992). For strains L6 and L8, the optimum growth temperature is reported to be 37 °C (Battaglia et al. 1994). Some strains can grow at the temperature up to 45 °C. Thus, the optimum temperature of leptospirilla is much higher than that of thiobacilli. This is evidenced by recently obtained data that the rate of growth and generation time in leptospirilla is more sensitive to temperature changes than in thiobacilli (Sand et al. 1992). Consequently, the temperature in natural biotopes can serve as a limiting factor for propagation of leptospirilla.

Leptospirilla are obligate acidophiles. The optimal pH for the growth of *L. ferrooxidans* VKM B-1339 was pH 2.5 (Markosyan 1972), whereas for strains isolated later, it was much lower. According to Battaglia and coworkers, the maximum growth rate for str. L8 has been observed at pH 1.8, and in str.

L6 – pH 1.3 (Battaglia et al. 1994). It is noteworthy that at pH 1.3, the growth in *At. ferrooxidans* is strongly inhibited (Merrettig et al. 1989; Sand et al. 1992; Battaglia et al. 1994). Thus, leptospirilla are more resistant to low pH values than thiobacilli.

For the strains 64 and 72 isolated from Alaverdi and Akhtala sulfide ores (Armenia), the optimum temperature was 37 °C. It is worth mentioning that iron-oxidizing activity in strain 64 remained at a high level at 40 °C and sharply reduced in strain 72. Both strains did not oxidize Fe^{2+} at 45 °C. The pH optimum for the growth of strains 72 and 64 was 2.0; the pH value of 1.4 was the lowest limit of bacterial growth. Bacterial growth and ferrous iron oxidation were inhibited by yeast extract (Vardanyan and Akopyan 2003; Vardanyan et al. 2013).

7.4.3 Oxidation of Iron and RISCs

Comparable activities of enzymes involved in RISCs oxidation, such as sulfite:cytochrome c oxidoreductase, thiosulfate dehydrogenase, rhodanese (thiosulfate-cyanide sulfurtransferase), sulfite:ferric ion oxidoreductase, as well as iron oxidase of isolated strains of *Sulfobacillus* spp., *Leptospirillum* spp., and *Acidithiobacillus* spp. bacteria isolated in Armenia, were studied.

As shown in Table 7.4, the activities of iron oxidase of *S. thermosulfidooxidans* str. 69 and str. 13Zn were comparable to that of *At. ferrooxidans* str. 61. High activity of iron oxidase was detected in *Leptospirillum* spp. bacteria. The activities of iron oxidase of *Leptospirillum* sp. str. ZC and str. Teg at 37 °C were 3.5–3.6 and 1.7–1.8 times higher than appropriate activities of *S. thermosulfidooxidans* str. 69 and *At. ferrooxidans* str. 61.

Although the studied strains were grown in the medium containing Fe^{2+} as the only energy source, sulfite:cytochrome c oxidoreductase and thiosulfate dehydrogenase activities were detected in cell-free extract of *S. thermosulfidooxidans* str. 69 and str. 13Zn and *At. ferrooxidans* str. 61. However, these strains almost showed no significant differences of sulfite:cytochrome c oxidoreductase activity (Vardanyan et al. 2015a, b).

It is worth to mention that sulfite oxidase and thiosulfate dehydrogenase were detected in cell-free extracts of *Leptospirillum* sp. str. ZC and str. Teg. According to literature data, isolated *S. thermosulfidooxidans* str. 69 showed significantly higher sulfite- and thiosulfate-oxidizing activities compared with *Sulfobacillus thermosulfidooxidans* VKM V-1269 (Krasil'nikova et al. 1998) and *Sulfobacillus sibiricus* strain N1 and strain SSO (Krasil'nikova et al. 2004). At the same time, there are no data on iron oxidase activity of the abovementioned bacteria. Sulfite:Fe (III) oxidoreductase activity has been previously described for *At. ferrooxidans* (Sugio et al. 1987). According to literature data, S. sibiricus also possesses sulfite:Fe (III) oxidoreductase activity. On the contrary *S. thermosulfidooxidans* VKM V-1269 doesn't show sulfite:Fe (III) oxidoreductase and sulfur:Fe (III) oxidoreductase activities.

The presence of sulfite:Fe (III) oxidoreductase in cells of *S. thermosulfidooxidans* str. 69 participating in oxidation of RISCs using Fe (III) as an electron acceptor will allow them like *Sulfobacillus sibiricus* to survive in condition of limited oxygen that occurs in bioleaching processes at high temperatures.

Table 7.4 Enzymatic activities of isolated *Sulfobacillus* spp., *Leptospirillum* spp., and *Acidithiobacillus* spp. bacteria (Vardanyan et al. 2015a, b)

Enzymes	*Enzymatic activity, μM/ mg protein min*						
	S .thermosulfidooxidans str. 69	*Sulfobacillus sp.* strain 13Zn	*At. ferrooxidans* str. 61	*Leptospirillum* sp. strain Teg		*Leptospirillum* sp. strain ZC	
T, °C	37	37	30	30	37	30	37
Sulfite: cytochrome c oxidoreductase	7.95	9.8	6.87	12.1	15.0	19.0	26.9
Thiosulfate dehydrogenase	0.17	0.67	0.34	0.38	0.45	0.78	1.4
Rhodanese	0	0	0	0	0	0	0
Sulfite:ferric ion oxidoreductase	-	-	0.0071	0.04	-	0.07	-
Iron oxidase	70.2	73.0	66.7	121.4	125.0	244.7	254.7

7.4.4 Carbon Metabolism

Moderately thermophilic bacteria of the genus *Sulfobacillus* are characterized by a flexible metabolism, which enables them to grow in autotrophic, mixotrophic, and heterotrophic conditions (Karavaiko et al. 1988). Nevertheless, stable growth of sulfobacilli is possible only in mixotrophic conditions, when elemental sulfur, reduced sulfur compounds, ferrous iron, and sulfide minerals are used as an energy source and some organic compounds and carbon dioxide serve as a source of carbon. In this connection, in order to elucidate the possible reasons for the absence of stable growth of these organisms under auto- and heterotrophic conditions, the enzyme systems of representatives of the genus *Sulfobacillus* responsible for growth in autotrophic conditions and ways of using organic compounds were studied.

Metabolism of carbon is studied in the most detail in *S. thermosulfidooxidans* VKM B-1269, *S. thermosulfidooxidans* subsp. asporogenes str. 41, and *S. acidophilus* (Zakharchuk et al. 1994; Karavaiko et al. 2001; Tsaplina et al. 2000; Norris et al. 1996).

7.4.4.1 Fixation of CO_2

The studies carried out have shown that the moderately thermophilic bacteria *S. thermosulfidooxidans* are able to assimilate $^{14}CO_2$ due to the oxidation of Fe^{2+} and, consequently, to grow autotrophically. Fixation of carbon dioxide in representatives of the genus *Sulfobacillus* occurs through the Calvin cycle (Wood and Kelly 1984; Krasilnikova et al. 1998; Zakharchuk et al. 2003). Activity of the key enzyme of Calvin cycle – ribulose bisphosphate carboxylase (RuBPCase) – was found in *S. thermosulfidooxidans* str. 41 and str. 1269 and *S. acidophilus* in all growth conditions. Comparative analyses showed a higher level of RuBPCase in cell-free extracts of *S. thermosulfidooxidans* subsp. asporogenes str. 41 (Table 7.5.). Probably the Calvin cycle for str. 41 is the main way of fixing CO_2. Activities of the RuBPCase in str. 41 and the other strains of sulfobacilli grown under mixotrophic conditions with glucose and Fe^{2+} were considerably lower and depended on the concentration of organic substances (Tsaplina et al. 2000; Krasilnikova et al. 1998; Wood and Kelly 1984; Zakharchuk et al. 2003). Unlike other representatives, activity of RuBPCase was not detected in heterotrophically growing cells of *S. thermosulfidooxidans* sp. thermotolerans str. K1 (Karavaiko et al. 2002).

It was established that fixation of CO_2 in *S. thermosulfidooxidans* subsp. asporogenes str. 41under all growth conditions could also be performed by the carboxylation of pyruvate and phosphoenolpyruvate (PEP). The activity of PEP-carboxylase increases significantly when organic compounds are added into the nutrient medium (mixotrophic conditions) (Table 7.6). Cells growing under heterotrophic conditions have the lowest activity of all carboxylases. The reactions of carboxylation of pyruvate and PEP resulting in the regeneration of oxaloacetate seem to be one of the mechanisms for providing the Krebs cycle with precursor amino acids.

In autotrophic and mixotrophic conditions, str. K1 and str. 1269 and S. sibiricus N1 were found to fix carbon dioxide with the help of PEP-carboxyltransferase (Krasilnikova et al. 1998; Karavaiko et al. 2002; Zakharchuk et al. 2003). The

Table 7.5 Comparative activities of Ribulose-bisphosphate carboxylase/oxygenase (RuBPCase) of strains *S. thermosulfidooxidans* at different growth conditions

	Growth conditions			
Bacterial strains	Autotrophic	Mixotrophic	Heterotrophic	References
S. thermosulfidooxidans subsp. asporogenes str.41	38.2	15.3	1.2	Tsaplina et al. (2000)
S. thermosulfidooxidans VKM V-1269	10.4	-	1.1	Krasil'nikova et al. (2001)
S. thermosulfidooxidans subsp. "thermotolerans" str. K1	4.3	2.1	N/A	Karavaiko et al. (2001)
S. sibiricus str. N1	12.8	7.0	4.8	Zakharchuk et al. (2003)

N/A-non available

Table 7.6 Activities of carboxylase *S. thermosulfidooxidans* str. 41 and str. K1 and *S. sibiricus* N1 at different growth conditions (Zakharchuk et al. 2003; Karavaiko et al. 2002; Tsaplina et al. 2000)

	Growth conditions								
	Autotrophic			Mixotrophic			Heterotrophic		
Enzymes	K1	41	N1	K1	41	N1	K1	41	N1
Ribulose-bisphosphate carboxylase/oxygenase (RuBPCase)	4,3	44.6	12.8	2,1	18.0	7.0	Nd	1.5	4.8
Pyruvate carboxylase	1.4	0.51	0.5	0.46	0.26	9.8	Nd	0.2	1.1
Phosphoenolpyruvate (PEP) carboxylase	3.5	8.7	1.8	1.52	20.7	0.6	1.34	0.2	2.2
PEP-carboxykinase	0.6	0	4.1	Nd	0	Nd	Nd	0	Nd
PEP-carboxyltransferase	3.1	0	0.4	0.92	0	0.7	Nd	0	Nd

activity of PEP-carboxykinase and PEP-carboxytransphosphorylase in str. 41 was not detected (Tsaplina et al. 2000) (Table 7.6).

Transition to mixotrophic conditions led to a decrease in the level of pyruvate carboxylase in the cells of strain K1, and in a heterotrophically growing culture, it was not detected at all (Karavaiko et al. 2002). Regardless of the growth conditions, the K1 strain also exhibited PEP-carboxylase activity, whereas the PEP-carboxykinase was detected only when growing in autotrophic conditions (Table 7.5).

Nevertheless, sulfobacilli, like other moderate thermophilic bacteria, are not capable of sustainable autotrophic growth (Wood and Kelly 1984; Dopson and Lindstrom 1999). It is considered that the reason for this is an ineffective mechanism for CO_2 fixation (Clark and Norris 1996). For this reason, sulfobacilli are often found in natural environments closely related to other thermoacidophilic bacteria, particularly *Acidimicrobium ferrooxidans*, which have a highly efficient CO_2 fixation mechanism and can supply them with the required amount of organic carbon (Clark and Norris 1996; Johnson 1998).

7.4.4.2 Tricarboxylic Acid Cycle

Tricarboxylic acid cycle enzymes: The cells of *S. thermosulfidooxidans* subsp. asporogenes str. 41 like *S. thermosulfidooxidans* VKM B-1269 are not capable of sustainable organotrophic growth on yeast extract or other organic substances. It is assumed that the cause is an ineffective mechanism for destruction of organic compounds. It has been revealed that there is no glyoxylate pathway in the *S. thermosulfidooxidans* VKM B-1269, and the tricarboxylic acid cycle in sulfobacilli is open at the level of α-ketoglutarate dehydrogenase and does not work completely (Krasilnikova et al. 2001; Zakharchuk et al. 1994, 2003; Tsaplina et al. 2000; Karavaiko et al. 2001). Studies have shown that the enzyme α-ketoglutarate dehydrogenase was not found in *S. thermosulfidooxidans* str. 41 (Table 7.7).

Therefore, it should be assumed that the final oxidation of organic substances in *S. thermosulfidooxidans* str. 41 occured via separate rations of TAC, as the full TAC cannot act due to the absence of α-ketoglutarate dehydrogenase. In the cells of sulfobacilli, one of the enzymes of the glyoxylate cycle – malate synthase – was also detected. However, the second enzyme, isocitrate dehydrogenase, is not detected independently of growth conditions (Zakharchuk et al. 1994; Tsaplina et al. 2000; Karavaiko et al. 2002).

Cells of str. 41 also showed the activity of one of the two enzymes of the glyoxylate cycle, malate synthase, whereas the second enzyme – isocitrate lyase – under no growth conditions was detected (Tsaplina et al. 2000).

Thus, comparing the obtained data with literature data, we can conclude that the TAC in str. 41, as *S. thermosulfidooxidans* VKM B-1269 and S. sibiricus, is not closed and is presented with individual reactions whose role is not only the final breakdown of organic compounds but also synthesis of organic compounds.

Summarizing the abovementioned, it can be concluded that the strains of sulfobacilli are characterized by a flexible metabolism that ensures their survival under extreme conditions, often characterized not only by the lack of organic substances but also by CO_2 and O_2.

Table 7.7 Activity of enzymes (nmol/min mg protein) of TAC and glyoxylate cycle in *S. thermosulfidooxidans* subsp. "asporogenes" str.41 under different growth conditions (Tsaplina et al. 2000)

	Growth conditions		
Enzymes	Autotrophic	Heterotrophic	Mixotrophic
Citrate synthase	13.6	8.7	19.8
Aconitase	146.8	17.6	18.8
Isocitrate dehydrogenase	24.4	2.2	17.6
α-Ketoglutarate dehydrogenase	0	0	0
Succinate dehydrogenase	46.7	64.0	56.1
Fumarase	30.0	105.0	58.2
Malate dehydrogenase	28.9	71.0	50.0
Isocitrate lyase	0	0	0
Malate synthase	6.4	9.6	5.2

7.4.5 Oxidation of Sulfide Minerals

It has been reported that *S. thermosulfidooxidans* VKM-V1269 oxidizes both separate sulfide minerals and their concentrates. According to the rate of oxidation of pyrite and other sulfide minerals, *S. thermosulfidooxidans* can 1.5 to 2.0 times and more exceed *At. ferrooxidans*. Moreover, this advantage of sulfobacilli is maintained even at 30–35 °C, which is not optimal for *S. thermosulfidooxidans* (Karavaiko et al. 1980, 1988).

S. thermosulfidooxidans is of particular interest for the leaching of gold-arsenic, copper-zinc, and other concentrates (Karavaiko et al. 1988). For the processing of refractory gold-bearing pyrite-arsenopyrite concentrates, the tank biooxidation technology is widely used. This technology is based on the oxidation of sulfide minerals by bacteria. As a result gold enclosed passes into the solution and becomes available for subsequent cyanidation. Tank bacterial leaching is usually carried out under mesophilic conditions using *At. ferrooxidans* or the association of *At. thiooxidans* and *L. ferrooxidans*. Studies have shown that moderately thermophilic bacteria *S. thermosulfidooxidans* are also capable of oxidizing a gravitational gold-containing pyrite-arsenopyrite concentrate under conditions of tank leaching at 50 °C. However, the bacterium exhibits high activity only at pulp densities not exceeding 8%. In order to leach this concentrate with moderately thermophilic bacteria at high pulp densities (20%), a variable temperature regime was used. According to this scheme, at the initial stage, the concentrate was oxidized at 30 °C by the association of *At. ferrooxidans* and *S. thermosulfidooxidans* and then at 42 °C with *S. thermosulfidooxidans*. The scheme used made it possible to increase the rate of arsenopyrite oxidation. The advantage of the proposed technology, according to the authors, is due to the nature of the interaction of mesophilic and moderately thermophilic bacteria. These interactions first of all are in the fact that at the initial stage of oxidation, the growth factors are synthesized by mesophilic bacteria, which are afterward used by *S. thermosulfidooxidans* (Norris et al. 1980; Karavaiko et al. 1980). In addition, it is assumed that exometabolites produced by *At. ferrooxidans* form chelates with arsenic, thereby reducing its toxicity. Thus, when using this scheme in the pulp, more favorable conditions for the growth of *S. thermosulfidooxidans* are created (Melamud et al. 1999).

In terms of practical application, it is important to identify the synthrophic relations that arise between *S. thermosulfidooxidans* and its accompanying bacteria – the thermotolerant form of *L. ferrooxidans*. *S. thermosulfidooxidans* in association with *L. ferrooxidans* exhibited equally good growth on medium containing copper-zinc-pyrite ore both in the presence of yeast extract and in its absence at 35–45 °C. According to the authors, the relationship between *S. thermosulfidooxidans* and *L. ferrooxidans* is of syntrophic nature, since both organisms have exhibited a higher growth rate and geochemical activity (Karavaiko et al. 1980).

Studies have shown that when *S. thermosulfidooxidans* MTFe-1 is cocultivated with *At. caldus* KU, the extent of iron leaching from arsenopyrite has reached 99.9%, compared to 36% observed with pure culture of *S. thermosulfidooxidans*. It is noteworthy that when using mixed cultures, no accumulation of elemental sulfur was observed, and the amount of Fe^{2+} was significantly lower. The role of *At. caldus* in the oxidation of arsenopyrite was estimated from the mechanism of its oxidation:

$$FeAsS + 7Fe^{3+} + 4H_2O \rightarrow H_3AsO_4 + 8Fe^{2+} + S^{o} + 5H^{+}$$

Proceeding from this equation, a hypothesis has been proposed, according to which *At. caldus* increases the rate of leaching of arsenopyrite by removing the sulfur accumulated on its surface, thereby making the mineral more accessible for further bacterial or chemical oxidation. The second possible mechanism for the effect of *At. caldus* on the leaching rate of arsenopyrite is associated with the release of growth factors. Earlier, in some representatives of acidophiles, the release of organic substances into the environment was observed (Borchewski 1967). From this point of view, the role of *At. caldus* in the association is to supply *S. thermosulfidooxidans* organic substances. In this case, the relationship between bacteria is symbiotic. It is assumed that with the growth of *S. thermosulfidooxidans*, the concentration of organic substances decreases, which, accumulating in the cytoplasm of chemolithotrophic acidophils, can cause inhibition of the growth (Alexander et al. 1987).

The third possible mechanism for the action of *At. caldus* on the leaching of arsenopyrite is the release of surfactants, which contributes to the dissolution of elemental sulfur. This mechanism underlies the action of the yeast extract as a wetting agent, leading to a dispersion of sulfur in the medium (Dopson and Lindstrom 1999).

The phenomenon of mutualism (mutual nutrition) was also observed in *S. thermosulfidooxidans* when co-grown with *Acidimicrobium ferrooxidans* (Clark and Norris 1996).

7.4.5.1 Oxidation of Pyrite

Studies of peculiarities of pyrite oxidation by *S. thermosulfidooxidans* str. 86 and str. 69 in pure and mixed cultures with *Leptospirillum* spp. bacteria showed that the strains were most active under mixotrophic conditions in the presence of 0.02% yeast extract (Table 7.8) (Vardanyan 1998).

The iron bioleaching rate under autotrophic conditions was significantly lower. The effectiveness of moderately thermophilic bacteria in the oxidation of pyrite without the addition of organic substances significantly increases when they are cocultivated with *Leptospirillum* spp. bacteria. It is worth mentioning that *Leptospirillum* spp. bacteria have shown relatively high efficiency in oxidation of pyrite (Table 7.8).

The activity of FeS_2 oxidation by tested bacteria and their associations correlates with the increase in biomass, which was estimated by the increase of protein and decrease of pH of the medium. Especially in case of the use of these associations, pH of the medium decreased to 1.1 and 0.95. It should be noted that at such extremely low pH values, the bacteria continued to actively grow and oxidize pyrite, which is proved by the presence of small amounts of Fe^{2+} (0.056 g/l) in the medium. Iron leached into the medium was exclusively in form of ferric iron (Fe (III)).

Table 7.8 Oxidation of pyrite by pure and mixed cultures *S. thermosulfidooxidans* and *Leptospirillum* spp. bacteria (FeS_2. 2 %, initial pH 2.0, T 37 °C, duration – 17 days) (Vardanyan 1998)

Bacteria used	Yeast extract	Leached, g/l		Protein, g/l	Final pH
		Fe^{3+}	*Fe^{2+}*		
Leptospirillum spp. str. 64	–	8120	756	0.06	1.25
S. thermosulfidooxidans str.86	–	1484	1232	0.02	1.4
S. thermosulfidooxidans str. 86[a]	+	2492	840	0.025	1.33
S. thermosulfidooxidans str. 86+ str. 64	–	11360	56	0.07	1.15
Leptospirillum spp. str. 72	–	9912	56	0.04	1.1
S. thermosulfidooxidans str. 69	–	2082	1260	0.01	1.25
S. thermosulfidooxidans str. 69[a]	+	3324	280	0.06	1.15
S. thermosulfidooxidans str. 69+ str. 72	–	12960	56	0.1	0.95

[a]*S. thermosulfidooxidans str. 86 and str. 69 were cultivated in mixotrophic condition*

Proceeding from the foregoing, it is assumed that an increase in the rate of pyrite oxidation in the presence of leptospirilla is due to the rapid regeneration of the ferric iron, which is a strong oxidant of pyrite (Vardanyan 1998). Similar data were obtained by Golovacheva who showed that in communities with thermotolerant form of *L. ferrooxidans*, *S. thermosulfidooxidans* can grow well at the temperature of 35–45 °C both in the absence and in the presence of a yeast extract in a medium containing copper-zinc-pyrite ore or pyrite, chalcopyrite, and arsenopyrite as a source of energy (Karavaiko et al. 1980).

The association of moderately thermophilic SIOB with thermotolerant *Leptospirillum* sp. bacteria is able to conduct the process of pyrite oxidation without the addition of organic substances at a rate of about two–three times greater, compared to the mixotrophically growing sulfobacilli. Application of this association allows to enhance pyrite leaching up to 92.7%. The advantage of this association is also the possibility of carrying out bacterial leaching at relatively high temperatures and low pH values.

7.4.5.2 Oxidation of Chalcopyrite

Chalcopyrite ($CuFeS_2$) is the most difficult substrate to be oxidized by chemolithotrophic bacteria (Fu et al. 2008). The dynamics of oxidation of $CuFeS_2$ with thermotolerant and moderately thermophilic SIOB and their associations is shown in Fig. 7.6. According to data presented, the greatest activity in chalcopyrite oxidation was shown by *S. thermosulfidooxidans* str. 86, growing under mixotrophic conditions in the presence of 0.02% yeast extract. Weak oxidation of $CuFeS_2$ was observed in *S. thermosulfidooxidans* str. 86 under autotrophic conditions, as well as *Leptospirillum* sp. str. 64 and *At. tandzuti* str. 5. It has been shown that the activity of str. 86 under autotrophic conditions increases when it is co-grown with a sulfur-oxidizing *At. tandzuti* str. 5. By the amount of leached iron, this association significantly exceeds the mixotrophically grown culture of *S. thermosulfidooxidans* str. 86 (Fig. 7.6).

Thus, the association of moderately thermophilic *S. thermosulfidooxidans* str. 86 and *Leptospirillum* sp. bacteria is several times more active than their monoculture in the oxidation of pyrite (Vardanyan and Vardanyan 2016). The use of a thermotolerant sulfur-oxidizing bacterium with moderately thermophilic bacteria makes it possible to oxidize pyrite and chalcopyrite without the addition of organic substances with the intensity observed in the growth of moderate thermophiles under mixotrophic conditions in the presence of a yeast extract (Vardanyan 1998, 2003).

7.4.5.3 Bioleaching of Refractory Gold-Bearing Ore

Biohydrometallurgical gold extraction is mainly used for the processing of arsenopyrite-pyrite concentrates with finely ingrained gold (Rawlings 1995). The ores of such type are the most appropriate objects for bioleaching, as the presence of pyrite, which has a higher electrode potential, enhances the rapid biooxidation of arsenopyrite and the almost complete exposure of gold for subsequent extraction by cyanidation. However, gold in some deposits, in particular, the Tandzut deposit (Armenia), is bound to pyrite, and the degree of its recovery by traditional cyanidation does not exceed 35%. The integrated biohydrometallurgical method is efficient for the processing of such ores. Taking into consideration that pyrite is the most sparingly oxidizable mineral, including bacterial oxidation, consequently, gold-bearing pyrite ores belong to a refractory type.

The main goal of our investigations was to study the biooxidation of gold-bearing pyrite ore of the Tandzut deposit (Armenia) at elevated temperatures using a moderate thermophilic bacterium from the genus *Sulfobacillus* and thermotolerant bacteria of the genus *Leptospirillum* (Vardanyan 1998; Vardanyan and Akopyan 2003). This ore displayed high contents of iron and sulfur and lacked arsenic. It also contained small amounts of copper, lead, molybdenum, and antimony and gold at a concentration of 1.0–2.0 g/t.

It was demonstrated that *Leptospirillum* sp. str. 64 at its optimal growth temperature oxidized pyrite 1.1–1.9-fold more actively than *S. thermosulfidooxidans* str. 86. Thus, *Leptospirillum* sp. str. 64 over 22-day cultivation leached approximately 64.1% of iron versus 34.1% by *S. thermosulfidooxidans* str. 86 (Table 7.9). In the case of *S. thermosulfidooxidans* subsp. asporogenes strain 86, the pyrite of Tandzut ore was oxidized at a constant low rate, whereas in the experiments with *Leptospirillum* sp. str. 64, a long lag phase (8 days) was observed followed by the active propagation of these bacteria and pyrite oxidation at a rate increasing the corresponding rate displayed by sulfobacilli (Fig. 7.7).

The use of *Leptospirillum* sp. str. 64 in association with *S. thermosulfidooxidans* str. 86 stimulated the bacterial growth and increased the rate of pyrite oxidation; consequently, the degree of its oxidation reached 98.4%. It is assumed that the

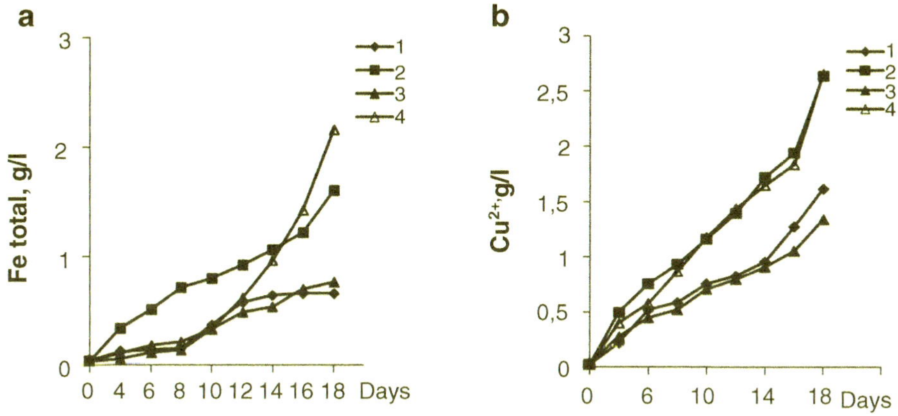

Fig. 7.6 Bioleaching of iron (**a**) and copper (**b**) during oxidation of chalcopyrite by SIOB: 1– *S. thermosulfidooxidans* str. 86, 2– str. 86*, 3 – str. 86 + *Leptospirillum* sp. str. 64, 4 – str. 86 + *Acidithiobacillus* sp. str. 5 (Vardanyan 2003)
*- *S .thermosulfidooxidans* str. 86 was cultivated under mixotrophic conditions

cocultivation led to syntrophic relationships between these bacteria. Presumably, *S. thermosulfidooxidans* str. 86 cells supplied *Leptospirillum* spp. bacteria with ferrous iron and, in turn, satisfied their demand in organic substances at the expense of exometabolites of leptospirilla (Vardanyan and Naghdalyan 2009). An increase in the ambient redox potential resulting from active iron oxidation by *Leptospirillum* cells and domination of Fe^{3+} over Fe^{2+} ions in the medium also enhanced the pyrite oxidation.

The use of thermophilic bacteria can exclude the need in cooling the reaction mass, which heats as a result of the exothermal oxidation reactions of sulfides, in particular, pyrite, to the temperatures exceeding the upper limit for mesophilic bacterial growth. The importance of involving leptospirilla in the leaching of pyrite ore is determined by their high adhesion to pyrite, high affinity for bivalent iron ions (KM = 0.25 mM versus 1.3 mM for *Acidithiobacillus ferrooxidans*), and low sensitivity to inhibition by trivalent iron (Ki = 42.8 mM versus 30 mM for *At. ferrooxidans*) (Norris and Kelly 1978; Norris et al. 1988). Laboratory studies have demonstrated that the bacteria *Leptospirillum* spp. are no less important in the bioleaching of iron-bearing ores and concentrates than *At. ferrooxidans* (Sand et al. 1992).

Molecular biological studies of the microbial population have demonstrated that *Leptospirillum* spp. are dominating bacteria in the oxidation of arsenopyrite concentrates in continuous reactors operating at 40 °C and pH 1.6 (Rawlings 1995; Rawlings et al. 1999). Our studies confirm that the association of moderate thermophilic sulfobacilli and thermotolerant *Leptospirillum* spp. can be successfully used for the bioprocessing of Tandzut and other refractory gold-bearing pyrite ores and concentrates.

Table 7.9 Bioleaching of pyrite containing gold-bearing ore from Tandzut deposit (Armenia) by SIOB and their association (PD –5%, initial pH 2.0, T – 37 and 50 °C, duration – 22 days) (Vardanyan and Naghdalyan 2009)

Culture used	Leached g/l	Fe, %	pH initial/final	Eh, mV
S. thermosulfidooxidans str. 86	3.5	34.1	2.0/1.2	680/710
Leptospirillum sp. str. 64	6.58	64.2	2.0/1.1	680/775
S. thermosulfidooxidans str. 86 + *Leptospirillum* sp. str. 64	10.1	98.5	2.0/1.25	680/815

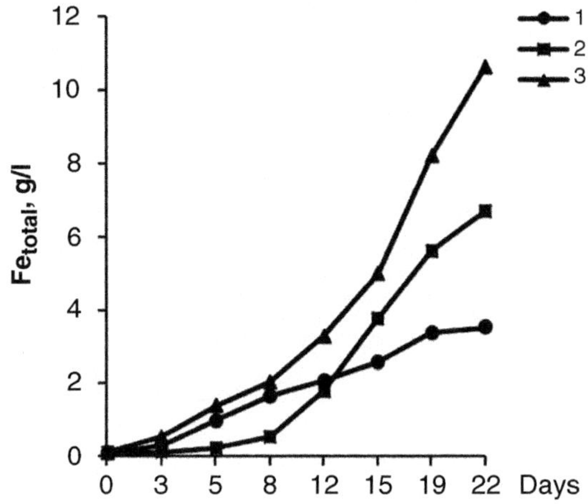

Fig. 7.7 Iron extraction dynamics during bioleaching of Tandzut ore by moderate thermophilic sulfur- and iron-oxidizing bacteria and their associations: 1– str. 86, 2 – str. 64, 3 – str. 86 + str. 64 (PD–5%, pH 2.0, T–37, and 50 °C for *S. thermosulfidooxidans* str. 86) (Vardanyan and Naghdalyan 2009)

7.5 Conclusion

Despite the extreme acidity and high concentration of various metals, sulfide ores of Armenia are characterized by a large variety of CB. Acidophilic IOB are widely represented, which can function in the temperature range of 10–50 °C. It is noteworthy that the microflora of copper, copper-molybdenum, and gold-polymetallic deposits are mainly represented by *At. ferrooxidans* and *S. thermosulfidooxidans*, whereas in the ores rich in pyrite (predominantly polymetallic deposits), *Leptospirillum* spp. bacteria dominate, which can be explained by their physiological characteristics, in particular, high resistance to low pH values and high concentrations of Fe^{3+}. As a result of studies carried out, new and efficient original strains of thermoacidophilic CB belonging to the genera *Sulfobacillus* and *Leptospirillum* have been isolated from sulfide ores of different mineralization.

It has been shown that the activities of iron oxidase in *Leptospirillum* spp. bacteria isolated in Armenia are considerably higher than appropriate activities of *S. thermosulfidooxidans* str. 69 and *At. ferrooxidans*. Although the studied strains of *S. thermosulfidooxidans* were grown in the medium containing Fe^{2+} as the only energy source, high level of sulfite:cytochrome c oxidoreductase and thiosulfate dehydrogenase activities involved in RISCs oxidation was detected in cells. The presence of sulfite:Fe (III) oxidoreductase in the cells of *S. thermosulfidooxidans* str. 69 participating in oxidation of RISCs using Fe (III) as an electron acceptor will allow them to survive under condition of limited oxygen that occurs in bioleaching processes at high temperatures.

Comparative analyses showed a higher level of RuBPCase in *S. thermosulfidooxidans* subsp. asporogenes str. 41 under all growth conditions. It is assumed that the Calvin cycle for str. 41 is the main way of fixing CO_2. It was established that fixation of CO_2 in *S. thermosulfidooxidans* subsp. asporogenes str. 41 under all growth conditions could also be performed by the carboxylation of pyruvate and phosphoenolpyruvate (PEP).

The cells of *S. thermosulfidooxidans* subsp. asporogenes str. 41 like *S. thermosulfidooxidans* VKM B-1269 are not capable of sustainable organotrophic growth on yeast extract or other organic substances. It is assumed that the cause is an ineffective mechanism for destruction of organic compounds. It has been revealed that there is no glyoxylate pathway in the str. 41 and the tricarboxylic acid cycle is open at the level of α-ketoglutarate dehydrogenase and does not work completely. Summarizing the obtained data, we can conclude that the TAC in *S. thermosulfidooxidans* subsp. asporogenes str. 41, as *S. thermosulfidooxidans* VKM B-1269 and S. sibiricus, is not closed and is presented with individual reactions whose role is not only the final breakdown of organic compounds but also the synthesis of organic compounds.

Summarizing the abovementioned, it can be concluded that the strains of sulfobacilli are characterized by a flexible metabolism that enables them to grow under autotrophic, mixotrophic, and heterotrophic conditions and ensures their survival under extreme conditions, often characterized not only by the lack of organic substances but also by CO_2 and O_2.

Thus, the association of moderately thermophilic *S. thermosulfidooxidans* str. 86 and *Leptospirillum* sp. bacteria is several times more active than their monoculture in the oxidation of pyrite. The use of thermotolerant iron or sulfur-oxidizing bacteria with moderately thermophilic sulfobacilli makes it possible to oxidize pyrite and chalcopyrite, respectively, without the addition of organic substances with the intensity observed in the growth of moderate thermophiles under mixotrophic conditions in the presence of a yeast extract. It is assumed that the cocultivation has led to syntrophic relationships between these bacteria. Our studies confirm that the association of moderate thermophilic sulfobacilli and thermotolerant *Leptospirillum* spp. bacteria can be successfully used for the bioprocessing of Tandzut and other refractory gold-bearing pyrite ores and concentrates.

References

Alexander B, Leach S, Ingledev WJ (1987) The relationship between chemiosmotic parameters and sensitivity to anions and organic acids in the acidophile Thiobacillus ferrooxidans. J Gen Microbiol 133:1171–1179

Baker BJ, Banfield JF (2003) Microbial communities in acid mine drainage. FEMS Microbiol Ecol 44:139–152

Battaglia F, Morin D, Garcia JL, Oliver P (1994) Isolation and study of two strains of Leptospirillum – like bacteria from a natural mixed population cultured on a cobaltiferrous pyrite substrate. Antonie van Leeawenhook 66:295–302

Bogdanova TI, Tsaplina IA, Kondrat'eva TF, Duda VI, Suzina NE, Melamud VS, Tourova TP, Karavaiko GI (2006) Sulfobacillus thermotolerans sp. Nov., a new thermotolerant, chemolithotrophic bacterium. Int J Sys Evol Microbiology 56:1039–1042

Bond PL, Druscel GK, Banfield JF (2000) Comparison of acid mine drainage microbial communities in physically and geochemically distinct ecosystems. Appl Environ Microbiol 66:4962–4971

Borchewski KM (1967) Keto acids as growth-limiting factors in autotrophic growth of Thiobacillus thiooxidans. J Bacteriol 93:597–599

Bridge TAM, Johnson DB (1998) Reduction of soluble iron and reductive dissolution of ferric iron-containing minerals by moderately thermophilic iron-oxidizing bacteria. Appl Environ Microbiol 64:2181–2218

Clark DA, Norris P (1996) Acidimicrobium ferrooxidans gen. nov., mixed-culture ferrous iron oxidation with Sulfobacillus species. Microbiology 142:785–790

Coram NJ, Rawlings DE (2002) Molecular relationship between two groups of the genus Leptospirillum and the finding that Leptospirillum ferriphilum sp. nov. dominates South African commercial biooxidation tanks that operate at 40 °C. Appl Environ Microbiol 68:838–845

Dew DW, Lawson EM, Broadhurst JL (1997) The BIOX process for biooxidation of gold-bearing ores or concentrates. In: Rawlings DE (ed) Biomining. Theory, microbes and industrial processes. Springer/Berlin Heilderberg, New York/Londes/Berlin, pp 45–80

Dopson M, Lindstrom EB (1999) Potential role of Thiobacillus caldus in Arsenopyrite Leaching. Appl Environ Microbiol 65:36–40

Dopson M, Lindstrom EB (2004) Analysis of community composition during moderately thermophilic bioleaching of pyrite, arsenical pyrite and chalcopyrite. Microb Ecol 48:19–28

Dufresne S, Bousquet J, Boissenot M, Guay R (1996) Sulfobacillus disulfidooxidans sp. nov., a new acidophilic, disulfide-oxidizing, gram-positive, spore-forming bacterium. Int J Syst Bacteriol 46:1056–1064

Edwards KJ, Goebel BM, Rodgers TM, Schrenk MO, Gihring TM, Cardona MM, Mcguire MM, Hamers RJ, Pace NR, Banfield JF (1999) Geomicrobiology of pyrite (FeS2) dissolution: case study at Iron Mountain, California. Geomicrobiol J 16:155–179

Edwards KJ, Bond PL, Gihring TM, Banfild J (2000) An Archaea iron-oxidizing extreme acidophilic important in acid mine drainage. Science 287:1796–1799

Fu B, Zhou H, Zhang R, Qiu G (2008) Bioleaching of chalcopyrite by pure and mixed cultures of Acidithiobacillus spp. and Leptospirillum ferriphilum. Int Biodeterior Biodegrad 62:109–115

Goebel BM, Stackebrandt E (1994) Cultural and phylogenetic analysis of mixed microbial populations found in natural and commercial bioleaching environments. Appl Environ Microbiol 60:1614–1621

Golovacheva RS (1976) Ultrastructural organization of Sulfobacillus thermosulfidooxidans cells and spores. Microbiologiya 48:681–689

Golovacheva RS (1979) Peculiarities of morphogenesis of Sulfobacillus thermosulfidooxidans. Microbiologiya 48:863–867

Golovacheva RS, Karavayko GI (1978) Sulfobacillus a new genera of thermophilic spore forming bacteria. Mikrobiologiya 47:815–822

Golovacheva RS, Golyshina OW, Karavaiko GI, Dorofeev AG, Pivovarova TA, Chernikh NA (1992) New Iron-oxidizing Bacteria Leptospirillum thermoferrooxidans sp. nov. Mikrobiologiya 61:1056–1064

Golyshina OV, Pivovarova TA, Karavaiko GI, Kondrat'eva TF, Moore ER, Abraham WR, Lansdorf H, Timmis K, Yakimov MN, Golyshin PN (2000) Ferroplasma acidophilum gen. nov., sp. nov., an acidophilic autotrophic, ferrous-iron-oxidizing cell-wall-lacking, mesophilic member of the Ferroplasmaceae fam. nov., comprising a distinct linage of the Archaea. Int J Syst Bacteriol 50:997–1006

Hallmann R, Friedrich A, Koops HP, Pommerening -Roser A, Rohde K, Zenneck C, Sand W (1993) Physiological characteristics of Thiobacillus ferrooxidans and Leptospirillum ferrooxidans and physiochemical factors influence in microbial metal leaching. Geomicrobiology 10:193–206

Harrison AP Jr (1986) Characteristics of Thiobacillus ferrooxidans and other iron-oxidizing bacteria, with emphasis on nucleic acid analyses. Biotechnol Appl Biochem 8:249–257

Harrison AP, Norris PR (1985) Leptospirillum ferrooxidans and similar bacteria: some characteristics and genomic diversity. FEMS Microbiol Lett 33:99–102

Hedrich S, Johnson DB (2013) Acidithiobacillus ferridurans sp. nov., an acidophilic iron-, sulfur- and hydrogen-metabolizing chemolithotrophic gammaproteobacterium. Int J Syst Evol Microbiol 63:4018–4025

Helle U, Onken U (1988) Continuous microbial leaching of a pyritic concentrate by Leptospirillum-like bacteria. Appl Microbiol Biotechnol 28:553–558

Hippe H (2000) Leptospirillum gen. nov. (ex Markosyan 1972) nom. rev. Including Leptospirillum ferrooxidans sp. nov. (ex Markosyan 1972), nom. rev. and Leptospirillum thermoferrooxidans sp. nov. System Evol Microbiol 50:501–503

Johnson DB (1995) Selective solid media for isolating and enumerating acidophilic bacteria. J Microbiol Methods 25:205–218

Johnson DB (1998) Biodiversity and ecology of acidophilic microorganisms. FEMS Microbiol Ecol 27:307–317

Johnson DB, Hallberg KB (2003) The microbiology of acidic mine waters. Res Microbiol 154:466–473

Johnson DB, Joulian C, d'Hugues P, Hallberg KB (2008) Sulfobacillus benefaciens sp. nov., an acidophilic facultative anaerobic Firmicute isolated from mineral bioleaching operations. Extremophiles 12:789–798

Johnson DB, Bacelar-Nicolau P, Okibe N, Thomas A, Hallberg KB (2009) Ferrimicrobium acidiphilum gen. nov., sp. nov. and Ferrithrix thermotolerans gen. nov., sp. nov.: heterotrophic, iron-oxidizing, extremely acidophilic actinobacteria. Int J Syst Evol Microbiol 59:1082–1089

Karavaiko GI, Golovacheva RS (1986) Aerobic thermophilic bacteria, oxidizing sulfur compounds and iron. Biology of thermophilic microorganisms. Nayka, Moscow, pp 35–47

Karavaiko GI, Kovalenko TV, Golovacheva RS (1980) Microbiological aspects of leaching copper from ores. In: Czegledi et al. (eds) Proceedings of the international conference on use of microorganisms in hydrometallurgy, B, Pecs. Hungary, December 4–6, pp 95–107

Karavaiko GI, Golovacheva RS, Pivovarova TA, Tzaplina IA, Vartanyan NS (1988) Thermophilic bacteria of the genus Sulfobacillus. In: Norris PR, Kelly DP (eds) Biohydrometallurgy. Proceedings of the International Symposium on biohydrometallurgy, Warwick, July 12–16. Kew: Science and Technology Letters, pp 29–41

Karavaiko GI, Bulygina ES, Tsaplina IA, Bogdanova TI, Chumakov KM (1990) Sulfobacillus thermosulfidooxidans: a new lineage of bacterial evolution. FEBS 261:8–10

Karavaiko GI, Tourova TP, Tsaplina IA, Bogdanova TI (2000) Investigation of the phylogenetic position of aerobic, moderately thermophilic Bacteria oxidizing Fe^{2+}, S^0, and sulfide minerals and affiliated to the Genus Sulfobacillus. Mikrobiologiya 69(6):732–735

Karavaiko GI, Krasil'nikova EN, Tsaplina IA, Bogdanova TI, Zakharchuk LM (2001) Growth and carbohydrate metabolism of Sulfobacilli. Mikrobiologiya 70:293–299

Karavaiko GI, Zakharchuk LM, Bogdanova TI, Egorova MA, Tsaplina IA, Krasil'nikova EN (2002) The enzymes of carbon metabolism in Thermotolerant Bacillar strain K1. Mikrobiologiya 71(6):755–761

Karavaiko GI, Dubinina GA, Kondrat'eva TF (2006) Lithotrophic microorganisms of the oxidative cycles of sulfur and iron. Microbiology 75:512–545

Kelly DP, Wood AP (2000) Reclassification of some species of Thiobacillus to the newly designated genera Acidithiobacillus gen. nov., Halothiobacillus gen. nov. and Thermithiobacillus gen. nov. Int J Syst Evol Microbiol 50:511–516

Kovalenko EV, Malakhova PT (1983) Spore forming iron oxidizing bacteria Sulfobacillus thermosulfidooxidans. Mikrobiologiya 52:962–966

Krasil'nikova E, Bogdanova T, Zakharchuk L, Tsaplina I, Karavaiko G (1998) Metabolism of reduced sulfur compounds in Sulfobacillus thermosulfidooxidans, strain 1269. Mikrobiologiya 67:156–164

Krasil'nikova EN, Tsaplina IA, Zakharchuk LM, Bogdanova TI (2001) Effects of exogenous factors on the activity of enzymes involved in carbon metabolism in thermoacidophilic bacteria of the genus Sulfobacillus. Prikl Biokhim Mikrobiol 37:418–423

Krasil'nikova E, Bogdanova T, Zakharchuk L, Tsaplina I (2004) Sulfur-metabolizing enzymes in thermoacidophilic bacteria Sulfobacillus sibiricus. Appl Biochem Microbiol 40:53–56

Markosyan GE (1972) A new Iron-oxidizing bacterium Leptospirillum ferrooxidans gen. nov. sp. nov. Biol Zh Armenii 25(2):26–29

Melamud VS, Pivovarova TA, Kondrat'eva TF, Karavaiko GI (1999) Study of oxidation by bacteria of a difficult-to-dress gold-containing pyrite-arsenopyrite concentrate under moderately thermophilic conditions. Prikl Biokhim Mikrobiol 35:182–189

Melamud VS, Pivovarova TA, Tourova TP, Kolganova TV, Osipov GA, Lysenko AM, Kondrat'eva TF, Karavaiko GI (2003) Sulfobacillus sibiricus sp. nov., a new moderately thermophilic bacterium. Miкrobiologiya 72:681–688

Merrettig U, Wlotzka P, Onken U (1989) The removal of pyritic sulphur from coal by Leptospirillum – like bacteria. Appl Microbiol Biotechnol 31:626–628

Norris PR (1983) Iron and mineral oxidation with Leptospirillum – like bacteria. In: Rossi G, Torma AE (eds) Recent progress in biohydrometallurgy. Associazione Mineraria Sarda, Iglesias, pp 43–59

Norris PS, Kelly DP (1978) In: Murr LF, Torma AE, Brierley JA (eds) Metallurgical application of bacterial leaching and related microbiological phenomena. Academic Press, New York, p 83

Norris PR, Brierley JA, Kelly DP (1980) Physiological characteristics of two facultatively-thermophilic mineral oxidizing bacteria. FEMS Microbiol Lett 7:119–122

Norris PR, Barr DW, Hinson D (1988) Iron and mineral oxidation by acidophilic bacteria: affinities for iron and attachment to pyrite. In: Norris PR, Kelly DP (eds) Biohydrometallurgy: proceedings of international symposium. Science and Technology Letters, Kew/Surrey, pp 43–59

Norris PR, Clark DA, Owen JP, Waterhouse S (1996) Characteristics of Sulfobacillus acidophilus sp. nov. and other moderately thermophilic mineral-sulfide-oxidizing bacteria. Microbiology 142:775–783

Norris PR, Burton NP, Foulis NAM (2000) Acidophiles in bioreactor mineral processing. Extremophiles 4:71–76

Okibe N, Gericke M, Hallberg KB, Johnson DB (2003) Enumeration and characterization of acidophilic microorganisms isolated from a pilot plant stirred-tank bioleaching operation. Appl Environ Microbiol 69:1936–1943

Olson GJ, Brierley JA, Brierley CL (2003) Bioleaching review part B: progress in bioleaching: applications of microbial processes by the minerals industries. Appl Microbiol Biotechnol 63:249–257

Pivovaarova TA, Markosyan GE, Karavaiko GI (1981) Morphogenesis and thin structure of Leptospirillum ferrooxidans. Mikrobiologiya 50:482–486

Pivovarova TA, Golovacheva RS (1985) Cytology, physiology and biochemistry of microorganisms important for hydrometallurgy. In: Karavaiko GI, Groudev SNM (eds) Biogeotechnology of metals. UNEP, Moscow, pp 27–55

Pizarro J, Jedlicki E, Orlellana O, Romero J, Espejo RT (1996) Bacterial populations in samples of bioleached copper ore as revealed by analysis of DNA obtained before and after cultivation. Appl Environ Microbiol 62:1323–1328

Rawlings DE (1995) In: Jerez CA, Vargas T, Toledo H, Wiertz JV (eds) Biohydrometallurgical processing. University of Chile, Santiago, pp 9–17

Rawlings DE (2002) Heavy metal mining using microbes. Ann Rev Microbiol 56:65–91

Rawlings DE, Johnson DB (2007) Biomining. Springer, Berlin-Heidelberg

Rawlings DE, Tributsch H, Hansford GS (1999) Reasons why 'Leptospirillum'-like species rather than Thiobacillus ferrooxidans are the dominant iron-oxidizing bacteria in many commercial processes for the biooxidation of pyrite and related ores. Microbiology 145:5–13

Sand W, Rohde K, Sobotke B, Zenneck C (1992) Evolution of Leptospirillum ferrooxidans for leaching. Appl Environ Microbiol 58:85–92

Sand W, Gehrke T, Hallmann R, Schippers A (1995) Sulfur chemistry, bio¢lm, and the (in)direct attack mechanism – a critical evaluation of bacterial leaching. Appl Microbiol Biotechnol 43:961–966

Sand W, Jozsa P-G, Kovacs Z-M, Săsăran N, Schippers A (2007) Long-term evaluation of acid rock drainage mitigation measures in large lysimeters. J Geochem Explor 92:205–211

Schippers A (2007) Microorganisms involved in bioleaching and nucleic acid-based molecular methods for their identification and quantification. In: Donati ER, Sand W (eds) Microbial processing of metal sulfides. Springer, Dordrecht, pp 3–33

Schippers A, Sand W (1999) Bacterial leaching of metal sulfides proceeds by two indirect mechanisms with thiosulfate or via Polysulfides and sulfur. Appl Environ Microbiol 65:319–321

Schippers A, Jozsa PG, Sand W (1996) Sulfur chemistry in bacterial leaching of pyrite. Appl Environ Microbiol 62:3424–3431

Schippers A, Rohwerder T, Sand W (1999) Intermediary sulfur compounds in pyrite oxidation: implications for bioleaching and biodepyritization of coal. Appl Microbiol Biotechnol 52:104–110

Senyushkin AA, Severina LO, Mityushina LL (1997) Capsule formation by Sulfobacillus thermosulfidooxidans cells growing under oligotrophic and mixotrophic conditions. Mikrobiologiya 66:455–461

Severina LO, Senyushkin AA, Karavaiko GI (1995) The structure and chemical composition of the S-layer in representatives of the genus Sulfobacillus. Mikrobiologiya 64:336–340

Steudel R (1996) Mechanism for the formation of elemental sulfur from aqueous sulfide in chemical and microbiological desulfurization processes. Ind Eng Chem Res 35:1417–1423

Sugio T, Mizunashi W, Inagaki K, Tano T (1987) Purification and some properties of sulfur: ferric ion oxidoreductase from Thiobacillus ferrooxidans. J Bacteriol 169:4916–4922

Suzina NF, Severina LO, Senyushkin AA, Karavaiko GI, Duda VI (1999) Ultrastructural Organization of Membrane System in Sulfobacillus thermosulfidooxidans. Mikrobiologiya 68:491–500

Tsaplina IA, Osipov GA, Bogdanova TI, Nedorezova TP, Karavaiko GI (1994) Fatty acid composition of lipids of thermoacidophilic bacteria from the genus Sulfobacillus. Mikrobiologiya 63:821–830

Tsaplina IA, Krasil'nikova EN, Zakharchuk LM, Egorova MA, Bogdanova TI, Karavaiko GI (2000) Carbon metabolism in Sulfobacillus thermosulfidooxidans subsp. asporogenes, strain 41. Mikrobiologiya 69:334–340

Turova TP, Poltoraus AB, Lebedeva IA, Bulygina ES, Tsaplina IA, Bogdanova TI, Karavaiko GI (1995) Phylogenetic position of Sulfobacillus thermosulfidooxidans: determination based on the 5S and 16SrRNA sequence analysis. Mikrobiologiya 64:366–374

Tyson GW, Lo I, Baker BJ, Allen EE, Hugenholtz P, Banfield JF (2005) Genome-directed isolation of the key nitrogen fixer Leptospirillum ferrodiazotrophum sp. nov. from an acidophilic microbial community. Appl Environ Microbiol 71:6319–6324

Vardanyan NS (1998) The effect of environmental factors on pyrite oxidation by Sulfobacillus thermosulfidooxidans. Biotekhnologiya 6:48–55

Vardanyan NS (2003) Oxidation of pyrite and chalcopyrite by mixed cultures of Sulfobacilli and other sulphur and iron oxidizing bacteria. Biotechnology 6:79–83

Vardanyan NS, Akopyan VP (2003) Leptospirillum-like bacteria and evaluation of their role in pyrite oxidation. Mikrobiologya 72:493–497

Vardanyan NS, Naghdalyan SZ (2009) Periodic bioleaching of refractory gold-bearing pyrite ore. Appl Biochem Mikrobiol 45(4):401–405

Vardanyan AK, Vardanyan NS (2016) Bioleaching of pyrite and chalcopyrite by new isolated thermotolerate sulfur-oxidizing bacteria Acidithiobacillus tandzuti sp.nov. Int J Sci Eng Res 7:203–207

Vardanyan AK, Vardanyan NS, Markosyan LM (2013) Peculiarities of adhesion and bioleaching of pyrite by new isolated Leptospirillum spp. Bacteria. Univ J Microbiol Res 1(2):22–25

Vardanyan A, Stepanyan S, Vardanyan N, Markosyan L, Sand W, Vera V, Zhang R (2015a) Study and assessment of microbial communities in natural and commercial bioleaching systems. Miner Eng 81:167–172

Vardanyan A, Vardanyan N, Markosyan L, Sand W, Vera M, Zhang R (2015b) Biofilm formation and extracellular polymeric substances (eps) analysis by new isolates of Leptospirillum, Acidithiobacillus and Sulfobacillus from Armenia. Adv Mater Res 1130:153–156

Vartanyan NS, Pivovarova TA, Tsaplina IA, Lysenko AM, Karavaiko GI (1988) A new thermoacidophilic bacterium belonging to the Sulfobacillus genus. Mikrobiologiya 57:268–274

Wisotzkey JD, Jurtsbuk P Jr, Fox GE et al (1992) Comparative sequence analysis on the 16S rRNA (rDNA) of Bacillus acidocaldarius, Bacillus acidoterrestris and Bacillus cycloheptanicus and proposal for creation of new genus, Alicyclobacillus gen. nov. Syst Bacteriol 42:263–269

Wood AP, Kelly DP (1984) Growth and sugar metabolism of a thermoacidophilic iron-oxidizing mixotrophic bacterium. J Gen Microbiol 130:1337–1348

Zakharchuk LM, Tsaplina IA, Krasil'nikova EN, Bogdanova TI, Karavaiko GI (1994) Carbon metabolism in Sulfobacillus thermosulfidooxidans strain 1269. Mikrobiologiya 63:573–580

Zakharchuk LM, Egorova M, Tsaplina IA, Bogdanova TI, Krasil'nikova EN, Melamud VS, Karavaiko GI (2003) Activity of the enzymes of carbon metabolism in Sulfobacillus sibiricus under various conditions of cultivation. Mikrobiologiya 72:62–626

Zhang R, Xia J, Peng J, Zhang Q, Zhang C, Nie Z, Qiu G (2010) A new strain Leptospirillum ferriphilum YTW315 for bioleaching of metal sulfides ores. Trans Nonferrous Met Soc China 20:135–141

Thermophilic and Halophilic Microorganisms Isolated from Extreme Environments of Turkey, with Potential Biotechnological Applications

8

Kemal Guven, Fatma Matpan Bekler, and Reyhan Gul Guven

Abstract

Turkey has a great number of different ecological areas, owning over 200 hot water resources and various hypersaline environments with a broad microbial diversity and opportunities for newly isolated microorganisms from extreme environments for many industrial applications. A variety of thermophilic and halophilic microorganisms in different regions of Turkey have been isolated and identified. The thermophilic bacterial members studied were *Anoxybacillus*, *Geobacillus*, *Bacillus*, *Brevibacillus*, and *Aeribacillus* belonging to the *Bacillaceae* family and the other thermophiles such as *Thermus* and *Thermomonas*, whereas the isolated halophilic microorganisms were mainly found to be members of the archaeal family *Halobacteriaceae* or grouped into bacterial phylum *Bacteroidetes*. In summary, the present study reviews on (1) isolating and identifying thermophiles and halophiles single or as community from various extreme habitats in Turkey based on conventional (morphological, physiological and biochemical tests) and/or molecular methods, (2) screening these extremophiles for industrially important enzymes, (3) studying other novel products and their use in other areas of biotechnology, and finally (4) discussing about the development strategies and the future perspectives on poorly studied extremophilic microorganisms in the country to fulfill future biotechnological and industrial demands.

K. Guven (✉)
Department of Molecular Biology and Genetics, Dicle University, Diyarbakir, Turkey
e-mail: kemalg@dicle.edu.tr

F. Matpan Bekler
Department of Biology, Dicle University, Diyarbakir, Turkey

R. Gul Guven
Department of Science and Mathematics Education, Dicle University, Diyarbakir, Turkey

D. Egamberdieva et al. (eds.), *Extremophiles in Eurasian Ecosystems: Ecology, Diversity, and Applications*, Microorganisms for Sustainability 8,
https://doi.org/10.1007/978-981-13-0329-6_8

Keywords
Thermophiles · Halophiles · Identification · Microorganisms · Biotechnological applications · Turkey

8.1 Introduction

Microbes in nature have served one of the largest and useful sources of many bioproducts including enzymes, polymers such as exopolysaccharides and polyhydroxyalkanoate, osmolytes, etc. Recently, biotechnology has increased its efforts to search for new organisms of practical use. Many microbial taxa need to be discovered and isolated from various extreme environmental samples all over the world. Extremophiles can survive in a variety of extreme conditions, which are classified as halophiles, thermophiles, psychrophiles, alkalophiles, acidophiles, barophiles, metalophiles and others depending on their adaptations to unusual environmental conditions.

Turkey has lots of different ecological areas, which possesses a broad microbial diversity. Turkey is a peninsular country surrounded by seas and also well known for its geothermal activity, and there are so many thermal springs all over the country. Therefore, there should be a great deal of opportunities for newly isolated microorganisms from extreme environments, including thermophilic and halophilic ones with numerous biotechnological applications. Because a search of extremophiles in the country is very recent, this potential has not been fully exploited.

It has been already easy to detect novel and rare microorganisms due to the improved classification methods based on the integrated use of phenotypic and genotypic data, also thanks to the molecular techniques developed most recently to expand the search for new bioproducts by exploring the diversity of microorganisms. Microbial diversity and novel molecular techniques, like genomics and metagenomics, are being utilised to discover new microbial enzymes and other bioproducts whose properties can be modified/improved by varying strategies, as well as using different bioinformatics tools. Microbial bioproducts with potential for biotechnological applications are obtained from a variety of bacterial groups including extremophiles, as well as mesophiles. Recent advances in modern biotechnology have led to great development in new bioproducts, through bioproduct applications; mainly enzymes are already well established. The application of novel biotechnology research in environmentally friendly bioprocesses is also rapidly expanding.

For many decades, the *Bacillaceae* family members have been good sources in biotechnological processes concerning whole cells or enzymes. In Turkey, the isolated and identified thermophilic members of the *Bacillaceae* family include *Anoxybacillus*, *Geobacillus*, *Bacillus*, *Brevibacillus*, *Aeribacillus*, etc. Moreover, isolated halophilic microorganisms were mainly found to be members of both bacteria and archaea.

In this chapter, an attempt has been made to review on the diversity of microorganisms isolated and identified in extreme environments of Turkey, which are hot water resources and hypersaline salterns, salt lakes or salt mines, as well as on the potential biotechnological applications of their industrially important enzymes and other novel products.

8.2 Thermal Springs Studied in Turkey

It is well known that hot water resource and geothermal region are the main thermophilic regions. Hot water resources are located in different parts of the world, due to volcanic activities. Turkey possesses many geothermal sources with varying typical temperatures and pH values. Figure 8.1 shows the map of all thermal water resources and those studied and documented within this review. The hot springs in Turkey, where the temperatures vary from 36 to 80 °C, are mostly rich in calcium and magnesium ions. It is well known that these springs are used for curing so many neurological, gynaecological, rheumatismal and dermatological diseases, as well as for digestive disorders and physical exhaustion (http://turkeyculture.org/). Although we are going to review on the studies carried out so far on the thermophilic microorganisms within this chapter, it should be mentioned that these environments have not yet been intensively studied in terms of microbiological diversity.

8.3 Thermophilic Microorganisms Isolated and Identified in Turkey

Microbial growth requires temperature as a vital parameter, and different temperature ranges are preferred by microorganisms to survive. Thus, microorganisms are grouped as psychrophiles, mesophiles, thermophiles and hyperthermophiles.

The family *Bacillaceae* currently consists of 62 genera and known as one of the largest bacterial families. The majority of *Bacillaceae* produce endospores and are Gram-positive, either rod-shaped (bacilli) or spherical (cocci), bacteria. *Bacillus*, *Anoxybacillus*, *Geobacillus*, *Brevibacillus*, etc. are classified into *Bacillaceae*, within the phylum *Firmicutes*, class *Bacilli* and order *Bacillales* (Mandic Mulec et al. 2016). The presence of thermophilic bacteria in the *Anoxybacillus*, *Bacillus*, *Geobacillus*, *Brevibacillus*, *Thermus* and *Aeribacillus* genera in thermal areas has been reported in Turkey (Belduz et al. 2003; Gul Guven et al. 2008; Inan et al. 2012; Poli et al. 2012; Bozoglu et al. 2013; Kacagan et al. 2015; Baltaci et al. 2016; Yildirim et al. 2017).

8.3.1 Anoxybacillus

Among bacilli members, unlike *Brevibacillus* and *Bacillus*, *Anoxybacillus* is a rather new genus that was recently proposed, known as aerotolerant anaerobes or

Fig. 8.1 The map of all geothermal areas in Turkey (blue circles) and those studied for microbial diversity within this review (red triangles), obtained and modified from documents of MTA (General Directorate of Mineral Research and Exploration of Turkey)

facultative anaerobes. The genus name *Anoxybacillus* means small rod living in the absence of oxygen (Pikuta et al. 2000). The members of the genus *Anoxybacillus* are Gram-positive, endospore-forming, rod-shaped (bacilli) bacteria, sizing 0.4–1.5 × 2.5–9.0 μm. Most *Anoxybacillus* species are moderately thermophilic having an optimal growth from 50 to 65 °C. *Anoxybacillus* cells are known to be alkalitolerant thermophile, which are suitable for most industrial applications. Since the first report of *Anoxybacillus*, this genus has been shown to serve as a possible choice in various applications involving lignocellulosic and starch biomasses, enzyme technology, waste treatment and also bioenergy manufacturing (Goh et al. 2013; Cihan and Yildiz 2016). As shown in Table 8.1, several important *Anoxybacillus* members have been identified, and their potential use in biotechnology has been evaluated in Turkey. In this section, the review of relevant literature will be presented in the chronological order.

Belduz et al. (2003) studied seven xylanolytic, thermophilic bacterial strains from mud and water samples from two major hot springs, namely, Gonen and Diyadin, located in the Turkish provinces of Balikesir and Agri, respectively. Among these strains, following morphological, biochemical and genetic analysis, *Anoxybacillus gonensis* was found to be a novel sporulating, rod-shaped, thermophilic bacterium (with an optimum temperature of 55–60 °C), growing on various carbon sources, such as xylose, glucose, starch and mannitol. It was also found to produce a high level of xylose isomerase.

Two moderately thermophilic (optimum temperature for growth, 50–55 °C) *Anoxybacillus* species were also isolated from Kestanbol and Ayder hot springs in Canakkale and Rize provinces, respectively, by Dulger et al. (2004). They were identified as *A. ayderensis* and *A. kestanbolensis* which were sporulating, Gram-positive, rod-shaped bacteria, growing on a variety of carbon sources including maltose, D-sucrose, D-glucose, D-mannose, D-mannitol, D-raffinose, D-fructose, D-xylose and L-arabinose.

In addition, Gul Guven et al. (2008) isolated a novel thermophilic Gram-positive strain KG8$^{(T)}$ from Taslidere hot spring in Batman. This strain was motile, spore-forming, aerobe, rod-shaped and occurring in pairs or filamentous. The growth range was between 35–65 °C (optimum temperature of 55 °C) and at pH 5.5–9.5 (optimum pH of 7.5). Because the strain KG8 was incapable to utilise most carbohydrates, this new subspecies was named as *A. kamchatkensis* subsp. *asaccharedens*. 16S rRNA gene sequence similarity, chemotaxonomic data and the results of biochemical and physiological tests, DNA–DNA hybridisation allowed phenotypic and genotypic differentiation of strain KG8 supporting the affiliation to the genus *Anoxybacillus*. This subspecies was found to be a good source of the enzyme amylase capable of utilising starch.

In a microbial diversity study, Adiguzel et al. (2009) identified 15 Gram-positive thermophilic bacteria isolated from Pamukcu (Balikesir), Sorgun (Yozgat), Ilica and Akdag (Erzurum) hot springs by using various methods including phenotypic, chemotaxonomic, 16S rRNA sequencing and rep-PCR genomic fingerprint profilings. They suggested that these profilings can be used as a reliable technique to identify thermophilic bacteria in the genera of *Bacillus*, *Anoxybacillus* and *Geobacillus* spp.

Table 8.1 *Anoxybacillus* members isolated and identified from various regions of Turkey and their potential use in biotechnology

Microorganisms	Strain code	GenBank accession number	Source	Growth temperature (°C)	Growth pH	Biotechnological applications and products	Reference
Anoxybacillus ayderensis	AB04^T	NCIMB 13972^T	Ayder hot spring/Rize	50–55	7.5–8.5	Xylose isomerase, esterase, fructose-1,6-bisphosphate aldolase, heavy-metal resistance	Dulger et al. (2004) and Belduz et al. (2015)
Anoxybacillus calidus	C161ab^T	FJ430012	Kizildere hot spring/Denizli	55	8.0-8.5	Amylase	Cihan et al. (2014)
Anoxybacillus flavithermus	M14, VI, R32, R40, M6.p1, 75B	–	Balcova geothermal site/ Izmir	30–70	9.0–10.0	Amylase, lipase	Yavuz et al. (2004)
Anoxybacillus flavithermus	–	–	Omer hot spring/ Afyonkarahisar	55	7.0	Amylase, metal biosorption	Ozdemir et al. (2011), Fincan et al. (2014) and Yener et al. (2017)
Anoxybacillus flavithermus	–	–	Gazligol hot spring/ Afyonkarahisar	60	7.0	Amylase	Ozdemir et al. (2012)

Anoxybacillus gonensis	G2^{T}	NCIMB 13933^{T}	Gonen hot spring/Balikesir	55–60	7.5–8.0	Xylose-utilising, esterase, thermozyme, amylase, pullulanase, cyclodextrinase, β-xylosidase, xylose isomerase, second-generation biofuel production, oligo-1,6-glucosidase, fructose-1,6-bisphosphate aldolase, α-galactosidase, α-glucosidase, β-galactosidase, mercury and arsenate reductase, glucose isomerase, CTP synthase, heavy-metal biosorption, ATPase activity of aluminium tolerance protein	Belduz et al. (2003), Colak et al. (2005), Faiz et al. (2007), Ertunga et al. (2007), Duran et al. (2009), Beris et al. (2011), Oztekin et al. (2011), Karaoglu et al. (2013), Sandalli et al. (2014), Lim et al. (2015) and Belduz et al. (2015)
Anoxybacillus gonensis	P39-89	FJ808725-26	Ilica hot spring/ Erzurum	55	7.5-10.0	Removal of textile dye, laccase	Adiguzel et al. (2009) and Yanmis et al. (2016)
Anoxybacillus gonensis	Z4		Ilica hot spring/ Erzurum	55	7.2	Amylase	Baltas et al. (2012)
Anoxybacillus gonensis	A4-O9	AY248707 KM596794	Diyadin hot spring/Agri	60	7.5	Amylase, cellulase	Colak et al. (2008), Genc et al. (2014), (2015) and Baltaci et al. (2016)
Anoxybacillus kamchatkensis subsp. *asaccharedens*	KG8$^{(T)}$	AM999779	Taslidere hot spring/Batman	55	7.5	Amylase	Gul Guven et al. (2008)

(continued)

Table 8.1 (continued)

Microorganisms	Strain code	GenBank accession number	Source	Growth temperature (°C)	Growth pH	Biotechnological applications and products	Reference
Anoxybacillus kaynarcensis	D1021^T	EU926955	Kaynarca hot spring/İzmir	60	7.0	Xylanase	Inan et al. (2013)
Anoxybacillus kaynarcensis	A1, A13	KC310452 KC310464	NED[1]	57±1	6.5–9.0	Amylase, cellulase	Yanmis et al. (2015)
Anoxybacillus flavithermus	A2	KC310453				Amylase	
Anoxybacillus rupiensis	A3	KC310454				Amylase	
Anoxybacillus gonensis	A5	KC310456				Amylase, cellulase	
Anoxybacillus kestanbolensis	K4^T	NCIMB 13971^T	Kestanbol hot spring/ Canakkale	50–55	7.5–8.5	Ribulokinase, protease	Dulger et al. (2004), Atasoy et al. (2012) and Tokgoz et al. (2014)
Anoxybacillus pushchinoensis	A8	AY248715	Diyadin hot spring/Agri	60	7.0–8.8	Xylanase	Kacagan et al. (2008)
Anoxybacillus salavatliensis	A343^T	EU326496	Salavatli hot spring/Aydin	60	8.0–9.0	α-Glucosidase	Cihan et al. (2011a)
Anoxybacillus sp.	PDF1	HQ875718	Seferihisar and Karakoc hot springs/Izmir	55	7.4	Carboxylesterase	Ay et al. (2011)
Anoxybacillus sp.	KP1	KC525949	Kopru hot spring/Agri	50	8.0	Amylase, protease, β-galactosidase	Matpan Bekler and Guven (2014) and Matpan Bekler et al. (2015c, 2017)
Anoxybacillus sp.	AH1	KP172526	Dargecit hot spring/Mardin	60	7.0–7.5	Amylase	Acer et al. (2015, 2016)

Anoxybacillus sp.	HBB 16 HBB 134	KR91195 GQ342689	Alangullu hot spring/Aydin	65	6.5–8.0	Lipase	Bakir and Metin (2015, 2017)
Anoxybacillus sp.	KB4	KU997674	Kusburnu hot spring/Agri	55–60	9.0–10.0	Amylase	Matpan Bekler (2016)
Anoxybacillus sp.	O20	KM668039	NED[2]	56	7.0	Amylase, cellulase, lipase	Genc et al. (2015) and Baltaci et al. (2016)
Anoxybacillus sp.	G18	FJ808715	Sorgun hot spring/Yozga	55–60	7.5–10.0	–	Adiguzel et al. (2009)
Anoxybacillus thermarum	A4	KC310455	Erzurum	55	7.0	Amylase, cellulase	Yanmis et al. (2015) and Baltas et al. (2016)
Anoxybacillus sp.	NB	FJ215763	Hisaralan hot spring/Balikesir	70	7.5	DNA polymerase I	Caglayan and Bilgin (2011)

NED: The samples collected from various hot springs and the source are **not exactly defined**

NED[1]: The samples collected from Kirsehir–Terma, Sirnak/Guclu Konak–Belkis Ana, Agri–Diyadin, Kayseri–Haciveli, Erzurum–Ilica, Urfa–Karaali, Rize–Ayder, Adiyaman–Kurucay and Ankara–Kizilcahamam

NED[2]: The samples collected from Bademli Bahce hot spring (Bursa), Sulusaray hot spring (Tokat), Sicak Cermik hot spring (Sivas), Bostanci hot spring (Balikesir) and Diyadin hot spring (Agri)

The results demonstrated that thermophilic bacterial strains collected from the hot springs were classified into three main clusters, one of which consisted three *Anoxybacillus* strains.

Cihan et al. (2011a) identified a moderate thermophilic bacilli, spore-forming, Gram stain-positive, facultative anaerobic, motile and α-glucosidase-producing novel *Anoxybacillus* species named *A. salavatliensis*, obtained from a high-temperature well-pipeline sediment in Salavatli town of Aydin, Turkey. In this study, the rep-PCR (BOX-PCR, (GTG)5-PCR) and ITS fingerprinting analyses were carried out between phylogenetically related species clustering of strain A343^T with its closely related *Anoxybacillus* species, as well as performing sodium dodecyl sulphate-polyacrylamide gel electrophoresis (SDS–PAGE) total protein profiles of relevant species. Growth of *A. salavatliensis* was observed at a range of 37–69 °C (optimum temperature of 60 °C) at a range of pH 5.5–9.5 (with optimum of 8.0–9.0) and able to grow on a variety of carbon sources. The biochemical tests showed the biotechnological potential of this bacterium concerning its enzymes, e.g. starch and gelatin utilisation; catalase, oxidase and amylase activities; and reduction of nitrate.

Nine of xylanolytic thermophilic microorganisms isolated from some hot springs located in the west of Turkey were found to belong to the genus *Anoxybacillus* on the basis of phenotypic characteristics and 16S rRNA gene sequence analysis (Inan et al. 2011a). Among these strains, a novel moderately thermophilic bacilli, endospore-forming, Gram-positive, motile, alkalitolerant strain D1021^T was described by Inan et al. (2013) from Kaynarca hot spring (water temperature, 60–100 °C) in Izmir Province of Turkey. The growth characteristics observed were temperature range of 35–70 °C (optimum 60 °C) and pH range of 6.0–10.0 (optimum pH of 7.0). The strain utilised a variety of carbon sources such as glucose, ribose, xylose, arabinose, maltose, melibiose and sucrose. They identified the strain as a new species and named *A. kaynarcensis* on the basis of phenotypic characteristics, phylogenetic and DNA–DNA hybridisation data, as well as *rpoB* gene analysis. In addition to 16S rRNA gene, the *rpoB* gene was shown to be successfully used as a genotyping approach to phylogenetic studies within *Anoxybacillus* (Inan et al. 2011b). The analysis of *rpoB* gene that is for the RNA polymerase beta subunit has been previously suggested in taxonomic studies of bacteria as an alternative of the 16S rRNA gene, containing conserved and variable regions (Da Mota et al. 2004).

Cihan (2013) has also studied 115 endospore-forming bacilli isolated from geothermal areas in Turkey by analysing 16S rDNA sequence analyses, as well as by ARDRA, ITS-PCR and rep-PCR. The isolates used in this study were collected from water, sediment, soil, stone and tree branch samples within ten hot springs and nine well pipelines of high temperatures, located in Aegean Region and Middle Anatolian Region. Among these strains, most widely distributed thermophiles belonged to the genus *Anoxybacillus* with 53 isolates. The isolated strains were grouped into eight phylogenetic lineages within the type strains of *A. flavithermus*, *A. salavatliensis*, *A. kamchatkensis* and *A. kamchatkensis* subsp. *asaccharedens*. In this study, the author also underlines the importance of *Anoxybacillus* isolates which produce biotechnologically valuable enzymes, the ability of carbohydrate degradation mainly amylolytic, glucosidic and proteolytic activities that made them superior in comparison to the remaining bacilli in the extreme habitats.

A novel amylase-producing thermophilic bacterial strain KP1 from the hot spring of Diyadin in Agri, Turkey, was isolated by Matpan Bekler and Guven (2014). The strain KP1 belonged to the genus *Anoxybacillus* on the basis of phylogenetic analysis by the sequence similarity of 16S rRNA gene by biochemical and physiological tests. Moreover, a thermophilic, starch-hydrolysing bacterium identified as *A. calidus* was isolated from soil near a thermal power plant near Kizildere (the water temperature of this geothermal region is between 195 and 212 °C) within Denizli province in Aegean Region of Turkey by Cihan et al. (2014). They also analysed the strain further by using the results of rep-PCR and ITS fingerprinting differentiating from related species of the genus *Anoxybacillus*. This novel species is facultatively anaerobic, rod-shaped, Gram-positive staining, motile and endospore-forming bacterium, which grows at a temperature range between 35 and 70 °C (with optimum 55 °C), at pH range of 6.5–9.0 (with optimum 8.0–8.5).

In another study, Acer et al. (2015) isolated α-amylase-producing thermophilic bacteria from the mud of Dargecit hot spring (water temperature and pH as 58 °C and 6.9, respectively) in Mardin Province of Turkey. The isolated strain AH1 was found to be a member of *Anoxybacillus* genus by characterising with the morphological, biochemical and physiological tests, in addition to the genetic analysis by 16S rRNA sequences. The analysis of 16S rRNA gene sequence showed that the most sequence similarity of the strain AH1 (DSM 23210^T) was to *A. flavithermus* subsp. *flavithermus* DSM 2641^T by 98.23%. The strain was Gram-positive, aerobe and spore-forming rod, which had an optimum growth temperature and pH values of 60 °C and 7.0–7.5, respectively. However, it was found to grow in a wide pH range (5.5–10.0), indicating that *Anoxybacillus* AH1 cells were alkaliphilic or alkalitolerant.

Belduz et al. (2015) have recently completed the study on genome sequences of thermophilic *A. ayderensis* AB04^T (=NCCB 100050^T = NCIMB 13972^T) which was isolated and described previously from the hot spring of Ayder in Rize Province of Turkey. The strain genome was 2,832,347 bp long and found to contain 2895 predicted genes as well as 103 RNA genes including 88 tRNAs, 14 rRNAs and 1 tmRNA. *A. gonensis* type strain G2^T (=NCCB 100040^T = NCIMB 13,933^T) isolated from Gonen hot springs in Turkey and identified previously was also studied by Lim et al. (2015) for its annotated and complete genome sequencing. It was presented that the total length of the genome was 2,803,668 bp, with 41.7% G + C content. Moreover, the genome comprised of 2934 protein-coding sequences, 62 pseudogenes, 2769 CDS, 78 tRNAs and 24 rRNAs.

An aerobic, motile, rod-shaped and Gram-positive thermophilic strain KB4 was isolated by Matpan Bekler (2016) from Kusburnu hot spring (the temperature and pH of the water 70 °C and 7.5) in Agri Province of Turkey. The strain growth was obtained at temperature of 55–60 °C, at pH 9.0–10.0 and at 3% (w/v) NaCl. The sequence analysis of 16S rRNA gene indicated that the KB4 strain was closely related to *A. pushchinoensis* K1$^{(T)}$ with a sequence similarity of 98.78%.

Most recently, Baltaci et al. (2016) have studied on the identification thermophilic bacteria from the water and sludge samples of different hot springs, including Sulusaray (50 °C, pH 7.2), Sicak Cermik (53 °C, pH 6.6), Bostanci (51 °C, pH 7.2), Bademli Bahce (58 °C, pH 6.6) and Diyadin (70 °C, pH 6.8). The identification was carried out by chemotaxonomic data from FAMEs, the biochemical, physiological and morphological tests and molecular methods (16S rRNA sequencing and GTG5-PCR). The strains were resembled mainly to three genera, namely, *Anoxybacillus*, *Bacillus* and *Aeribacillus*, according to 16S rRNA sequencing results. However, a strain designated as O20 exhibited 97% similarity to *A. kaynarcensis*, while the other (O9) was resembled to *Anoxybacillus gonensis* with a similarity rate of 99%.

8.3.2 Geobacillus

Considering many bacilli genus, similar to *Anoxybacillus*, *Geobacillus* is also a relatively new genus that was recently proposed by Nazina et al. (2001), of which members are thermophilic, endospore-forming and aerobic, grow at a temperature range between 35 and 78 °C and are also widespread in many geographical areas (Poli et al. 2011; Cihan et al. 2011b).

There have been some reports on members of *Geobacillus* genus isolated from hot springs within Turkey and on their industrially important enzymes and new products (Table 8.2). The first report was carried out by Canakci et al. 2007 a decade ago. They isolated 16 Gram-positive rods from water and mud samples of Dikili–Bergama Kaynarca hot spring in province of Izmir and Camkoy Camur, Omerbeyli and Alangullu hot springs in the province of Aydin. The water temperatures of the hot springs studied varied between 70 and 130 °C. Based on 16S rRNA gene sequence analysis, all of 16 isolates resembled *Geobacillus* species by ≥97%, and most of them were found to produce xylanase and arabinofuranosidase enzymes.

Adiguzel et al. (2009) also studied as many as 15 thermophilic bacilli isolated from several different hot springs in Turkey and then characterised and identified by using molecular methods such as (GTG) 5-PCR cluster analysis. *G. pallidus* sub-cluster was comprised of the strains G19A, P112, P66 and P161 with similarity ratios of ≥82%, ≥80% and ≥ 81%, respectively. Second cluster included four strains of *G. toebii* and *G. stearothermophilus* (M66A, Ah22, G5A and G7 and a reference strain of *G. thermodenitrificans*), with lower similarity ratio (≥76%).

In a very comprehensive study, Coleri et al. (2009) isolated 451 thermophilic bacilli from 42 different hot springs and high-temperature power plants of different locations in the provinces of Ankara, Denizli, Aydin, Manisa, Nevsehir and Izmir belonging to different geographical regions of Turkey. The water temperatures of these geothermal waters were in the range 60–90 °C and pH of 6.0–9.0. Sixty-seven isolates showed a high amylase activity. All isolates were Gram-positive, rod-shaped, endospore-forming, motile, catalase-positive bacteria. Four thermophilic bacilli strains, F84b, A333, F84a and E134, producing α-glucosidase at significant levels were chosen for further experiments. The 16S rRNA gene sequence analysis showed that all isolates chosen belonged to the genus *Geobacillus*. *Geobacillus* spp.

Table 8.2 *Geobacillus* members isolated and identified from various regions of Turkey and their potential use in biotechnology

Microorganisms	Strain code	GenBank Accession number	Source	Growth temperature (°C)	Growth pH	Biotechnological applications and products	Reference
Geobacillus anatolicus	C4	AF411064	Hisaralan hot spring/Balikesir	70	7.5	DNA polymerase I	Caglayan and Bilgin (2012)
Geobacillus bogazici	NT	AY323205	Hisaralan hot spring/Balikesir	70	7.5	DNA polymerase I	Caglayan and Bilgin (2011)
Geobacillus kaue	NB	AF411066	Gonen hot spring/Balikesir	70	7.5	DNA polymerase I	Caglayan and Bilgin (2011)
Geobacillus pallidus	G19A	FJ808716	Sorgun hot spring/Yozgat	55–60	7.5–10.0	–	Adiguzel et al. (2009)
	P66, P112	FJ808718, FJ808721	Akdag hot spring/Erzurum				
Geobacillus pallidus	AH23, M21, G19 P26-45	EU9355598, EU9355594, EU9355595, EU9355591-93	Pasinler hot spring/Erzurum	55	7.5–8.5	Amylase	Adiguzel et al. (2011)
Geobacillus sp.	R30, R72, R21, V90		Balcova geothermal site/Izmir	30–70	9.0–10.0	Amylase, lipase	Yavuz et al. (2004)
Geobacillus sp.	DB2	KX184730	Dibekli hot spring/Agri	60	6.0–6.5	Amylase	Matpan Bekler (2017)
Geobacillus sp.	TF16	KY013619	Germencik hot spring/Aydin	55		Endo-xylanase, phytase	Cakmak and Saglam Ertunga (2017) and Sirin et al. (2017)
Geobacillus sp.	P161	FJ808724	Ilica hot spring/Erzurum	55–60	7.5–10.0		Adiguzel et al. (2009)

(continued)

Table 8.2 (continued)

Microorganisms	Strain code	GenBank Accession number	Source	Growth temperature (°C)	Growth pH	Biotechnological applications and products	Reference
Geobacillus spp.	I2, IT4.1 11, IT3, 9.1, 4.5, I4.3		NED[1]	55	5.5-8.5	Xylanase, arabinofuranosidase	Canakci et al. (2007)
Geobacillus stearothermophilus	AH22	FJ808712	Ilica hot spring/ Erzurum	55	7.2	Lipase	Adiguzel et al. (2009) and Ekinci et al. (2016)
Geobacillus subterraneus subsp. *aromaticivorans*	Ge1^{T}	CIP 110341^{T}	Guclukonak hot spring/Sirnak	60	9.0	β-Galactosidase, lipase	Poli et al. (2012)
Geobacillus thermodenitrificans	A333	EU326497	Germencik hot spring/Aydin	60	6.8	α-Glucosidase	Cihan et al. (2009)
Geobacillus thermodenitrificans subsp. *calidus*	F84b$^{(T)}$	NCIMB 14582$^{(T)}$	Kizilcahamam hot spring/Ankara	60	8.0	α-Glucosidase	Cihan et al. (2011b)
Geobacillus toebii	G5A, G7	FJ808713, FJ808714	Sorgun hot spring/Yozgat	55–60	7.5–10.0	–	Adiguzel et al. (2009)
	M66A	FJ808717	Pamukcu hot spring/Balikesir				

NED: The samples collected from various hot springs and the source is **not exactly defined**
NED[1] : The samples collected from Dikili–Bergama Kaynarca hot spring and Camkoy Camur, Omerbeyli, and Alangullu hot springs in the provinces of Izmir and Aydin in Turkey

F84a, A333 and F84b strains were determined as extracellular enzyme producers. Moreover, seven thermophilic *Geobacillus* strains were isolated from the hot springs named Hisaralan and Gonen in Turkey by Caglayan and Bilgin (2011) to study novel DNA polymerases. They cloned and sequenced the complete coding sequences of the polA genes (2637 bp) of these *Geobacillus* species, encoding DNA polymerase I with a molecular weight of 99 kDa.

Cihan et al. (2011b) also isolated a thermophilic, endospore-forming, facultatively anaerobic, rod-shaped and motile bacterial strain F84b$^{(T)}$ from well-pipeline sediment sample with a high temperature in Kizilcahamam, Turkey. The growth was observed at temperatures between 45 and 69 °C (optimum 60 °C) and pH ranging of 7.0–8.5 (optimum 8.0). Strain F84b$^{(T)}$ was found to produce α-glucosidase, growing on various carbon sources. The G + C content of genomic DNA was 49.6 mol %. The 16S rRNA gene sequence analysis of the strain F84b$^{(T)}$ displayed a high relatedness to *G. subterraneus* (99.3%) and to *G. thermodenitrificans* (99.8%) with DNA hybridisation values of 29.1% and 74.3%, respectively. In this study, rep-PCR and the intergenic 16S–23S rRNA gene fingerprinting profiles as well as the physiological and biochemical methods helped to differentiate strain F84b$^{(T)}$ from *G. thermodenitrificans*. For this reason, F84b$^{(T)}$ strain is assigned as a new subspecies and named *G. thermodenitrificans* subsp. *calidus* (=NCIMB 14582$^{(T)}$ = DSM 22629$^{(T)}$).

In a more recent study, a new thermophilic rod-shaped, endospore-forming, Gram-positive, alkaliphilic and motile *Geobacillus* strain was isolated from the mud of Guclukonak hot spring in Sirnak city in the southeast region of Turkey (Poli et al. 2012). The temperature and pH of the muddy water of the hot spring were 60 °C and 6.9, respectively. Growth of the isolate was observed at the temperature range of 30–65 °C (with optimum of 60 °C) and at pH range between 5.5 and 10.0 (optimum pH of 9.0). The strain was able to utilise starch and gelatine, the ONPG activity, positive for lipase, phosphatase, catalase and urease. 16S rRNA gene sequence studies in comparison to other members showed that the isolate belonged to the genus *Geobacillus*. The genomic DNA G + C content of the strain was 52.0%. The DNA–DNA hybridisation results showed that the representative strain Ge1^{T} closely related to *G. subterraneus*, *G. thermodenitrificans*, *G. thermocatenulatus*, *G. vulcani* and *G. thermoleovorans* were 69.3%, 57%, 37%, 27% and 26%, respectively. Chemotaxonomic analyses of FAME and other conventional tests allowed phenotypic and genotypic differentiation of this strain to be assigned as a novel subspecies named as *G. subterraneus* subsp. *aromaticivorans* Ge1^{T} (DSM 23066 ^{T} = CIP 110341^{T}), due to utilising typical hydrocarbons such as n-decane and squalene. RAPD–PCR using both OPR2 and GTG5 primers is also used for comparison, and the fingerprint profiles of the strain were clearly different from those produced by its closest relatives as well as *G. subterraneus*.

8.3.3 *Bacillus, Brevibacillus, Aeribacillus* and the Other Thermophiles

Literature review showed that there have been a few other species belonging to *Bacillus*, *Brevibacillus*, *Aeribacillus* and *Thermus* genus isolated from the thermal waters in different areas of Turkey and studied for their biotechnological potential (Table 8.3). Among these, *Brevibacillus* is reclassified from *Bacillus brevis* (first described in 1900) more recently as the type species (*Brevibacillus brevis*) of a new genus (Shida et al. 1996; Inan et al. 2012, 2016).

Members of *Bacillus* genus are well known to be widespread all over the world in various extreme and geographical areas including hot springs of Turkey. Adiguzel et al. (2009) studied several hot springs in different provinces of Turkey isolating 15 thermophilic bacteria and observed three clusters containing *Anoxybacillus*, *Geobacillus* and *Bacillus* strains by classification using 16S rRNA sequences and rep-PCR profiling techniques and showed that P151, P100, P79 and P130 strains were resembled to strains of *B. licheniformis* and *B. pumilus*.

A thermostable metalloprotease-producing bacterial strain KG5 was isolated from the mud of Kos hot spring in Bingol. The strain KG5 which was facultative anaerobic, motile, rod-shaped and Gram-positive and possessing central and oval endospores, grown optimally at 40 °C, was found to be a strain of *B. cereus* defined by phenotypic characterisation and by gene sequence analysis of 16S rRNA (Gul Guven 2007; Ahmetoglu et al. 2015). In another study carried out in a hot spring named Taslidere (water temperature of 78 °C) in Batman located in the southeast of Turkey, a thermotolerant rod-shaped bacterial strain was isolated and deposited as DSM 18503. The growth temperature range of this thermo-alkalitolerant strain KG9 was determined as 30–55 °C and pH of 5.0–10.5. The isolate was found to be a member of the species *B. licheniformis* identified by the analysis of morphological, biochemical and physiological characteristics, as well as by 16S rRNA gene sequence similarities. It was revealed that the 16S rRNA gene sequence of the strain KG9 was 99.9% similar to that of *B. licheniformis* strain DSM 13 (Gul Guven 2007; Matpan Bekler et al. 2015a). Matpan Bekler et al. (2015b) also studied on the isolation, identification and enzyme production of the strain *B. licheniformis* DV3, which were isolated from the water of Davut hot spring in Diyadin township of Agri Province (water temperature 78 °C, pH 7.7), in northeastern Turkey. The strain *B. licheniformis* DV3 was identified by biochemical, morphological tests and 16S rRNA sequence analysis.

Additionally, Adiguzel et al. (2011) revealed a population of *B. licheniformis* and *Aeribacillus pallidus* in Pasinler hot spring, Erzurum Province in Turkey, carrying out a study using the analyses of 16S rRNA gene sequences, FAME and BOX PCR fingerprint profiles. Nine different bacterial strains selected based on biochemical, physiological and morphological tests were classified into two phenotypic groups; the first group represented by four strains was identified as *B. licheniformis*, while the second group represented by five strains was identified as *A. pallidus*. In a similar study, more than ten strains were isolated from different geothermal areas of Turkey, five of which were found to belong to *A. pallidus*, six strains were close to

Table 8.3 The members of *Bacillus*, *Brevibacillus*, *Aeribacillus* and the other thermophiles isolated and identified and their potential use in biotechnology, from various regions of Turkey

Microorganisms	Strain code	GenBank accession number	Source	Growth temperature (°C)	Growth pH	Biotechnological applications and products	Reference
Bacillus licheniformis	A1	EU314720	Diyadin hot spring/Agri	60	7.0	Chitinase	Sandalli et al. (2008)
Bacillus licheniformis	G3B, G1, P38, AH2	EU9355596, EU9355597, EU9355592, EU9355599	Pasinler hot spring/ Erzurum	55	7.5–8.5	Amylase	Adiguzel et al. (2011)
Bacillus licheniformis	A10	KC310461	Ilica hot spring/ Erzurum	57	6.5–9.0	Amylase, protease, cellulase	Yilmaz et al. (2014) and Yanmis et al. (2015)
Bacillus licheniformis	A7, A9	KC310458, KC310460,	NED	57±1	6.5-9.0	Amylase, cellulase	Yanmis et al. (2015)
Bacillus licheniformis	O12, O16	KM596797, KM596801	NED	56	7.0	Amylase, protease, lipase, cellulase	Baltaci et al. (2016)
Bacillus licheniformis	P79	FJ808719	Akdag hot spring/ Erzurum	55–60	7.5–10.0	–	Adiguzel et al. (2009)
	P100, P151	FJ808720, FJ808723	Pamukcu hot spring, Balikesir				
Bacillus licheniformis	KG9	DSM 18503	Taslidere hot spring/ Batman	30–55	5.0–10.5	β-Galactosidase, heavy-metal removal	Gul Guven, (2007), Matpan Bekler et al. (2015a) and Alkan et al. (2015)

(continued)

Table 8.3 (continued)

Microorganisms	Strain code	GenBank accession number	Source	Growth temperature (°C)	Growth pH	Biotechnological applications and products	Reference
Bacillus licheniformis	DV3	HQ864575	Davut hot spring/Agri	55	7.0	Amylase, protease	Matpan Bekler et al. (2015b)
Bacillus pumilus	P130	FJ808722	Pamukcu hot spring/ Balikesir	55–60	7.5–10.0		Adiguzel et al. (2009)
Bacillus pumilus	O2, O7, O8, O14	KM596787, KM596792, KM596793, KM596799	NED[2]	56	7.0	Amylase, protease, lipase, cellulase	Baltaci et al. (2016)
Bacillus sp.	KG5	KP318029	Kos Hot Spring/ Bingol	40	7.5	Metalloprotease	Gul Guven (2007) and Ahmetoglu et al. (2015)
Bacillus sp.	–	–	Alangullu hot spring/Aydin	65	6.0	Esterase	Bakir Ateslier and Metin (2006)
Bacillus subtilis	–	–	Diyarbakir	37	7.0	Amylase	Ozdemir et al. (2011)
Bacillus sp.	ANT-6		Adana	20–55		Amylase	Arikan et al. (2003)
Bacillus thermoamylovorans	O10, O15, O21	KM596795, KM596800, KT364469	NED[2]	56	7.0	Amylase, protease, lipase	Baltaci et al. (2016)
Bacillus thermoamylovorans	A11	KC310462	NED[1]	57±1	6.5–9.0	Amylase, cellulase	Yanmis et al. (2015)
Bacillus thermolactis	A6	KC310457					

Aeribacillus pallidus	O1, O3, O5, O6, O11	KM668038, KM596788, KM596790, KM596791, KM596796	NED[2]	56	7.0	Amylase	Baltaci et al. (2016)
Aeribacillus pallidus	P26	EU935591	Pasinler hot spring/ Erzurum	55	7.5–8.5	Catalase, peroxidase, removal of some textile dyes	Taslimi et al. (2013)
Aeribacillus pallidus	C10	KC333049	Belkisana hot spring/Sirnak	55	9.0	Protease	Yildirim et al. (2017)
Brevibacillus aydinogluensis	PDF25[T]	NR_117986	Karakoc hot spring/İzmir	55	7.0	–	Inan et al. (2012)
Brevibacillus sp.	Z1	KC292196	Diyadin hot spring/Agri	59 ± 1	6.5-8.5	Amylase, cellulase, laccase, removal of some textile dyes	Bozoglu et al. (2013) and Yanmis et al. (2015)
Brevibacillus gelatini	PDF4[T]	KP899808	Camkoy hot spring/Aydin	40	7.0	Gelatinase	Inan et al. (2016)
Thermomonas hydrothermalis	O4	KM596789	NED[2]	56	7.0	Amylase, lipase	Baltaci et al. (2016)
Thermus anatoliensis	MT1[T]	NCCB100425[T]	Buharkent geothermal area/Aydin	65	7.5	β-Glucosaminidase; the valine, leucine and cystine aminopeptidases; lipase (C14); α-galactosidase; α-glucosidase; alkaline; and acid phosphatases	Kacagan et al. (2015)

NED: The samples collected from various hot springs and the source is **not exactly defined**
NED[1]: The samples collected from Kirsehir–Terma, Sirnak/Guclu Konak–Belkis Ana, Agri–Diyadin, Kayseri–Haciveli, Erzurum–Ilica, Urfa–Karaali, Rize–Ayder, Adiyaman–Kurucay and Ankara–Kizilcahamam
NED[2]: The samples collected from Bademli Bahce hot spring (Bursa), Sulusaray hot spring (Tokat), Sicak Cermik hot spring (Sivas), Bostanci hot spring (Balikesir) and Diyadin hot spring (Agri)

B. pumilus and two strains resembled to *B. licheniformis* (all at a similarity ratio of 99%). However, three isolates were very close to *B. thermoamylovorans* (≥99%), and two belonged to *Anoxybacillus* genus. In addition, one of the isolates belonged to *Thermomonas hydrothermalis* (with 99% similarity). These isolated thermophilic bacteria were also evaluated for their capability to produce enzymes such as protease, lipase, cellulase and amylase (Baltaci et al. 2016).

Inan et al. (2012) isolated two moderately thermophilic, Gram-positive, endospore-forming, rod-shaped, motile bacteria, designated as PDF25^T and PDF30 from water and mud samples of Karakoc hot spring (the water temperature is around 60–70 °C) in Izmir Province. Cells grow at a temperature range of 35–65 °C (optimum of 55 °C) and pH 6.0–0 (optimum pH of 7.0), hydrolysing casein, starch, gelatin and ONPG (o-nitrophenyl-beta-D-galactoside). The strain PDF25^T was identified as *Brevibacillus aydinogluensis* characterised by gene sequence analysis of 16S rRNA. It was demonstrated that both strains were the members of the genus *Brevibacillus*; the strain PDF25^T had a high sequence similarity to *Brevibacillus thermoruber* DSM 7064 T (98.5%). DNA–DNA hybridisation results displayed 58% relatedness between the strain PDF25^T and *B. thermoruber* DSM 7064^T. The G + C content of genomic DNA was 56.09 mol %. Based on phenotypic and genetic characterisation (particularly by the analysis of the sequence of hypervariable (HV) region), the strain PDF25^T distinguished as a novel species of the genus *Brevibacillus*. Furthermore, Inan et al. (2016) isolated two moderately thermophilic, Gram-positive, endospore-forming, motile, rod-shaped bacteria from Camkoy hot spring Aydin Province, Turkey. The strain designated as PDF4^T had a DNA G + C content of 51.7 mol %, and DNA–DNA hybridisation of strain PDF4^T and type strains of the closely related species displayed less than 60% relatedness. For the type strain PDF4^T (=NCCB 100559^T = DSM 100115^T), the species name of *Brevibacillus gelatini* sp. nov. was proposed.

In a recent study, Kacagan et al. (2015) have isolated a Gram-negative, aerobic rods, nonmotile, catalase, urease and oxidase-positive bacterium (strain MT1^T) from Buharkent hot water in Aydin city of Turkey (water temperature and pH were 88 °C and 6.5, respectively). The strain hydrolysed starch and gelatin, as well as possessing a variety of enzyme activities including β-glucosaminidase, valine, leucine, and cysteine aminopeptidases, lipase (C14), α-galactosidase, α-glucosidase, acid and alkaline phosphatases, which need to be exploited for possible uses in biotechnology. The isolated strain was found to grow at temperature range of 45–80 °C (with optimum of 65 °C) and at pH 5.5–10.5 (with optimum of 7.5). The comparison of16S rRNA gene sequence similarity values between strainMT1^T and other *Thermus* species revealed highest similarity of 96.92% to *T. islandicus* PRI 383^T, followed by *T. arciformis* TH92^T (96.48%) and *T. composti* K-39^T (95.73%). The G + C content of strain MT1^T genomic DNA was 69.6 mol %. On the basis of various methodologies using a polyphasic approach, strain MT1^T was suggested as a novel species named as *Thermus anatoliensis*. The type strain is MT1^T (=NCCB 100425 ^{T} = LMG 26880^T).

8.4 Biotechnological Importance of Thermophiles Isolated in Turkey

8.4.1 Possible Applications Related to Enzyme Industry

Thermozymes have been widely used in many industrial applications as they are well known to possess thermal tolerance and stability to harsh industrial processes at very high temperatures (Demirjian et al. 2001). Microbes and their enzymes are used in a wide range of biotechnological applications such as detergent, fine chemicals, pharmaceutical, bioremediation, food, leather, paper and textile industry. The most important enzymes for industry are lipases, carboxylesterases, cellulases, xylanases, pectinases, amylases, galactosidases and proteases. Thermophilic bacilli are the natural source of most thermostable enzymes. As can be seen in Tables 8.1 and 8.3, a number of thermozymes from thermophilic bacilli isolated in Turkey are as follows: those belonging to *Bacillus* species such as amylase (Arikan et al. 2003; Ozdemir et al. 2011; Matpan Bekler et al. 2015b), chitinase (Sandalli et al. 2008), metalloprotease (Matpan Bekler et al. 2015b; Ahmetoglu et al. 2015) and β-galactosidase (Matpan Bekler et al. 2015a) and to the *Anoxybacillus* species such as α-amylase (Matpan Bekler and Guven 2014; Acer et al. 2015, 2016), xylanase (Kacagan et al. 2008; Inan et al. 2013), glucosidase (Cihan et al. 2011a), glucose isomerase (Karaoglu et al. 2013), ribulokinase (Tokgoz et al. 2014), esterase (Colak et al. 2005; Faiz et al. 2007; Ay et al. 2011), lipase (Bakir and Metin 2015 and 2017), aldolase (Ertunga et al. 2007), CTP synthase (Sandalli et al. 2014), β-galactosidase (Matpan Bekler et al. 2017) and protease (Matpan Bekler et al. 2015c) which have been well characterised. There have been also several studies on the novel enzymes of thermophile *Geobacillus* sp. isolated from Turkey (see Table 8.2), including xylanase (Canakci et al. 2012; Cakmak and Saglam Ertunga 2017), arabinofuranosidase (Canakci et al. 2007), α-glucosidase (Cihan et al. 2009, 2011b), and DNA polymerase I (Caglayan and Bilgin 2011, 2012). Kocabiyik and Erdem (2002) also studied on alkaline proteases produced by various thermoacidophilic archaeal and bacterial strains growing optimally around pH 2.0–5.0, which were originally isolated from acidic hot springs in various hydrothermal sites in Turkey. Here, we summarise on various thermostable enzymes and their characterisation in studied thermophiles that are isolated in hot springs or geothermal areas in Turkey (Tables 8.1, 8.2 and 8.3).

It is very clear that *Anoxybacillus* species are most studied microorganism in hot springs of Turkey, in terms of both identification and their thermostable enzymes. For example, Colak et al. (2005) reported on *A. gonensis* G2 secreting an esterase responsible for the degradation of poly-3-hydroxybutyrate (P3HB). The optimum enzyme parameters were pH 7.5 and 60 °C. The enzyme activity was enhanced by Ca^{2+} indicating to be a cofactor which is a characteristic for lipases/esterases. The esterase activity is inhibited by the metal chelating agent ethylenediaminetetraacetic acid (EDTA), supporting its metalloenzyme characteristic. A similar study was carried out by Faiz et al. (2007) on a thermostable esterase in a novel thermophile, *A. gonensis* A4, capable to degrade tributyrin. The extracellular enzyme had a molecular weight of 62 kDa. The optimum pH and temperature values for the esterase of strain A4 were 5.5 and 60–80 °C, respectively. They also showed that the enzyme esterase had serine residue in active site and –SH groups were found to be essential for enzyme activity. In addition, a gene encoding a thermostable carboxylesterase from *Anoxybacillus* sp., PDF1, was cloned in *Escherichia coli* BL21. The molecular mass of purified recombinant protein was about 26 kDa as determined by SDS–PAGE. The enzyme showed activity under a wide pH (pH 5.0–10.0) and temperature range (25–90 °C) with optimum temperature and pH values of 60 °C and 8.0, respectively. The inhibition tests on carboxylesterase of *Anoxybacillus* sp. PDF1 revealed that it possesses a serine residue in active site and –SH groups in specific sites, required for its activity (Ay et al. 2011). Lipases and esterases are well known to catalyse many reactions such as esterification, interesterification, alcoholysis or acidolysis, used for fat and oil industry, in the synthesis of flavour esters for food industry, for the synthesis of fine chemicals in the pharmaceutical industry.

Bakir and Metin (2015) isolated 201 thermophilic bacteria from a hot spring in Alangullu/Aydin (50 °C). Among these, 22 isolates exhibited lipase activity. However, the strain HBB 134 having a maximum 16S rRNA sequence similarity (99%) with *Anoxybacillus flavithermus* was found to be the best lipase-producing isolate, which was the first report for a lipase production in the genus *Anoxybacillus*. In another study, the authors isolated another thermophilic lipase-producing bacterium, namely, *Anoxybacillus* sp. HBB16, showing 16S rDNA sequence similarity of 96% with *A. flavithermus*. The maximum activity of the alkaline lipase occurred at 55 °C and pH 9.5 (Bakir and Metin 2017). Lipases (triacylglycerol acylhydrolase; EC 3.1.1.3) are biotechnologically important enzymes which catalyse the hydrolysis of mono-, di- and triacylglycerides to glycerol and free fatty acids at an oil–water interface.

Another thermostable enzyme, which has a biotechnological importance, is amylase, used in many other industrial areas, such as in removing food and starch stains in dry cleaning, in the textile, starch and food industry, and in the purification of apple and pear juice, in the detergent and pharmaceutical industries. Matpan Bekler and Guven (2014) carried out a study on a novel α-amylase produced by a newly isolated thermophilic bacterial strain, namely, *Anoxybacillus* KP1 from the Diyadin hot spring (water temperature 50 °C, pH 7.4) in Agri Province, in northeastern Turkey. Maximal activity of the α-amylase was observed at the pH and temperature of 8.0 (pH range at 6.0–10.0) and 60 °C, respectively. The α-amylase production increased in the presence of 2% (w/v) soluble starch, some nitrogen sources and Mn^{2+}. The enzyme was calcium-independent, considerably stable at a range of pH and temperature, which may be advantageous in industrial applications for food processing and traditional brewing, where the temperatures could denature the enzymes after fermentation. Moreover, an extracellular α-amylase production by a novel thermophilic *Anoxybacillus* sp. AH1 from Dargecit hot springs in Turkey was investigated in the presence of many different media containing a variety of carbon and nitrogen sources. It was also found that α-amylase from *Anoxybacillus* sp. AH1 was thermostable and Ca^{2+} dependent (Acer et al. 2015). In a more recent study,

Acer et al. (2016) purified this α-amylase from *Anoxybacillus* sp. AH1 and determined the molecular mass as 85 kDa. The enzyme had the optimum temperature and pH values of 60 °C and 7.0, respectively. The enzyme activity was seen to increase by various detergents, Mg^{2+} and Ca^{2+}, but there was a significant inhibition by metal ion chelators 1,10-phenanthroline and EDTA. In addition, the activity of α-amylase was enhanced by dithiothreitol (DTT) and β-mercaptoethanol, but it was inhibited by p-chloromercuribenzoic acid (PCMB), indicating the presence of one essential cysteine residue at least in the enzyme active site. The strain AH1 α-amylase inhibition by phenylmethylsulfonyl fluoride (PMSF) also indicated the importance of the seryl hydroxyl group in the catalysis of this enzyme.

A thermophilic *Anoxybacillus ayderensis* $AB04^T$ that was isolated from the Ayder hot spring was found to possess a number of glycoside hydrolases (GHs) which are of importance for carbohydrate-related industries. The GHs of *A. ayderensis* $AB04^T$ were compared to those of other sequenced *Anoxybacillus* spp. genomes, where 14 GH enzyme genes encoded in the genome that belong to GH families 1, 10, 13, 31, 32, 51, 52 and 67 were detected. It was predicted that nine GH enzymes were active on α-chain polysaccharides (pullulanase, α-amylase, α-glucosidase, CDase and oligo-1,6-glucosidase), while the other five GH enzymes act specificity on β-linked polysaccharides (i.e., xylan and cellulose). Those uniquely present were endo-1,4-β-xylanase (NCB I locus ID: KIP 21668) and α-glucuronidase (KIP 21917). Despite the GHs, other *A. ayderensis* $AB04^T$ enzyme genes coding for industrially important enzymes were esterase, aldolase and xylose isomerase. Particularly, xylose isomerase (EC 5.3.1.5) catalyses the isomerisation of glucose to fructose and of xylose to xylulose, which is important in the industry of high-fructose corn syrup production. Two esterases (KIP 19922 and KIP 21735) were detected in the strain $AB04^T$ genome, which had 96.0% and 96.3% amino acid sequence similarity with the esterase from *A. gonensis* $G2^T$ and *Anoxybacillus* sp. PDF-1, respectively. Moreover, *A. ayderensis* $AB04^T$ contains two aldolases, KIP 21451 and KIP 21450 (Belduz et al. 2015).

Complete genome sequencing of *Anoxybacillus gonensis* type strain $G2^T$ (=NCCB 100040^T = NCIMB $13{,}933^T$) isolated previously from Gonen hot springs showed that this strain consisted various carbohydrases, such as pullulanase (AKS39285) and α-amylase (NCBI locus ID: AKS37565), which are valuable for starch hydrolysis in industry, cyclodextrinase (AKS37561) used in cyclodextrin-related research and also xylose isomerase (AKS39170) and β-xylosidase (AKS39172) which are good candidates for second-generation biofuel production. They also reported on some other novel enzymes of *A. gonensis* $G2^T$ and their potential use in biotechnology, such as β-galactosidase (AKS39183), α-galactosidase (AKS39187), oligo-1,6-glucosidase (AKS37459) and α-glucosidase (AKS37566). This strain was found to produce many other well-studied enzymes with biotechnological importance, including fructose-1,6-bisphosphate aldolase and carboxylesterase (Lim et al. 2015). Moreover, in the thermophile *A. gonensis* $G2^T$, a new glucose isomerase (GI) was described by Karaoglu et al. (2013), which is particularly suitable for the production of high-fructose corn syrup in the food industry. The gene encoding this enzyme was cloned and expressed in *E. coli*. The purified recombinant enzyme, with a molecular weight of approximately 50 kD determined by SDS–PAGE and MALDI–TOF analysis, had an optimal activity at 85 °C and pH 6.5. In a study carried out by Sandalli et al. (2014), a novel CTP synthase gene of *A. gonensis* G2 was cloned, expressed and characterised. The thermophilic cytidine-5′-triphosphate synthase (EC 6.4.3.2) gene (pyrG) was 1590 bp long encoding a protein with 530 amino acids, possessing a molecular weight of 59.5 kDa. The CTP synthase amino acid sequence showed a similarity of 90%–94% with *Bacillus* sp.

A ribulokinase from *Anoxybacillus kestanbolensis* AC26Sari isolated from the hot spring mud (Camkoy in Canakkale province, Turkey) was studied, cloned and expressed in *E. coli* BL21 (Tokgoz et al. 2014). The ribulokinase of the strain AC26Sari was found to have 99% DNA and 99% amino acid identity with ribulokinase of *A. flavithermus* WK1, while 90% DNA and 96% amino acid identity with *Geobacillus thermodenitrificans* NG80–2 ribulokinase. The purified enzyme had a molecular mass about 61 kD, as determined by SDS–PAGE, and was found to be active at a wide temperature (50–75 °C) and pH (pH 5.0–10.0) range, with optimum temperature of 60 °C and an optimum pH of 9.0. The activity of purified enzyme was strongly inhibited by Zn^{2+} but enhanced by Mg^{2+}, though the ribulokinase from *A. kestanbolensis* AC26Sari did not require any other metallic cations for its activity. This was the first report to characterise a thermophilic ribulokinase of thermophilic bacteria. L-ribulokinase is unusual among kinases because it is known to phosphorylate all four 2- ketopentoses (L- or D-xylulose and L- or D-ribulose).

Xylanases are well known to be important in biotechnology increasing the nutritional quality of animal feed and in the textile fibre recovery and used for industrial wastes in pulp and paper industry, as well as in the clarification of fruit juices, wine, etc. A thermophilic, xylanolytic bacterium isolated from the Diyadin hot springs was identified as *Anoxybacillus pushchinoensis* strain A8 by sequence similarity of 16S rRNA gene and DNA–DNA hybridisation studies. The extracellular xylanase had a molecular mass of approximately 83 kDa. The maximal activity obtained for the enzyme was pH 6.5 (stable over a broad pH range of 6.5–11 for 24 h) and 55 °C (stable at temperature between 50 and 60 °C up to 24 h). The enzyme was found to be an exo-acting xylanase (Kacagan et al. 2008). In another study, Inan et al. 2013 found that *Anoxybacillus kaynarcensis* produced xylanase activity, and the zymogram analysis of SDS–PAGE revealed apparent molecular weights between 100 and 150 kDA, with the optimum temperature and pH values of 65 °C and 7.0–9.0, respectively.

Proteases, particularly thermostable ones, have been used for a long time for many industrial applications. In search of proteases, Matpan Bekler et al. (2015c) studied a novel extracellular alkaline protease (EC 3.4.21–24, 99) in thermophilic *Anoxybacillus* sp. KP1 strain. The purified enzyme had a molecular weight of 106 kDa using SDS–PAGE, which was stable at pH 9.0 and at 50–60 °C for 1 h. Some chemicals such as Triton X-100, Tween 80, Ca^{2+} and Cu^{2+} increased the activity of the enzyme, while EDTA and PMSF inhibited proteolytic activity, suggesting that the enzyme was a serine alkaline protease. They also stated that the detergent stability (residual activity between 73% and 82%) was an important feature for their industrial applications, such as detergent industry.

The genus *Geobacillus* has also drawn attention due to their potential use in biotechnology. Canakci et al. (2007) investigated on thermophilic xylanase and arabinofuranosidase activities in the isolated 16 Gram-positive bacilli which belonged to the genus *Geobacillus* from Dikili–Bergama Kaynarca hot spring (Izmir Province) and Camkoy Camur, Omerbeyli and Alangullu hot springs in Aydin Province in Turkey. They reported that seven of the isolates had both arabinofuranosidase and xylanase activities, while four of them had only xylanase and the other five isolates had none of both activities. The xylanase of isolates 3.3, 7.1 and 9.1 had the highest optimum temperature of 80 °C, while the isolates AO4, AO17, 7.2, 9.1 and 9.2 had the highest optimum pH of 8. The optimum temperature for arabinofuranosidase activity for isolates 7.2, AO4, AC15 and 12 was 75 °C, whereas only isolate AC15 had the lowest pH of 5.5.

A xylanase-encoding gene from *Geobacillus* sp. 7.1, isolated from the hot spring of Dikili–Bergama Kaynarca, was cloned and sequenced, followed by overexpression in *E. coli* and purification. Extracellular xylanase having a molecular weight of 47 kDa had the optimum pH and temperature values of 8.0 and 75 °C, respectively. The xylanase had the most sequence similarity (93%) with the enzyme from *G. thermodenitrificans* NG80–2. It was found that the enzyme carried a catalytic domain which belonged to the glycoside hydrolase family 10 (GH10), exhibiting an excellent pH stability. The enzyme did not have cellulase activity, whereas degraded xylan in an endo-fashion (Canakci et al. 2012). In addition, Cakmak and Saglam Ertunga (2017) have recently studied on cloning, expression, immobilisation and characterisation of an endo-xylanase and its industrial applications in *Geobacillus* sp. TF16 collected from the Germencik Omerbeyli hot spring in Aydin. The molecular weight of the recombinant enzyme was found to be a single band of 39.8 kDa on SDS–PAGE. The immobilised enzyme compared to free enzyme showed an increase in optimum temperature from 55 to 65 °C. The optimum temperature for the free enzyme was pH 8.5, whereas immobilised enzyme displayed a higher activity in the pH range 6.0–8.5. The endo-xylanase was shown to have importance for use in biotechnology as it was capable of releasing the reducing sugar from juice and poultry feed and oven spring in bakery.

A study was performed on the purification and characterisation of novel DNA polymerases of *Geobacillus kaue* strain NB isolated from Gonen and Hisaralan hot springs in Turkey. It was shown that the optimum values for the enzymatic activity of *G. kaue* polI was 70 °C and pH 7.5–8.5. In addition, polyamines stimulated the polymerisation activity of the enzyme. Three-dimensional structure of polI showed that all functionally important regions were conserved in the polymerase active site computed using homology modelling (Caglayan and Bilgin 2011).

The thermophilic chitinases that degrade chitin, the most abundant renewable natural resource after cellulose, have a wide range of biotechnological applications. A chitinase gene (chiB65) in *Bacillus licheniformis* A1 obtained from Diyadin hot spring was cloned and expressed in *E. coli* and then sequenced. The purified recombinant protein was analysed on SDS–PAGE using the fluorogenic substrate

4-methylumbelliferyl β-D-N,N′-diacetylchitobioside, having a molecular weight of approximately 71 kD. The optimum values for the enzyme were pH 6.0 and temperature of 65 °C, though it was stable at a pH range of 5.0–9.0 for 4 h at 65 °C and 24 h at room temperature (Sandalli et al. 2008).

Thermophilic amylases are well known to be used for hydrolysis of starch to produce glucose and related chemicals in industry. From this point of view, an alkaline, thermostable α-amylase-producing *Bacillus* sp. ANT-6 was identified by Arikan et al. (2003). The enzyme had an optimum activity at 80 °C and pH 10.5. The relative molecular mass of the enzyme was found as 94.5 kDa. A *Bacillus subtilis* strain isolated from soil samples in Diyarbakir, Turkey, was also studied for its thermostable α-amylase. The effects of many parameters such as incubation time, different culture media, carbon and nitrogen sources and various starches, flours, detergents and other chemicals on the production of α-amylase were studied. The purified enzyme was found to be Ca-dependent, having the optimum pH and temperature of 6.0 and 60 °C, respectively (Ozdemir et al. 2011).

Ahmetoglu et al. (2015) studied on a novel extracellular protease produced by *Bacillus* sp. KG5 isolated from Kos hot spring (Bingol, Turkey). The molecular weight of purified enzyme was approximately 48 kDa by both native and SDS–PAGE and was not a serine–protease as PMSF did not have an inhibitory effect on protease activity. The enzyme showed maximum activity at pH of 7.0–7.5. It was also determined that the protease was thermostable, particularly fully stable in the Ca^{2+} presence at 50 °C even after 120 min. It is clear that thermostability of proteases is a critical feature required for industrial applications such as leather processing and detergent. Proteases are also used in many applications such as bioremediation, biotransformation and biosynthesis, brewing, food, meat, dairy industries and diagnostics. In a newly isolated thermophilic *Bacillus licheniformis* DV3, extracellular thermostable α-amylase and protease were studied. The optimum temperature and pH values for both extracellular enzymes were 70 °C and 7.0 for the α-amylase, respectively, while it was 10.0 and 50 °C for the protease, respectively. The α-amylase activity was enhanced in the presence of Mn^{2+}, inhibition was obtained in the presence of Ca^{2+} indicating to be a member of calcium–independent amylases. The protease activity increased in the presence of Ca^{2+} and Zn^{2+}, whereas the activity was decreased by EDTA and PMSF, indicating that the enzyme was a metallo- and serine protease (Matpan Bekler et al. 2015b).

A thermostable β-galactosidase from a thermo- and alkalitolerant KG9 strain belonging to *Bacillus licheniformis* isolated from Taslidere hot spring in Batman (Turkey) was cloned, expressed in *E. coli* and characterised. Due to genomic sequence similarity of *B. licheniformis* strain KG9 to that of *B. licheniformis* strain DSM 13 (99.9% identity), PCR primers based on four putative β-galactosidase genes in the genome of strain DSM 13 were employed for the isolation of the corresponding β-galactosidase genes from KG9 strain. The molecular masses of β-galactosidases I, II, III and IV were calculated as 30, 79, 74 and 79 kDa, respectively, using sequencing data. Similarly, the number of identified β-galactosidase genes in strain KG9 was four, and three genes were expressed in *E. coli* as intracellular and active. Among these three, the authors purified and characterised the recombinant β-galactosidase III, having the optimal pH and temperature of 6.0 and 60 °C, respectively. The purified enzyme analysed on SDS–PAGE displayed one single band with a molecular weight of ~75 kDa (Matpan Bekler et al. 2015a). It has been also claimed that the characteristic such as thermostability makes this recombinant β-galactosidase favourable in the application of β-galactosidase in dairy and food processes involving hydrolysis of lactose in order to enhance the digestibility of milk or to improve the functional characteristics of milk products, etc.

8.4.2 Applications Related to Environmental Biotechnology

The use of thermophiles and their bioproducts in environmental biotechnology is well known such as biohydrogen production, bioconversion of lignocellulose to hydrogen, conversion of glycerol to lactate, conversion of D-xylose into ethanol, biodegradation of dyes or petroleum hydrocarbons, recovery of heavy metals, etc. The thermophiles and their products are resistant to harsh conditions in industrial applications by supplementing or replacing traditional chemical processes (Mehta et al. 2016).

The H_2-producing bacteria, closely affiliated to genus *Thermoanaerobacterium* determined by PCR–DGGE profiling, were isolated from hot spring of Hisaralan in Balikesir Province, Turkey. H_2 bioproduction was accompanied by production of acetate, butyrate, ethanol and lactate. It was found that H_2 production was maximum at the temperature range from 49.6 to 54.8 °C (Karadag et al. 2016).

The dyes are commonly used in different industrial fields such as textile, food, cosmetics and paper; on the other hand, they cause health and environmental problems. The enzyme called laccase (oxidoreductase) has ability to oxidise the compounds associated with both phenolic and nonphenolic lignin and to deoxidise the pollutants resistant to biodegradation, for example, used in the removal of textile dyes, phenols and detoxification of wastes. In a study carried out by Yanmis et al. (2016), an extracellular laccase from *Anoxybacillus gonensis* P39 (Gen Bank No:FJ808725) isolated from Ilica hot spring, Erzurum Province, was purified with a molecular weight of 40 kDa on SDS–PAGE and with optimum pH and temperature values of 5.0 and 60 °C. Bozoglu et al. (2013) also studied on the purification and characterisation of a laccase (with molecular mass 93 and 110 kDa) and its possible use in removal of textile dyes, from a new thermophilic strain of *Brevibacillus* sp. (Z1) isolated from Diyadin hot springs in Agri Province of Turkey. The evaluation of laccase in both studies for possible use in bioremediation process of some textile dyes showed that the laccase reduced the amount of several dyes such as the Reactive Black 5, Fuchsine, Allura Red and Acid Red 37 in wastewater.

Lim et al. (2015) found that the *Anoxybacillus gonensis* $G2^T$ consisted arsenate reductase (AKS38388) and three mercury (AKS37713, AKS38377, AKS38379) genes in its genome, showing that the $G2^T$ strain may be used in heavy metal bioremediation. The analysis of *A. ayderensis* $AB04^T$ genome showed the presence of at least six heavy metal resistance genes, four of which were mercury resistance (mer) operons (KIP 20706 and KIP 20408) and mercuric reductases, catalysing the reduction of Hg^{2+} to Hg 0 (KIP 19952 and KIP 20409). Other two genes were arsenate reductase (KIP 20402) and arsenic efflux pump protein (KIP 20401). Beris et al. (2011) also reported on an aluminium tolerance gene (*G2alt*) and the effects of environmental conditions on its biological functioning in the thermophilic $G2^T$ strain. The *G2alt* gene was 666 bp long and encoded a protein of 221 amino acids. The amino acid sequence of the protein with ATPase activity was highly similar to proteins which are responsible for aluminium resistance.

There has been an emphasis given to the utilisation of microorganisms, so as thermophiles for their great metal ion absorption ability from aqueous solutions. Duran et al. (2009) used *A. gonensis* which was immobilised on Diaion HP-2MG as a new biosorption system for the enrichment of various metals prior to the atomic absorption spectrometric analysis. More recently, a thermophilic haloalkalitolerant bacterial strain named KG9 was newly isolated and identified as a close member of *Bacillus licheniformis* which was also evaluated for possible use in environmental technology by Alkan et al. (2015) as a new biosorbent for preconcentrating Cd(II), Ni(II) and Cu(II) prior to flame atomic absorption spectrometric (FAAS) analysis. The strain (KG9) immobilised on Amberlite XAD-4 was used for the measurement of toxic metal ions in real samples such as the Tigris river and drinking water and in mushrooms. The optimum parameters such as eluent type and volume, amount of adsorbent, pH, sample solution volume, sample solution flow rate and matrix interference effect on the metal ion retention were investigated for the analyte quantitative recovery.

8.5 Hypersaline Environments of Turkey

Turkey, especially Central Anatolia, is rich for hypersaline environments. Tuz Lake which is the largest salt lake in central Turkey occupies a depression in the dry central plateau of Turkey, located in 105 km northeast of Konya city and 120 km south of Ankara. The lake stays shallow (1–2 m) and has a total surface area of 1665 km^2, with a length of 90 km and a width of 35 km within a closed basin. The water salt concentration reaches up to 33%. In summer, when the lake dries out, a 30-cm layer of salt forms due to the evaporation. The lake and the salterns provide a main source of solar salt: 73% of the salt consumption of Turkey, meaning that the lake produces more than 200 million tons of salt (Birbir and Sesal 2003; Mutlu et al. 2008). Moreover, Camalti Saltern is the biggest artificial marine solar saltern in Turkey. It is a multipond system consisting of 182 ponds covering 58 km^2 and located about 38°35′N and 26°57′E on the east cost of the Aegean Sea. Sea salt extraction has been carried out in the area since 1863 (Mutlu and Guven 2015).

8.6 Halophilic Microorganisms Isolated from Extreme Environments of Turkey and Their Possible Use in Biotechnology

8.6.1 Halophilic Archaea and Bacteria from Hypersaline Environments of Turkey

A distinct class of extremophiles is halophiles which means that salt is required for them to survive. Halophilic microorganisms are found in various hypersaline environments including crystalliser ponds, saline sand and soils, marine environments, solar lakes and hypersaline lakes. Microorganisms adapted to life at very high salt concentrations are widely spread, both within the archaeal and the bacterial domain which are well known to comprise a well-defined, aerobic or facultatively anaerobic microorganisms (Ozcan et al. 2007; Mutlu and Guven 2015). There have been several studies recently on extremely halophilic communities in various hypersaline environments such as salterns, salt lakes or salt mines in Turkey, for the aim of identification and/or possible biotechnological uses (Table 8.4).

Salt lakes and the solar salt contain huge numbers of prokaryotes, mainly extremely halophilic *Archaea* of the family *Halobacteriaceae* (Birbir et al. 2007). Typical characteristics of the family *Halobacteriaceae* members are having different shades of red as colony colour, various morphological types from rods, cocci, to extremely pleomorphic, and growing at 25% NaCl concentration (Grant et al. 2001). Moreover, all isolates are known to comprise ether-bound membrane lipids as well as being resistant to antibiotics that target the bacterial peptidoglycan (Ozcan et al. 2007). Both metabolic diversity and biotechnological potential have been found in halophilic and halotolerant microorganisms (Tatar et al. 2016).

Birbir and Sesal (2003) studied on extremely halophilic microorganism communities in Sereflikochisar Salt Lake in central Turkey. A research on microbial diversity was carried out in this area due to being a main source of solar salt for food and hide and also due to the economic importance. In total, 82 extremely halophilic aerobic strains from six salt and three brine samples were detected, indicating a diverse bacterial community, 32 of which were randomly selected strains. Most cells of the strains stained Gram-negative and motile. Optimum growth was observed at 40 °C, at a pH of 7.5 and in the presence of 25% (w/v) NaCl. The results of morphological, biochemical and physiological characteristics of the isolates and antibiotic sensitivities were used to distinguish *Archaebacteria* and *Eubacteria* in the lake. It was also demonstrated that the lake accommodated a fairly wide diversity of halophilic species producing industrial enzymes such as cellulases, β-galactosidases, lipases and gelatinases.

Extremely halophilic archaea are well known to survive in the hypersaline conditions such as salt mines or salt lakes. In a hypersaline environment, namely, the Ayvalik saltern, seven extremely halophilic archaea were isolated by Elevi et al. (2004). The characterisation of halophilic strains was based on the conventional methods, including polar lipid composition, exoenzyme production, protein profiles, plasmid size and number. The Ayvalik saltern is also a very important resource in the country, and the produced salt is widely used in a variety of industrial processes such as food preservation and curing of hides (cheese, pickles, tomato, paste, fish, etc.), as well as in leather industries in Turkey. It also provides a good habitat for halophilic microorganisms. All studied isolates needed at least 15% (w/v) NaCl concentration in the medium to grow. The optimum growth for seven red halophilic *Archaea* strains were salt concentrations ranging 20–25% (w/v) NaCl at 39 °C. The isolates characterised belonged to the archaeal family *Halobacteriaceae*. The red colour (based on α-bacterioruberin derivatives) observed due to the extremely halophilic characteristic of the cultures was evidence for the presence of *Archaea* species. Due to the lipid compositions, triglycosyl diether as glycolipid of four isolates (strains R1–R4) assigned them to the genus *Haloarcula*, while strains R5–R7 containing sulphated diglycosyl diether instead resembled to *Halorubrum saccharovorum*.

A study was also conducted on the microbial diversity in the hypersaline Tuz Lake and its salterns, Kaldirim and Kayacik, located in Central Anatolia, Turkey. This study presented the results on diversity of extremely halophilic *Archaea*. Twenty-seven different strains belonged to the family *Halobacteriaceae*, which are known to be aerobic and possess red or pink pigments, and were characterised based on colony pigmentation, phenotypic characteristics, polar lipid compositions and antibiotic sensitivities. Moreover, gene sequence analysis of 16S rRNA of the isolates was performed, and the phylogenetic analysis revealed that the isolated strains are mostly assigned to the genera *Halorubrum*, *Halobacterium* and *Haloarcula*. In particular, the most dominant genus in Lake samples was *Haloarcula*, while *Halorubrum* members were detected in Tuz Lake and the saltern samples of Kaldirim, and the species of *Halobacterium* were obtained from Tuz Lake and the Kayacik saltern. All archaeal strains possessed hydrolytic enzymes (cellulases, amylases, proteases and others), used in food, detergent and leather industries (Birbir et al. 2007).

Ozcan et al. (2007) reported on the diversity of archaeal strains isolated from water and soil samples of six hypersaline locations in the provinces of Ankara (salt lake, 45.667% salinity), Denizli (Aci Lake, 0.265% salinity), Konya (Bolluk Lake, 48.452% salinity), Kayseri (Tuzla Lake), Kirsehir (Seyfe Lake) and Burdur (Salda Lake, 1.114% salinity). By analyses of morphological and biochemical properties, sensitivity to different antibiotics, plasmids and total lipid composition as well as comparisons of 16S rRNA gene sequences (1388 bp), thirty-three strains were characterised, which all belonged to the family *Halobacteriaceae*. All isolates were found to be Gram-negative, catalase- and oxidase-positive and possessing pink to red colony colour. By phylogenetic analyses, these isolates were clustered into nine genera, namely, *Halomicrobium* (one isolate), *Halalkalicoccus* (one isolate), *Haloterrigena* (three isolates), *Haloferax* (three isolates), *Natrialba* (four isolates), *Natronococcus* (four isolates), *Haloarcula* (four isolates), *Natrinema* (five isolates) and *Halorubrum* (eight isolates).

The prokaryotic diversity in a hypersaline Tuz Lake, Turkey, was also demonstrated by Mutlu et al. (2008). The authors studied microbiota in this lake by using the methodology of FISH, denaturing gradient gel electrophoresis of PCR-amplified

Table 8.4 Isolation, identification and possible biotechnological uses of halophilic microorganisms from various hypersaline environments in Turkey

Microorganisms	Strain code	GenBank accession number	Source	Growth temperature (°C)	Growth pH	Biotechnological applications and products	Reference
Halobacterium sp.	3KYS1, 3TL6, 2KYS1, 5TL6, 3TL4, 2TL9	DQ352855, DQ352856, DQ352857, DQ352858, DQ352859, DQ352860	Kaldırım and Kayacık salt lake	40	7.5	Amylases, cellulases, proteases	Birbir et al. (2007)
Halobacterium sp.	1SL3, 2SL3, 3SL6, 6SL6, 2SL8, 3SL8, 3SL9	–	Şereflikochisar Salt Lake	40	7.5	Cellulases, gelatinases, lipases and β-galactosidases	Birbir and Sesal (2003)
Halobacterium salinarum	1TK2, 1TK3	–	Tuzköy salt mine	40	6.5–7.5	Amylase, lipase, β-galactosidase, cellulase	Birbir et al. (2004)
Haloarcula vallismortis	3TK1, 3TK2, 3TK3, 2TK2, 2TK3, 4TK1, 4TK3					Amylase, lipase, cellulase	
Natrinema	2TK1, 5TK1					Amylase, lipase	
Halorubrum sodomense	1TK1					Amylase, lipase, DNase, cellulase	

(continued)

Table 8.4 (continued)

Microorganisms	Strain code	GenBank accession number	Source	Growth temperature (°C)	Growth pH	Biotechnological applications and products	Reference
Haloquadratum and *Salinibacter*	1B-8B, 4A-17A	EF459702, EF459730	Tuz Lake	37	–	–	Mutlu et al. (2008)
Halobacillus	EM7, EM14, EM33	KF499274, KF499281, KF499299	Salt-affected soil of Erzurum	30	7.0–11.0	Amylase	Orhan and Gulluce (2015)
Halobacteriaceae	R1–R7		Ayvalik saltern	39	7.2	Amylase, gelatinase	Elevi et al. (2004)
Halobacillus sp.	C-22	HM037269	Camalti Saltern/Izmir	37	4.5	Decolourisation of textile dyes	Demirci et al. (2011)
Haloferax sp.	C24, C27	JX067385, JX067386	Camalti Saltern/Izmir	37	6.5–7.5	PAH-degrading enzymes	Erdogmus et al. (2013)
Halobacterium sp.	C37	JX067388					
Halorubrum sp.	C41, C43, C46, C50, C51	JX067389, JX067390, JX067391, JX067392, JX067393					
Halalkalicoccus	C15	DQ373058	Salda Lake/Burdur	37	7.0	Amylase	Ozcan et al. (2007, 2009)
Haloferax	A440,	DQ309084,	Salt Lake/Ankara			Amylase, gelatinase	
	A317,	DQ309085,	Seyfe Lake/Kirsehir				
	D107	DQ373057					

Halorubrum	F23A,	DQ309087,	Bolluk Lake/Konya			Gelatinase	
	C35,	DQ373061,					
	F42A,	DQ309086,	Salda Lake/Burdur				
	A29,	DQ309088,	Salt Lake/Ankara				
	B36,	DQ373060,	Aci Lake/Denizli				
	A87B,	DQ309089,	Seyfe Lake/Kirsehir				
	D10,	DQ373059,					
	F100	DQ309090					
Haloarcula	A191	DQ309091	Salt Lake/Ankara			Amylase, gelatinase, lipase, esterase	
Halomicrobium	A337,	DQ309092,	Salt Lake/Ankara				
	A283,	DQ309093,	Aci Lake/Denizli				
	B44A,	DQ373062,					
	A43	DQ309094					
Halomonas smyrnensis	AAD6^T	DQ131909	Camalti/Izmir	37	5.5–8.5	Amylase, exopolysaccharide, levan production	Poli et al. (2009), Sam et al. (2011), Sogutcu et al. (2012), Poli et al. (2013), Ates et al. (2013) and Calimlioglu and Arga (2016)
Halomonas sp.	AAD12	GU397429	Camalti/Izmir	37	7.0	Hydroxyectoine production, Amylase	Uzyol et al. (2012) and Ozturk et al. (2015)

(continued)

Table 8.4 (continued)

Microorganisms	Strain code	GenBank accession number	Source	Growth temperature (°C)	Growth pH	Biotechnological applications and products	Reference
Halomonas sp.	LM1C	KT588 664	Pamukkale/ Denizli	21.6	6.9	Lipase, esterase	Gutiérrez Arnillas et al. (2016)
Halomonas halophila	C13,	HM0372678,	Camalti marine solar saltern/Izmir	37	6.5–7.5	–	Mutlu and Guven (2011, 2015)
	C18,	KF863791,					
	C20	KF863792					
Halobacillus sp.	C12,	KF863788,					
	C17,	KF863790,					
	C25	KF863793,					
Virgibacillus marismortui	C15	KF863789					
Halolamina sp.	YKT1	KU051660	Yozgat salt mine	37	7.5	–	Kurt Kizildogan et al. (2017)
Streptomonospora tuzyakensis	$BN506^T$	KCTC 29210^T	Tuz (Salt) Lake, Konya	37	6.0–12.0	–	Tatar et al. (2016)

16S rRNA genes, and by comparisons of the clone library 16S rRNA gene sequences. Interestingly, the number of *Archaea* members in the community was three times more than those of *Bacteria* detected by FISH. The archaeal members were dominantly clustered into the square *Haloarchaea* of the Walsby group, while bacterial members dominantly grouped into *Bacteroidetes*, such as *Salinibacter ruber*-related phylotypes. It is well known that *Bacteroidetes* species are widespread in various hypersaline environments. The comparison between 16S rRNA sequences from the Tuz Lake bacterial strains and those from other hypersaline environments showed a 'halophilic branch' within the *Bacteroidetes* phylum that clustered together.

Moreover, a natural reserve in Sasali, Izmir, in the Aegean Region was studied for the isolation and identification of halophiles from several pond soil samples with salt contents in the range 30–50% and pH in the range of 6.5–7.5. The isolated strains designated as AAD6^T, AAD4, AAD17 and AAD21 were Gram-negative, exopolysaccharide-producing and moderately halophilic bacteria which grew at an optimum of 10% (w/v) NaCl. The G + C compositions of the genomic DNAs of AAD21 AAD17, AAD4 and AAD6^T were 62.6, 62.8, 63.3 and 63.0 mol %, respectively. Sequence comparisons of 16S rRNA gene between the strain AAD6^T and most related species indicated that the strain AAD6^T was close to *Halomonas salina* F8-11^T (99.4% similarity) and *Halomonas halophila* CCM 3662^T (99.4%), and the mean values of DNA–DNA hybridisation between the representative strain AAD6^T and the most related species mentioned above were calculated as 40.8 and 39.6%, respectively. On the basis of the phenotypic, phylogenetic and genomic properties presented above, the strain AAD6^T represents a novel species of the genus *Halomonas* and thus named *Halomonas smyrnensis* (=DSM 21644^T = JCM 15723^T). *H. smyrnensis* was rod-shaped and formed circular and slightly irregular colonies with cream-yellowish colour and was also different from all closely related species of the genus *Halomonas* in terms of hydrolysing starch and casein. This novel species also produced a higher yield of exopolysaccharide named levan which was previously described as repeating unit comprised of beta (2,6)-D-fructofuranosyl residues (Poli et al. 2009, 2013). Moreover, whole genome sequencing of *H. smyrnensis* AAD6^T was succeeded by Sogutcu et al. 2012.

Orhan and Gulluce (2015) have recently mentioned about the importance of halophilic and halotolerant microorganisms in salt-affected soils, which may possess basic enzyme activities that can enhance nutrient cycling and fertility in soil. This study was carried out in salt-affected soil of Erzurum Province in the East Anatolian Region of Turkey. Forty-five bacterial strains were isolated and characterised by phenotypic and phylogenetic techniques. The strains isolated from salt-affected soils belonged to 16 different genera, as follows: *Bacillus* (19 strains), *Staphylococcus* (3 strains), *Halobacillus* (4 strains), *Zhihengliuella* (2 strains), *Oceanobacillus* (2 strains), *Halomonas* (1 strain), *Nesterenkonia* (2 strains), *Promicromonospora* (2 strains), *Jeotgalibacillus* (2 strains), *Planococcus* (2 strains), *Virgibacillus* (1 strain), *Terribacillus* (1 strain), *Thalassobacillus* (1 strain), *Marinibacillus* (1 strain), *Gracilibacillus* (1 strain) and *Microbacterium* (1 strain). They claimed that the characterised strains in salt-affected soils had high salt tolerance and significant enzyme activities which may be used for improvement of agricultural soils.

Another recent study to identify the bacterial diversity of Camalti marine solar saltern has been carried out by Mutlu and Guven 2015. The total salt concentrations and the pH values of samples collected from this area were measured between 6% and 32% and pH 6.5 and 7.5, respectively. The bacterial communities of Camalti Saltern were characterised by molecular techniques that included the analysis of PCR-amplified fragments of 16S rRNA gene by the denaturing gradient gel electrophoresis. They identified a total of 42 isolates at the genus/species level, and 17 of them belonged to the *Bacteria* domain. All of bacterial strains were phylogenetically related to *Halomonas*, *Halobacillus* and *Virgibacillus* genus. 16S rRNA sequence analysis of the clones by ARDRA method showed that most (85%) of the bacterial clones were the members of *Salinibacter* genus within the *Bacteroidetes*.

A novel halophilic actinobacterium was isolated from Tuz Lake soil sample in Konya by Tatar et al. (2016). The isolate designated as BN506^T was associated with members of the genus *Streptomonospora* based on morphological and chemotaxonomic properties. Moreover, analysis of 16S rRNA gene sequence and DNA–DNA relatedness showed that strain BN506^T was a member of a new species of the *Streptomonospora* genus, named as *Streptomonospora tuzyakensis* (= DSM 45930^T = KCTC 29210^T). The 16S rRNA gene sequence similarities between strain BN506^T and related species showed close relation to *S. halophila* YIM 91355^T (98.1%) and *S. arabica* S186^T (97.9%), with also DNA relatedness values of 41.0 ± 3.5% and 25.2 ± 3.6%, respectively. The genomic DNA G + C content was detected as 71.1 mol %. The isolate was aerobic, Gram-positive, nonmotile actinomycete. The aerial mycelium of the species was white and found to grow at 4–20% NaCl (w/v) and between temperatures of 28 and 37 °C (optimally 37 °C in 10% (w/v) NaCl) and between a pH range of 6.0–12.0.

A global transcriptome analysis has been recently conducted by Kurt Kizildogan et al. (2017) in an extremely halophilic archaeon, *Halolamina* sp. YKT1, isolated from a sample in Yozgat salt mine, in order to explore the molecular mechanisms leading to the high salt tolerance. It was found that the salt tolerance of the YKT1 strain involves the up-regulation of genes related with osmoprotectant solutes, membrane transporters, oxidative stress proteins, CRISPR–Cas systems and iron metabolism. This comprehensive transcriptome analysis however showed that the genes that encode the proteins involved in translation, transcription, DNA replication and DNA repair were downregulated.

8.6.2 Biotechnological Applications of Halophilic Microorganisms Isolated in Turkey

There are promising studies on halophilic bacteria and archaea, as they have ability for producing biochemicals, which possess a significant potential use in industrial and environmental technology (Oren 2010). Furthermore, halophilic bacteria have ability to produce biopolymers that are used in industrial and medical technology

(Table 8.4). For instance, levan is an extracellular biopolymer produced by a moderately halophilic bacterium *Halomonas* species (Poli et al. 2009, 2013; Ates et al. 2013). Moreover, exopolysaccharide-producing halophile, namely, *Halomonas* sp. AAD6 (DQ131909), was isolated from Camalti Saltern Area in Turkey. The strain cultivated on agro-industrial waste produced exopolysaccharides which had a potential use as an alternative and easily biodegradable polyelectrolytes, compared to synthetic ones that are commonly in use and contain toxic and carcinogenic monomers (Sam et al. 2011). The activated sludge culture supplemented with a salt tolerant, *Halobacter halobium*, was utilised for saline wastewater treatment in order to alleviate salt inactivation effects in a biodisc contactor (Kargi and Dincer 1998).

A moderate halophile identified as *Halomonas* sp. AAD12 from salt sediments in Camalti Saltern Area in Turkey was isolated by Ozturk et al. (2015), pointing out that it was a promising candidate as a hydroxyectoine producer. *Halomonas* sp. AAD12 was found to adapt stress conditions by changing its osmolyte accumulation ratio and the fluidity of membrane to prevent the effects of stress. A number of moderate halophiles are known to change the accumulated concentrations of the osmoprotectants ectoine, hydroxyectoine and proline to protect its cytoplasm during stress exposure, such as oxygen limitation, temperature and salinity. These molecules are desired for a variety of uses in biotechnology, for protection of enzymes against different stress factors such as freeze-thawing, freeze-drying and heating and also as a protection for the healthy cell desiccation during chemotherapy and for medical use as a molecular chaperon for Alzheimer's disease, as well as for preservation of cardiac death donors (DCD) livers.

An application field of halophilic bacteria has been the use for biodegradation of dyes. For example, a report was carried out about the use of halophilic bacterium isolated from water and soil samples of a solar sea saltern (Camalti, Turkey) in environmental technology, especially in textile industry for the decolourisation of some of azo–metal complex dyes. Among these, only one bacterium identified by 16S rRNA gene sequence analysis as *Halobacillus* sp. C-22 (99% sequence similarity) was determined as resistant against two dyes, which are Lanaset Brown B and Lanaset Navy R. Following exposure to Lanaset Brown B, the bacterium decolourised the dye at a high absorbance ratio (96.12%) after 78th h, while Lanaset Navy R was significantly decolourised in 10 min by 46.67% and 60.66% at the third hour (Demirci et al. 2011).

Another application field was the use of archaeal isolates in environmental technology for degradation of polyaromatic hydrocarbons (PAHs). Erdogmus et al. (2013) isolated some archaeal strains from the Camalti Saltern (Turkey) and identified them by 16S rRNA gene sequences as *Halobacterium salinarum*, *Halobacterium piscisalsi*, *Halorubrum ezzemoulense*, *Haloarcula hispanica*, *Haloarcula* sp., *Haloferax* sp. and *Halorubrum* sp., which were found to degrade PAHs (namely, naphthalene, p-hydroxybenzoic acid, pyrene and phenanthrene) to use for the carbon and energy sources. Recently, halophilic microorganisms have been also used for biological treatment of highly saline wastewaters containing aromatic hydrocarbons. Acikgoz and Ozcan (2016) isolated a total of 103 halophilic *Archaea* from different parts of Turkey to study phenol biodegradation. The aromatic compound phenol is known to be toxic produced after various industrial activities. The maximum phenol degradation capacity was obtained with the strain A235, among all strains studied on liquid and solid media with 20% (w/v) NaCl and phenol as the only carbon source.

The novel industrially important enzymes isolated and characterised from halophiles, which are stable in harsh conditions such as thermal, salt, alkaline and organic solvent stability, may well present advantages in different industrial processes (Souza 2010; Kumar et al. 2012). Although a review on halophilic proteins and their applications have been already reviewed by Calimlioglu and Arga (2016) in general, there are not enough studies on halophilic enzymes from microorganisms isolated from saltern areas in Turkey. As an example, extremely halophilic microorganism communities comprising of *Archaebacteria* and *Eubacteria* isolated in Sereflikochisar Salt Lake located in central Turkey were studied by Birbir and Sesal (2003). It was also demonstrated that a fairly wide diversity of halophilic species were found to produce industrial enzymes such as lipases, gelatinases, cellulases and β-galactosidases.

Another study was carried out on a thermostable amylase produced by moderately halophilic microorganism, namely, *Halomonas* sp. strain AAD21, isolated from the Camalti Saltern located in Izmir Province. On the basis of morphological and biochemical characteristics and 16S rRNA gene sequence analysis, the strain was assigned to the genus *Halomonas*. The strain was found to grow at wide salt concentration range of 3–20% (w/v) NaCl, with optimum of 10%. The optimum temperature and pH of the α-amylase were determined as 50 °C and 7.0, respectively. The α-amylase from *Halomonas* sp. AAD21 was found to be thermostable, as 70% of original enzyme activity was retained during 120 min of incubation at 90 °C, which claimed to be a good candidate for use in severe process conditions of starch hydrolysis or detergent industry (Uzyol et al. 2012).

Ozcan et al. (2009) screened as many as 118 halophilic archaeal strains in search of lipolytic activity, five of which were selected and further characterised to determine the effects of salt, temperature and pH at various ranges on the optimum esterase and lipase activities. The highest hydrolytic production was determined for the strains grown at a special medium containing 25% NaCl and 1% arabic gum. The maximum activity of esterase was observed at temperature 60–65 °C, pH 8–8.5 and at 3–4.5 M NaCl, while the highest activity of lipase was determined at temperatures between 45 and 65 °C, pH of 8.0, and NaCl range of 3.5–4 M, indicating the presence of the temperature-tolerant and salt-dependent archaeal lipolytic enzymes. The results also showed that the strains had a higher esterase activity compared to lipase activity.

A most recent study has been published on new bacterial sources of halophilic lipases. It has been highlighted on the Turkish and Spanish hypersaline biotopes to be a suitable source of halophilic microorganisms producing lipolytic enzymes, which are from two different points in salterns of Parque Natural de las Lagunas de La Mata y Torrevieja (Spain) and from Pamukkale (Turkey). Three strains growing at NaCl concentration greater than 15% were capable to synthesise lipolytic enzymes, though one of them identified as *Halomonas* sp. LM1C was demonstrated to have high enzyme production levels. Subsequently this strain was used, and the highest lipase production was obtained at pH 6.9 and 21.6 °C. The optimum values for the enzyme-biocatalysed hydrolysis were determined as neutral pH and 29 °C. The extracellular lipase displayed a high salt tolerance, which claimed to pose the economic advantages in industrial applications (Gutiérrez-Arnillas et al. 2016).

8.7 Future Perspective

Industrial biotechnology is a key technology for future economic development. Thus, for developing countries such as Turkey, there is a need to expand the research in biotechnology field. Since Turkey owns different ecological areas, i.e. surrounded by seas, salt lakes and many hot springs with a broad microbial diversity including extremophiles, there are a great deal of opportunities for newly isolated microorganisms from extreme environments for their use in biotechnological applications. From this point of view, it seems that the search of extremophiles in the country is very recent, and this potential needs to be fully exploited.

There is a limited data on archaea isolated from extreme environments in Turkey, which need to be considered in many investigations. Particularly, their roles as source of enzymes from extremophile archaea (thermostable DNA polymerases, amylase, galactosidases and pullulanases) have a wide range of potential uses and also known to be very stable in organic solvents, providing an advantage in use for environmentally friendly processes. For instance, acidophilic archaea give a promise in mineral processing for the extraction of several metals such as gold and copper, as Turkey is known to possess gold and copper mines.

Recent advances in molecular genetic tools for extreme microorganisms lead to their use for metabolic engineering for the production of chemicals and fuels. The advantages and drawbacks of using extremophiles as industrial hosts need to be discussed further with perspectives on future developments in this emerging technology (Loder et al. 2017). Today, in most countries, there is a trend towards cheap, renewable and readily available biomass in the production of various chemicals utilising extremophiles and their enzymes. It is a fact that Turkey is one of the top agriculture-producing countries in the world. As a result, there are potential and existing applications of both thermophiles and thermostable enzymes on conversion of raw materials containing carbohydrate into the desired products in industrial biotechnology. Moreover, the studies on thermophiles have extended to energy biotechnology such as biofuel, biohydrogen and ethanol production. Microorganisms isolated can be explored for the production of next-generation biofuels by the use of the carbohydrate fraction in lignocellulosic material (Turner et al. 2007).

It has been already suggested that at higher temperatures the ethanol production would make the process design easy, and the engineered progeny of *Thermoanaerobacter mathranii*, *Thermoanaerobacterium saccharolyticum* and *Geobacillus thermoglucosidasius* now forms the platform for new biotechnology companies (Taylor et al. 2009; Olson et al. 2015).

On the other hand, the potentials of halophiles and their bioproducts have been extensively reviewed (Ventosa et al. 1998; Madern et al. 2000; Oren 2010; DasSarma and DasSarma 2017), as well as genetic tools for manipulation of moderately halophilic bacteria with promising applications in biotechnology (Vargas and Nieto 2004). Halophilic enzymes are important candidates for use in biotransformation reactions in harsh industrial environments such as cosmetic, agrochemical, textile, detergent, paper, fuel, energy and pharmaceutical industries, which all require the organic solvent addition, very high temperatures, high salt concentration, low water level, high pH levels, etc. Despite some studies which have been already performed on halophilic microorganisms in different regions of Turkey, halophiles still need a special interest in terms of isolation and characterisation of new species producing desirable biocatalysts and biomolecules to fulfill future biotechnological and industrial demands. Most recently, biosynthesis of nanoparticles using extreme microorganisms has emerged as rapidly developing research area in green nanotechnology in the world as forming an alternative for conventional chemical and physical methods, which should also be taken into consideration.

References

Acer O, Pirinccioglu H, Matpan Bekler F, Gul Guven R, Guven K (2015) *Anoxybacillus* sp. AH1, an amylase producing thermophilic bacterium isolated from Dargecit hot spring. Biologia 70(7):853–862

Acer O, Matpan Bekler F, Pirinccioglu H, Gul Guven R, Guven K (2016) Purification and characterization of thermostable and detergent-stable α-amylase from *Anoxybacillus* sp. AH1. Food Technol Biotechnol 54(1):70–77

Acikgoz E, Ozcan B (2016) Phenol biodegradation by halophilic archaea. Int Biodeterior Biodegrad 107:140–146

Adiguzel A, Ozkan H, Baris O, Inan K, Gulluce M, Sahin F (2009) Identification and characterization of thermophilic bacteria isolated from hot springs in Turkey. J Microbiol Methods 79:321–328

Adiguzel A, Inan K, Sahin F, Arasoglu T, Gulluce M (2011) Molecular diversity of thermophilic bacteria isolated from Pasinler hot spring (Erzurum, Turkey). Turk J Biol 35(3):267–274

Ahmetoglu N, Matpan Bekler F, Acer O, Guven G, Guven K (2015) Production, purification and characterisation of thermostable metallo-protease from newly isolated *Bacillus* sp. KG5. Eurasia J Biosci 9:1–11

Alkan H, Gul Guven R, Guven K, Erdogan S, Dogru M (2015) Biosorption of Cd^{2+}, Cu^{2+}, and Ni^{2+} ions by a thermophilic haloalkalitolerant bacterial strain (KG9) immobilized on amberlite XAD-4. Pol J Environ Stud 24(5):1903–1910

Arikan B, Unaldi N, Coral G, Colak O, Aygan A, Gulnaz O (2003) Enzymatic properties of a novel thermostable, thermophilic, alkaline and chelator resistant amylase from an alkaliphilic *Bacillus* sp. isolate ANT-6. Process Biochem 38:1397–1403

Atasoy P, Canakci S, Belduz AO (2012) Purification, characterization and industrial application of the alkaline protease (akap) from *Anoxybacillus kestanbolensis* K4. New Biotechnol 29S:118

Ates O, Arga KY, Toksoy Oner E (2013) The stimulatory effect of mannitol on levan biosynthesis: lessons from metabolic systems analysis of *Halomonas smyrnensis* AAD6. Biotechnol Prog 29(6):1386–1397

Ay F, Karaoglu H, Inan K, Canakci S, Belduz AO (2011) Cloning, purification and characterization of a thermostable carboxylesterase from *Anoxybacillus* sp. PDF1. Protein Expr Purif 80:74–79

Bakir Ateslier ZB, Metin K (2006) Production and partial characterization of a novel thermostable esterase from a thermophilic *Bacillus* sp. Enzyme MicrobTechnol 38(5):628–635

Bakir ZB, Metin K (2015) Screening for industrially important enzymes from thermophilic bacteria; selection of lipase-producing microorganisms and optimization of culture conditions. European J Biotechnol Biosci 3(6):43–48

Bakir ZB, Metin K (2017) Production and characterization of an alkaline lipase from thermophilic *Anoxybacillus* sp. HBB16. Chem Biochem Eng Q 31(3):303–312

Baltaci MO, Genc B, Arslan S, Adiguzel G, Adiguzel A (2016) Isolation and characterization of thermophilic bacteria from geothermal areas in Turkey and preliminary research on biotechnologically important enzyme production. Geomicrobiol J 34(1):53–62

Baltas N, Dincer B, Kolayli S, Ekinci AP, Adiguzel A, Gulluce M (2012) Purification and characterization of extracellular alpha-amylase from *Anoxybacillus gonensis* Z4 strain. Turk J Biochem 37(1):140

Baltas N, Dincer B, Ekinci AP, Kolayli S, Adiguzel A (2016) Purification and characterization of extracellular α-amylase from a thermophilic *Anoxybacillus thermarum* A4 strain. Braz Arch Biol Technol 59:e16160346

Belduz AO, Dulger S, Demirbag Z (2003) *Anoxybacillus gonensis* sp. nov., a moderately thermophilic, xylose-utilizing, endospore-forming bacterium. Int J Syst Evol Microbiol 53:1315–1320

Belduz AO, Canakci S, Chan KG et al (2015) Genome sequence of *Anoxybacillus ayderensis* AB04^{T} isolated from the Ayder hot spring in Turkey. Stand Genomic Sci 10:70

Beris FS, De Smet L, Karaoglu H, Canakci S, Van Beemuen J, Belduz AO (2011) The ATPase activity of the G2alt gene encoding an aluminium tolerance protein from *Anoxybacillus gonensis* G2. J Microbiol 49:641–650

Birbir M, Sesal C (2003) Extremely halophilic bacterial communities in Sereflikochisar Salt Lake in Turkey. Turk J Biol 27(7):7–21

Birbir M, Ogan A, Calli B, Mertoglu B (2004) Enzyme characteristics of extremely halophilic archaeal community in Tuzkoy Salt Mine, Turkey. World J Microbiol Biotechnol 20:613–621

Birbir M, Calli B, Mertoglu B, Bardavid RE, Oren A, Ogmen MN, Ogan A (2007) Extremely halophilic archaea from Tuz Lake, Turkey, and the adjacent Kaldirim and Kayacik salterns. World J Microbiol Biotechnol 23:309–316

Bozoglu C, Adiguzel A, Nadaroglu H, Yanmis D, Gulluce M (2013) Purification and characterization of laccase from a newly isolated thermophilic *Brevibacillus* sp. (Z1) and its application of removal textile dyes. Res J Biotechnol 8(9):56–66

Caglayan M, Bilgin N (2011) Cloning and sequences analysis of novel DNA polymerases from thermophilic *Geobacillus* species isolated from hot springs in Turkey: characterization of a DNA polymerase I from *Geobacillus kaue* strain NB. Appl Biochem Biotechnol 165:1188–1200

Caglayan M, Bilgin N (2012) Temperature dependence of accuracy of DNA polymerase I from *Geobacillus anatolicus*. Biochimie 94(9):1968–1973

Cakmak U, Saglam Ertunga N (2017) Gene cloning, expression, immobilization and characterization of endo-xylanase from *Geobacillus* sp. TF16 and investigation of its industrial applications. J Mol Catal B: Enzym 133:288–298

Calimlioglu B, Arga KY (2016) Proteins from halophilic bacteria: purification and their applications. In: Protein purification: principles and trends. iConcept Press Hong Kong isbn:978-1-922227-40-9

Canakci S, Inan K, Kacagan M, Belduz AO (2007) Evaluation of arabinofuranosidase and xylanase activities of *Geobacillus* spp. isolated from some hot springs in Turkey. J Microbiol Biotechnol 17:1261–1270

Canakci S, Cevher Z, Inan K, Tokgoz M, Bahar F, Kacagan M, Fulya AS, Belduz AO (2012) Cloning, purification and characterization of an alkali-stable endoxylanase from thermophilic *Geobacillus* sp. 71. World J Microbiol Biotechnol 28:1981–1988

Cihan AC (2013) Taxonomic classification of *Anoxybacillus* isolates from geothermal regions in Turkey by 16S rRNA gene sequences and ARDRA, ITS-PCR, Rep-PCR analyses. Pol J Microbiol 62(2):149–163

Cihan AC, Yildiz ED (2016) General characteristics, taxonomic studies and biotechnological importance of the genus *Anoxybacillus*. Commun Fac Sci Univ Ank Series C 25(1-2):56–82
Cihan A, Ozcan B, Cokmus C (2009) Characterization of thermostable glucosidases from newly isolated *Geobacillus* sp. A333 and thermophilic bacterium A343. World J Microbiol Biotechnol 25:2205–2217
Cihan AC, Ozcan B, Cokmus C (2011a) *Anoxybacillus salavatliensis* sp. nov., an α-glucosidase producing, thermophilic bacterium isolated from Salavatli, Turkey. J Basic Microbiol 51(2):136–146
Cihan AC, Ozcan B, Tekin N, Cokmus C (2011b) *Geobacillus thermodenitrificans* subsp. *calidus*, subsp. nov., a thermophilic and α-glucosidase producing bacterium isolated from Kizilcahamam, Turkey. J Gen Appl Microbiol 57(2):83–92
Cihan AC, Cokmus C, Koc M, Ozcan B (2014) *Anoxybacillus calidus* sp. nov., a thermophilic bacterium isolated from soil near a thermal power plant. Int J Syst Evol Microbiol 64:211–219
Colak A, Sisik D, Saglam N, Guner S, Canakci S, Belduz AO (2005) Characterization of a thermoalkalophilic esterase from a novel thermophilic bacterium, *Anoxybacillus gonensis* G2. Bioresour Technol 96:625–631
Colak A, Col M, Canakci S, Belduz AO, Omarov I (2008) Investigation of extracellular highly thermostable starch hydrolyzing activity from a novel thermophilic bacterium *Anoxybacillus gonensis* A4. Asian J Chem 20:1577–1587
Coleri A, Cokmus C, Ozcan B, Akkoc N, Akcelik M (2009) Isolations of α-glucosidase-producing thermophilic bacilli from hot springs of Turkey. Microbiology 78(1):56–66
Da Mota FF, Gomes EA, Paiva E, Rosado AS, Seldin L (2004) Use of *rpoB* gene analysis for identification of nitro-genfixing *Paenibacillus* species as an alternative to the 16S rRNA gene. Lett Appl Microbiol 39:34–40
DasSarma S, DasSarma P (2017) Halophiles. In: eLS. Wiley, Chichester. http://www.els.net. https://doi.org/10.1002/9780470015902.a0000394.pub4
Demirci A, Mutlu MB, Guven A, Korcan E, Guven K (2011) Decolorization of textile azo-metal complex dyes by a halophilic bacterium isolated from Camalti Saltern in Turkey. CLEAN 39(2):177–184
Demirjian DC, Moris-Varas F, Cassidy CS (2001) Enzymes from extremophiles. Curr Opin Chem Biol 5:144–151
Dulger S, Demirbag Z, Belduz AO (2004) *Anoxybacillus ayderensis* sp. nov. and *Anoxybacillus kestanbolensis* sp. nov. Int J Syst Evol Microbiol 54:1499–1503
Duran C, Bulut VN, Gundogdu A, Soylak M, Belduz AO, Beris FS (2009) Biosorption of heavy metals by *Anoxybacillus gonensis* immobilized on diaion HP-2MG. Sep Sci Technol 44(2):335–358
Ekinci AP, Dincer B, Baltas N, Adiguzel A (2016) Partial purification and characterization of lipase from *Geobacillus stearothermophilus* AH22. J Enzyme Inhib Med Chem 31(2):325–331
Elevi R, Assa P, Birbir M, Ogan A, Oren A (2004) Characterization of extremely halophilic Archaea isolated from the Ayvalik Saltern, Turkey. World J Microbiol Biotechnol 20:719–725
Erdogmus SF, Mutlu B, Korcan SE, Guven K, Konuk M (2013) Aromatic hydrocarbon degradation by halophilic archaea isolated from Camalti Saltern, Turkey. Water Air Soil Pollut 224:1449
Ertunga NS, Colak A, Belduz AO, Canakci S, Karaoglu H, Sandalli C (2007) Cloning, expression, purification and characterization of fructose-1,6-bisphosphate aldolase from *Anoxybacillus gonensis* G2. J Biochem 141:817–825
Faiz O, Colak A, Saglam N, Canakci S, Belduz AO (2007) Determination and characterization of thermostable esterolytic activity from a novel thermophilic bacterium *Anoxybacillus gonensis* A4. J Biochem Mol Biol 40:588–594
Fincan SA, Enez B, Ozdemir S, Matpan Bekler F (2014) Purification and characterization of thermostable α-amylase from thermophilic *Anoxybacillus flavithermus*. Carbohydr Polym 102:144–150
Genc B, Baltaci O, Nadaroglu H, Adiguzel A (2014) A novel cellulase from Anoxybacillus gonensis A4 isolated from geothermal area in Turkey. In: Molecular biology conference, 10–12 September, 2014, Turkey, p 47

Genc B, Nadaroglu H, Adiguzel A, Baltaci O (2015) Purification and characterization of an extracellular cellulase from *Anoxybacillus gonensis* O9 isolated from geothermal area in Turkey. J Environ Biol 36:1319–1324

Goh KM, Kahar UM, Chai YY, Chong CS, Chai KP, Ranjani V, Illias RM, Chan KG (2013) Recent discoveries and applications of *Anoxybacillus*. Appl Microbiol Biotechnol 97:1475–1488

Grant WD, Kamekura M, McGenity TJ, Ventosa A (2001) Order I. Halobacteriales. In: Garrity GM (ed) Bergey's manual of systematic bacteriology V.I, the Archaea and deeply branching and phototrophic bacteria, 2nd edn. Springer, New York, p 294–333. http://turkeyculture.org/

Gul Guven R (2007) Isolation and identification of bacteria from hot springs, and β-galactosidase purification from *Alicyclobacillus acidocaldarius* subsp. *rittmanii*. PhD thesis, Dicle University, Diyarbakir

Gul Guven R, Guven K, Poli A, Nicolaus B (2008) *Anoxybacillus kamchatkensis* subsp. *asaccharedens* subsp. nov., a thermophilic bacterium isolated from a hot spring in Batman. J Gen Appl Microbiol 54:337–354

Gutiérrez-Arnillas E, Rodríguez A, Sanromán MA, Deive FJ (2016) New sources of halophilic lipases: Isolation of bacteria from Spanish and Turkish saltworks. Biochem Eng J 109:170–177

Inan K, Belduz AO, Canakci S (2011a) Isolation and characterization of xylanolytic new strains of *Anoxybacillus* from some hot springs in Turkey. Turk J Biol 35:529–542

Inan K, Bektas Y, Canakci S, Belduz AO (2011b) Use of *rpoB* sequences and rep-PCR for phylogenetic study of *Anoxybacillus* species. J Microbiol 49:782–790

Inan K, Canakci S, Belduz AO, Sahin F (2012) *Brevibacillus aydinogluensis* sp. nov., a moderately thermophilic bacterium isolated from Karakoc hot spring. Int J Syst Evol Microbiol 62:849–855

Inan K, Belduz AO, Canakci S (2013) *Anoxybacillus kaynarcensis* sp. nov., a moderately thermophilic, xylanase producing bacterium. J Basic Microbiol 53:410–419

Inan K, Ozer A, Guler HI, Belduz AO, Canakci S (2016) *Brevibacillus gelatini* sp. nov., isolated from a hot spring. Int J Syst Evol Microbiol 66:712–718

Kacagan M, Canakci S, Sandalli C, Inan K, Colak DN, Belduz AO (2008) Characterization of a xylanase from a thermophilic strain of *Anoxybacillus pushchinoensis* A8. Biologia 63:599–606

Kacagan M, Inan K, Canakci S, Guler HI, Belduz AO (2015) *Thermus anatoliensis* sp. nov., a thermophilic bacterium from geothermal waters of Buharkent, Turkey. J Basic Microbiol 55:1367–1373

Karadag D, Makinen AE, Efimova E, Puhakka JA (2016) Thermophilic biohydrogen production by an anaerobic heat treated-hot spring culture. Bioresour Technol 100:5790–5795

Karaoglu H, Yanmis D, Ay F, Celik A, Canakci S, Belduz AO (2013) Biochemical characterization of a novel glucose isomerase from *Anoxybacillus gonensis* G2T that displays a high level of activity and thermal stability. J Mol Catal B Enzym 97:215–224

Kargi F, Dincer AR (1998) Saline wastewater treatment by halophile-supplemented activated sludge culture in an aerated rotating biodisc contactor. Enzym Microb Technol 22:427–433

Kocabiyik S, Erdem B (2002) Intracellular alkaline proteases produced by thermoacidophiles: detection of protease heterogeneity by gelatin zymography and polymerase chain reaction (PCR). Bioresour Technol 84:29–33

Kumar S, Karan R, Kapoor S, Singh SP, Khare SK (2012) Screening and isolation of halophilic bacteria producing industrially important enzymes. Braz J Microbiol 43:1595–1603

Kurt Kizildogan A, Abanoz B, Okay S (2017) Global transcriptome analysis of *Halolamina* sp. to decipher the salt tolerance in extremely halophilic archaea. Gene 601:56–64

Lim YL, Chan KG, Ee R, Belduz AO, Canakci S, Kahar UM, Yaakop AS, Goh KM (2015) Complete genome of the potential thermozyme producer *Anoxybacillus gonensis* G2^{T} isolated from the Gonen hot springs in Turkey. J Biotechnol 212:65–66

Loder AJ, Zeldes B, Conway JM, Counts JA, Straub CT, Khatibi PA, Lee LL, Vitko NP, Keller M, Rhaesa AM, Rubinstein GM, Scott IM, Lipscomb GL, Adams MWW, Kelly M (2017) Extreme thermophiles as metabolic engineering platforms: strategies and current perspective. In: Wittmann C, Liao JC (eds) Industrial biotechnology: microorganisms. Wiley-VCH Verlag GmbH & Co. KGaA, Weinheim. https://doi.org/10.1002/9783527807796.ch14

Madern D, Ebel C, Zaccai G (2000) Halophilic adaptation of enzymes. Extremophiles 4:91–98
Mandic Mulec I, Stefanic P, van Elsas J (2016) Ecology of *Bacillaceae*. In: Driks A, Eichenberger P (eds) The bacterial spore: from molecules to systems. ASM Press, Washington, DC, pp 59–85
Matpan Bekler F (2016) Isolation and identification of a novel thermo-alkalophilic *Anoxybacillus* sp. strain KB4 from Kusburnu hot spring in Turkey. BioDiCon 9(2):169–179
Matpan Bekler F (2017) Physio-biochemical characterization and 16S rRNA sequencing of thermophilic *Geobacillus* sp. strain DB2 from Dibekli hot spring, Turkey. Iğdır Univ J Inst Sci Tech 7(1):63–71
Matpan Bekler F, Guven K (2014) Isolation and production of thermostable α-amylase from thermophilic Anoxybacillus sp. KP1 from Diyadin hot spring in Ağri. Turkey Biologia 69(4):419–427
Matpan Bekler F, Stougaard P, Guven RG, Guven K, Acer O (2015a) Isolation and cloning of extracellular thermostable β-galactosidases from a newly isolated thermophilic Bacillus licheniformis KG9. Cell Mol Biol 61(3):71–78
Matpan Bekler F, Acer O, Guven K (2015b) Co-Production of thermostable, calcium-independent α-amylase and alkali-metallo protease from newly isolated *Bacillus licheniformis* DV3. Innov Rom Food Biotechnol 16:21–30
Matpan Bekler F, Acer O, Guven K (2015c) Production and purification of novel thermostable alkaline protease from *Anoxybacillus* sp. KP1. Cell Mol Biol 61(4):118–125
Matpan Bekler F, Yalaz S, Acer O, Guven K (2017) Purification of thermostable β-galactosidase from *Anoxybacillus* sp. KP1 and estimation of combined effect of some chemicals on enzyme activity using semiparametric errors in variables model. Fresenius Environ Bull 26(3):2253–2261
Mehta R, Singhal P, Singh H, Damle D, Sharma AK (2016) Insight into thermophiles and their wide-spectrum applications. 3. Biotech 6(1):81
Mutlu MB, Guven K (2011) Detection of prokaryotic microbial communities of Çamaltı Saltern, Turkey, by fluorescein in situ hybridization and real-time PCR. Turk J Biol 35:687–695
Mutlu MB, Guven K (2015) Bacterial diversity in Camalti Saltern, Turkey. Pol J Microbiol 64(1):37–45
Mutlu MB, Martinez-Garcia M, Santos F, Pena A, Guven K, Anton J (2008) Prokaryotic diversity in Tuz Lake, a hypersaline environment in Inland Turkey. FEMS Microbiol Ecol 65:474–483
Nazina TN, Tourova TP, Poltaraus AB, Novikova EV, Grigoryan AA, Ivanova AE, Lysenko AM, Petrunyaka VV, Osipov GA, Belyaev SS, Ivanov MV (2001) Taxonomic study of aerobic thermophilic bacilli: Descriptions of *Geobacillus subterraneus* gen. nov., sp. nov. and *Geobacillus uzenensis* sp. nov. from petroleum reservoirs and transfer of *Bacillus stearothermophilus*, *Bacillus thermocatenulatus*, *Bacillus thermoleovorans*, *Bacillus kaustophilus*, *Bacillus thermoglucosidasius* and *Bacillus thermodenitrificans* to *Geobacillus* as the new combinations *G. stearothermophilus, G. thermocatenulatus*, *G. thermoleovorans*, *G. kaustophilus*, *G. thermoglucosidasius* and *G. thermodenitrificans*. Int J Syst Evol Microbiol 51:433–446
Olson DG, Sparling R, Lynd LR (2015) Ethanol production by engineered thermophiles. Curr Opin Biotechnol 33:130–141
Oren A (2010) Industrial and environmental applications of halophilic microorganisms. Environ Technol 31(8-9):825–834
Orhan F, Gulluce M (2015) Isolation and characterization of salt-tolerant bacterial strains in salt-affected soils of Erzurum, Turkey. Geomicrobiol J 32(6):521–529
Ozcan B, Ozcengiz G, Coleri A, Cokmus C (2007) Diversity of halophilic archaea from six hypersaline environments in Turkey. J Microbiol Biotechnol 17(5):745–752
Ozcan B, Ozyilmaz G, Cokmus C, Caliskan M (2009) Characterization of extracellular esterase and lipase activities from five halophilic archaeal strains. J Ind Microbiol Biotechnol 36:105–110
Ozdemir S, Matpan F, Guven K, Baysal Z (2011) Production and characterization of partially purified extracellular thermostable α-amylase by *Bacillus subtilis* in submerged fermentation (SmF). Prep Biochem Biotechnol 41:365–381

Ozdemir S, Matpan F, Okumus V, Dundar A, Ulutas MS, Kumru M (2012) Isolation a thermophilic *Anoxybacillus flavithermus* sp. nov. and production of thermostable α-amylase under solid-state fermentation (SSF). Ann Microbiol 62:1367–1375

Oztekin Z, Ay F, Karaoglu H, Colak DN, Belduz AO (2011) Purification and characterization of W137F/V184T double mutations on *Anoxybacillus gonensis* G2 xylose isomerase. Curr Opin Biotechnol 22:87

Ozturk HU, Sariyar Akbulut B, Ayan B, Poli A, Denizci AA, Utkan G, Nicolaus B, Kazan D (2015) Moderately halophilic bacterium *Halomonas* sp. AAD12: a promising candidate as a hydroxyectoine producer. J Microb Biochem Technol 7:262–268

Pikuta E, Lysenko A, Chuvilskaya N, Mendrock U, Hippe H, Suzina N, Nikitin D, Osipov G, Laurinavichius K (2000) *Anoxybacillus pushchinensis* gen. nov., sp. nov., a novel anaerobic, alkaliphilic, moderately thermophilic bacterium from manure, and description of *Anoxybacillus flavithermus* comb. nov. Int J Syst Evol Microbiol 50:2109–2117

Poli A, Kazak H, Gurleyendag B, Tommonaro G, Pieretti G, Oner ET, Nicolaus B (2009) High level synthesis of levan by a novel *Halomonas* species growing on defined media. Carbohydr Polym 78:651–657

Poli A, Laezza G, Gul Guven R, Orlando P, Nicolaus B (2011) *Geobacillus galactosidasius* sp. nov., a new thermophilic galactosidase producing bacterium isolated from compost. Syst Appl Microbiol 34:419–423

Poli A, Guven K, Romano I, Pirinccioglu H, Gul Guven R, Euzeby JP, Matpan F, Acer O, Orlando P, Nicolaus B (2012) *Geobacillus subterraneus* subsp. *aromaticivorans* subsp. nov., a novel thermophilic and alkaliphilic bacterium isolated from a hot spring in Sirnak, Turkey. J Gen Appl Microbiol 58:437–446

Poli A, Nicolaus B, Denizci AA, Yavuzturk B, Kazan D (2013) *Halomonas smyrnensis* sp. nov., a moderately halophilic, exopolysaccharide-producing bacterium. Int J Syst Evol Microbiol 63:10–18

Sam S, Kucukasik F, Yenigun O, Nicolaus B, Toksoy Oner E, Yukselen MA (2011) Flocculating performances of exopolysaccharides produced by a halophilic bacterial strain cultivated on agro-industrial waste. Bioresour Technol 102(2):1788–1794

Sandalli C, Kacagan M, Canakci S, Belduz AO (2008) Cloning, expression, purification and characterisation of a thermostable chitinase from *Bacillus licheniformis* A1. Ann Microbiol 58:245

Sandalli C, Saral A, Ulker S, Karaoglu H, Belduz AO, Copur Cicek A (2014) Cloning, expression, and characterization of a novel CTP synthase gene from *Anoxybacillus gonensis* G2. Turk J Biol 38:111–117

Shida O, Takagi H, Kadowaki K, Komagata K (1996) Proposal for two new genera, *Brevibacillus* gen. nov. and *Aneurinibacillus* gen. nov. Int J Syst Bacteriol 46:939–946

Sirin Y, Yildirim Akatin M, Colak A, Saglam Ertunga N (2017) Dephytinization of food stuffs by phytase of *Geobacillus* sp. TF16 immobilized in chitosan and calcium-alginate. Int J Food Prop. https://doi.org/10.1080/10942912.2016.1261151

Sogutcu E, Emrence Z, Arikan M, Cakiris A, Abaci N, Toksoy Oner E, Ustek D, Arga KY (2012) Whole genome sequencing, assembly and annotation of *Halomonas smyrnensis*, a levan producing halophilic bacterium. EMBnet.journal 18.A. POSTERS. 89

Souza PMD (2010) Application of microbial α-amylase in industry-A review. Braz J Microbiol 41(4):850–861

Taslimi P, Adiguzel A, Nadaroglu H, Bozoglu C, Gulluce M (2013) Removal of some textile dyes from aqueous solution by using a catalase-peroxidase from *Aeribacillus pallidus* (P26). J Pure Appl Microbiol 7(4):2629–2640

Tatar D, Guven K, Inan K, Cetin D, Belduz AO, Sahin N (2016) *Streptomonospora tuzyakensis* sp. nov., a halophilic actinomycete isolated from saline soil. Antonie Van Leeuwenhoek 109(1):35–41

Taylor MP, Eley KL, Martin S, Tuffin MI, Burton SG, Cowan DA (2009) Thermophilic ethanologenesis: future prospects for second-generation bioethanol production. Trends Biotechnol 27(7):398–405

Tokgoz M, Inan K, Belduz AO, Gedikli O, Canakci S (2014) Cloning, purification, and characterization of a thermophilic ribulokinase from *Anoxybacillus kestanbolensis* AC26Sari. Turk J Biol 38:633–639

Turner P, Mamo G, Nordberg Karlsson E (2007) Potential and utilization of thermophiles and thermostable enzymes in biorefining. Microb Cell Factories 6:9. https://doi.org/10.1186/1475-2859-6-9

Uzyol KS, Sariyar Akbulut B, Denizci AA, Kazan D (2012) Thermostable α-amylase from moderately halophilic *Halomonas* sp. AAD21. Turk J Biol 36:327–338

Vargas C, Nieto JJ (2004) Genetic tools for the manipulation of moderately halophilic bacteria of the family Halomonadaceae. In: Balbás P, Lorence A (eds) Methods in Molecular Biology, Recombinant Gene Expression: Reviews and Protocols, vol 267, 2nd edn. Humana Press Inc., Totowa, pp 183–208

Ventosa A, Márquez MC, Garabito MJ, Arahal DR (1998) Moderately halophilic gram-positive bacterial diversity in hypersaline environments. Extremophiles 2:297–304

Yanmis D, Baltaci MO, Gulluce M, Adiguzel A (2015) Identification of thermophilic strains from geothermal areas in Turkey by using conventional and molecular techniques. Res J Biotechnol 10(1):39–45

Yanmis D, Adiguzel A, Nadaroglu H, Gulluce M, Demir N (2016) Purification and characterization of laccase from thermophilic *Anoxybacillus gonensis* P39 and its application of removal textile dyes. Rom Biotechnol Lett 21:11485–11496

Yavuz E, Gunes H, Harsa S, Yenidunya AF (2004) Identification of extracellular enzyme producing thermophilic bacilli from Balcova (Agamemnon) geothermal site by ITS rDNA RFLP. J Appl Microbiol 97:810–817

Yener I, Varhan Oral E, Dolak I, Ozdemir S, Ziyadanogullari R (2017) A new method for preconcentration of Th(IV) and Ce(III) by thermophilic *Anoxybacillus flavithermus* immobilized on Amberlite XAD-16 resin as a novel biosorbent. Ecol Eng 103:43–49

Yildirim V, Baltaci MO, Ozgencli I, Sisecioglu M, Adiguzel A, Adiguzel G (2017) Purification and biochemical characterization of a novel thermostable serine alkaline protease from *Aeribacillus pallidus* C10: a potential additive for detergents. J Enzyme Inhib Med Chem 32(1):468–477

Yilmaz B, Baltaci MO, Yildirim V, Sisecioglu M, Adiguzel A (2014) Purification and Characterization of Thermostable Alkaline Protease Enzyme from *Bacillus licheniformis* A10. In: MolBiolCon2014. 10–12 September/Turkey, p48

Hypersaline Environments of Iran: Prokaryotic Biodiversity and Their Potentials in Microbial Biotechnology

9

Atefeh Safarpour, Mohammad Ali Amoozegar, and Antonio Ventosa

9.1 Introduction

Our planet is called The Blue Planet because about 70% of its surface is covered with seawater/water. Each liter of the seawater contains approximately 35 g of different salts, and sodium chloride (NaCl) is the major salt in most seawaters. This amount of salt in seawater has not been a limitation for microorganisms to live in such habitats (Libes 2011). An extensive diversity of microorganisms is found in seawaters, and this diversity is similar to the freshwaters. Salt concentration of several places in the world is higher than seawaters. The increase in salt concentrations reduces the number of present organisms, where only halophilic or halotolerant ones can survive in such hypersaline environment. These halophilic and halotolerant microorganisms can be found in all three domains of life: Archaea, Bacteria, and Eukarya. Survival of macroorganisms seems to be impossible in salt concentrations more than 20%. Hypersaline environments are widespread in all parts of the world. Natural salt lakes, hypersaline soils, salt wetlands, salt travertines, underground deposits of rock salt or salt mines, artificial salt lakes (e.g., solar salterns for NaCl production from seawater), coastal lagoons, and even salted food products are examples of hypersaline environments (Oren 2002). Generally hypersaline environments are divided in two major groups based on their origins: thalassohaline and athalassohaline environments. Thalassohaline environments have originated from

A. Safarpour · M. A. Amoozegar (✉)
Extremophiles Laboratory, Department of Microbiology, Faculty of Biology and Center of Excellence in Phylogeny of Living Organisms, College of Science, University of Tehran, Tehran, Iran
e-mail: amoozegar@ut.ac.ir

A. Ventosa
Department of Microbiology and Parasitology, Faculty of Pharmacy, University of Sevilla, Sevilla, Spain

D. Egamberdieva et al. (eds.), *Extremophiles in Eurasian Ecosystems: Ecology, Diversity, and Applications*, Microorganisms for Sustainability 8,
https://doi.org/10.1007/978-981-13-0329-6_9

seawater and include marine salterns, some saline soils, and some lakes like Great Salt Lake. Athalassohaline environments, however, are not originated from the sea; and they can be found in all continents where they include saline soil and lakes along soda lakes (Ventosa and Arahal 2009). A distinct majority of halophilic microorganisms have called these saline environments "home," and their survival depends on different salts of these highly saline environments, especially NaCl (Ma et al. 2010). Like any other saline environments, salt lakes and other salt bodies of water are classified into thalassohaline and athalassohaline. Thalassohaline lakes have resulted from the evaporation of seawater, and usually their ionic composition is similar to seawater, and therefore, NaCl is the dominant and most abundant salt in these lakes. The pH of these lakes is usually around 7–8; thus, several halophilic microorganisms prefer to live in thalassohaline lakes. The Great Salt Lake in Utah, USA, is an example of thalassohaline lakes. Although this lake is not connected with the sea, it has originated from the evaporation of a salt lake from ice age, Bonneville, and as the water is similar to seawater, it has been classified as a thalassohaline lake. The salinity of Great Salt Lake is about 30 and 12% in the north and south arms, respectively (Oren 2011). The Dead Sea is the most famous example of athalassohaline body of water in the world. Total salt concentration of the Dead Sea is about 35% but sodium is not the dominant ion of this environment, where the concentration of divalent ions magnesium and calcium is much higher (Bardavid et al. 2007). The main anions of the Dead Sea are bromide and chloride, and the pH of this athalassohaline environment is about 6. The predominant microbial strains that live in the Dead Sea are magnesium-tolerant ones which acquired low amount of sodium (Buchalo et al. 1998). Some of athalassohaline environments are alkaline with pH about 9.7–10. Examples of such environments are Mono Lake in California and Lake Magadi in Kenya (Javor 1989). Near the bottom of the Red Sea, the Mediterranean Sea, and the Gulf of Mexico, hypersaline brines have been found, and some microbial communities are found in depth of 1.5–3.5 km under the water surface (Hallsworth et al. 2007). Great diversity of microorganisms exists that can grow in salt concentrations up to saturation amount of NaCl (>300 g/l), and because of their pigments, they could be detected with naked eyes. The most common halophilic strains that could be found in all salt-saturated brines are unicellular alga *Dunaliella salina*, the square archaeon *Haloquadratum walsbyi*, and the red bacterium *Salinibacter ruber* (Oren 2002). Almost all of archaeal strains from the phylum *Euryarchaeota* have the optimal growth in presence of salt concentrations above 15%, and surprisingly many of them don't have the ability to live in salt concentrations below 10% (Savage et al. 2008). On the other hand, halophilic bacteria are characterized. They belong to several phyla including the *Cyanobacteria*, the *Gammaproteobacteria*, the *Firmicutes*, and the *Bacteroidetes* (Makhdoumi-Kakhki et al. 2012a, b, c, d). Also, several eukaryotic microorganisms and even macroorganisms are found in hypersaline environments. *Artemia*, the brine shrimp, is the most frequent macroorganism in hypersaline environments with the ability to live in salt concentrations more than 15%. In case of eukaryotic halophilic microorganisms, *Dunaliella*, the green algae, is the most important and well-studied one. Survival of several heterotrophic microorganisms in hypersaline environments depends on this autotrophic alga. Furthermore, its red β-carotene pigments increase its importance in biotechnology (de Lourdes Moreno et al. 2012). In general, halophiles have several applications in industry and biotechnology including food industry, medicine, depleting heavy metals and toxins, petroleum industry, detergents, and textile industry. Furthermore, these microorganisms have the great ability to produce novel bioactive molecules (Yin et al. 2015). In this chapter we describe different saline environments of Iran and discuss the studies about halophilic microorganism's diversity in these environments. Furthermore, we focus on studies which exhibited the biotechnological potential and/or application of these native halophilic and halotolerant microorganisms.

9.2 Hypersaline Lakes and Wetlands of Iran

Iran is a country with continental climates. Large parts of this country, especially in central and southern parts, consist of deserts. One of the most important features of Iranian deserts is that they are salty. The presence of salt in different places of Iran varied in amount from low percentages to saturated concentrations. Also, different types of saline environments including saline and hypersaline soils, wetlands, and permanent or seasonal lakes exist in Iran (Breckle 2002). These saline environments have two aspects of importance for mankind life. An old aspect is that these deserts are a great reservoir of food and raw materials for agriculture, industry, and medicine. As we know these places are rich of important compounds, like sodium chloride, sodium sulfate, calcite, and selenite and important elements, like magnesium, manganese, lithium, boron, and tungsten (Shadrin and Oren 2015; Nissenbaum 1993). Saline lakes and wetlands are found everywhere in Iran, and most of them are seasonal lakes and only have water in winter and spring, and with increasing sunlight, they become dry. Usually, these environments have no water from May to October, and during this time, their salinity reaches to its maximum amount. Based on their geographical position, surface of these seasonal lakes is covered with millimeters to centimeters of salt. The most frequent salt in all Iranian lake is NaCl, but in Meighan wetland, sodium sulfate is the predominant one (Ghadimi and Ghomi 2013). In recent years, several studies have been done on isolated microorganisms from different hypersaline lakes of Iran, and they were categorized in two groups: first, the ecologic and taxonomic studies, and second, the studies on biotechnological applications of native microorganisms. With its great variation of ecologic regions, Iran is a hotspot for biodiversity studies, and unfortunately, a high number of its native species are exposed to human threats. Studies on the biodiversity of Iranian microorganism have started from two decades ago; thus most of these studies are new, and several of them are about isolation, identification, and taxonomic investigations of new strains from hypersaline lakes and drawing a biologic map for these regions. Among these studies, there are good investigations from Gomishan wetland, Urmia Lake, and seasonal lakes like Aran-Bidgol and Incheh Borun. Up to now more than 50 new eukaryotic (mold and yeast) and prokaryotic (archaea and bacteria) strains in taxonomic level of species, genus, and family were isolated and characterized from hypersaline environments of Iran including two new families of bacteria and yeasts (*Soortiaceae* and *Fereydouniaceae*) (Amoozegar et al. 2017; Nasr et al. 2014), five genera of archaea (Amoozegar et al. 2012; Makhdoumi-Kakhki et al. 2012a, c; Mehrshad et al. 2015, 2016), six genera of bacteria (Zarparvar et al. 2014; Amoozegar et al. 2014a, c, e; Shahinpei et al. 2014a; Munoz et al. 2016), two new genera from actinomycetes (Nikou et al. 2015b, 2017), five species of molds (Arzanlou et al. 2016; Crous et al. 2014; Hyde et al. 2016), eight species of archaea (Amoozegar et al. 2013b, 2014c, d, 2015; Corral et al. 2015, 2016; Rasooli et al. 2017a, b; Naghoni et al. 2017a, b), 22 new species of bacteria (Amoozegar et al. 2008, 2009a, b, 2013a, 2014b, f, 2016a, b, c; Bagheri et al. 2012, 2013a, b; Didari et al. 2012, 2013; Makhdoumi-Kakhki et al. 2012b, Mehrshad et al. 2013, Sanchez-Porro et al. 2009, 2010; Shahinpei et al. 2014a, b), and two new species of actinomycetes (Nikou et al. 2014, 2015a). As shown in Table 9.1, Aran-Bidgol salt lake was the origin of most of these novel taxa. In the following sections we introduce saline environments of Iran separately and discuss the studies on the biodiversity and biotechnological applications of their isolated microorganisms. Finally, in Table 9.2, we summarize the biotechnological applications of halophilic microorganisms isolated from different saline environments of Iran. Study of enzymes from these microorganisms was the major biotechnological approach in almost all regions.

9.2.1 Urmia Lake

9.2.1.1 Geographical Characteristic of Urmia Lake

Urmia Lake with ancient name of Chichast is the largest permanent, inland, hypersaline lake of Iran which is located in northwest of this country (Fig. 9.1). The most important water suppliers of Urmia Lake are Zarineh River, Simineh River, Talkhe River, and Aji Chai River. The main ions of the lakes are cations like sodium, magnesium, potassium, calcium, and lithium and anions including chloride, sulfate, and bicarbonate (Eimanifar and Mohebbi 2007). This ecosystem was registered in the Ramsar Convention on Wetlands as a wetland of international importance; also Urmia Lake has been selected as 1 of the 59 biosphere reserves by UNESCO (Asem et al. 2014, 2016). In previous years the amount of the lake's water reached to 14×10^9 m^2, and its average depth was about 6 m, but now the amount of its water is about 3×10^9 m^2 with an average depth of ˃1 m; therefore its water is approximately salt saturated. The shrimp, *Artemia*, is the sole macroorganism found on the lake (Shadkam et al. 2016).

9.2.1.2 Microbiology and Biodiversity of Microorganisms in Urmia Lake

In recent years several studies had been carried out on microbial life of Urmia Lake. In a study on the biodiversity of microorganisms of the lake, Barin et al. (2015) reported that the increase in salinity levels was not the main reason behind microbial biomass declination in the nearby saline soils. It was also shown that microbial

Table 9.1 New halophilic taxa from different saline and hypersaline environments of Iran

Domain		Taxonomic level	Species name	Isolated from	Reference
Eukaryotes	Mold	Species	*Aspergillus iranicus*	Urmia Lake	Arzanlou et al. (2016)
		Species	*Aspergillus urmiensis*	Urmia Lake	Arzanlou et al. (2016)
		Species	*Emericellopsis persica*	Urmia Lake	Crous et al. (2014)
		Species	*Neocamarosporium chichastianum*	Urmia Lake	Hyde et al. (2016)
	Yeast	Family	*Fereydounia khargensis*	Kharg Island	Nasr et al. (2014)
Bacteria		Family	*Soortia roseihalophila*	Badab-Soort travertine spring	Amoozegar et al. (2017)
		Genus	*Aliicoccus persicus*	Aran-Bidgol salt lake	Amoozegar et al. (2014e)
		Genus	*Aquibacillus halophilus*	Aran-Bidgol salt lake	Amoozegar et al. (2014a)
		Species	*Oceanobacillus halophilus*	Aran-Bidgol salt lake	Amoozegar et al. (2016b)
		Genus	*Alteribacillus bidgolensis*	Aran-Bidgol salt lake	Didari et al. (2012)
		Species	*Bacillus iranensis*	Aran-Bidgol salt lake	Bagheri et al. (2012)
		Species	*Oceanobacillus limi*	Aran-Bidgol salt lake	Amoozegar et al. (2014b)
		Species	*Oceanobacillus longus*	Aran-Bidgol salt lake	Amoozegar et al. (2016a)
		Species	*Bacillus persicus*	Aran-Bidgol salt lake	Didari et al. (2013)
		Species	*Bacillus halosaccharovorans*	Aran-Bidgol salt lake	Mehrshad et al. (2013)
		Species	*Bacillus salsus*	Aran-Bidgol salt lake	Amoozegar et al. (2013a)
		Species	*Lentibacillus persicus*	Aran-Bidgol salt lake	Sanchez-Porro et al. (2010)
		Species	*Ornithinibacillus halophilus*	Aran-Bidgol salt lake	Bagheri et al. (2013b)
		Species	*Marinobacter persicus*	Aran-Bidgol salt lake	Bagheri et al. (2013a)
		Genus	*Salinivenus iranica*	Aran-Bidgol salt lake	Makhdoumi-Kakhki et al. (2012b)
		Species	*Salinivenus lutea*	Aran-Bidgol salt lake	Makhdoumi-Kakhki et al. (2012b)
		Genus	*Limimonas halophila*	Aran-Bidgol salt lake	Amoozegar et al. (2013c)
		Species	*Bacillus persepolensis*	Howz Soltan salt lake	Amoozegar et al. (2009b)
		Species	*Piscibacillus halophilus*	Howz Soltan salt lake	Amoozegar et al. (2009a)
		Species	*Thalassobacillus cyri*	Howz Soltan salt lake	Amoozegar et al. (2009a)

(continued)

Table 9.1 (continued)

Domain	Taxonomic level	Species name	Isolated from	Reference
	Species	*Salinivibrio proteolyticus*	Bakhtegan Lake	Amoozegar et al. (2008)
	Species	*Salinifilum proteinilyticum*	Meighan wetland	Nikou et al. (2017)
	Genus	*Salininema proteolyticum*	Meighan wetland	Nikou et al. (2015b)
	Species	*Alloactinosynnema iranicum*	Incheh Borun wetland	Nikou et al. (2014)
	Genus	*Salinithrix halophila*	Incheh Borun wetland	Zarparvar et al. (2014)
	Species	*Nocardia halotolerans*	Incheh Borun wetland	Nikou et al. (2015a)
	Genus	*Salinispirillum marinum*	Gomishan wetland	Shahinpei et al. (2014a)
	Species	*Aliidiomarina iranensis*	Gomishan wetland	Amoozegar et al. (2016c)
	Species	*Aliidiomarina sedimenti*	Gomishan wetland	Shahinpei et al. (2017)
	Species	*Cyclobacterium halophilum*	Gomishan wetland	Shahinpei et al. (2014b)
	Species	*Pseudomonas salegens*	Gomishan wetland	Amoozegar et al. (2014f)
Archaea	Genus	*Halosiccatus urmianus*	Urmia Lake	Mehrshad et al. (2015)
	Genus	*Halovarious luteus*	Urmia Lake	Mehrshad et al. (2016)
	Species	*Halorubrum halodurans*	Aran-Bidgol salt lake	Corral et al. (2016)
	Species	*Halorubrum persicum*	Aran-Bidgol salt lake	Corral et al. (2015)
	Species	*Halovivax certinus*	Aran-Bidgol salt lake	Amoozegar et al. (2015)
	Species	*Halorientalis persicus*	Aran-Bidgol salt lake	Amoozegar et al. (2014c)
	Species	*Halovivax limisalsi*	Aran-Bidgol salt lake	Amoozegar et al. (2014d)
	Species	*Halopenitus malekzadehii*	Aran-Bidgol salt lake	Amoozegar et al. (2013b)
	Genus	*Halovenus aranensis*	Aran-Bidgol salt lake	Makhdoumi-Kakhki et al. (2012a)
	Genus	*Halopenitus persicus*	Aran-Bidgol salt lake	Amoozegar et al. (2012)
	Genus	*Haloarchaebious iranensis*	Aran-Bidgol salt lake	Makhdoumi-Kakhki et al. (2012c)
	Species	*Natrinema soli*	Meighan wetland	Naghoni et al. (2017a, b)
	Species	*Natronoarchaeum persicum*	Meighan wetland	Rasooli et al. (2017a, b)

Table 9.2 Biotechnological potentials of halophilic and halotolerant microorganisms isolated from different saline and hypersaline environments of Iran

Cell type	Microorganisms	Biotechnological potentials	Site of isolation	Reference
Eukaryotes	Microalgae	Carotenoid production	Urmia Lake	Heidari et al. (2000) and Fazeli et al. (2006)
		β-carotene production	Maharloo Lake	Nikookar et al. (2004, 2005, 2013)
		Biodiesel production	Maharloo Lake	Rasoul-Amini et al. (2014)
Prokaryotes	Archaea	Arsenic accumulation	Urmia Lake	Taran (2011)
		Bacterioruberin production	Urmia Lake	Asgarani et al. (2014)
		Carotenoid production	Urmia Lake	Naziri et al. (2014) and Hamidi et al. (2014)
		Biosurfactant production	Namakdan Lake	Jadidi et al. (2014)
		Laccase activity	Urmia Lake	Siroosi et al. (2016)
	Bacteria	Antineoplastic enzyme production	Urmia Lake	Shirazian et al. (2016)
		Para-amino acetanilide degradation	Urmia Lake	Heris et al. (2014a)
		Carotenoid production	Urmia Lake	Hamidi et al. (2012)
		Amylopullulanase activity	Aran-Bidgol salt lake	Siroosi et al. (ad)
		Amylase activity	Aran-Bidgol salt lake	Shafiei et al. (2010, 2011, 2012)
		Production of acetone, butanol, and ethanol	Aran-Bidgol salt lake	Amiri et al. (2016)
		Chromate reduction	Aran-Bidgol salt lake	Amoozegar et al. (2007)
		Amylase activity	Howz Soltan salt lake	Fahimeh et al. (2013)
		Biodegradation of glyphosate herbicide	Howz Soltan salt lake	Sharifi et al. (2015)
		Bioconversion of ferulic acid to vanilic acid	Howz Soltan salt lake	Ashengroph and Nahvi (2014)

(continued)

Table 9.2 (continued)

Cell type	Microorganisms	Biotechnological potentials	Site of isolation	Reference
		Production of bioactive compounds	Maharloo Lake	Hashemi et al. (2014)
		Protease activity	Maharloo Lake	Shahbazi and Karbalaei-Heidari (2012)
		Biodegradation of polycyclic aromatic hydrocarbons	Maharloo Lake	Kafilzadeh et al. (2007b) and Kafilzadeh and Behzadi (2015)
		Proteolytic activity	Maharloo Lake	Ghasemi et al. (2011)
		Production of protease	Bakhtegan Lake	Amoozegar et al. (2008)
		Protease activity	Bakhtegan Lake	Karbalaei-Heidari et al. (2007, 2008)
		Biodegradation of polycyclic aromatic hydrocarbons	Tashk Lake	Kafilzadeh et al. (2007a)
		Poly-beta-hydroxybutyrate production	Gavkhooni wetland	Ramezani et al. (2015)
		Biodegradation of xenobiotic compounds	Gavkhooni wetland	Azarbaijani et al. (2016)
		Bio-absorption of cesium	Gavkhooni wetland	Bakhshi et al. (2007)
		Mercury reduction	Gavkhooni wetland	Noroozi et al. (2017)
		Production of keratinolytic protease	Incheh Borun wetland	Khoshnevis et al. (2014)
		Laccase activity	Incheh Borun wetland	Rezaei et al. (2014)
		Precipitation of calcium carbonate	Badab-Soort travertine	Khansha et al. (2016)
		Resistant to gamma radiation	Lut Desert	Shirsalimian et al. (2016)
		Decreasing agent against drought and saline stress	Eshtehard wetland	Talebi et al. (2013)

Fig. 9.1 The Urmia Lake in the northwest of Iran. The color of the lake is red in some regions (up left), and some salt crystals can be observed in this lake (up right). Red brines are found beneath salt layers of the lake (bottom left). The salt cressets are present beside the lake (bottom right)

stress indices such as cis to trans and saturated to unsaturated conversion of cell membrane fatty acids increased with salinity. Furthermore, microbial communities were altered due to high saline conditions, where they found more fungi and Gram-negative bacteria compared to bacteria and Gram-positive ones, respectively (Barin et al. 2015). In 2014, a study on the biodiversity of cultivable microorganisms of Urmia Lake reported that the number of cultivable microorganisms in water and soil of the lake were 6×10^4 and 5×10^6 cell/ml, respectively. The cultivable bacteria of the lake belong to the following phyla: *Firmicutes*, *Proteobacteria*, and *Actinobacteria* with percentages of 78.6%, 21.4%, and 1.8%, respectively (Kashi et al. 2014). Another report was about archaeal diversity of Urmia Lake. In this study 14 cultivable archaeal genera were reported from this lake, and the genera *Halorubrum* and *Haloarcula* with percentages of 48 and 14.5%, respectively, were the most frequent ones; On the other hand, culture-independent studies showed that the genus *Halonotius* with a percentage of 44% was the predominant archaea of the lake (Farahani et al. 2014). *Halosiccatus urmianus* and *Halovarious luteus* are two new halophilic archaea which were recently isolated from this lake (Mehrshad et al. 2015; Mehrshad et al. 2016). Also, four new fungal species from eukaryotic world were isolated from this lake. These new species are *Aspergillus iranicus*, *Aspergillus urmiensis*, *Emericellopsis persica*, and *Neocamarosporium chichastianum* (Arzanlou et al. 2016; Crous et al. 2014; Hyde et al. 2016).

9.2.1.3 Biotechnological Studies on Urmia Lake's Microorganisms

In recent years several studies have been focused on biotechnological applications of isolated microorganisms from Urmia Lake (Table 9.2). A study on hydrolytic enzymes of bacterial strains isolated from the lake reported that Gram-positive bacteria have more ability to produce hydrolytic enzymes than Gram-negative bacteria. The percentages of hydrolytic enzymes produced were in order from highest to lowest inulinase, DNase, xylanase, lipase, amylase, pullulanase, protease, cellulase, and pectinase. The genus *Halobacillus* from Gram-positive and the genus *Halomonas* from Gram-negative bacteria had the highest percentages number in enzyme-producing strains. The genus *Halobacillus* produced cellulase, protease, amylase, pectinase, and inulinase, and the genus *Halomonas* produced inulinase, pullulanase, and xylanase. The genus *Thallasobacillus* produced amylase, DNase, and inulinase, and the genus *Marinobacter* did not produce any hydrolytic enzyme (Babavalian et al. 2014). Recently, a laccase enzyme which is alkaline-chloride tolerant was purified from a *Bacillus* strain from Urmia Lake. Laccases are multicopper oxidases of different aromatic or inorganic substrates. These enzymes have various biotechnological applications like azo dye decolorization in textile industry. This purified laccase had a molecular weight of 180 kDa and was active in presence of NaCl with 800 mM concentration, and that's why this laccase is unique among bacterial laccases. This was the first case of a halotolerant bacterial laccase to be reported, which had been isolated from hypersaline environments (Siroosi et al. 2016). The productive ability of halotolerant bacterial strains on antineoplastic enzymes like L-asparaginase and L-glutaminase was assayed. These enzymes were used for patients with acute lymphoblastic leukemia. A moderate halophile bacterium from the genus *Bacillus* had the highest production of L-asparaginase while the strain belonging to the genus *Salicola* had the highest production of L-glutaminase (Shirazian et al. 2016). In regard of bioaccumulation of arsenic, a novel halophilic archeon from Urmia Lake, *Haloarcula* sp. IRU1, exhibited an efficiency of 60.89%. This feature was obtained at 40 °C, pH 8, and 90 mg/L $NaAsO_2$ (Taran 2011). *Marinobacter* sp. TBZ23 isolated from Urmia Lake had the potential to biodegrade para-amino acetanilide in the presence of 14% NaCl (Heris et al. 2014a). Also, it was reported that *Halomonas* sp. TBZ9 from this permanent lake is capable of reducing Fe III (Heris et al. 2014b). The tolerance capacity of extremely halophilic archaeon, *Haloferax radiotolerans*, isolated from this lake against the effects of ultraviolet light (UV) and ^{60}Co r-rays had been investigated. It was shown that, in comparison with a radioresistant strain of *Escherichia coli*, *E. coli* B/r, *Haloferax radiotolerans* was more resistant when exposed to DNA-damaging agents. This study was the first report of radio resistance ability in archaeal strains (Asgarani et al. 2006). Several reports were about pigments of halophilic microorganisms of Urmia Lake. In one study, it was exhibited that the main pigment of the halophilic archaeon, *Haloarcula* sp. IRU1, from Urmia Lake is bacterioruberin (Asgarani et al., 2014). The other study was focused on the carotenoid production by a novel halophilic bacterial strain *Marinobacter* sp. TBZ112. The results exhibited that the carotenoid produced by this strain is monodemetyl spirilloxanthin (Hamidi et al. 2012). Also, halophilic archaeon, *Halorubrum* sp. TBZ126, isolated from the lake showed high production of different carotenoids including bacterioruberin, lycopene, and β-carotene (Naziri et al. 2014; Hamidi et al. 2014). Some *Dunaliella* strains were isolated from Urmia Lake. The ability of these strains to produce carotenoid in presence of salt and irradiance stress was investigated (Heidari et al. 2000). Moreover, it was exhibited that *Dunaliella tertiolecta* DCCBC26 from Urmia Lake has the ability to produce β-carotene (Fazeli et al. 2006).

9.2.2 Aran-Bidgol Salt Lake

9.2.2.1 Geographical Characteristic of Aran-Bidgol Salt Lake

Aran-Bidgol hypersaline lake, also known as Qom salt lake or Namak Lake, is the largest seasonal playa of Iran which is salt saturated in all seasons. The lake looks like a triangle between Tehran, Qom, and Semnan provinces (Fig. 9.2). It has a surface area of about 2.4×10^3 km^2. Water only covers 40 km^2 of its surface during spring, and its depth is between 45 cm and 1 m. The surface of the Aran-Bidgol Lake is covered by salt, and the depth of this salt layer varies between 5 and 55 m. Colorful salt layers can be seen in this lake. The colors seen in salt layers are cyan blue, brown, white, green, pink, and gray or black from up to down (Fig. 9.2). The array of colors in laminated layers resembles the typical layers of the marine salters of Salin-de-Giraud (Oren 2011). This inland lake is a thalassohaline lake, and rainfalls and seasonal rivers are the most important water suppliers of it. Colorful brines of the lake with biologic colors of green, orange, red, black, and brown on hexagonal salt layers create a unique picturesque of the lake, from September to November.

9.2.2.2 Microbiology and Biodiversity of Microorganisms in Aran-Bidgol Salt Lake

In 2012, a study on biodiversity of Aran-Bidgol Lake exhibited that the number of prokaryotic population of the lake is about $3–4 \times 10^7$ cells/ml, which is higher than those of the microbial populations of the seas; thus it was revealed that this lake is an active and efficient ecosystem. According to FISH analysis, the proportion of bacteria to archaea in this ecosystem was 1:2–1:3, which was unexpected due to the high salinity of the lake. Culture-independent studies revealed that *Halorubrum* and *Salinibacter* were the most frequent genera of the domains Archaea and Bacteria, respectively. The study exhibited that Aran-Bidgol Lake is an active and complete ecosystem which contains autotrophs like *Cyanobacteria* and purple sulfur bacteria of the genus *Halorhodospira* and all kinds of heterotrophs. In general, the classes of *Bacterioidetes* and *Halobacteria* from bacterial and archaeal domains are the predominant ones in this lake (Makhdoumi-Kakhki et al. 2012d). As shown in Table 9.1, up to now 16 new bacterial species or genera were isolated and identified from this lake including moderately halophilic bacteria, *Aliicoccus persicus* (Amoozegar et al. 2014e), *Aquibacillus halophilus* (Amoozegar et al. 2014a),

Fig. 9.2 Aran-Bidgol salt lake with colorful brines and salts of it (top and bottom left). This salt lake with its great area is an important reservoir of salt (bottom right). Array of colors in salt layers of Aran-Bidgol salt lake is similar to laminated layers of the marine salters of Salin-de-Giraud (Oren 2011)

Oceanobacillus halophilus (Amoozegar et al. 2016b), *Alteribacillus bidgolensis* (Didari et al. 2012), *Bacillus iranensis* (Bagheri et al. 2012), *Oceanobacillus limi* (Amoozegar et al. 2014b), *Oceanobacillus longus* (Amoozegar et al. 2016a), *Bacillus persicus* (Didari et al. 2013), *Bacillus halosaccharovorans* (Mehrshad et al. 2013), *Bacillus salsus* (Amoozegar et al. 2013a), *Lentibacillus persicus* (Sanchez-Porro et al. 2010), *Ornithinibacillus halophilus* (Bagheri et al. 2013b), and *Marinobacter persicus* (Bagheri et al. 2013a) along extremely halophilic bacteria, *Salinibacter luteus* and *Salinibacter iranicus* (Makhdoumi-Kakhki et al. 2012b) which recently have been renamed as *Salinivenus lutea* and *Salinivenus iranica* (Munoz et al. 2016) and also *Limimonas halophila* (Amoozegar et al. 2013c). Furthermore, nine new taxa from archaea were isolated and identified from this lake, all of them belong to the class *Halobacteria* including *Halorubrum halodurans* (Corral et al. 2016), *Halorubrum persicum* (Corral et al. 2015), *Halovivax certinus* (Amoozegar et al. 2015), *Halorientalis persicus* (Amoozegar et al. 2014c), *Halovivax limisalsi* (Amoozegar et al. 2014d), *Halopenitus malekzadehii* (Amoozegar et al. 2013b), *Halovenus aranensis* (Makhdoumi-Kakhki et al. 2012a), *Halopenitus persicus* (Amoozegar et al. 2012), and *Haloarchaebious iranensis* (Makhdoumi-Kakhki et al. 2012c).

9.2.2.3 Biotechnological Studies on Aran-Bidgol Salt Lake's Microorganisms

There are several studies about biotechnological applications of microorganisms from Aran- Bidgol salt lake (Table 9.2). In 2013, Babavalian et al. reported the hydrolytic activity of the enzymes produced by halophilic bacterial strains of the lake. In this study, the hydrolytic enzyme activity of 83 moderately halophilic bacterial strains from Aran-Bidgol Lake was examined. The results showed that the most frequent enzymes in Gram-positive strains were DNases, inulinases, pullulanases, and cellulases while Gram-negative bacteria had a great ability to produce lipases. In this study seven strains exhibited a mixed activity of six different enzymes which revealed a high potential of the lake ecosystem in biotechnological applications. Furthermore, two bacterial genera *Salicola* and *Salinicoccus* showed the highest production for lipase and cellulase, respectively (Babavalian et al. 2013). The hydrolytic enzymes from archaeal strains of this lake were also investigated (Makhdoumi-Kakhki et al. 2011). Pectinase activity was not found in any of the 293 strains of the study, but DNase, amylase, lipase, inulinase, pullulanase, protease, cellulase, chitinase, and xylanase activity was observed, and several strains showed more than one enzyme activity. *Halorubrum*, *Haloarcula*, and *Natrinema* had the most enzyme activity while *Halovivax* and *Natronomonas* did not have any hydrolytic activity at all. These enzymes had been produced as response to stress or extreme conditions, and most of the strains are polyextremophiles. The presence of distinct enzymes in halophilic bacteria and archaea is highly valuable in industry and economy (Makhdoumi-Kakhki et al. 2011). In 2014, an amylopullulanase enzyme had been purified from the halophilic archaeon, *Halorubrum*, isolated from Aran-Bidgol salt lake. It was the first time that the presence of this enzyme had been reported in halophilic microorganisms. Maximum activity of this enzyme was at 3–4 M salt, pH 7, and 40 °C. The molecular weight of it was 140 kDa and had activity in presence of nonpolar organic solvent, which is really valuable for industrial processes (Siroosi et al. 2014). One of the most important microorganisms isolated from Aran-Bidgol salt lake was *Nesterenkonia* sp. strain F, which exhibited notable functions in biotechnology. In 2011 the draft genome of this strain was obtained (Sarikhan et al. 2011). Three amylase enzymes from this strain have been purified with molecular weight of 57, 100 and 110 kDa. They had their maximum activity at pH 6.5–7.5 and 40 °C. Besides, they were active at 0–4 M concentration of salt and

Fig. 9.3 Howz Soltan salt lake. Some regions of the lake are dry (left) while other regions have water (right) in waterfall seasons

tolerated polar and nonpolar organic solvents. One of these amylases had the ability to hydrolyze starch which made it very important in biotechnology (Shafiei et al. 2010, 2011, 2012). Furthermore, Amiri et al. (2016) reported that *Nesterenkonia* sp. strain F had the ability to produce acetone, butanol, and ethanol (ABE) under aerobic conditions. This was the first report of ABE production from a wild microorganism that does not belong to class *Clostridia*. Also, this study was the first report of butanol production from a halophilic bacterium under aerobic conditions. Through fermentation with 50 g/l initial glucose concentration, 66 mg/l of butanol and 291 mg/l of ethanol were produced by this strain (Amiri et al. 2016). Also, it was reported that the halophilic bacterium *Nesterenkonia* sp. strain MF2 from this salt lake had the ability to live in up to 600 mM of chromate. Further studies showed that under aerobic conditions this isolate reduced 0.2 mM soluble Cr (VI) into nontoxic insoluble Cr (III) after 24 h. In the presence of different amounts of salt, this chromate reduction ability had remained (Amoozegar et al. 2007). An enzyme with tellurite and nitrate reduction ability was purified from *Salinicoccus iranensis* strain QW6 isolated from Aran-Bidgol salt lake. This enzyme had three subunits with molecular weights of 135, 63, and 57 kDa. The optimum activity of tellurite removal was observed at pH 7.5 and 5% of NaCl (Alavi et al. 2014).

9.2.3 Howz Soltan Salt Lake

Howz Soltan salt lake is a small (24 km^2) seasonal salt lake which is located in the border of Dasht-e Kavir in Qom province, Iran (Fig. 9.3). The lake is also known as Saveh Lake and Shahi Lake. Howz Soltan salt lake consists of two separate hollow. The western hollow is Howz Soltan and the eastern hollows is Howz Morreh (Fig. 9.4) which are connected through a small stream. During winter and spring, water fills Howz Morreh first, and then the excessive amount pours into Howz Soltan. Major water suppliers of this salt lake are rainfalls and some rivers like Shoor River and Ghare Chay River. This salt flat is located 710 m higher than sea

Fig. 9.4 Howz Morreh salt lake. This lake has water in almost all seasons (up left and bottom right), and colorful plants could be found nearby (up right and bottom left)

level with an annual rainfall of 100–120 mm. The surface of the lake is covered by hexagonal salt layers, but in cold and rainy seasons, a thin layer of water coats it. The frequent ions of the lake are Cl^-, Na^+, SO_4^{2-}, K^+, Mg^{2+}, Ca^{2+}, and CO_3^{2-} as sodium chloride, sodium sulfate, potassium chloride, and magnesium chloride are the main salt of the lake. The pH of the lake varies among 6.5 to 8.2 so it is neutral to moderate alkaline. The amount of water salinity in Howz Soltan is 25 to 28% and in dry seasons reaches to saturation (Babavalian et al. 2011; Rohban et al. 2009).

In recent years some studies have been carried out on the biodiversity of microorganisms in Howz Soltan salt lake. As a result, three new endospore-forming Gram-positive bacterial strains were isolated from this lake. These novel species were *Bacillus persepolensis* (Amoozegar et al. 2009b), *Piscibacillus halophilus* (Amoozegar et al. 2009a) and *Thalassobacillus cyri* (Sanchez-Porro et al. 2009). Furthermore, two halophilic microalgae from the genus *Dunaliella* including *D. parva* and *D. viridis* were isolated from this lake (Sedghi et al. 2016). A strain of *Dunaliella salina* was also isolated from Howz Soltan, and it had the ability to produce carotenoids and protein in the presence of different pH and salt concentrations (Tavallaie et al. 2015). Furthermore, the archaeon *Halobacterium salinarum* has been reported in Howz Soltan salt lake (Hassanshahian and Mohamadian 2011).

As shown in Table 9.2, some studies have been carried out on biotechnological applications of microorganisms from Howz Soltan salt lake. In 2009, a report about hydrolytic enzyme activity of halophilic bacteria from this lake was published. In this study 231 bacterial strains were assayed for production of 10 hydrolytic enzymes. Lipase activity was the most encountered enzymatic activity in these strains. Amylase, protease, inulinase, xylanase, cellulase, pullulanase, DNase, and pectinase activity were also reported from these strains. Gram-positive strains produced more efficient enzymes, whereas the genus *Salicola* from Gram-negative bacteria had the ability to produce more efficient lipases. The genera *Gracilibacillus*, *Virgibacillus*, *Thalassobacillus*, and *Halobacillus* were more capable of producing hydrolytic enzymes than others (Rohban et al. 2009). Additionally, it was reported that 18 halo-thermophilic strains were isolated from Howz Soltan lake where three of them exhibited amylase activity (Fahimeh et al. 2013). In other study, it was reported that the bacterial strain *Salinicoccus* sp. from the lake had the capability to biodegrade glyphosate herbicide (Sharifi et al. 2015). Furthermore, bioconversion of ferulic acid to vanilic acid by resting cells of *Halomonas salina* HSL5 isolated from Howz Soltan has been reported. The results showed that it can act as a biocatalyst for biological production of vanilic acid (Ashengroph and Nahvi 2014).

9.2.4 Maharloo, Tashk, and Bakhtegan Lakes

Three hypersaline seasonal lakes are located in Fars province, near the historical city of Shiraz, in the south of Iran. These lakes are Maharloo Lake, Tashk Lake, and Bakhtegan Lake. Maharloo Lake (Fig. 9.5) is a seasonal lake which only has water

Fig. 9.5 Maharloo Lake. The pink color of the lake gives it a unique and beautiful picturesque (top right). Not only the water (up left and bottom right) but also the plants near the lake (bottom left) have this pink color

Fig. 9.6 Bakhtegan Lake. This lake is located near the historical city of Shiraz, in Fars province of Iran. During winter and spring, this lake has water (bottom left and middle) while the surface of it is covered by a layer of salt during summers (up left). Different shapes of salts can be observed in the lake (up right and bottom right)

in winter and spring, and the depth of its water reaches to 3 m. In summers it is completely dry and converts from a lake to a salt marsh. The pink color of its water is a result of residing unicellular microalgae like *Dunaliella* and halophilic archaeal strains. Its area is about 600 km^2 with the width of 15 km^2, and Soltan Abad River and Khoshk River are the main water suppliers. Na^+, Cl^-, SO_4^{2+}, K^+, Mg^{2+}, and Ca^{2+} are the most frequent ions of Maharloo Lake. Tashk and Bakhtegan lakes are twin seasonal salt lakes. Tashk Lake is located in the north, and Bakhtegan Lake (Fig. 9.6) is located in the south, and a stream connects them to each other. During summer

Bakhtegan is a saturated salt marsh while its surface is covered with water in winter and spring. Sodium chloride and sodium sulfate are the main salts of the Bakhtegan Lake (Sajedipour et al. 2017). Kor River is the main water supplier of Tashk Lake which directly feeds it. Later, the extra water enters into Bakhtegan Lake (Eskandari et al. 2016).

Some biotechnological studies were carried out on microorganisms from Maharloo Lake (Table 9.2). In one of these studies, it was reported that two halophilic isolates from the lake produced bioactive compounds. These strains belonged to the species *Bacillus licheniformis* and *Bacillus subtilis*. It was discovered that the mentioned bioactive molecules have a glycoprotein structure and *Staphylococcus aureus*, *Aspergillus niger*, and *Mucor* sp. were sensitive, whereas *Pseudomonas aeruginosa*, *Escherichia coli*, and *Bacillus cereus* were resistant to these bioactive molecules. This study revealed the potential of halophilic bacteria from Maharloo Lake for developing new drugs (Hashemi et al. 2014). A novel extracellular protease with a molecular weight of 21 KDa was purified from a *Salinivibrio* sp. strain MS-7 isolated from the lake. This serine metalloprotease had optimal activity at 50 °C, pH 8.0, and 0.5 M NaCl (Shahbazi and Karbalaei-Heidari, 2012). Furthermore, two studies revealed that the bacterial strains from the genus *Pseudomonas* isolated from this lake have the ability to biodegrade polycyclic aromatic hydrocarbons (PAHs) in the presence of 6% NaCl (Kafilzadeh et al. 2007b; Kafilzadeh and Behzadi 2015). Also, a screening survey was carried out on bacterial strains isolated from Maharloo Lake. In this report, 16 isolates showed proteolytic activity, and all of them had optimal growth in 7–15% NaCl. Gram-positive bacteria showed higher proteolytic activity, and *Bacillus* sp. BCCS041 was the best proteolytic strain (Ghasemi et al. 2011). Several studies were focused on microalga strains isolated from this lake which belong to the genus *Dunaliella*. Most of these studies are about growth and β-carotene production of *D. salina* in the presence of different factors and harsh situations like copper toxicity, osmotic shock, manganese, iron and sulfur starvation, phytohormones, and ammonium nitrate nutrition (Zarei et al. 2016; Nikookar et al. 2004, 2005, 2013; Montazeri-Najafabady et al. 2016; Shaker et al. 2017; Mousavi et al. 2016). The ability of biodiesel formation by *Dunaliella* strain isolated from Maharloo Lake was also investigated (Rasoul-Amini et al. 2014).

Bakhtegan Lake was also a subject of microbiological studies. In a study on the microbial diversity of Bakhtegan Lake, it was reported that four archaeal genera from the orders *Halobacteriales* and *Haloferacales* were found in this lake, including *Halobacterium*, *Haloarcula*, *Halococcus*, and *Haloferax*. Among these, *Halobacterium* and *Haloferax* had the highest and lowest frequency, respectively. Also, it was reported that four genera of bacteria were found in this lake including *Pseudomonas*, *Flavobacterium*, *Micrococcus*, and *Bacillus* (Kafilzadeh et al. 2007c). A Gram-negative halophilic strain, *Salinivibrio proteolyticus* was isolated from this lake which is highly capable of producing halothermotolerant alkaliphilic protease (Amoozegar et al. 2008). Two proteases with molecular weights of 31 and $\geq$ 43 kDa were purified from this strain. These proteases were resistant to organic solvents and temperature while having activity in a wide spectrum of pH, temperature, and salt. They were active even in 4 M concentration of salt. The enzymes had

Fig. 9.7 Gavkhooni wetland in the central region of Iran. This wetland is almost salt saturated in all seasons

their maximum activity at temperature of 55 and 65 °C; hence, this makes them very important in biotechnological approaches. One of these enzymes has been cloned in *Escherichia coli* (Karbalaei-Heidari et al. 2007, 2008).

Another study showed the biodegradation of polycyclic aromatic hydrocarbons (PAHs) by bacteria isolated from Tashk Lake. In this report *Pseudomonas* sp. was the sole bacterium that degraded PAH optimally in the presence of 6% of NaCl (Kafilzadeh et al. 2007a).

9.2.5 Gavkhooni Wetland

Gavkhooni (Fig. 9.7), with area of 470 km^2, is an international wetland which is located in the central region of Iran in Isfahan province. This salty wetland with a salinity of 30% is the terminal basin of the Zayandehrod River. The depth of it reaches to 1 m in springs while it is often dry during summer. Gavkhooni wetland was registered as an international wetland by Ramsar Convention in 1957 (Taghavi et al. 2013).

In a study on microbial isolation from Gavkhooni wetland, 161 isolates showed the ability to accumulate lipid inclusions in their intracellular space. All of strains were moderately halophilic or halotolerant. One of them, a Gram-negative strain

Oceanimonas sp. GK1, produced the highest amount of this inclusion in almost all examined culture conditions. Further studies clarified that inclusions were poly-beta-hydroxybutyrate (PHB). It was reported that in the presence of 5% sucrose and 0.5% peptone, this strain accumulated PHB at 35 °C, pH 8.0, and 5% NaCl with a efficiency of 75% (Ramezani et al. 2015). Whole-genome sequencing of *Oceanimonas* sp. GK1 revealed that the genome of this strain consisted of a single circular chromosome with 3,514,537 base pair length and also two plasmids with 8462 and 4245 base pair length (Yeganeh et al. 2012). Further analysis revealed that some virulence genes like ZOT, RTX toxin, thermostable hemolysin, lateral flagella, and type IV pili are present in its genome. These genes have a role in infection caused by other pathogenic bacteria and also in adhesion and biofilm formation (Yeganeh et al. 2015). Also, it had been exhibited that this halotolerant strain has high ability to biodegrade xenobiotic compounds such as phenol. This strain uses phenol as its carbon source via the *ortho*-cleavage pathway in the citrate cycle. Besides, further studies showed that this strain had strong adaptation to harsh environments and that genes encoding carbohydrate active enzymes are rare in its genome (Azarbaijani et al. 2016).

Isolation of *Dunaliella tertiolecta* sp. ABRIINW-G3, a new strain of *Dunaliella tertiolecta* from this wetland had also been reported (Hosseinzadeh Gharajeh et al. 2012). The cesium bio-absorption had been reported from halophilic and halotolerant bacterial strain isolated from soil samples near Gavkhooni wetland. It was shown that halotolerant strains had higher ability for bio-absorbing cesium than the halophilic strains, with averages of 33.1 and 15.6 mg/gdw, respectively (Bakhshi et al. 2007). On the other hand, it was reported that the halophilic bacterium *Bacillus firmus* MN8, isolated from this wetland, had the ability of reducing mercury. Also, it was shown that this strain had *merA* gene and its mercuric reductase had the optimum activity at pH and temperature of 7.5 and 35 °C, respectively, while its activity in 1.5 M concentration of NaCl was 50%. This strain was assumed as a excellent choice for bioremediation of mercury-contaminated environments (Noroozi et al. 2017).

9.2.6 Meighan Wetland

Meighan wetland or Meighan desert wetland is a seasonal hypersaline wetland with an area of 1.2×10^3 km^2, which is located in Markazi province of Iran near Arak city, 1700 m above sea level. The depth of its water reaches to 1.5 m in some seasons. This environment is the largest reservoir of sodium sulfate in Iran. The climate of the wetland is warm and dry, like the Mediterranean climate. The annual amount of rainfalls in this region is about 300 mm. The highest and lowest reported temperature of Meighan wetland are 44 and − 33 °C, and like other saline environments of Iran, the ions Na^+, Cl^-, and SO_4^{+2} are abundant in this wetland.

Metagenomic analysis and culture dependent studies had been carried out on the microbial diversity of Meighan wetland recently (Naghoni et al. 2017a, b). Based on these results, 48 archaeal and 57 bacterial strains were isolated from this wetland, and dominant archaeal and bacterial strain distribution was similar in culture-dependent and culture-independent studies.

Recently two new halophilic actinomycetes were isolated from Meighan wetland. These new genera are *Salininema proteolyticum* and *Salinifilum proteinilyticum* (Nikou et al. 2015b, 2017). Also, two new halophile archaea, *Natrinema soli* and *Natronoarchaeum persicum*, were isolated from this wetland (Naghoni et al. 2017a, b; Rasooli et al. 2017a, b). Furthermore, in a recent study, ten chemolithoautotrophic, haloalkaliphilic sulfur-oxidizing strains belonging to the genus *Thioalkalivibrio* were isolated from this wetland (Makzum et al. 2017).

9.2.7 Incheh Borun and Gomishan Wetlands

The north of Iran, with 700 km length, is divided in two different regions with two different ecosystems. One region in south of Caspian Sea is covered by rainforests with high humidity while the other part in southeast of Caspian Sea is dry with sporadic saline lands. These saline environments include ecosystems like brackish to hypersaline wetlands with neutral to alkaline pH and multiple mud volcanoes. There are few biological studies from these regions, and microbiological studies have been carried out only on two ecosystems of this region: moderate alkaline brackish Gomishan wetland and hypersaline Incheh Borun wetland (Nouri et al. 2008).

9.2.7.1 Incheh Borun Wetland

Incheh Borun wetland (Fig. 9.8) is a hypersaline wetland in north of Iran, near the border with Turkmenistan Republic. This thalassohaline wetland has a salinity of about 23–28%, and its pH varies between 2.8 and 6.8. Eastern part of the wetland is affected by wastewater of a iodine extraction factory, and therefore its pH is lower than other parts of the wetland. Cl^-, Na^+, Ca^{2+}, Mg^{2+}, and K^+ are the most frequent ions of Incheh Borun wetland (Zarparvar et al. 2016).

Biodiversity studies about prokaryotic life of this wetland showed that the number of cultivable microorganisms is 2.1×10^6 cells/ml. Those from the bacterial domain belonged to the phyla *Firmicutes*, *Proteobacteria*, and *Actinobacteria*. Forty percent of the bacteria were halophilic and the remaining 60% were halotolerant. The most frequent halophilic strains belonged to the genera *Marinobacter* and *Halomonas* and most halotolerant belonged to the genera *Bacillus*, *Dietza*, *Oceanobacillus*, and *Kocuria* (Zarparvar et al. 2016). In the archaeal domain, the genera *Haloarcula* and *Halostagnicola* had the highest and lowest abundancy, respectively. The frequency order of archaeal genera of the wetland were *Haloarcula*, *Halorubrum*, *Haloferax*, *Halobellus*, *Halogeometricum*, *Halobacterium*, *Halolaminia*, *Halorhabdius*, and *Halostagnicola* (Rasooli et al. 2016). Up to now three new bacterial taxa at the level of genus or species have been isolated from Incheh Borun wetland. These novel taxa are *Alloactinosynnema iranicum* (Nikou et al. 2014), *Salinithrix halophila* (Zarparvar et al. 2014), and *Nocardia halotolerans* (Nikou et al. 2015a). All of them belong to the phylum *Actinobacteria*. Isolation and purification of a protease enzyme had been reported from *Salicola* sp., which was isolated from Incheh Borun wetland. This

Fig. 9.8 Incheh Borun wetland. This wetland has two different picturesques during the summer (up) and the winter (bottom right and middle). Salt crusts are visible beside the wetland (bottom left)

keratinolytic protease had the capability of producing 86 μg/ml keratin from 1 g pretreated feather (Khoshnevis et al. 2014). The purification of laccase enzyme from *Chromohalobacter* sp. from Incheh Borun wetland was also reported. The purified laccase had a molecular weight of about 60 kDa and showed optimal activity at 3 M NaCl, pH 8.0, and 45 °C (Rezaei et al. 2014).

9.2.7.2 Gomishan Wetland

Gomishan wetland is located two meters lower than sea level and has the area of about 1.7×10^3 km^2. This wetland consists of salt marshes with few amount of water and is connected to the Caspian Sea, so its hydrological features are directly influenced by the sea. Typically, the depth of this wetland is 1 m and reaches to 2.5 m near the sea. In 2001, this wetland was registered in the List of Wetlands of International Importance as declared in the Ramsar Convention (Saba et al. 2016). The salinity and pH of Gomishan brackish alkaline wetland are between 2 to 4% and 7.2 to 9.3, respectively. The ions of the wetland in order from higher to lower are Cl^-, Na^+, SO_4^{2-}, Mg^{2+}, Ca^{2+}, HCO_3^-, and K^+ (Shahinpei et al. 2013).

Microbial studies of Gomishan wetland showed that 23% of the isolated prokaryotes are polyextremophiles and haloalkaliphiles (Shahinpei et al. 2013). These strains belong to the following genera: *Idiomarina*, *Halomonas*, *Halobacillus*, and *Bacillus*, and the following phyla, *Actinobacteria*, *Bacteroidetes*, *Firmicutes*, and

Proteobacteria. *Firmicutes* and Gram-positive endospore-forming strains were the predominant ones, followed by reperesentatives of, the phylum *Proteobacteria* and the class *Gammaproteobacteria*. The genera *Altererythrobacter*, *Caenispirillum*, *Erythrobacter*, *Martelella*, *Nesiotobacter*, *Stappia*, and *Thalassospira* from the *Alphaproteobacteria* and the genus *Achromobacter* from the *Betaproteobacteria* were also detected in this wetland. More than 50% of isolated strains had lipase activity while DNase activity was very rare in these strains. These results varied from other studies of enzyme activity in other saline and hypersaline environments of Iran (Shahinpei et al. 2013). Up to now, three diatom species belonging to the family *Bacillariophyceae* were isolated from Gomishan wetland. These new strains are *Fallacia pygmaea*, *Halamphora coffeiformis*, and *Navicula veneta* (Saba et al. 2016). Also, five new bacterial taxa at the genus or species level have been described from this wetland, including the haloalkaliphilic microorganisms *Salinispirillum marinum* (Shahinpei et al. 2014a), *Aliidiomarina iranensis* (Amoozegar et al. 2016c), and *Aliidiomarina sedimenti* (Shahinpei et al. 2017) and the halophilic species, *Cyclobacterium halophilum* (Shahinpei et al. 2014b) and *Pseudomonas salegens* (Amoozegar et al. 2014f).

9.2.8 Badab-Soort Travertine Spring

Badab-Soort (Fig. 9.9) is a travertine-maker spring which is located in Mazandaran province in the north of Iran. As a result of calcium carbonate accumulation on this spring, Badab-Soort travertine had been created. It has two different spring heads which varied in characteristics and colors and sediments. As shown in Fig. 9.9, Badab-Soort with its natural unique features is a suitable environment to microbiological studies. In this region the relationship between microorganisms and their surroundings is really notable, because several microorganisms are responsible for calcium carbonate precipitation in travertines. Five strains were isolated from Badab-Soort travertine which were capable of calcium carbonate precipitation. One of these strains had the highest (45.6 mg/ml) amount of calcium carbonate precipitation (Khansha et al. 2016). *Soortia roseihalophila*, a Gram-negative bacterium, has been isolated from Badab-Soort travertine spring and belongs to the novel family *Soortiaceae* (Amoozegar et al. 2017).

9.2.9 Lut Desert

Lut Desert is located in south east of Iran and has a very hot and dry climate. This desert with an area of 2×10^5 km^2 is the 25th largest desert of the world. In 2005, NASA estimated that the temperature of the Gandom Beryan region of Lut Desert is about 70.7 °C, and this was the hottest registered temperature on terrestrial areas of the Earth (Aghanabati 2017). Despite the typical hot and dry climate, a permanent saline river exists in this desert, called Shoor River, and has a length of 2×10^2 km where it stretches from north to south of the Lut Desert. It is the sole

Fig. 9.9 Badab-Soort travertine. Old (up left and bottom right) and new (up right and bottom left) spring and travertine of Badab-Soort

permanent river of the whole region. The pH of the river is neutral, and sodium chloride, sodium sulfate, and potassium chloride are frequent in this river. The average salt concentration in this region is about 15%; thus it is a good habitat for halophilic microorganisms (Yazdi et al. 2014). Some extreme halophilic archaeal strains have been isolated from Shoor River recently. These isolated strains belong to the genera *Haloterigena*, *Natrialba*, and *Natrinema*. These strains tolerated gamma radiations between 2 and 6 kGy, and *Natrialba* sp. strain MS17 had the highest level of tolerance, with 6 kGy (Shirsalimian et al. 2017). Furthermore, the halotolerant actinobacterium *Kocuria polaris* strain A10, isolated from Gandom Beryan area of Lut Desert, exhibited resistance toward gamma radiation up to 4 kGy and remained viable after desiccation for 28 days (Shirsalimian et al. 2016).

9.2.10 Other Saline Environments of Iran

In addition to the mentioned environments, several saline environments like Hamun Lake, Lipar Lake, Bezangan Lake, Eshtehard wetland, Jazmorian wetland, Kafter wetland, Shahdad wetland, Namakdan Lake, Sirjan River, and Behesht-e Masoumeh

Fig. 9.10 Behesht-e Masoumeh wetland. This wetland is located in the central region of Iran, near the city of Qom. Colorful algae and plants are abundant in and around the wetland (up right and bottom). Oil could be found in this wetland (up left)

wetland (Fig. 9.10) exist in Iran. Some sporadic studies have been carried out on microbial life of these environments. For example, in a study two species of *Dunaliella,* identified from Sirjan River and their response to salinity, were examined and compared (Nezhad and Mansouri 2016). Also, in another study a halophilic archaeal strain, Pars Q2, isolated from Namakdan Lake in Qeshm Island showed the ability to produce a biosurfactant when crude oil was its sole carbon source. Furthermore, this strain was able to use molasses and glycerol as its carbon and energy source (Jadidi et al. 2014). Besides, two halophilic exopolysaccharide-producing strains were

isolated from Eshtehard wetland (Fig. 9.11) in Alborz province. These two bacteria were used as decreasing agent against drought and saline stress in order to increase the wheat crops. 16S rRNA analysis showed that these strains are close to *Bacillus subtilis* sub sp. *inaquosorum* and *Marinobacter lipolyticus* sp. The inoculation of these bacterial strains into soil resulted in dried and fresher roots with higher shoot weight. Furthermore, this inoculation increased germination rate and percent

Fig. 9.11 Eshtehard salt marsh. During dry seasons the wetland seems completely white (up left and bottom right), and white brines could be observed around the wetland (up right). Salt crusts are visible on the surface of the marsh (bottom left)

of wheat germination (Talebi et al. 2013). In another study the phylogenetic diversity of cultivable bacteria of Bezangan Lake in northeast of Iran was examined. The study showed the isolation of 51 Gram-positive and 15 Gram-negative strains. Furthermore, 30 different isolates were selected for further studies. These strains belonged to several phyla including *Beta-* and *Gammaproteobacteria*, *Bacteroidetes*, and *Firmicutes*. The Gram-negative strains belonged to the genera *Luteibacter*, *Xanthomonas*, *Varivorax*, *Collimonas*, and *Flavobacterium* while the Gram-positive belonged to the genera *Bacillus*, *Fictibacillus*, *Staphylococcus*, and *Paenibacillus*. *Pseudomonas* was the predominant genus. It was shown that the hydrolytic enzymes were the same in both Gram-negative and Gram-positive bacteria (Shahnavaz and Ghasemzadeh 2015).

9.3 Conclusion

Iran is a country of distinct and variable climates. North of Iran is very humid with frequent floods while the south of Iran is dry, and the main areas of southwest of this country are covered by deserts. Salt lakes and other saline environments are found frequently in Iran, and in previous sections, we describe some important ones. Some of them have been widely studied where some have been less studied. Besides, some are currently under investigation. Among hypersaline environments of Iran, Aran-Bidgol salt lake and Urmia Lake are the most significant ones. Thus, there have been more studies on their microbial diversity. Considering the distinct spectrum of biodiversity in them, there have been many studies focused on the biotechnological applications of the residing microorganisms. Individual studies on other hypersaline environments have shown that there are lots of opportunities to examine the biodiversity of them. With unique and distinctive characteristics of these environments, it won't be unexpected to isolate microorganisms with better and more important biotechnological abilities.

References

Aghanabati SA (2017) A brief review of the geology of West Asia. In: Ancient West Asian civilization. Springer, Singapore, pp 39–50

Alavi S, Amoozegar MA, Khajeh K (2014) Enzyme (s) responsible for tellurite reducing activity in a moderately halophilic bacterium, *Salinicoccus iranensis* strain QW6. Extremophiles 18:953–961

Amiri H, Azarbaijani R, Yeganeh LP, Fazeli AS, Tabatabaei M, Salekdeh GH, Karimi K (2016) *Nesterenkonia* sp. strain F, a halophilic bacterium producing acetone, butanol, and ethanol under aerobic conditions. Sci Rep 6:18408–18418

Amoozegar MA, Ghasemi A, Razavi MR, Naddaf S (2007) Evaluation of hexavalent chromium reduction by chromate-resistant moderately halophile, *Nesterenkonia* sp. strain MF2. Process Biochem 42:1475–1479

Amoozegar MA, Schumann P, Hajighasemi M, Fatemi AZ, Karbalaei-Heidari HR (2008) *Salinivibrio proteolyticus* sp. nov., a moderately halophilic and proteolytic species from a hypersaline lake in Iran. Int J Syst Evol Microbiol 58:1159–1163

Amoozegar MA, Sanchez-Porro C, Rohban R, Hajighasemi M, Ventosa A (2009a) *Piscibacillus halophilus* sp. nov., a moderately halophilic bacterium from a hypersaline Iranian lake. Int J Syst Evol Microbiol 59:3095–3099

Amoozegar MA, Sánchez-Porro C, Rohban R, Hajighasemi M, Ventosa A (2009b) *Bacillus persepolensis* sp. nov., a moderately halophilic bacterium from a hypersaline lake. Int J Syst Evol Microbiol 59:2352–2358

Amoozegar MA, Makhdoumi-Kakhki A, Fazeli SAS, Azarbaijani R, Ventosa A (2012) *Halopenitus persicus* gen. nov., sp. nov., an archaeon from an inland salt lake. Int J Syst Evol Microbiol 62:1932–1936

Amoozegar MA, Didari M, Bagheri M, Fazeli SAS, Schumann P, Spröer C, Ventosa A (2013a) *Bacillus salsus* sp. nov., a halophilic bacterium from a hypersaline lake. Int J Syst Evol Microbiol 63:3324–3329

Amoozegar MA, Makhdoumi-Kakhki A, Mehrshad M, Fazeli SAS, Ventosa A (2013b) *Halopenitus malekzadehii* sp. nov., an extremely halophilic archaeon isolated from a salt lake. Int J Syst Evol Microbiol 63:3232–3236

Amoozegar MA, Makhdoumi-Kakhki A, Ramezani M, Nikou MM, Fazeli SAS, Schumann P, Ventosa A (2013c) *Limimonas halophila* gen. nov., sp. nov., an extremely halophilic bacterium in the family *Rhodospirillaceae*. Int J Syst Evol Microbiol 63:1562–1567

Amoozegar MA, Bagheri M, Didari M, Mehrshad M, Schumann P, Spröer C, Ventosa A (2014a) *Aquibacillus halophilus* gen. nov., sp. nov., a moderately halophilic bacterium from a hypersaline lake, and reclassification of *Virgibacillus koreensis* as *Aquibacillus koreensis* comb. nov. and *Virgibacillus albus* as Aquibacillus albus comb. nov. Int J Syst Evol Microbiol 64:3616–3623

Amoozegar MA, Bagheri M, Makhdoumi-Kakhki A, Didari M, Schumann P, Spröer C, Ventosa A (2014b) *Oceanobacillus limi* sp. nov., a moderately halophilic bacterium from a salt lake. Int J Syst Evol Microbiol 64:1284–1289
Amoozegar MA, Makhdoumi-Kakhki A, Mehrshad M, Fazeli SAS, Spröer C, Ventosa A (2014c) *Halorientalis persicus* sp. nov., an extremely halophilic archaeon isolated from a salt lake and emended description of the genus *Halorientalis*. Int J Syst Evol Microbiol 64:940–944
Amoozegar MA, Makhdoumi-Kakhki A, Mehrshad M, Riazi S, Ventosa A (2014d) *Halovivax limisalsi* sp. nov., an extremely halophilic archaeon from hypersaline mud. Int J Syst Evol Microbiol 64:3422–3426
Amoozegar MA, Bagheri M, Makhdoumi-Kakhki A, Didari M, Schumann P, Nikou MM, Ventosa A (2014e) *Aliicoccus persicus* gen. nov., sp. nov., a halophilic member of the Firmicutes isolated from a hypersaline lake. Int J Syst Evol Microbiol 64:1964–1969
Amoozegar MA, Shahinpei A, Sepahy AA, Makhdoumi-Kakhki A, Seyedmahdi SS, Schumann P, Ventosa A (2014f) *Pseudomonas salegens* sp. nov., a halophilic member of the genus Pseudomonas isolated from a wetland. Int J Syst Evol Microbiol 64:3565–3570
Amoozegar MA, Makhdoumi-Kakhki A, Mehrshad M, Rasooli M, Fazeli SAS, Spröer C, Ventosa A (2015) *Halovivax cerinus* sp nov, an extremely halophilic archaeon from a hypersaline lake. Int J Syst Evol Microbiol 65:65–70
Amoozegar MA, Bagheri M, Makhdoumi A, Mehrshad M, Didari M, Schumann P, Ventosa A (2016a) *Oceanobacillus longus* sp. nov., a moderately halophilic bacterium isolated from a salt lake. Int J Syst Evol Microbiol 66:4225–4230
Amoozegar MA, Bagheri M, Makhdoumi A, Nikou MM, Fazeli SAS, Schumann P, Ventosa A (2016b) *Oceanobacillus halophilus* sp. nov., a novel moderately halophilic bacterium from a hypersaline lake. Int J Syst Evol Microbiol 66:1317–1322
Amoozegar MA, Shahinpei A, Fazeli SAS, Schumann P, Spröer C, Ventosa A (2016c) *Aliidiomarina iranensis* sp. nov., a haloalkaliphilic bacterium from a coastal-marine wetland. Int J Syst Evol Microbiol 66:2099–2105
Amoozegar MA, Khansha J, Mehrshad M, Fazeli SAS, Ramezani M, Rafael R, Ventosa A (2017) *Soortia roseihalophila* gen. nov., sp. nov., a new taxon in the order Balneolales isolated from a travertine spring, and description of *Soortiaceae* fam. nov. Int J Syst Evol Microbiol 67:113–120
Arzanlou M, Samadi R, Frisvad JC, Houbraken J, Ghosta Y (2016) Two novel Aspergillus species from hypersaline soils of the National Park of Lake Urmia, Iran. Mycol Prog 15:1081–1092
Asem A, Eimanifar A, Djamali M, De los Rios P, Wink M (2014) Biodiversity of the hypersaline Urmia Lake National Park (NW Iran). Diversity 6:102–132
Asem A, Eimanifar A, Wink M (2016) Update of "biodiversity of the Hypersaline Urmia Lake National Park (NW Iran)". Diversity 8:6–9
Asgarani E, Shirzad M, Soudi MR, Shahmohammadi HR, Falsafi T (2006) Study on the resistance of *Haloferax Radiotolerans*, an extreme halophilic Archaebacterium from Urmia Lake against ultraviolet (UV) light and 60Co Gama-rays. J Nucl Sci Technol 36:13–18
Asgarani E, Safari N, Zare D (2014) Partial purification and characterization of carotenoid pigment from *Haloarcula* sp. IRU-I, an extremely halophilic archaeon isolated from Urmia Salt Lake. J Mol Cell Res 26:326–338
Ashengroph M, Nahvi I (2014) Use of resting cells of halomonas salina HSL5 as biocatalyst for biological vanillic acid production. Cell Mol Res J 1:1–13
Azarbaijani R, Yeganeh LP, Blom J, Younesi H, Fazeli SA, Tabatabaei M, Salekdeh GH (2016) Comparative genome analysis of Oceanimonas sp. GK1, a halotolerant bacterium with considerable xenobiotics degradation potentials. Ann Microbiol 66:703–716
Babavalian H, Amoozegar MA, Pourbabaee AA, Moghaddam MM, Shakeri F (2013) Isolation and identification of moderately halophilic bacteria producing hydrolytic enzymes from the largest hypersaline playa in Iran. Microbiology 82:466–474
Babavalian H, Amoozegar MA, Zahraei S, Rohban R, Shakeri F, Moghaddam MM (2014) Comparison of bacterial biodiversity and enzyme production in three hypersaline lakes; Urmia, Howz-Soltan and Aran-Bidgol. Indian J Microbiol 54:444–449

Bagheri M, Didari M, Amoozegar MA, Schumann P, Sanchez-Porro C, Mehrshad M, Ventosa A (2012) *Bacillus iranensis* sp. nov., a moderate halophile from a hypersaline lake. Int J Syst Evol Microbiol 62:811–816

Bagheri M, Amoozegar MA, Didari M, Makhdoumi-Kakhki A, Schumann P, Spröer C, Ventosa A (2013a) *Marinobacter persicus* sp. nov., a moderately halophilic bacterium from a saline lake in Iran. Antonie Van Leeuwenhoek 104:47–54

Bakhshi S, Sooudi MR, Ghafourian H, Ahmadi (2007) Cesium biosorption in halophilic and halotolerant bacteria and investigation of effects of environmental factors on biosorption. Res J Univ Isfahan 1:15–30

Bagheri M, Amoozegar MA, Schumann P, Didari M, Mehrshad M, Spröer C, Ventosa A (2013b) *Ornithinibacillus halophilus* sp. nov., a moderately halophilic, gram-stain-positive, endospore-forming bacterium from a hypersaline lake. Int J Syst Evol Microbiol 63:844–848

Bardavid RE, Mana L, Oren A (2007) Haloplanus natans gen. nov., sp. nov., an extremely halophilic, gas-vacuolate archaeon isolated from Dead Sea–Red Sea water mixtures in experimental outdoor ponds. Int J Syst Evol Microbiol 57:780–783

Barin M, Aliasgharzad N, Olsson PA, Rasouli-Sadaghiani M (2015) Salinity-induced differences in soil microbial communities around the hypersaline Lake Urmia. Soil Res 53:494–504

Breckle SW (2002) Salt deserts in Iran and Afghanistan. Barth & Böer (eds.), Sabkha Ecosys. pp 109–22

Buchalo AS, Nevo E, Wasser SP, Oren A, Molitoris HP (1998) Fungal life in the extremely hypersaline water of the Dead Sea: first records. Proceedings of the Royal Society of London B: Biol Sci 265:1461–1465

Corral P, Rafael R, Sánchez-Porro C, Amoozegar MA, Papke RT, Ventosa A (2015) *Halorubrum persicum* sp. nov., an extremely halophilic archaeon isolated from sediment of a hypersaline lake. Int J Syst Evol Microbiol 65:1770–1778

Corral P, Rafael R, Sánchez-Porro C, Amoozegar MA, Papke RT, Ventosa A (2016) *Halorubrum halodurans* sp. nov., an extremely halophilic archaeon isolated from a hypersaline lake. Int J Syst Evol Microbiol 66:435–444

Crous PW, Wingfield MJ, Schumacher RK, Summerell BA, Giraldo A, Gené J, Garethjones EB (2014) Fungal planet description sheets: 281–319. Persoonia-Mol Phyl Evol Fungi 33:212–289

de Lourdes MM, Sánchez-Porro C, García MT, Mellado E (2012) Carotenoids' production from halophilic bacteria. In: Microbial Carotenoids from Bacteria and Microalgae. Humana Press, New York, pp 207–217

Didari M, Amoozegar MA, Bagheri M, Schumann P, Spröer C, Sánchez-Porro C, Ventosa A (2012) *Alteribacillus bidgolensis* gen. nov., sp. nov., a moderately halophilic bacterium from a hypersaline lake, and reclassification of *Bacillus persepolensis* as *Alteribacillus persepolensis* comb. nov. Int J Syst Evol Microbiol 62:2691–2697

Didari M, Amoozegar MA, Bagheri M, Mehrshad M, Schumann P, Spröer C, Ventosa A (2013) *Bacillus persicus* sp. nov., a halophilic bacterium from a hypersaline lake. Int J Syst Evol Microbiol 63:1229–1234

Eimanifar A, Mohebbi F (2007) Urmia Lake (Northwest Iran): a brief review. Saline syst 3:5–13

Eskandari H, Borji M, Khosravi H, Nakhaee Nejadfar S, Eskandari H (2016) Change detection of Bakhtegan and Tashk Basin during 2001–2013. Int J Forest Soil Erosion (IJFSE) 6:67–71

Fahimeh M, Nima B, Majid B (2013) Isolation, characterization and identification of amylase producing halothermophilic isolates from Howz Soltan Lake, Iran. Afr J Microbiol Res 7:4483–4490

Farahani H, Amoozegar MA, Asadi B, Mehrshad M, Yakhchali B, Motamedi S (2014) Taxonomic classification of newly isolated archaeal strain from Urmia hypersaline lake. Iranian J Public Health 43:147–148

Fazeli MR, Tofighi H, Samadi N, Jamalifar H (2006) Effects of salinity on β-carotene production by *Dunaliella tertiolecta* DCCBC26 isolated from the Urmia salt lake, north of Iran. Bioresour Technol 97:2453–2456

Ghadimi F, Ghomi M (2013) Assessment of the effects of municipal wastewater on the heavy metal pollution of water and sediment in Arak Mighan Lake, Iran. J Tethys 1:205–214

Ghasemi Y, Rasoul-Amini S, Ebrahiminezhad A, Kazemi A, Shahbazi M, Talebnia N (2011) Screening and isolation of extracellular protease producing bacteria from the Maharloo Salt Lake. Iranian J Pharma Sci 7:175–180

Hallsworth JE, Yakimov MM, Golyshin PN, Gillion JL, D'Auria G, de Lima Alves F, Harris G (2007) Limits of life in $MgCl_2$-containing environments: chaotropicity defines the window. Environ Microbiol 9:801–813

Hamidi M, Hejazi M, Nazemyieh H, Maleki B, Hejazi M, Naziri D, Hassanzadeh S (2012) Study of carotenoid production by *marinobacter* sp. TBZ112 isolated from Urmia lake in northwestern of Iran. Res Pharmac Sci 7:S492

Hamidi M, Abdin MZ, Nazemyieh H, Hejazi MA, Hejazi MS (2014) Optimization of total carotenoid production by *Halorubrum* sp. TBZ126 using response surface methodology. J Microb Biochem Technol 6:286–294

Hashemi T, BaseriSalehi M, Bahador N (2014) Isolation of halophilic bacteria from Maharlu salt Lake-Iran and their evaluation for the production of bioactive compounds. Int J Mol Clinical Microbiol 1:365–370

Hassanshahian M, Mohamadian J (2011) Isolation and characterization of *Halobacterium salinarum* from saline lakes in Iran. Jundishapur J Microbiol 4:59–66

Heidari R, Riahi H, Saadatmand S (2000) Effects of salt and irradiance stress on photosynthetic pigments and proteins in Dunaliella salina teodoresco. J Sci I R Iran 1:1173–1178

Heris YS, Hajizadeh N, Vahed SZ, Vallipour J, Amin M (2014a) Bacterial degradation of para-amino acetanilide by *Marinobacter* sp. TBZ23, isolated from Urmia Lake in Iran. Trends life Sci 3:1–9

Heris YS, Hajizadeh N, Vahed SZ, Vallipour J, Hejazi MA, Golabi SM, Hejazi MS (2014b) Fe (III) reduction by *Halomonas* sp. TBZ9 and *Maribobacter* sp. TBZ23, isolated from Urmia Lake in Iran. Advan Environ Biol 1:59–66

Hosseinzadeh Gharajeh N, Hejazi MA, Nazeri S, Barzegari A (2012) Characterization of an indigenous isolate, *Dunaliella tertiolecta* ABRIINW-G3, from Gavkhooni salt marsh in Iran based on molecular and some morpho-physiological attributes. J Agric Sci Technol 14:1579–1590

Hyde KD, Hongsanan S, Jeewon R et al (2016) Fungal diversity notes 367-491: taxonomic and phylogenetic contributions to fungal taxa. Fungal Divers 80:1–270

Jadidi N, Roozbehani B, Saadat A (2014) Application of organic surfactants to recover hydrocarbons from oil Sludges. J Environ Friendly Proc 2:253–260

Javor B (1989) Hypersaline, Alkaline Lakes. In: Hypersaline Environments. Springer, Berlin/Heidelberg, pp 292–312

Kafilzadeh F, Behzadi SS (2015) Characterization and isolation and anthracene degrading gram negative bacteria from the contaminated water and sediments in Maharlou Lake. J Environ Sci Technol 16:513–519

Kafilzadeh F, Javid H, Mohammadi H (2007a) Isolation decomposing bacteria of polycyclic aromatic hydrocarbons (PAHs) from Tashk lake and examination of salt concentration effect on them. Iranian Scientific Fisheries J 16:103–112

Kafilzadeh F, Mohammadi H, Kadivar E (2007b) Isolation of bacteria degrading polycyclic aromatic hydrocarbons (PABs) from the Maharloo Lake and the effect of salt contration on PAB degradation. J School Public Health 5:67–77

Kafilzadeh F, Javid H, Kargar M (2007c) Isolation of halophilic and halotolerant microorganisms from the Bakhtegan lake and the effect of physicochemical factors on their frequency. Water Wastewater 18:81–67

Kakhki AM, Amoozegar MA, Khaledi EM (2011) Diversity of hydrolytic enzymes in haloarchaeal strains isolated from salt lake. Int J Environ Sci Technol 8:705–714

Karbalaei-Heidari HR, Ziaee AA, Schaller J, Amoozegar MA (2007) Purification and characterization of an extracellular haloalkaline protease produced by the moderately halophilic bacterium, *Salinivibrio* sp. strain AF-2004. Enzym Microb Technol 40:266–272

Karbalaei-Heidari HR, Ziaee AA, Amoozegar MA, Cheburkin Y, Budisa N (2008) Molecular cloning and sequence analysis of a novel zinc-metalloprotease gene from the *Salinivibrio* sp. strain AF-2004 and its extracellular expression in *E. coli*. Gene 408:196–203

Kashi FJ, Owlia P, Amoozegar MA, Yakhchali B, Kazemi B (2014) Diversity of cultivable micro-organism in the eastern part of Urmia Salt Lake, Iran. J Microbiol Biotechnol Food Sci 4:36–43

Khansha J, Amoozegar MA, Ranjbaran M (2016) Studying of biomineralization on Badab Sourt travertine deposit spring, Kiasar, Mazandaran Province. Biol J Microorganism 5:141–158

Khoshnevis N, Rezaei S, Samaei NA, Khoshayand MR, Faramarzi MA (2014) Production and characterization of a novel halophilic protease produced by the isolate *Salicola* sp. Mol Biol Res Comm 3:193–193

Libes S (2011) Introduction to marine biogeochemistry. In: pp:27–29. Academic, Cambridge, MA

Ma Y, Galinski EA, Grant WD, Oren A, Ventosa A (2010) Halophiles 2010: life in saline environments. Appl Environ Microbiol 76:6971–6981

Makhdoumi-Kakhki A, Amoozegar MA, Ventosa A (2012a) *Halovenus aranensis* gen. nov., sp. nov., an extremely halophilic archaeon from Aran-Bidgol salt lake. Int J Syst Evol Microbiol 62:1331–1336

Makhdoumi-Kakhki A, Amoozegar MA, Ventosa A (2012b) *Salinibacter iranicus* sp. nov. and *Salinibacter luteus* sp. nov., isolated from a salt lake, and emended descriptions of the genus *Salinibacter* and of *Salinibacter ruber*. Int J Syst Evol Microbiol 62:1521–1527

Makhdoumi-Kakhki A, Amoozegar MA, Bagheri M, Ramezani M, Ventosa A (2012c) *Haloarchaeobius iranensis* gen. nov., sp. nov., an extremely halophilic archaeon isolated from a saline lake. Int J Syst Evol Microbiol 62:1021–1026

Makhdoumi-Kakhki A, Amoozegar MA, Kazemi B, PaiC L, Ventosa A (2012d) Prokaryotic diversity in Aran-Bidgol salt lake, the largest hypersaline playa in Iran. Microbes Environ 27:87–93

Makzum S, Amoozegar MA, Dastgheib SM (2017) Isolation and identification of obligately chemolithoautotrophic, haloalkaliphilic bacterium *Thioalkalivibrio* sp. strain EMA and optimizing its thiosulfate removal activity in haloalkaliphilic condition. Biol J Microorganisms 2:2–17

Mehrshad M, Amoozegar MA, Didari M, Bagheri M, Fazeli SAS, Schumann P, Ventosa A (2013) *Bacillus halosaccharovorans* sp. nov., a moderately halophilic bacterium from a hypersaline lake. Int J Syst Evol Microbiol 63:2776–2781

Mehrshad M, Amoozegar MA, Makhdoumi A, Rasooli M, Asadi B, Schumann P, Ventosa A (2015) *Halovarius luteus* gen. nov., sp. nov., an extremely halophilic archaeon from a salt lake. Int J Syst Evol Microbiol 65:2420–2425

Mehrshad M, Amoozegar MA, Makhdoumi A, Fazeli SAS, Farahani H, Asadi B, Ventosa A (2016) *Halosiccatus urmianus* gen. nov., sp. nov., a haloarchaeon from a salt lake. Int J Syst Evol Microbiol 66:725–730

Montazeri-Najafabady N, Negahdaripour M, Salehi MH, Morowvat MH, Shaker S, Ghasemi Y (2016) Effects of osmotic shock on production of β-carotene and glycerol in a naturally isolated strain of Dunaliella salina. J Appl Pharma Sci 6:160–163

Mousavi P, Morowvat MH, Montazeri-Najafabady N, Abolhassanzadeh Z, Mohagheghzadeh A, Hamidi M, Ghasemi Y (2016) Investigating the effects of phytohormones on growth and β-carotene production in a naturally isolates stain of *Dunaliella salina*. J Appl Pharma Sci 6:164–171

Munoz R, Rosselló-Móra R, Amann R (2016) Revised phylogeny of Bacteroidetes and proposal of sixteen new taxa and two new combinations including *Rhodothermaeota* phyl. nov. Syst Appl Microbiol 39:281–296

Naghoni A, Emtiazi G, Amoozegar MA, Rasooli M, Etemadifar Z, Shahzadeh Fazeli SA, Minegishi H, Ventosa A (2017a) *Natronoarchaeum persicum* sp. nov., a haloarchaeon isolated from a hypersaline lake. Int J Syst Evol Microbiol 67:3339–3343

Naghoni A, Emtiazi G, Amoozegar MA, Silvia Cretoiu M, Stal LJ, Etemadifar Z, Shahzadeh Fazeli SA, Bolhuis H (2017b) Microbial diversity in the hypersaline Lake Meyghan, Iran. Sci Rep 7:11522

Nasr S, Soudi MR, Fazeli SAS, Nguyen HD, Lutz M, Piątek M (2014) Expanding evolutionary diversity in the Ustilaginomycotina: *Fereydouniaceae* fam. nov. and *Fereydounia* gen. nov., the first urocystidalean yeast lineage. Mycol Prog 13:1012–1015

Naziri D, Hamidi M, Hassanzadeh S, Tarhriz V, Zanjani BM, Nazemyieh H, Hejazi MS (2014) Analysis of carotenoid production by *Halorubrum* sp. TBZ126; an extremely halophilic archeon from Urmia Lake. Adv Pharm Bull 4:61–67

Nezhad FS, Mansouri H (2016) Identification of two species of the green algae *Dunaliella* and comparing their response to salinity. Indian J Plant Physiol 21:44–49

Nikookar K, Moradshahi A, Kharati M (2004) Influence of salinity on the growth, pigmentation and ascorbate peroxidase activity of *Dunaliella salina* isolated from Maharlu salt lake in Shiraz. Iranian J Sci Tech (Sciences) 28:117–125

Nikookar K, Moradshahi A, Hosseini L (2005) Physiological responses of *Dunaliella salina* and *Dunaliella tertiolecta* to copper toxicity. Biomol Eng 22:141–146

Nikookar K, Rowhani L, Mohsenzadeh S, Kholdebarin B (2013) Growth and pigment development of *Dunaliella salina* Teod. in response to ammonium nitrate nutrition. Mol Biol Res Comm 2:73–79

Nikou MM, Ramezani M, Amoozegar MA, Fazeli SAS, Schumann P, Spröer C, Ventosa A (2014) *Alloactinosynnema iranicum* sp. nov., a rare actinomycete isolated from a hypersaline wetland, and emended description of the genus *Alloactinosynnema*. Int J Syst Evol Microbiol 64:1173–1179

Nikou MM, Ramezani M, Amoozegar MA, Rasooli M, Harirchi S, Fazeli SAS, Ventosa A (2015a) *Nocardia halotolerans* sp. nov., a halotolerant actinomycete isolated from saline soil. Int J Syst Evol Microbiol 65:3148–3154

Nikou MM, Ramezani M, Amoozegar MA, Rasouli M, Fazeli SAS, Schumann P, Ventosa A (2015b) *Salininema proteolyticum* gen. nov., sp. nov., a halophilic rare actinomycete isolated from wetland soil, and emended description of the family Glycomycetaceae. Int J Syst Evol Microbiol 65:3727–3733

Nikou MM, Ramezani M, Harirchi S, Makzoom S, Amoozegar MA, Fazeli SA, Schumann P, Ventosa A (2017) *Salinifilum* gen. nov., with description of *Salinifilum proteinilyticum* sp. nov., an extremely halophilic actinomycete isolated from Meighan wetland, Iran, and reclassification of Saccharopolyspora aidingensis as *Salinifilum aidingensis* comb. nov. and *Saccharopolyspora ghardaiensis* as *Salinifilum ghardaiensis* comb. nov. Int J Syst Evol Microbiol 67:4221–4227

Nissenbaum A (1993) Utilization of Dead Sea asphalt throughout history. Rev Chem Eng 9:365–384

Noroozi M, Amoozegar MA, Pourbabaee AA, Naghavi NS, Nourmohammadi Z (2017) Isolation and characterization of mercuric reductase by newly isolated halophilic bacterium, *Bacillus firmus* MN8. Global J Environ Sci Manag 3:4–13

Nouri J, Karbassi AR, Mirkia S (2008) Environmental management of coastal regions in the Caspian Sea. Int J Environ Sci Technol 5:43–52

Oren A (2002) Diversity of halophilic microorganisms: environments, phylogeny, physiology, and applications. J Ind Microbiol Biotechnol 28:56–63

Oren A (2011) Ecology of halophiles. In: Extremophiles Handbook. Springer, Tokyo, pp 343–361

Ramezani M, Amoozegar MA, Ventosa A (2015) Screening and comparative assay of polyhydroxyalkanoates produced by bacteria isolated from the Gavkhooni Wetland in Iran and evaluation of poly-β-hydroxybutyrate production by halotolerant bacterium *Oceanimonas* sp. GK1. Ann Microbial 65:517–526

Rasooli M, Amoozegar MA, Sepahy AA, Babavalian H, Tebyanian H (2016) Isolation, identification and extracellular enzymatic activity of culturable extremely halophilic archaea and bacteria of Inche Boroun wetland. Int Letters Nat Sci 56:40–51

Rasooli M, Naghoni A, Amoozegar MA, Mirfeizi L, Nikou MM, Fazeli SA, Minegishi H, Ventosa A (2017a) *Natrinema soli* sp. nov., a novel halophilic archaeon isolated from a hypersaline wetland. Int J Syst Evol Microbiol 67:2142–2147

Rasooli M, Naghoni A, Amoozegar MA, Mirfeizi L, Nikou MM, Fazeli SAS, Ventosa A (2017b) *Natrinema soli* sp. nov., a novel halophilic archaeon isolated from a hypersaline wetland. Int J Syst Evol Microbiol 67:2142–2147

Rasoul-Amini S, Mousavi P, Montazeri-Najafabady N, Mobasher MA, Mousavi SB, Vosough F, Ghasemi Y (2014) Biodiesel properties of native strain of *Dunaliella salina*. Int J Renewable Energy Res (IJRER) 4:39–41

Rezaei S, Shahverdi AR, Faramarzi MA (2014) An extremely halophilic laccase from the saline water isolate *Chromohalobacter* sp. Mol Biol Res Commu 3:260–260

Rohban R, Amoozegar MA, Ventosa A (2009) Screening and isolation of halophilic bacteria producing extracellular hydrolyses from Howz Soltan Lake, Iran. J Ind Microbiol Biotechnol 36:333–340

Saba F, Noroozi M, Ghahremaninejad F, Sedghi M, Papizadeh M (2016) Isolation, purification and identification of three diatom species (*Bacillariophyceae*) from Gomishan wetland (N. Iran) using phylogeny and silica cell wall ultra-structure analysis. Rostaniha 17:28–39

Sajedipour S, Zarei H, Oryan S (2017) Estimation of environmental water requirements via an ecological approach: a case study of Bakhtegan Lake, Iran. Ecol Eng 100:246–255

Sanchez-Porro C, Amoozegar MA, Rohban R, Hajighasemi M, Ventosa A (2009) *Thalassobacillus cyri* sp. nov., a moderately halophilic gram-positive bacterium from a hypersaline lake. Int J Syst Evol Microbiol 59:2565–2570

Sanchez-Porro C, Amoozegar MA, Fernandez AB, Ramezani M, Ventosa A (2010) *Lentibacillus persicus* sp. nov., a moderately halophilic species isolated from a saline lake. Int J Syst Evol Microbiol 60:1407–1412

Sarikhan S, Azarbaijani R, Yeganeh LP, Fazeli AS, Amoozegar MA, Salekdeh GH (2011) Draft genome sequence of *Nesterenkonia* sp. strain F, isolated from Aran-Bidgol salt Lake in Iran. J Bacteriol 193:5580–5580

Savage KN, Krumholz LR, Oren A, Elshahed MS (2008) *Halosarcina pallida* gen. nov., sp. nov., a halophilic archaeon from a low-salt, sulfide-rich spring. Int J Syst Evol Microbiol 58:856–860

Sedghi M, Saba F, Noroozi M (2016) Isolation, morphological and molecular identification of Dunaliella species in Hoz-e Soltan and Aran-va-Bidgol salt lakes (Iran). Rostaniha 17:70–77

Shadkam S, Ludwig F, van Vliet MT, Pastor A, Kabat P (2016) Preserving the world second largest hypersaline lake under future irrigation and climate change. Sci Total Environ 559:317–325

Shadrin N, Oren A (2015) Past, present and future of saline lakes: research for global sustainable development. Chin J Oceanol Limnol 33:1349–1353

Shafiei M, Ziaee AA, Amoozegar MA (2010) Purification and biochemical characterization of a novel SDS and surfactant stable, raw starch digesting, and halophilic α-amylase from a moderately halophilic bacterium, *Nesterenkonia* sp. strain F. Process Biochem 45:694–699

Shafiei M, Ziaee AA, Amoozegar MA (2011) Purification and characterization of an organic-solvent-tolerant halophilic α-amylase from the moderately halophilic *Nesterenkonia* sp. strain F. J Ind Microbiol Biotechnol 38:275–281

Shafiei M, Ziaee AA, Amoozegar MA (2012) Purification and characterization of a halophilic α-amylase with increased activity in the presence of organic solvents from the moderately halophilic *Nesterenkonia* sp. strain F. Extremophiles 16:627–635

Sharifi Y, Pourbabaei AA, Javadi A, Abdolmohammad MH, Saffari M, Morovvati A (2015) Biodegradation of glyphosate herbicide by Salinicoccus spp isolated from Qom Hoze-soltan lake, Iran. Environ Health Eng Manag J 2:31–36

Shahbazi M, Karbalaei-Heidari HR (2012) A novel low molecular weight extracellular protease from a moderately halophilic bacterium *Salinivibrio* sp. strain MS-7: production and biochemical properties. Mol Biol Res Comm 1:45–56

Shahinpei A, Amoozear MA, Akhavan SA (2013) Isolation and identification of poly-extremophilic alkalophilic, halophilic and halotolerant bacteria from alkaline thalassohaline Gomishan wetland. J Taxonomy Biosistematics 5:79–100

Shahinpei A, Amoozegar MA, Fazeli SAS, Schumann P, Ventosa A (2014a) *Salinispirillum marinum* gen. nov., sp. nov., a haloalkaliphilic bacterium in the family '*Saccharospirillaceae*. Int J Syst Evol Microbiol 64:3610–3615

Shahinpei A, Amoozegar MA, Sepahy AA, Schumann P, Ventosa A (2014b) *Cyclobacterium halophilum* sp. nov., a marine bacterium isolated from a coastal-marine wetland. Int J Syst Evol Microbiol 64:1000–1005

Shahinpei A, Amoozegar MA, Fazeli SAS, Schumann P, Spröer C, Ventosa A (2017) *Aliidiomarina sedimenti* sp. nov., a haloalkaliphilic bacterium in the family *Idiomarinaceae*. Int J Syst Evol Microbiol 67:2087–2092

Shahnavaz B, Ghasemzadeh F (2015) Phylogenetic diversity of cultivable bacteria in Lake Bazangan and their ecological role. Cell Mol Res 28:360–370

Shaker S, Morowvat MH, Ghasemi Y (2017) Effects of sulfur, Iron and manganese starvation on growth, β-carotene production and lipid profile of *Dunaliella salina*. J Young Pharmacists 9:1–5

Shirazian P, Asad S, Amoozegar MA (2016) The potential of halophilic and halotolerant bacteria for the production of antineoplastic enzymes: L-asparaginase and L-glutaminase. EXCLI J 15:268–279

Shirsalimian MS, Sepahy AA, Amoozegar MA, Kalantar SM, Dabbagh R (2016) Isolation of a mesophilic and halotolerant strain of *Kocuria polaris* from Gandom Beryan area in the Lut Desert of Iran, moderately resistant to gamma radiation and desiccation. Biosci Biotech Res Asia 13:2343–2350

Shirsalimian MS, Amoozegar MA, Sepahy AA, Kalantar SM, Dabbagh R (2017) Isolation of extremely halophilic Archaea from a saline river in the Lut Desert of Iran, moderately resistant to desiccation and gamma radiation. Microbiology 86:403–411

Siroosi M, Amoozegar MA, Khajeh K, Fazeli M, Rezaei MH (2014) Purification and characterization of a novel extracellular halophilic and organic solvent-tolerant amylopullulanase from the haloarchaeon, *Halorubrum* sp. strain Ha25. Extremophiles 18:25–33

Siroosi M, Amoozegar MA, Khajeh K (2016) Purification and characterization of an alkaline chloride-tolerant laccase from a halotolerant bacterium, *Bacillus* sp strain WT. J Mol Catalysis B: Enzym 134:89–97

Taghavi L, Tayebi S, Tayebi S, Karimian B (2013) Geochemistry application of major and trace elements for analysis of palaeo climate of north part of Gavkhooni wetland. J Wetland Ecobiology 5:53–62

Talebi AM, Pourbabaee AA, Shorafa M (2013) Effect of halophilic exopolysaccharide-producing Bacteria on wheat growth under drought and saline stresses. Iranian J Soil Res 27:97–106

Taran M (2011) Utilization of petrochemical wastewater for the production of poly (3-hydroxybutyrate) by *Haloarcula* sp. IRU1. J Hazard Mater 188:26–28

Tavallaie S, Emtyazjoo M, Rostami K, Kosari H, Assadi MM (2015) Comparative studies of β-carotene and protein production from *Dunaliella salina* isolated from Lake Hoze-soltan, Iran. J Aqua Food Prod Tech 24:79–90

Ventosa A, Arahal DR (2009) Physico-chemical characteristics of hypersaline environments and their biodiversity. Extremophiles 2:1–6

Yazdi A, Emami MH, Shafiee SM (2014) Dasht-e lut in Iran, the most complete collection of beautiful geomorphological phenomena of desert. Open J Geo 4:249–261

Yeganeh LP, Azarbaijani R, Sarikhan S, Mousavi H, Ramezani M, Amoozegar MA, Salekdeh GH (2012) Complete genome sequence of *Oceanimonas* sp. GK1, a halotolerant bacterium from Gavkhouni Wetland in Iran. J Bacteriol 194:2123–2124

Yeganeh LP, Azarbaijani R, Mousavi H, Fazeli SA, Amoozgar MA, Salekdeh GH (2015) Genome-wide analysis of *Oceanimonas* sp. GK1 isolated from Gavkhouni Wetland (Iran) demonstrates presence of genes for virulence and pathogenicity. Cell J (Yakhteh) 17:451–460

Yin J, Chen JC, Wu Q, Chen GQ (2015) Halophiles, coming stars for industrial biotechnology. Biotechnol Adv 33:1433–1442

Zarei M, Mobasher MA, Morowvat MH, Mousavi P, Montazeri-Najafabady N, Hajighahramani N, Ghasemi Y (2016) Effects of menthone and piperitone on growth, chlorophyll a and β-carotene production in *Dunaliella salina*. J App Pharma Sci 6:215–219

Zarparvar P, Amoozegar MA, Nikou MM, Schumann P, Ventosa A (2014) *Salinithrix halophila* gen. nov., sp. nov., a halophilic bacterium in the family *Thermoactinomycetaceae*. Int J Syst Evol Microbiol 64:4115–4119

Zarparvar P, Amoozegar MA, Babavalian H, Reza FM, Tebyanian H, Shakeri F (2016) Isolation and identification of culturable halophilic bacteria with producing hydrolytic enzyme from Incheh Broun hypersaline wetland in Iran. Cell Mol Bio (Noisy-le-Grand, France) 62:12–31

Halotolerant and Halophilic Microbes and Their Environmental Implications in Saline and Hypersaline Lakes in Qinghai Province, China

10

Hongchen Jiang, Jianrong Huang, and Jiang Yang

Abstract

Abundant and diverse halophilic and halotolerant microbes exist in the lakes of Qinghai Province, China. However, it is poorly known about their roles in the biogeochemical cycling of carbon, nitrogen, and sulfur elements and how their ecological functions respond to environmental conditions. The purpose of this chapter is to summarize the diversity/community composition and ecological implications of halophilic and halotolerant microbes and their responses to environmental conditions in the Qinghai lakes. Halophilic and halotolerant microbes in the Qinghai lakes are important mines for exploring new taxonomic units, and they are extensively involved in ecological functions related to biogeochemical cycling of carbon, nitrogen, and sulfur elements. The halophilic and halotolerant microbes in global saline and hypersaline lakes may contribute higher fraction to global carbon flux than expected. So reappraisals are to be conducted on microbial roles in biogeochemistry. In addition, biomarkers (ancient DNA or lipids) derived from halophilic and halotolerant microbial functional groups can be employed to reconstruct the paleoenvironmental conditions of the lakes.

Keywords

Qinghai · Saline/hypersaline lakes · Microbes · Diversity · Function

H. Jiang (✉) · J. Huang · J. Yang
State Key Laboratory of Biogeology and Environmental Geology, China University of Geosciences, Wuhan, Hubei Province, China
e-mail: jiangh@cug.edu.cn

D. Egamberdieva et al. (eds.), *Extremophiles in Eurasian Ecosystems: Ecology, Diversity, and Applications*, Microorganisms for Sustainability 8,
https://doi.org/10.1007/978-981-13-0329-6_10

10.1 Distribution of Saline and Hypersaline Lakes in Qinghai Province, China

Qinghai Province is located in the northern Qinghai-Tibetan Plateau, the third pole of the world. It possesses hundreds of lakes with a total of 12,856 km^2 of surface area (~ 380 lakes each with an area of >1 km^2) and a total of 2.247×10^{11} m^3 of water (Zheng et al. 2002). Most (about 85% of the total lake surface area in Qinghai Province) of the Qinghai lakes are saline (salinity < 35 g/L) and hypersaline (salinity > 35g/L) with neutral or slightly alkaline pH (Zheng et al. 2002). These saline and hypersaline lakes are inhabited by abundant and diverse microbes (Dong et al. 2006; Jiang et al. 2007; Jiang et al. 2006; Wu et al. 2006; Yang et al. 2016a; Yang et al. 2016b), which could be classified into halophilic and halotolerant classes on the basis of their salt requirement and/or tolerance (Oren 2008).

The community composition and diversity of halotolerant and halophilic microbes in the Qinghai lakes were ever extensively investigated with the use of cultivation-dependent and cultivation-independent technologies. In this chapter, summaries will be given on the major findings of microbial diversity, function, and potential ecological significance in the saline and hypersaline lakes in Qinghai Province.

10.2 Microbial Diversity in Saline and Hypersaline Lakes of Qinghai Province, China

During the past decade, cultivation-dependent and cultivation-independent microbial works were extensively performed in seven lakes of Qinghai Province (Table 10.1). Salinity of the studied lakes ranged from 0.8% to salt saturation, and pH was neutral to slightly alkaline (Table 10.1). Below summaries of the microbial research work with the use of cultivation and molecular techniques will follow.

10.2.1 Cultivation-Based Bacterial Diversity

The surface sediments from seven Qinghai lakes were employed in enrichment and isolation experiments with eight types of culture media (Table 10.2). A total of 646 strains were obtained, and they were affiliated with 4 bacterial phyla, 7 classes, 18 orders, 39 families, 89 genera, and 210 species. The four phyla were Firmicutes, Proteobacteria, Actinobacteria, and Bacteroidetes, each accounting for 43.2%, 30.9%, 18.3%, and 7.6% of the obtained strains, respectively. The Firmicutes and Proteobacteria were dominant phyla among the retrieved bacterial strains. The dominant genera consisted of *Halomonas*, *Bacillus*, *Halobacillus*, *Kocuria*, *Planococcus*, *Nesterenkonia*, *Dietzia*, *Salegentibacter*, and *Marinobacter*, and they accounted for about 60% of the obtained strains (Fig. 10.1). In addition, some genera (e.g., *Salinimicrobium*, *Halolactibacillus*, *Fontibacter*, *Mameliella*, *Trichococcus*, *Terribacillus*, *Stappia*) each consisted of only one species (strain). At the 98%

Table 10.1 Carbon fixation rates in the waters and surface sediments of the Qinghai-Tibetan lakes

Lakes	GPS location (N/E) of sampling sites	Salinity (g/L)	pH	Chla (ug/L)	DOC in lake water (mg/L)	Light-dependent C fixation in water (ug C L^{-1} h^{-1})	SD	Light-dependent C fixation in surface sediment (ug C g^{-1} h^{-1})	SD	Dark C fixation of the surface sediment (ug C g^{-1} h^{-1})	SD
Erhai lake	36°33.4′/100°43.3′	0.8	8.7	19.0	27.2	104.05	3.36	36.68	8.38	0.29	0.54
Qinghai lake	36°33.3′/100°37.5′	15.9	9.1	0.2	132.8	55.02	18.15	92.58	30.62	0.79	0.89
Gahai lake 1	36°43.8′/100°40.3′	30.0	8.5	nd	25.6	nd	nd	nd	nd	nd	nd
Tuosu lake	37°11.6′/96°53.3′	29.0	8.9	1.4	50.0	29.13	7.46	262.36	135.70	3.33	1.82
Gahai lake 2	37o08.2′/97°34.6′	71.7	7.9	1.0	31.6	50.08	1.22	25.90	14.67	0.11	0.34
Xiaochaidan lake	37°27.2′/95°30.6′	139.4	8.3	3.4	44.8	73.50	17.31	21.02	2.96	0.87	0.93
Chaka lake	36°45.1′/99°04.8′	341.9	7.6	4.9	306.0	7.90	2.59	2.20	1.48	0.00	0.06
Dabuxun lake	37 °05.1′/95°06.8	371.1	7.0	nd	nd	nd	nd	nd	nd	nd	nd

Note: *SD:* standard deviation; *nd:* not determined

Table 10.2 Media recipes employed for the cultivation work in the Qinghai lake samples

Media type	#	Recipe (/L)	Ref
Media supplemented with saccharides as carbon substrates	1	NaCl, 0.2g; glucose, 0.5g; tryptone, 0.01g; 10mL of (Na_2SO_4•$10H_2O$ 0.02g; $MgSO_4$•$7H_2O$, 0.02g; KBr 0.02, K_2HPO_4 0.01g; KH_2PO4, 0.01g; $CaCl_2$, 0.02g; $NaHCO_3$, 0.02g; KNO_3, 0.01g; H_2O, 100mL)	This study
	2	NaCl, 0.2g; yeast extract, 0.5g; tryptone, 0.01g; 10mL of (Na_2SO_4•$10H_2O$, 0.02g; $MgSO_4$•$7H_2O$, 0.02g; KBr, 0.02, K_2HPO_4, 0.01g; KH_2PO_4, 0.01g; $CaCl_2$, 0.02g; $NaHCO_3$, 0.02g; KNO_3, 0.01g; H_2O, 100mL);	This study
Media made on the basis of the physiochemical properties of the studied lakes	3	Na_2SO_4•$10H_2O$, 0.01g; K_2HPO_4, 0.01g; $CaCO_3$, 0.002g; KCl, 0.02g, $FeSO_4$, 0.002g; NaCl, 0. 2g; KNO_3, 0.02g; $MgSO_4$•$7H_2O$, 0.005g; NaF, 0.001g; KBr, 0.1g; H_3BO_3, 0.002g; peptone, 0.1g	This study
	4	$NaHCO_3$, 0.005g; $MgCl_2$, 0.05g; $CaCl_2$, 0.01g; $ZnSO_4$•$7H_2O$, 0.001g; $FeCl_3$•$6H_2O$, 0.003g; $MnCl_2$, 0.001g; $CaCl_2$, 0.002g; Na_2SO_4•$10H_2O$, 0.01g; KBr 0.1; $MgCl_2$, 0.005g; yeast extract, 1g; casein acids hydrolysate, 0.75g	This study
Published media for halophilic bacteria	5	KNO_3, 2g; $MgSO_4$•$7H_2O$, 0.05g; K_2HPO_4, 2g; $CaCl_2$, 1g; $FeSO_4$, 10mg; glucose, 10g; casein acids hydrolysate, 0.3g	Tang et al. (2007)
	6	NaC1, 2g; KCl, 3g; K_2HPO_4, 1g; KNO_3, 1g; $MgCl_2$, 5g; $MnCl_2$•$4H_2O$, 0.02g; $ZnSO_4$, 0.07g; $FeSO_4$•$7H_2O$, 0.02g; glycerol, 5g; fucose, 5g; asparagine, 0.5g; vitamin B1, 0.2mg; inositol, 0.5mg; vitamin C, 0.2mg	Guan et al. (2013)
	7	1/10 of ASW medium (SO_4, 4g, KCl, 0.68g; KBr, 0.1g; H_3BO_3, 0.025g; $MgCl_2$, 5.4g; $CaCl_2$•$2H_2O$, 1.5g; $SrCl_2$•$6H_2O$, 0.024g; $NaHCO_3$, 0.2g; Na_2HPO_4, 0.04g; NH_4Cl_2, 0.5g; NaF, 0.002g, peptone, 5.0g; yeast extract, 1.0g; pH 8.0)	Tarhriz et al. (2013)
	8	NaCl, 28.13g; KCl, 0.77g; $CaCl_2$•$2H_2O$, 1.60g; $MgCl_2$•$6H_2O$, 4.80 g; $NaHCO_3$, 0.11g; $MgSO_4$•$7H_2O$, 3.50g	This study

cutoff (Stackebrandt 2006), 79 of the obtained strains could be classified as potential new species affiliated with 44 potential new taxonomic units. So far the validly characterized microbes only occupy about 1–10% of the global population of prokaryotes (Schleifer 2004). Thus it is reasonable to sepculate that the Qinghai lakes are inhabited by abundant new taxonomic units considering that the high ratio (12%, 79 out of 646) of culturable bacteria belonged to potential new taxa.

The diversity and composition (dominant and unique taxa) of the culturable bacterial population varied among the saline/hypersaline lakes of Qinghai Province (Fig. 10.2). For example, in the freshwater EHL, 47 strains were obtained (accounted for 7.3% of the total retrieved isolates), and they were dominated by *Paenibacillus*, *Bacillus*, *Pseudomonas*, and *Dietzia*. None of the EHL isolates were related to

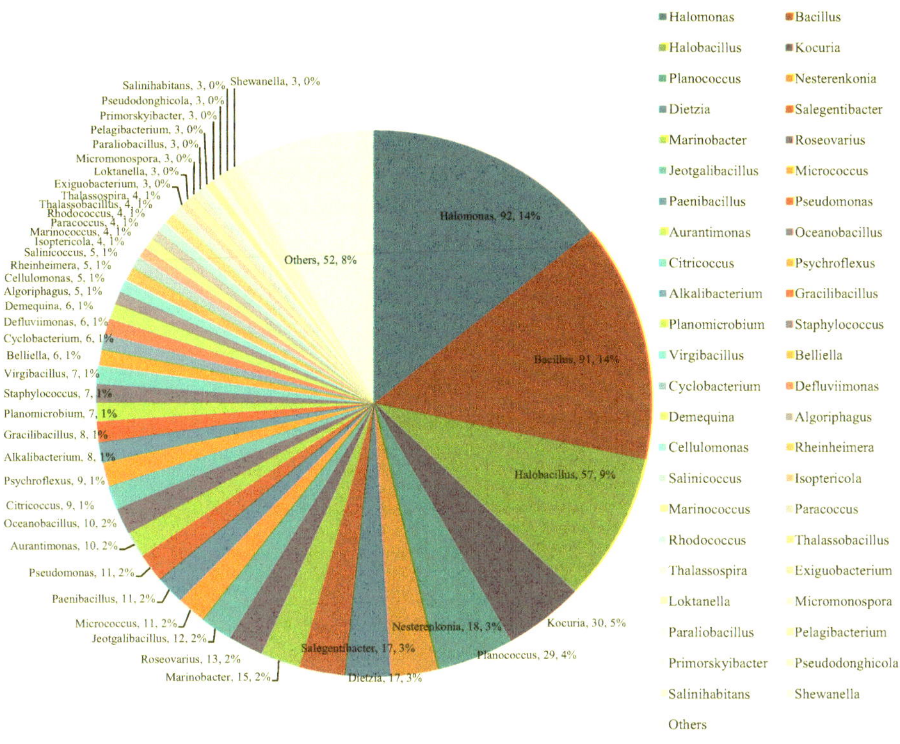

Fig. 10.1 Genus affiliation of the obtained isolates in the sediments of saline and hypersaline lakes in Qinghai Province

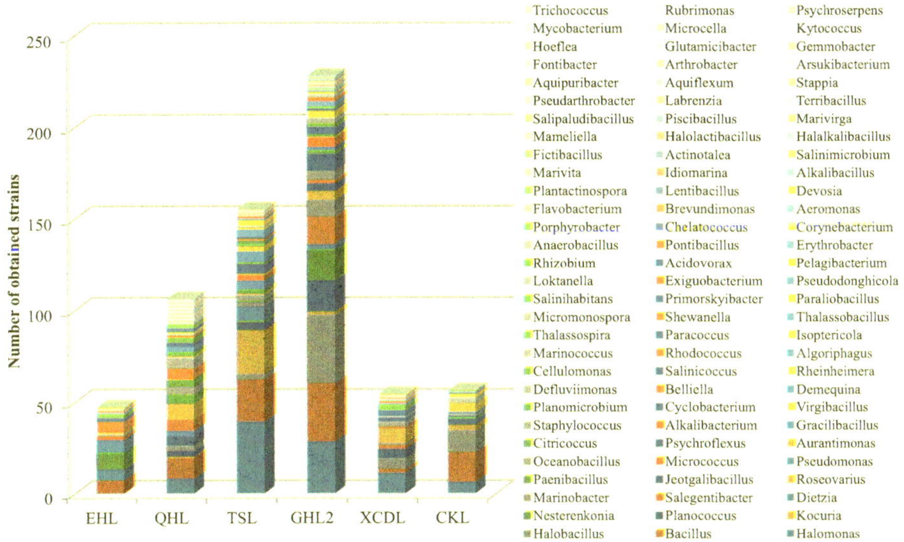

Fig. 10.2 Genus-level diversity of culturable bacteria among the saline and hypersaline lakes in Qinghai Province

halophilic taxa, such as *Halomonas*, *Halobacillus*, and *Salinicoccus*. In contrast, a total of 106 strains were retrieved from the saline QHL (salinity 1.4%) and TSL (salinity 2.6%), accounting for 16.4% of the total obtained strains, with *Bacillus*, *Aurantimonas*, *Halomonas*, and *Micrococcus* being dominant. In the TSL 155 strains were obtained (accounting for 23.9% of the total obtained strains) with *Halomonas*, *Kocuria*, *Bacillus*, and *Dietzia* being dominant. A total of 282 strains (accounting for 43.7% of total) were obtained from the mid-salinity hypersaline GHL2 and XCDL with dominance of *Halobacillus*, *Bacillus*, *Halomonas*, *Planococcus*, *Nesterenkonia*, *Roseovarius*, and *Salegentibacter*. For the salt CKL, a total of 56 strains (8.7% of total) were obtained with the dominance of *Bacillus*, *Halobacillus*, *Halomonas*, and *Virgibacillus* (Fig. 10.2). So it can be seen that halophilic species (e.g., *Halomonas*, *Halobacillus*) became dominant in the culturable bacterial population with increasing salinity of the studied lakes. The higher number and diversity of obtained isolates in the GHL and TSL suggested that bacteria in hypersaline lakes with mid-salinity might be of higher culturability.

10.2.2 Cultivation-Independent Bacterial and Archaeal Diversity

The bacterial and archaeal diversity of the waters and surface sediments of the lakes were ever reported elsewhere (Jiang et al. 2009a; Jiang et al. 2007, 2008, 2010b, 2016a, b; Liu et al. 2017; Wu et al. 2006). Generally speaking, the diversity and community composition of bacteria and archaea differed between waters and sediments of the lakes in Qinghai Province (Fig. 10.3), which may be ascribed to the different physicochemical properties between waters and sediments (Yang et al. 2016a). Bacteria in the Qinghai lakes were dominated by the Proteobacteria (including α-, β-, γ-, and δ-subgroups), which accounted for 21–61% of total

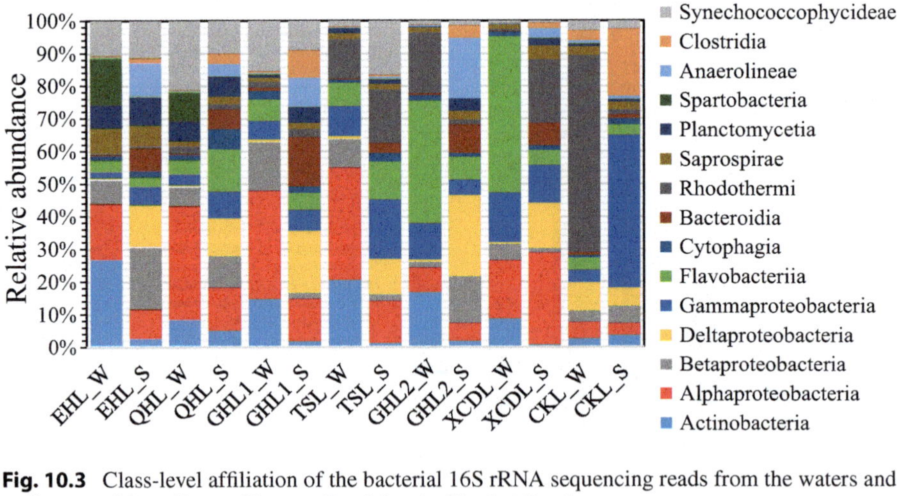

Fig. 10.3 Class-level affiliation of the bacterial 16S rRNA sequencing reads from the waters and sediments of the saline and hypersaline lakes in Qinghai Province

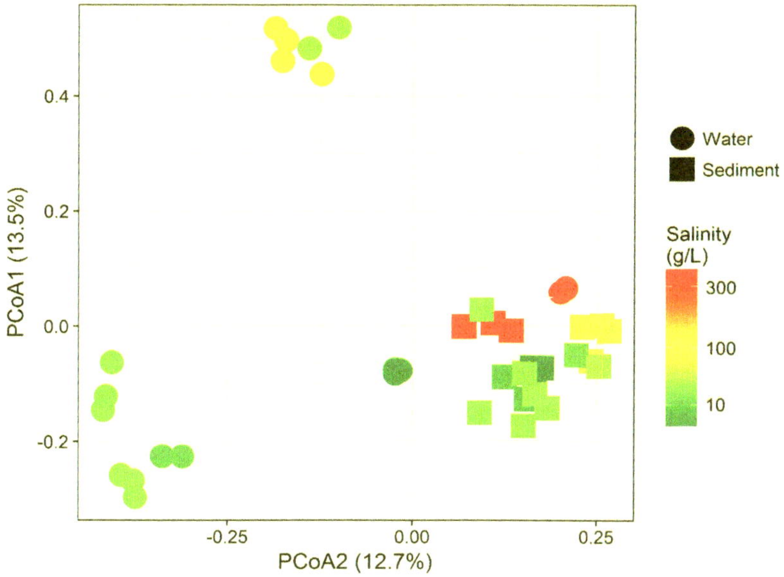

Fig. 10.4 Bray-Curtis dissimilarity-based principal coordinate analysis among the bacterial communities among of the studied samples

bacterial sequencing reads of each sample for both waters and sediments (Fig. 10.4). Such dominance of Proteobacteria was consistent with the abovementioned cultivation data. The aquatic archaeal communities mainly consisted of marine benthic groups (MBGs, such as MBG-B, MBG-C, and MBG-D)/ Bathyarchaeota and Euryarchaeota (mainly halophilic archaea) in the Qinghai lakes (Jiang et al. 2009a). The Euryarchaeota and Woesearchaeota were dominant in the sedimental archaeal communities of the Qinghai saline/hypersaline lakes, and the Euryarchaeota mainly consisted of halophilic archaea (i.e., *Halobacteria*) (Fig. 10.5).

It is notable that the Thaumarchaeota are present in all the examined saline/ hypersaline lakes. All known species affiliated with Thaumarchaeota are capable of aerobically oxidizing ammonia to nitrite, the first and rate-limiting step in nitrification (Pester et al. 2011; Stieglmeier et al. 2014). So the widespread distribution of Thaumarchaeota in the studied Qinghai lakes suggested that ammonia-oxidizing archaea could exist under a full range of salinity (from freshwater to salt saturation) (Hu et al. 2010; Jiang et al. 2009b) and they might play important roles in nitrification in saline/hypersaline environments. However, the ecological function of Thaumarchaeota in the Qinghai lakes still awaits further investigation.

In addition, the aquatic and sedimentary archaeal communities were dominated by deep-sea archaeal groups (e.g., marine benthic groups B, C, and D) in Qinghai lake (Jiang et al. 2008). These marine benthic groups of archaea were

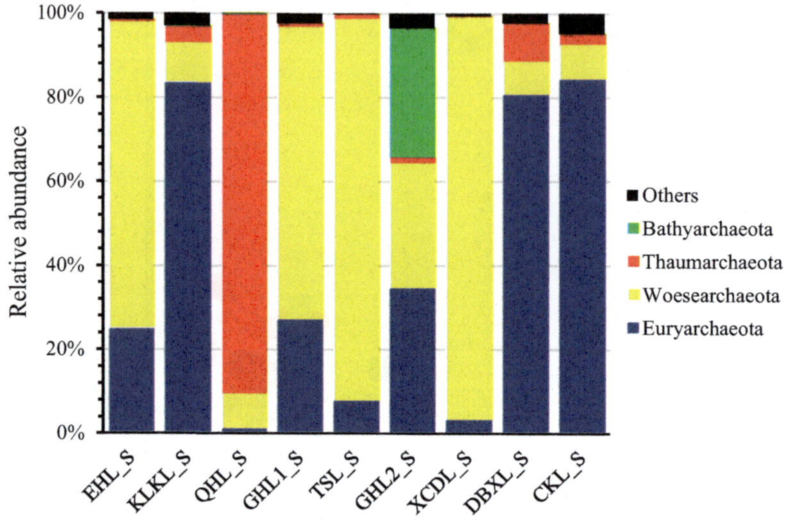

Fig. 10.5 Phylum-level affiliation of the archaeal 16S rRNA sequencing reads from the sediments of the saline and hypersaline lakes in Qinghai Province

extensively observed in deep ocean environments where methane hydrate and/or high content of methane was present and was thus thought to be involved in methane cycling (Inagaki et al. 2006). The dominance of deep-sea archaeal groups in Qinghai lake suggested that such microbial groups were not limited to ocean environments (Jiang et al. 2008). However, their metabolic functions are still to be surveyed (Fig. 10.6).

The microbial communities in the Qinghai lakes were very sensitive to salinity, and the community composition varied with salinity. For example, in low-salinity environment, bacterial and archaeal communities were dominated by halotolerant bacteria and Crenarchaeota, respectively, while, halophilic archaea (i.e., *Halobacteria*) were dominant in high-salinity environments (Crenarchaeota completely disappeared when salinity was larger than 28%) (Jiang et al. 2007, 2009a). Salinity was more important than geographical distance in shaping microbial distribution among the Qinghai lakes of different salinity (Yang et al. 2016b).

Generally microbes in natural environments do not exist individually but live in the form of community. They have to interact with each other either through synergy/symbiosis or competition and thus form a complex microbial network, which could also be influenced by environmental factors. To reveal the coexistence of microbial communities, phylogenetic molecular ecological networks (pMEN) of the Qinghai saline/hypersaline lakes were constructed according to the methods described previously (Zhou et al. 2011). The pMEN results showed that the microbial network size and connectivity varied among the saline/hypersaline lakes of different salinity. The connectivity composition of the pMEN nodes was significantly distinct (Bray-Curtis dissimilarity >0.65) among the Qinghai lakes of different salinity. The number of the pMEN nodes and connectivity decreased with the

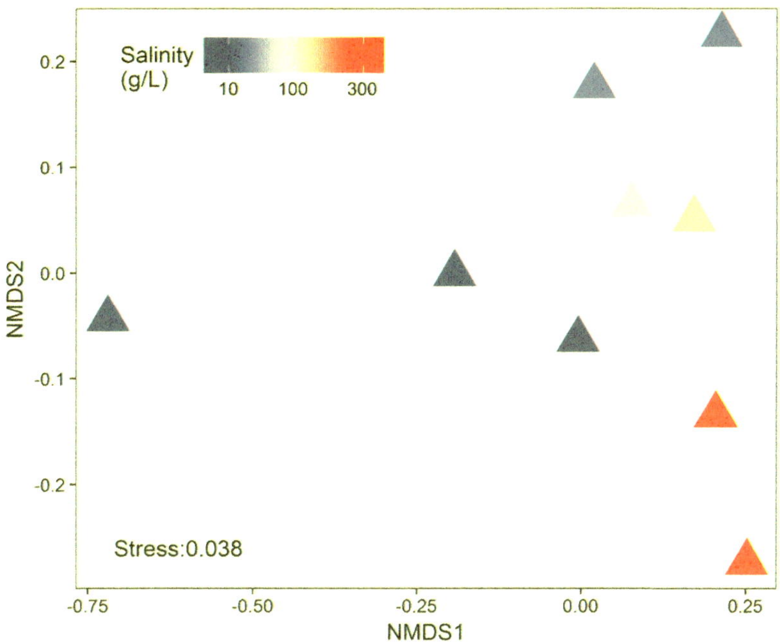

Fig. 10.6 Nonmetric dimensional scaling (NMDS) ordination according to Bray-Curtis dissimilarities among the archaeal communities of the studied lake samples

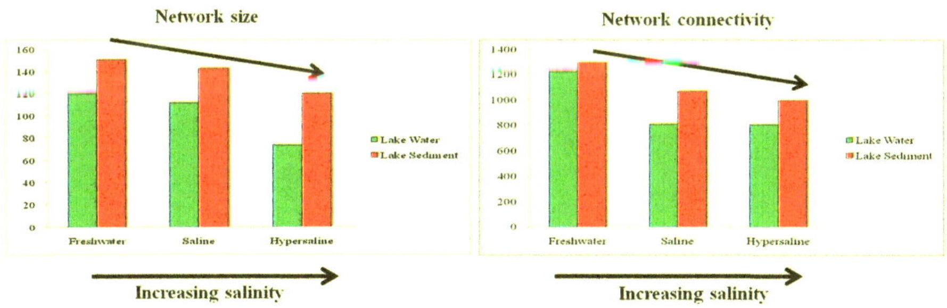

Fig. 10.7 Variation of network size and connectivity with salinity. Note: network size indicates the number of node OTUs; network connectivity indicates the number of connections between one node and other nodes in a network

increasing salinity of the Qinghai lakes (Fig. 10.7), suggestive of the decrease of the functional individuals within the microbial communities and of the interactions (synergy or competition) among individuals with increasing salinity of the lakes. This may be due to the fact that microbes require more energy to cope with osmotic pressure (Oren 2011). So it is reasonable to observe the salinity influence on the pMEN structure complexity of the microbial communities.

10.3 Response of Microbial Functional Groups to Environmental Variables and Ecological Importance

In saline and hypersaline lakes, microbes require to pay energy to maintain the osmotic pressure of their cells, and thus their functional activities are commonly restricted by salinity (Oren 2011). Our previous studies showed that most of the known microbial functional groups (e.g., carbon fixation, organic matter decomposition, ammonia oxidation, and sulfur oxidation and reduction) were widely distributed in the Qinghai lakes with a wide range of salinity and their diversity, population composition, and ecological functions were significantly influenced by salinity and/or other environmental variables (Table 10.3). For example, as the important primary producers (carbon fixation) in various aquatic ecosystems (e.g., salinity and hypersaline lakes), Cyanobacteria and eukaryotic algae were widespread in waters and surface sediments of the Qinghai saline and hypersaline lakes, and their population composition varied with salinity (Liu et al. 2016b; Yang et al. 2018). It is notable that geographic distance presented a predominant role in affecting the benthic algal distribution in comparison to aquatic alga among the Qinghai lakes, which could be ascribed to the fact that benthic alga are difficult (relative to their aquatic counterparts) to travel among lakes (Yang et al. 2018).

In order to assess the microbial carbon fixation rates in the waters and sediments of the Qinghai lakes, in situ incorporation of inorganic carbon experiments was performed with the use of radiolabeled bicarbonate ($NaH^{14}CO_3$) according to the method described elsewhere (Boyd et al. 2014). The results showed that the photosynthetic carbon-fixing rates were 7.9–104.1 μg C L^{-1} h^{-1} and 2.2–262.3 μg C g^{-1} h^{-1} for the waters and surface sediments of the Qinghai lakes, respectively. The log-transformed photosynthesis carbon-fixing rates showed significant ($R^2 = 0.651$ and 0.772 for water and sediment, respectively) linear negative correlation with salinity (Fig. 10.8). This suggested that salinity was a predominant factor influencing photosynthetic carbon fixation in the Qinghai lakes. Annual carbon fixed in the Qinghai lakes was computed from the obtained carbon fixation rate (using the average values of 53.3 μg C L^{-1} h^{-1} and 73.5 μg C g^{-1} h^{-1} for water and surface sediments, respectively) assuming that carbon fixation could take place with similar or higher rate to/than the measured season (middle September) in 9 months (210 days, ice-free season) per year and 10 h (with sunlight) per day. A total of 12,856 km^2 surface area and a total of 2.247 × 1011 m^3 lake water (Zheng et al. 2002) were employed for the calculation of total fixed carbon. The resulting total amount of fixed carbon could reach up to 0.233 Pg (10^{15} gram) in the Qinghai lakes, which is amazingly huge considering the small ratio of surface and water volume of the Qinghai lakes to that of the ocean (Moran et al. 2016). This suggested that global saline and hypersaline lakes may contribute higher carbon fixation than expected, which however awaits further investigation (Table 10.4).

In addition, the populations and functions of microbes related to organic carbon degradation also varied with lake salinity. For example, microbial communities in charge of organic carbon degradation were different at different salinity: in freshwater lakes, the genera of *Acinetobacter*, *Pseudomonas*, *Sphingomonas*, and

Table 10.3 Summary of microbial studies related to functional groups involved in biogeochemistry of C, N, and S elements

Microbial functional groups	Methods employed	Diversity and/or composition of microbial populations and their response to environmental variables	Refs
Cyanobacteria and algae	Illumina sequencing	Aquatic algal community composition and the relative abundance ratio of Cyanobacteria to algae varied with salinity The benthic algal community compositions showed significant correlation with many environmental (e.g., dissolved inorganic and organic carbon, illumination intensity, total nitrogen and phosphorus, turbidity, and water temperature) and spatial factors	Liu et al. (2016b) and Yang et al. (2018)
AAPB	Cloning	Population composition varied with salinity; the *pufM* gene diversity showed significant correlation with salinity, N and P availability, TOC and/or DOC, and HCO_3^- / CO_3^- ions; the ratio between particle-attached and free-living AAPB was positively correlated with increasing salinity	Jiang et al. (2010a). Jiang et al. (2009c), and Liu et al. (2017)
COX	Cloning	Population composition varied with salinity; the composition structure showed significant correlation with salinity, pH, and major ions	Yang et al. (2013b)
Organic matter decomposition	Cultivation, cloning and Illumina sequencing	The diversity and carbon utilization of cultivable bacteria were affected be salinity	Liu et al. (2017) and Liu et al. (2016a)
AOB and AOA	Cloning	The diversity and ratio of AOA to AOB were affected by salinity	Hu et al. (2010) and Jiang et al. (2009b)
n-DAMO/ anammox	Cloning	Nitrite-dependent anaerobic methane-oxidizing (n-DAMO) bacteria and anaerobic ammonia oxidizing (anammox) can coexist in lakes with salinity up to 84 g/L	Yang et al. (2012)
SOX	Cloning	Population composition varied with salinity	Yang et al. (2013a)
SRB	Cloning	The diversity of SRB was affected by salinity; SRB were involved in dolomite formation	Deng et al. (2010); Yang et al. (2013)c

Tolumonas were the major contributors, in contrast with genera of *Marinobacterium*, *Marinobacter*, *Nitrincola*, *Vibrio*, and *Halomonas* in saline lakes (unpublished data). In response to increasing salinity, microbial species within one genus showed decreasing capability of carbon degradation and decreasing number of organic carbon types for cell growth or energy (Liu et al. 2017); microbial species within one

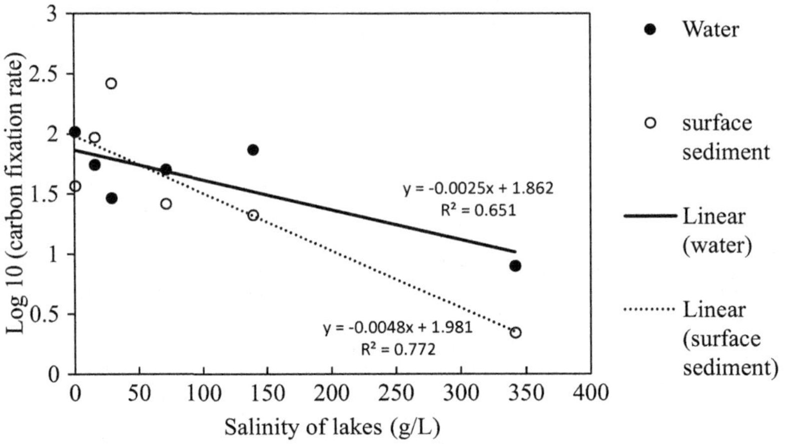

Fig. 10.8 Linear relationship between the log-transformed rates of benthic algal carbon fixation and salinity among the lake samples

genus but from different lakes showed negative correlation between the number of their carbon sources and organic carbon degradation ability and the salinity of lakes where they were from (Liu et al. 2017).

10.4 Implications for Paleoenvironments in Qinghai Province

The content of organic matter in lake sedimentary records can reflect the variations in climate and environment in ancient time (Yang et al. 2015). A 6-m-long sediment core was collected from Qinghai Lake, which could cover the sedimentary records during the past 18,000 years (Fig. 10.9). The total organic carbon (TOC) content was highest (>3.5%) in the section at the 1.65–4.20 m depth, lowest (<3%) below the 4.20 m depth, and moderate (3–3.5%) above the 1.65 m depth. Bacterial community composition varied in response to TOC content in the sediment core: in the sediments with highest TOC content, the relative of Euryarchaeota reads was higher than other samples, while OP9 was more abundant in the sediment with moderate TOC content than other samples. In contrast, Firmicutes and Proteobacteria were more abundant in the sediments with lowest TOC content than other samples (Fig. 10.10). Such microbial community variation along with geochemical conditions (i.e., TOC content) resulted from microbial adaptation to environmental evolution, which has certain implications for sedimentary environment.

In addition, some functional microbes (e.g., algae, ammonia-oxidizing archaea) require light, oxygen, and/or other specific environmental variables (e.g., salinity, nutrition level) for their metabolic activities. Their abundance, diversity, and community composition sensitively respond to variation in environmental conditions

Table 10.4 Profiles of age, total organic carbon (TOC) content, and basic mineral composition of the studied sediment samples

Sample	Age (years BP)	TOC(%)	Quartz	Calcite	Aragonite	Muscovite 2M1	Kaolinite	Chalcosine	Pyrite	Dolomite	Rankinite
QHLS009	401	4.2	*****	***	***	**	***	bdl	bdl	bdl	**
QHLS205	4606	2.0	*****	**	***	*	*	**	bdl	bdl	bdl
QHLS211	4710	2.3	*****	bdl	****	**	bdl	**	bdl	bdl	*****
QHLS217	4815	3.2	*****	***	***	*	bdl	**	bdl	bdl	bdl
QHLS223	4923	3.0	*****	*****	**	**	**	*	bdl	bdl	bdl
QHLS257	5583	1.9	*****	**	****	**	**	bdl	bdl	bdl	bdl
QHLS277	6012	2.6	*****	*****	***	*	*	**	bdl	bdl	bdl
QHLS291	6332	4.4	*****	***	***	bdl	**	bdl	bdl	bdl	bdl
QHLS311	6818	3.6	*****	*****	****	**	bdl	*	bdl	bdl	bdl
QHLS415	10018	3.9	****	*****	**	*	bdl	*	bdl	bdl	bdl
QHLS417	10093	4.4	nm	nm	nm	nm	nm	nm	nm	nm	nm
QHLS421	10243	3.2	nm	nm	nm	nm	nm	nm	nm	nm	nm
QHLS425	10396	3.1	****	**	*	*	bdl		*	*****	bdl
QHLS427	10473	3.0	nm	nm	nm	nm	nm	nm	nm	nm	nm
QHLS433	10708	2.1	*****	**	**	*	*	*	bdl	bdl	bdl
QHLS439	10949	3.1	nm	nm	nm	nm	nm	nm	nm	nm	nm
QHLS459	11789	1.9	nm	nm	nm	nm	nm	nm	nm	nm	nm
QHLS485	12980	nd	*****	**	*	*	*	*	bdl	bdl	bdl

BP before present, *nd* not determined, *bdl* below detection limit

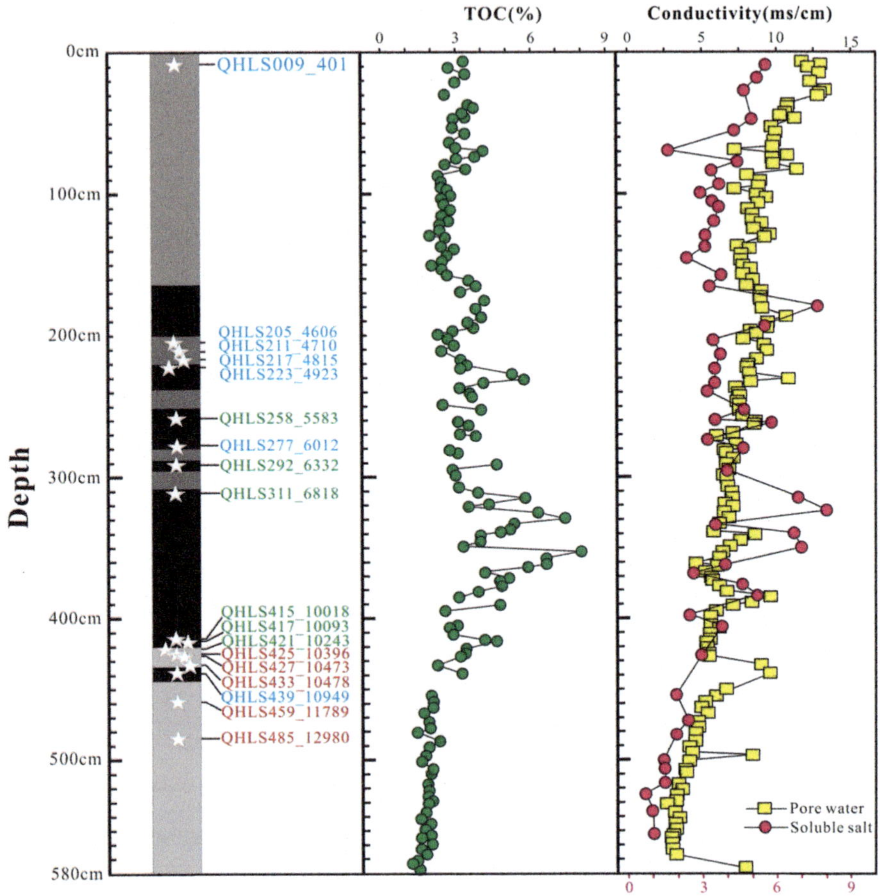

Fig. 10.9 Profiles of TOC content and conductivity of the sediment core collected from Qinghai Lake. Note: the stars indicate the subsamples for DNA work. The subsamples were coded as follows for the example of QHLS205_4606: QHLS, Qinghai lake sediment; 205, the depth of the sample in the unit of centimeters (cm); 4606, the age of the subsample dated with ^{14}C isotope

(e.g., temperature, salinity, nutrition level). Metabolic activities of such functional microbes will end once they are buried in dark and anaerobic sediments. However, their DNA and/or lipids (e.g., glycerol dialkyl glycerol tetraethers, commonly abbreviated as GDGTs, core membrane lipids from bacteria and archaea) could be preserved for some period of time and therefore can be employed to reconstruct paleoenvironmental conditions such as temperature (Wang et al. 2015; Wu et al. 2013), nutrition level (Yang et al. 2015), ancient water level (Wang et al. 2014, 2015), and water salinity (Wang et al. 2013) of ancient lakes.

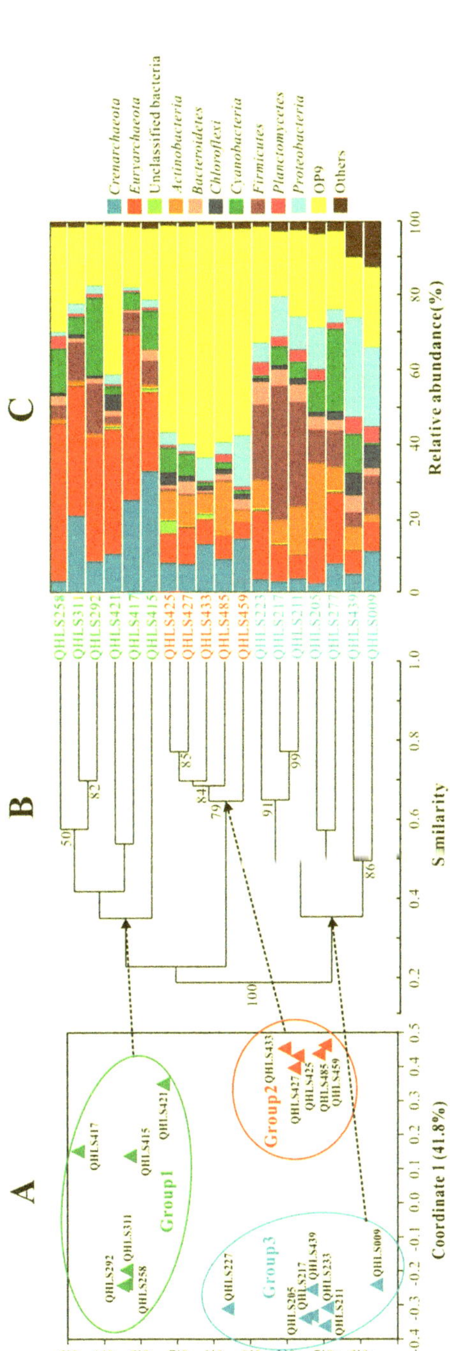

Fig. 10.10 A panel: canonical correlation analysis (CCA) of bacterial and archaeal distributions among the subsamples. Clustering (B panel) and relative abundance (C panel) of major phylogenetic groups as determined by 16S rRNA Illumina sequencing

10.5 Closing Remarks and Future Perspectives

The results summarized in this chapter suggest that the Qinghai saline and hypersaline lakes are inhabited by abundant and diverse halophilic and halotolerant microorganisms, among which a large portion belong to potential new taxonomic units and thus await further explorations for new taxa or metabolic functions. These microbes of different ecological functions are enigmatically involved in the biogeochemical cycling of carbon, nitrogen and sulfur elements and are of great ecological importance. State-of-the-art molecular techniques (e.g., single-cell genomics, metagenomics, metatranscriptomics, stable isotope probing, proteomics) should be integrated with traditional cultivation studies and stoichiometric biogeochemistry in future microbial studies, which could help understand the biogeochemistry in saline/hypersaline lakes.

Acknowledgments This research was supported by grants from the National Natural Science Foundation of China (Grant Nos. 41422208, 41521001, 41672331, 41672337, and 41602346), the 111 Program (State Administration of Foreign Experts Affairs & the Ministry of Education of China, grant B18049), the Fundamental Research Funds for National University (China University of Geosciences—Wuhan), and State Key Laboratory of Biogeology and Environmental Geology.

References

Boyd ES, Hamilton TL, Havig JR, Skidmore ML, Shock EL (2014) Chemolithotrophic primary production in a subglacial ecosystem. Appl Environ Microbiol 80:6146–6153

Deng S, Dong H, Lv G, Jiang H, Yu B, Bishop ME (2010) Microbial dolomite precipitation using sulfate reducing and halophilic bacteria: results from Qinghai Lake, Tibetan Plateau, NW China. Chem Geol 278:151–159

Dong H, Zhang G, Jiang H, Yu B, Chapman L, Lucas C, Fields M (2006) Microbial diversity in sediments of saline Qinghai Lake, China: linking geochemical controls to microbial ecology. Microb Ecol 51:65–82

Guan T, Teng Y, Che Z, Zhang L, Zhang X, Xing Y, Liu P, Zhang L (2013) Comparison of isolation media for Actinobacteria from LopNur salt lake. Biotechnology 23:56–60 (*in Chinese with English abstract*)

Hu A, Yao T, Jiao N, Liu Y, Yang ZAO, Liu X (2010) Community structures of ammonia-oxidising archaea and bacteria in high-altitude lakes on the Tibetan Plateau. Freshwat Biol 55:2375–2390

Inagaki F, Nunoura T, Nakagawa S, Teske A, Lever M, Lauer A, Suzuki M, Takai K, Delwiche M, Colwell FS, Nealson KH, Horikoshi K, D'Hondt S, Jorgensen BB (2006) Biogeographical distribution and diversity of microbes in methane hydrate-bearing deep marine sediments on the Pacific Ocean margin. Proc Natl Acad Sci U S A 103:2815–2820

Jiang H, Deng S, Huang Q, Dong H, Yu B (2010a) Response of aerobic anoxygenic phototrophic bacterial diversity to environment conditions in saline lakes and Daotang River on the Tibetan Plateau, NW China. Geomicrobiol J 27:400–408

Jiang H, Dong H, Deng S, Yu B, Huang Q, Wu Q (2009a) Response of archaeal community structure to environmental changes in lakes on the Tibetan Plateau, northwestern China. Geomicrobiol J 26:289–297

Jiang H, Dong H, Yu B, Liu X, Li Y, Ji S, Zhang CL (2007) Microbial response to salinity change in Lake Chaka, a hypersaline lake on Tibetan plateau. Environ Microbiol 9:2603–2621

Jiang H, Dong H, Yu B, Lv G, Deng S, Berzins N, Dai M (2009b) Diversity and abundance of ammonia-oxidizing archaea and bacteria in Qinghai Lake, northwestern China. Geomicrobiol J 26:199–211

Jiang H, Dong H, Yu B, Lv G, Deng S, Wu Y, Dai M, Jiao N (2009c) Abundance and diversity of aerobic anoxygenic phototrophic bacteria in saline lakes on the Tibetan plateau. FEMS Microbiol Ecol 67:268–278

Jiang H, Dong H, Yu B, Ye Q, Shen J, Rowe H, Zhang C (2008) Dominance of putative marine benthic Archaea in Qinghai Lake, northwestern China. Environ Microbiol 10:2355–2367

Jiang H, Dong H, Zhang G, Yu B, Chapman LR, Fields MW (2006) Microbial diversity in water and sediment of Lake Chaka, an athalassohaline lake in northwestern China. Appl Environ Microbiol 72:3832–3845

Jiang H, Huang Q, Deng S, Dong H, Yu B (2010b) Planktonic actinobacterial diversity along a salinity gradient of a river and five lakes on the Tibetan Plateau. Extremophiles 14:367–376

Liu W, Jiang H, Yang J, Wu G (2017) Gammaproteobacterial diversity and carbon utilization in response to salinity in the lakes on the Qinghai-Tibetan Plateau. Geomicrobiol J. https://doi.org/10.1080/01490451.01492017.01378951

Liu W, Yang J, Wu G, Zhang G, Jiang H (2016a) Diversity of cultivable bacteria based on different carbon sources in the sediments of lakes in northern Qing-Tibetan Plateau. Journal Salt Lake Research 24:92–102 (*in Chinese but with English abstract*)

Liu X, Hou W, Dong H, Wang S, Jiang H, Wu G, Yang J, Li G (2016b) Distribution and diversity of Cyanobacteria and eukaryotic algae in Qinghai–Tibetan lakes. Geomicrobiol J 33:860–869

Moran MA, Kujawinski EB, Stubbins A, Fatland R, Aluwihare LI, Buchan A, Crump BC, Dorrestein PC, Dyhrman ST, Hess NJ, Howe B, Longnecker K, Medeiros PM, Niggemann J, Obernosterer I, Repeta DJ, Waldbauer JR (2016) Deciphering Ocean carbon in a changing world. Proc Natl Acad Sci U S A 113:3143–3151

Oren A (2008) Microbial life at high salt concentrations: phylogenetic and metabolic diversity. Saline Systems 4:2–2

Oren A (2011) Thermodynamic limits to microbial life at high salt concentrations. Environ Microbiol 13:1908–1923

Pester M, Schleper C, Wagner M (2011) The Thaumarchaeota: an emerging view of their phylogeny and ecophysiology. Curr Opin Microbiol 14:300–306

Schleifer KH (2004) Microbial diversity: facts, problems and prospects. Syst Appl Microbiol 27:3–9

Stackebrandt E (2006) Taxonomic parameters revisited: tarnished gold standards. Microbiol Today 6:152–155

Stieglmeier M, Alves RJ, Schleper C (2014) The phylum Thaumarchaeota., in: Rosenberg E, DeLong EF, Lory S, Stackebrandt E, Thompson F (Eds.), The Prokaryotes. Springer, pp. 347–362

Tang S, Jiang Y, Zhi X, Lou K, Li W, Xu L (2007) Isolation methods for halophilic and halotolerant actinomycetes. Microbiology China:390–392

Tarhriz V, Thiel V, Nematzadeh G, Hejazi MA, Imhoff JF, Hejazi MS (2013) *Tabrizicola aquatica* gen. nov. sp. nov., a novel alphaproteobacterium isolated from Qurugöl Lake nearby Tabriz city, Iran. Antonie Van Leeuwenhoek 104:1205–1215

Wang H, Dong H, Zhang CL, Jiang H, Liu Z, Zhao M, Liu W (2015) Deglacial and Holocene archaeal lipid-inferred paleohydrology and paleotemperature history of Lake Qinghai, northeastern Qinghai–Tibetan Plateau. Quat Res 83:116–126

Wang H, Dong H, Zhang CL, Jiang H, Zhao M, Liu Z, Lai Z, Liu W (2014) Water depth affecting thaumarchaeol production in Lake Qinghai, northeastern Qinghai-tibetan plateau: implications for paleo lake levels and paleoclimate. Chem Geol 368:76–84

Wang H, Liu W, Zhang CL, Jiang H, Dong H, Lu H, Wang J (2013) Assessing the ratio of archaeol to caldarchaeol as a salinity proxy in highland lakes on the northeastern Qinghai–Tibetan Plateau. Org Geochem 54:69–77

Wu QL, Zwart G, Schauer M, Agterveld MPK, Bahn MW (2006) Bacterioplankton community composition along a salinity gradient of sixteen high-mountain lakes located on the Tibetan Plateau, China. Appl Environ Microbiol 72:5478–5485

Wu X, Dong H, Zhang CL, Liu X, Hou W, Zhang J, Jiang H (2013) Evaluation of glycerol dialkyl glycerol tetraether proxies for reconstruction of the paleo-environment on the Qinghai-Tibetan Plateau. Org Geochem 61:45–56

Yang J, Jiang H, Dong H, Hou W, Li G, Wu G (2015) Sedimentary archaeal amoA gene abundance reflects historic nutrient level and salinity fluctuations in Qinghai Lake, Tibetan Plateau. Sci Rep 5:18071

Yang J, Jiang H, Dong H, Wu G, Hou W, Zhao W, Sun Y, Lai Z (2013a) Abundance and diversity of sulfur-oxidizing bacteria along a salinity gradient in four Qinghai-Tibetan lakes, China. Geomicrobiol J 30:851–860

Yang J, Jiang H, Dong H, Wu G, Hou W, Zhao W, Sun Y, Lai Z (2013b) Diversity of carbon monoxide-oxidizing bacteria in five lakes on the Qinghai-Tibet Plateau, China. Geomicrobiol J 30:758–767

Yang J, Jiang H, Liu W, Wang B (2018) Benthic algal community structures and their response to geographic distance and environmental variables in the Qinghai-Tibetan lakes with different salinity. Front Microbiol 9:578

Yang J, Jiang H, Sun Y, Wu G, Hou W, Lai Z, Dong H (2013c) Abundance and diversity of sulfate-reducing bacteria in Qinghai-Tibetan lakes. Journal of Salt Lake Research 21:7–13 (*In Chinese but with English abstract*)

Yang J, Jiang H, Wu G, Hou W, Sun Y, Lai Z, Dong H (2012) Co-occurrence of nitrite-dependent anaerobic methane oxidizing and anaerobic ammonia oxidizing bacteria in two Qinghai-Tibetan saline lakes. Frontiers of Earth Science 6:383–391

Yang J, Jiang H, Wu G, Liu W, Zhang G (2016a) Distinct factors shape aquatic and sedimentary microbial community structures in the lakes of Western China. Front Microbiol 7:1782

Yang J, Ma L, Jiang H, Wu G, Dong H (2016b) Salinity shapes microbial diversity and community structure in surface sediments of the Qinghai-Tibetan Lakes. Sci Rep, in press

Zheng X, Zhang M, Xu X, Li B (2002) Salt lakes in China. Beijing Science Press, Beijing

Zhou J, Deng Y, Luo F, He Z, Yang Y (2011) Phylogenetic molecular ecological network of soil microbial communities in response to elevated CO_2. MBio 2:e00122–e00111

Soil Salinity and Microbes: Diversity, Ecology, and Biotechnological Potential

11

Dilfuza Egamberdieva, Kakhramon Davranov, and Stephan Wirth

Abstract

Soil salinity is a severe problem worldwide to crop production and ecosystems because it disturbed soil biological processes and microbial functioning. The adverse effects of salt stress on soil microbial activity and populations have been studied extensively. The understanding of the adaptive properties of soil microbes makes it possible to use them in restoring abandoned salt-affected lands. Salt-tolerant microorganisms are essential components in carbon, nitrogen, and phosphorus cycling. There is evidence that these microbes play an important role in soil biochemical processes and nutrient cycle, improving plant stress tolerance and nutrient acquisition through their ability to fix atmospheric nitrogen, solubilize phosphate or by enhancing decomposition of plant residues. These stress tolerant microbes have a great biotechnological potential to improve productivity and plant health on saline soils, or under arid conditions.

Keywords

Salinity · Drought · Soil microbes · Diversity · Nutrient cycle · Plant nutrients

D. Egamberdieva (✉)
Leibniz Centre for Agricultural Landscape Research, Müncheberg, Germany

Faculty of Biology, National University of Uzbekistan, Tashkent, Uzbekistan

K. Davranov
Faculty of Biology, National University of Uzbekistan, Tashkent, Uzbekistan

S. Wirth
Leibniz Centre for Agricultural Landscape Research, Müncheberg, Germany

D. Egamberdieva et al. (eds.), *Extremophiles in Eurasian Ecosystems: Ecology, Diversity, and Applications*, Microorganisms for Sustainability 8,
https://doi.org/10.1007/978-981-13-0329-6_11

11.1 Introduction

Climate change has resulted in the reduction of available land resources used for agricultural production, thus leading to pose a potential threat to global food security (El-Beltagy and Madkour 2012; Mahalingam 2015). Among abiotic factors, soil salinity is increasing annually, and more than 50% of lands devoted to agriculture are salinated (Jadhav et al. 2010; FAO 2010).

According to previous reports, 77 million hectares of land used for crop cultivation were affected by salinity (Brady and Well 2008). Salinization is common in arid and semiarid areas where rainfall is low and salt contents increase in top soil (Egamberdieva et al. 2008; Bui 2013). Furthermore, inappropriate application of chemical fertilizers, irrigation of agricultural lands, and use of saline groundwater have resulted in secondary salinization (de Wit et al. 2011). There are two types of salinity, primary and secondary salinization. Primary salinity results from the excessive weathering of rocks, intrusion of salt water from the sea along the coast, or improper drainage. Secondary salinity results from the activities of humans, which include excess use of saline water for irrigation of agricultural lands or inappropriate application of chemical fertilizers (Lakhdar et al. 2009). For example, the Aral Sea, once the fourth largest inland body of water in the world, has been steadily decreasing in size, because of cotton production accompanied by intense irrigation of arable fields (Egamberdiyeva et al. 2007). The dry seabed exposed to weathering further increased soil salinization and desertification around the region (Ragab and Prudhomme 2002). The area of saline irrigated lands in Uzbekistan amounted to 2399,7 th/ha, including low-saline lands, 1317,7 th/ha; medium-saline lands, 665,2 th/ha; and strong-saline lands, 416,5 th/ha. Saline soils with high availability of soluble salts affect plant growth at various growth stages leading to yield loss and also impact ion compositions at maturity. Higher soil salinity hampers the growth of several crop plants mainly because of reduction in the osmotic potential of the soil solution and specific ion effects leading to nutritional imbalance (Ahmad 2010). The exposure of plants to salt stress brings alterations in several major physiological plant processes like photosynthesis, protein synthesis, respiration, water uptake, as well as energy and lipid metabolisms (Abd_Allah et al. 2017).

Soil salinity not only inhibits plant growth, or disturbs physiological properties, but also affects the microbiome which plays an important role for soil productivity (Hashem et al. 2015), especially the maintenance of food webs and nutrient turnover. Soil microbial diversity is largely determined by the plant rhizosphere which is considered as hot spot for microbial colonization and proliferation due to nutrient rich exudates (Renella et al. 2006). Investigating the microbial population, physiology, ecology, and functions in arid saline environments is important in developing understanding that new agricultural technologies will have upon plant-microbe interactions, nutrient transformations, and plant succession. Several abiotic factors including pH, soil nutrients, and moisture have been proposed as main factors limiting microbial activity and abundance (Lauber et al. 2009; Egamberdieva et al. 2010a, 2011). The soil microbial community is the most active in upper soil layers and therefore strongly affected by the accumulation of soluble salts (Kapur et al. 2010). Thus, negative effects of salinity on soil microbial activity, including microbial respiration or enzyme activities, were reported in many studies (Yuan et al. 2007; Setia et al. 2010). There are several studies reporting negative correlations between soil microorganisms and salt content of soils (Mavi et al. 2012; Luo et al. 2017). The reduction of the diversity of microbiota in salt-affected soils was reported previously, e.g., by Lijuan et al. (2017) who observed *Planctomycetes*, *Bacteroidetes*, *Proteobacteria*, *Actinobacteria*, *Chloroflexi*, *Gemmatimonadetes*, and *Acidobacteria* as dominant groups in salt-affected soils of Gansu Province, China. The understanding of the adaptive properties of soil microbes makes it possible to use them in restoring abandoned salinated lands.

The soil microbes have adaptive abilities to survive in harsh environmental conditions, including drought, salinity, and extreme temperatures (Turan et al. 2014; Egamberdieva et al. 2016, 2017c). They produce several specific metabolites which help microbes to withstand abiotic stresses (Sorty et al. 2016). Furthermore, many studies have reported that salt-tolerant bacteria enhance plant growth, physiological properties, and crop yield (Singh and Jha 2016; Cho et al. 2015; Ahmad et al. 2015). The mode of action of PGPR that have beneficial effects on plant growth and physiological processes includes nitrogen fixation, modulation of antioxidant enzymes, antimicrobial properties, and production of osmotic compounds, cell wall degrading enzymes, ACC deaminase, hydrogen cyanide, or siderophores solubilizing minerals (Asaf et al. 2017; Hashem et al. 2016; Egamberdieva 2009; Egamberdieva et al. 2017a; Shahzad et al. 2016). In this review, we discuss the diversity, ecology, physiology, and biotechnological potential of microbes from saline environments.

11.2 Soil Microbes and Microbial Diversity

11.2.1 Microbial Populations

A huge multitude of different microorganism is commonly found in the soil including bacteria, fungi, actinomycetes, protozoa, and algae (Egamberdieva 2012; Mendes et al. 2013). The most common type of soil microbes is bacteria that have great effect on many processes such as organic matter oxidation, hydrolysis, and degradation, and these in turn determine the natural cycles of carbon, nitrogen, phosphorus, and other elements (Renella et al. 2006; Gougoulias et al. 2014). In addition, they help the roots take up nutrients, recycle nutrient elements into the ecosystems from atmosphere or mineral reserves, break down detritus, release mineral elements in soluble forms, and protect the roots from pathogens (Egamberdieva et al. 2010b; Forchetti et al. 2010). Extreme conditions may reduce the activity of some microorganisms but will stimulate that of others, and in many extreme environments, microorganisms are the crucial contributors to nutrient cycling. Research has shown that by measuring the size and activity of soil microbial communities, we can assess soil degradation and the effects of management designed to reverse the degradation. The salt-tolerant microorganisms are important components of carbon, nitrogen, and phosphorus cycling. In an earlier report, Belimov et al. (1995) described such soil microorganisms as oligonitrophilic, ammonifying, nitrifying, and denitrifying bacteria that play an important role in the transformation of mineral nutrients.

The extreme environment is also characterized by low N concentration, which is an essential element to all forms of life. Nitrogen fixation is carried out only by prokaryotes which may be symbiotic or free living, and oligonitrophilic bacteria have the ability to fix N_2. In nitrogen-poor soils under extreme environmental conditions, they are widely distributed compared to other soil. In salt-affected soils, the nitrogen fixation capability of bacteria is suppressed due to high pH (Butale et al. 2010; Kouas et al. 2010).

Arid calcareous soils of Uzbekistan are characterized by very low nitrogen content, where oligonitrophilic bacteria are dominant. The highest density of oligonitrophilic bacteria was observed under alfalfa and cotton during summer and the lowest in winter (Fig. 11.1). The salt-affected soils of Uzbekistan are nitrogen deficient soils, and thus the abundance of oligonitrophilic bacterial strains are higher than other microbial populations. Bacterial density was the lowest in winter and increased gradually through spring and autumn. The differences of microbial populations varied with soil depth with higher amounts in the 0–10 and 10–20 cm soil depth than at the 20–30 cm depth regardless of plant type. Moradi et al. (2011) studied free-living diazotrophs and total bacterial populations in saline soils and observed a negative effect of salinity on total bacterial population. In their study, the free-living diazotrophic and total heterotrophic bacterial populations were significantly higher in non-salinated soil compared to salt-affected soil. Among the soil microbes, *Corynebacterium* and *Enterobacter* genera were moderate halophiles, and *Bacillus* and *Agrobacterium* genera were extreme halophiles.

The nitrogen mineralization and immobilization by microbes are one of the key components of the soil nitrogen cycle. The organic nitrogenous compounds are decomposed by microbial enzymes to form ammonia (NH_3), and thus the amount of plant available N in soils is increased through those processes (Luo et al. 2017). Ammonifying bacterial populations were predominant groups in soil attached to plant roots compared to soil without plant cultivation. Marked effects were found to have taken place on the ammonifying bacteria populations under tomato and wheat grown in saline arid soils (Fig. 11.2). Luo et al. (2017) observed a decreased population of ammonifying bacteria, nitrifying bacteria, and denitrifying bacteria under saline alkaline soil of China. The differential effects of plant species on soil microbial communities were related to root exudates which contain different nutrients (Marschner et al. 2004; Bertin et al. 2003). The highest density of ammonifying bacteria in soil under various crops was observed during spring and summer which indicate the importance of plantation of salt-affected soils. A year which was characterized with hot summer and less precipitation affected soil microbial population negatively (Wallenstein et al. 2010).

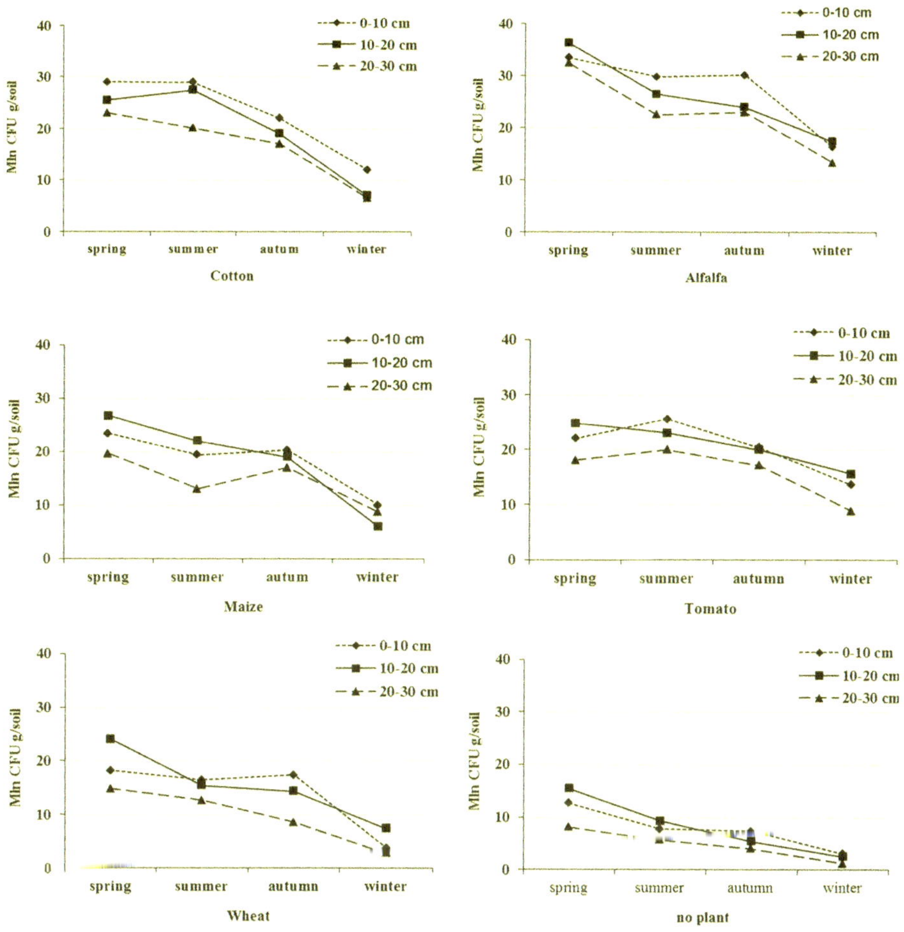

Fig. 11.1 The number (CFU) of oligonitrophilic bacteria under cotton, alfalfa, maize, tomato, wheat, and fallow. The samples were taken in spring, summer, autumn, and winter (0–30 cm soil depth)

In nitrification process, NH_3 or $NH_4{}^+$ is oxidized to nitrite (NO_2^-) and then to nitrate (NO_3^-). NO_3^- is readily taken up by plants, and because of its negative charge, it moves freely through the soil. Nitrification can also prevent nitrogen losses in soils where NH_3 volatilization is a major factor (Bock et al. 1986). The oxidation of NH_3 and NH_4^+ to NO_2^- is carried out in most soils by species of *Nitrosomonas,* while the oxidation of NO_2^- to NO3 $^-$ is carried out by *Nitrobacter* (Bock et al. 1986). Luo et al. (2017) reported a decrease of nitrifying bacteria in salt-affected rhizosphere soil of cotton. A significant inhibition of soil microbial population was also found by Yuan et al. (2007).

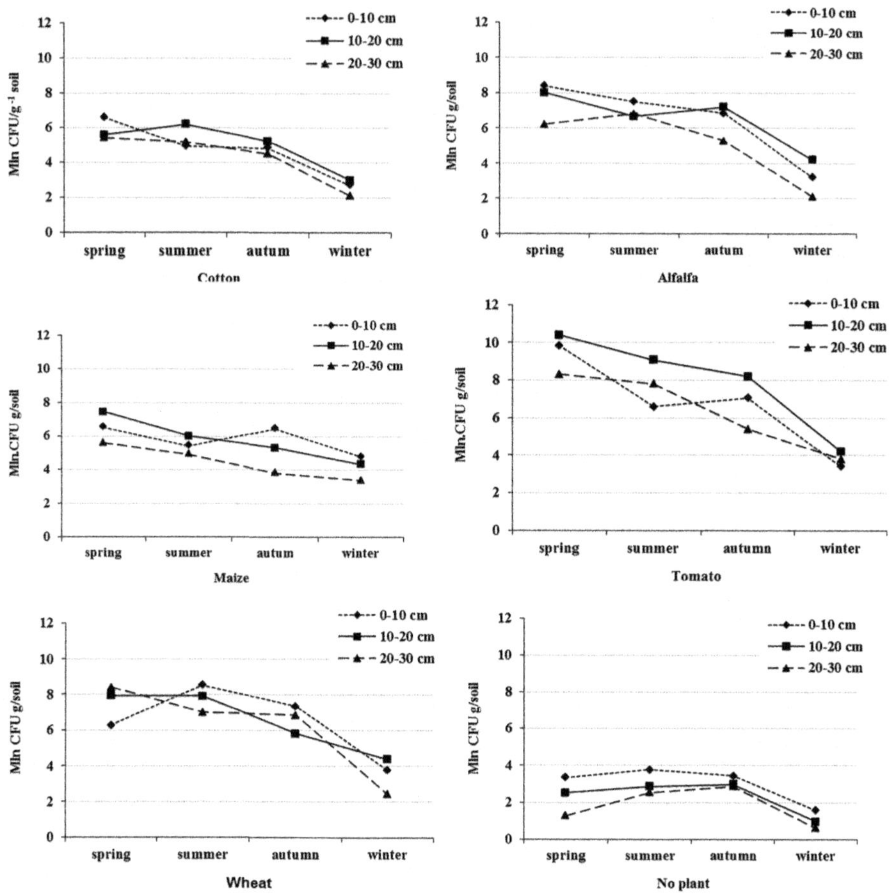

Fig. 11.2 The number (CFU) of ammonifying bacteria under cotton, alfalfa, maize, tomato, wheat, and fallow. The samples were taken in spring, summer, autumn, and winter (0–30 cm soil depth)

11.2.2 Microbial Diversity

External abiotic factors, such as soil type, soil temperature and moisture, or salinity, induce effects on soil microbial communities and diversity (Mishra et al. 2012). Canfora et al. (2014) studied microbial diversity of a natural saline soil located in Sicily and found diverse bacterial phyla. Dominant phyla were *Acidobacteria, Actinobacteria, Bacteroidetes, BRC1, Chlorobi, Chloroflexi, Cyanobacteria, Deferribacteres, Firmicutes, Gemmatimonadates, Nitrospira, Planctomycetes, Proteobacteria, Spirochaetes, Tenericutes, Verrucomicrobia*, and *WS3*. In another study, Yang et al. (2016) found in low-saline soil higher percentages of *Proteobacteria, Acidobacteria, Chloroflexi*, and *Gemmatimonadetes* and lower percentages of

Firmicutes and *Verrucomicrobia*, whereas high saline soil contained mostly *Loktanella* and *Kordiimonas*.

Research on salt-affected soils in India showed a presence of *Pseudomonas* species belonging to *P. aeruginosa*, *P. pseudoalcaligenes*, *P. alcaligenes*, *P. fluorescens*, *P. putida*, *P. stutzeri*, *P. mendocina*, *P. mallei*, *and P. diminuta* (Rangarajan et al. 2002). In salinated soil of Uzbekistan, several salt-tolerant *Pseudomonas* strains were identified as *Pseudomonas putida*, *P. extremorientalis*, *P. chlororaphis*, and *P. aurantiaca*. The isolates tolerated up to 5% of NaCl (Egamberdieva and Kucharova 2009). In these salinated Uzbekistan soil, other species were found including *Mycobacterium phlei* and *Mycoplana bullata* (Egamberdiyeva and Höflich 2003); however, a majority of salt-tolerant bacteria belong to the genus *Bacillus* which is well adapted in arid saline soils because of their capability to produce endospores (Egamberdiyeva and Höflich 2003). Among them *B. laevolacticus*, *B. amyloliquefaciens*, *B. polymyxa*, *B. subtilis*, and *B cereus* were isolated from cotton, tomato, and wheat grown (Egamberdiyeva 2005). Panosyan et al. (2018) reported about the bacterial community composition in saline-alkaline soil located in Ararat Plain (Armenia) and described the dominance of *Firmicutes* populations. The salt-tolerant bacterial diversity showed that 41.2% belong to *Halobacillus*, 23.5% to *Piscibacillus*, 23.5% to *Bacillus*, and 11.8% to *Virgibacillus*. Other researchers have isolated a number of microbial strains from *Sesbania cannabina* grown in saline environments of Rudong County, China, and they are belonging to the genera *Ensifer*, *Agrobacterium*, *Neorhizobium*, *and Rhizobium*. Among them *Ensifer meliloti* and *Neorhizobium huautlense* were the dominant species (Li et al. 2016b). Moreover, Mishra et al. (2012) observed diverse bacterial genera in Bhitarkanika mangrove soil including *Azotobacter*, *Bacillus*, *Desulfotomaculum*, *Desulfovibrio*, *Desulfomonas*, *Klebsiella*, *Methylococcus*, *Micrococcus*, and *Pseudomonas*.

In other studies, the most predominant phylum in saline soils of India was *Proteobacteria*, which included *Ectothiorhodospiraceae*, *Azospira*, *Stenotrophomonas*, *Thiobacillus*, *Levilinea*, *Desulfobacteraceae*, *Thioalkalivibrio*, and *Rhodocylales* (Sah et al. 2014). Li et al. (2016a) studied the bacterial communities in salt-alkali soils under mulberry and soybean. They found dominant taxonomic groups to be *Proteobacteria*, *Acidobacteria*, *Actinobacteria*, *Chloroflexi*, *Bacteroidetes*, *Planctomycetes*, and *Gemmatimonadetes*.

11.3 Beneficial Microbes

Salt-affected soils are a source for diverse microorganisms that adapted in extreme environmental conditions (Sorty et al. 2016; Egamberdieva et al. 2017b). The plant species grown in salt-affected soils are also associated with a microbial community which competes for nutrient and niches in the rhizosphere (Hashem et al. 2015, 2017). The plant development is sensitive to salt stress and results in an inhibition of the root system growth, nutrient absorption, and physiological processes (Hashem et al. 2016; Parray et al. 2016). In addition, the amount of reactive oxygen species (ROS) which lead to oxidative stress in plants (Ahmad et al. 2015) is increased.

The use of beneficial microbes to control plant diseases and to stimulate plant growth has been considered to be a viable alternative and environmentally friendly method (Egamberdieva et al. 2011). The bacteria that exert beneficial effects on plant growth and physiology are known as plant growth-promoting rhizobacteria (PGPR) and have been reported widely and are used for many crops (Lugtenberg and Kamilova 2009). There is some evidence that soil microbes have the potential to help alleviate abiotic stresses and improve plant growth and physiological properties under extreme environmental conditions (Table 11.1). Thus, the mechanisms utilized by plant growth-promoting bacteria have been reported by several authors (Cho et al. 2015; Nadeem et al. 2014). Application of bacterial inoculants such as *Azotobacter, Azospirillum, Bacillus, and Pseudomonas* has improved growth, nutrient uptake, and yield of cotton under various climatic conditions (Anjum et al. 2007; Park et al. 2011; Egamberdieva et al. 2013; Bharti et al. 2016). For example, Li-Hua et al. (2016) observed an increase of plant biomass, contents of proline, organic acids, amino acids, soluble sugars, as well as of enzyme activities such as peroxidase and superoxide dismutase as well as of root vigor in case of *Suaeda salsa* L. after inoculation with *Trichoderma harzianum.* Also microbial inoculant also increased population of fungi, bacteria, and actinomycetes under saline soil condition of China. The strains of *Bacillus subtilis* isolated from India showed phosphate solubilizing ability and improved seed yield and content of essential oil of fennel (*Foeniculum vulgare* Mill.). In addition, they improved the uptake of soil nutrients such as C, N, P, and K by plants, under saline soil condition of India (Mishra et al. 2016). In other study halotolerant *Enterobacter cloacae* strain (HSNJ4) stimulated seed germination, plant growth, nutrient uptake, and chlorophyll content of canola under salt stress condition (Li et al. 2017). Moreover, a modulation of physiological properties of plants by bacterial inoculation was observed, e.g., the concentration of proline, indole-3-acetic acid (IAA), and antioxidant enzyme activity were increased. Rajput et al. (2013) found that salt-tolerant *Planococcus rifietoensis* strain stimulated growth and yield of wheat under salinity stress. The strain was able to tolerate up to 65 g/L NaCl salinity in the medium and produce IAA, enzymes, and compatible solutes. The strain *Curtobacterium flaccumfaciens* E108 isolated from *Hordeum secalinum* increased plant biomass and development in barley under salt stress condition (Cardinale et al. 2015). The plant growth, physiological properties, and symbiotic performance of host chickpea with rhizobia under saline soil were improved by salt-tolerant *B. subtilis* NUU4 (Egamberdieva et al. 2017b) (Fig. 11.3).

Soil salinity inhibited root and shoot growth and nutrient acquisition of cotton, whereas bacterial inoculation helped plants to withstand salt stress (Egamberdiyeva and Höflich 2004). The cotton plants treated with bacterial inoculants *P. alcaligenes* PsA15, *P. denitrificans* PsD6, and *A. simplex* 50 significantly increased root dry weight. The concentrations of N, P, and K in root and shoot tissues (13–42%) were increased by bacterial inoculants compared to untreated plants under salinated soil of Syrdarya province. In another study, the plant growth of wheat grown under saline soil was increased by IAA produced by bacteria of *Pseudomonas putida, P. extremorientalis, P. chlororaphis*, and *P. aurantiaca* (Egamberdieva and Kucharova 2009). An increased plant biomass of wheat was also reported by inoculation of

Table 11.1 Some examples of root associated microorganisms and their capability to improve plant growth under stress conditions

Microorganisms	Plant	Beneficial effect	References
Trichoderma harzianum	Seepweed	Plant biomass, proline, enzyme activity of peroxidase and superoxide dismutase	Li-Hua et al. (2016)
Bacillus subtilis	Fennel	Plant growth, nutrient uptake, seed yield, essential oil content	Mishra et al. (2016)
Enterobacter cloacae	Canola	Plant growth, nutrient uptake and chlorophyll content, and antioxidant enzyme activity	Li et al. (2017)
Planococcus rifietoensis	Wheat	Plant growth, nutrient uptake, yield	Rajput et al. (2013)
Bacillus amyloliquefaciens	Chinese cabbage, radish, tomato	Plant growth, nutrient uptake	Kim et al. (2017)
Serratia plymuthica Stenotrophomonas rhizophila, *Pseudomonas fluorescens*, *P. extremorientalis*, *P. fluorescens* P	Cucumber	Phytohormone and proline modulation, plant growth	Egamberdieva et al. (2011)
P. denitrificans PsD6, *M. bullata* MpB46, *A. tumescens*	Pea	Plant growth, nutrient uptake	Egamberdiyeva and Höflich (2004)
Mycobacterium phlei, *Mycoplana bullata*	Wheat	Plant growth, nutrient uptake	Egamberdiyeva and Höflich (2003)
Klebsiella sp.	Wheat	Plant growth, nutrient uptake	Singh et al. (2015)
Pseudomonas putida, *P. extremorientalis*, *P. chlororaphis*, *P. aurantiaca*	Wheat	Plant growth, nutrient uptake	Egamberdieva and Kucharova (2009)
Curtobacterium flaccumfaciens,	Barley	Plant biomass	Cardinale et al. (2015)
Bacillus subtilis, *Arthrobacter* sp.	Wheat	Plant growth, nutrient uptake	Upadhyay et al. (2012)
Enterobacter sp.	Wheat	Plant growth, nutrient uptake	Sorty et al. (2016)
Serratia marcescens	Maize	Plant growth, stress tolerance, nutrient uptake	Lavania and Nautiyal (2013)
Aspergillus fumigatus	Chickpea	Plant growth, stress tolerance, nutrient uptake	Khan et al. (2011)
Bacillus subtilis	Thal tree	Plant growth, stress tolerance, nutrient uptake	Hashem et al. (2016)

Fig. 11.3 The effect of salt-tolerant *Bacillus subtilis* NUU4 with the combination of *Mesorhizobium ciceri* IC53 on chickpea growth

other bacterial strains such as *Bacillus subtilis* and *Arthrobacter* sp. by Upadhyay et al. (2012).

Hidri et al. (2016) reported an increased plant growth of *Sulla carnosa* by *Pseudomonas* sp. that was able to produce IAA under salt stress. In another study, *Bacillus amyloliquefaciens* produced ABA and stimulated plant growth of rice under saline conditions (Shahzad et al. 2016). Similar results were observed by ABA produced by *Bacillus aryabhattai* which increased plant growth, nutrient acquisition, nodule formation, and drought stress tolerance in soybean (Park et al. 2017). The salt-tolerant *Bacillus amyloliquefaciens* H-2-5 isolated from Korean soil stimulated the growth of Chinese cabbage, radish, tomato, and mustard plants under salt stress (Kim et al. 2017). The strain was able to produce gibberellins and abscisic acid and showed phosphate-solubilizing activity. In earlier studies salt-tolerant strains of *Serratia plymuthica* RR-2-5-10, *Stenotrophomonas rhizophila* e-p10, *Pseudomonas fluorescens* SPB2145, *P. extremorientalis* TSAU20, and *P. fluorescens* PCL1751 reduced cucumber root rot caused by *Fusarium solani* and stimulated cucumber growth and fruit yield on saline soil in greenhouse conditions (Egamberdieva et al. 2011). Egamberdiyeva and Höflich (2004) reported an increased plant biomass and N, P, and K uptake of pea by *P. denitrificans* PsD6, *M. bullata* MpB46, und *A. tumescent* under saline soil conditions in Uzbekistan. In other study *Mycobacterium phlei* MbP18 and *Mycoplana bullata* MpB46 isolated from salinated soil were found to significantly increase root and shoot growth of winter wheat in a nutrient-poor soil (Egamberdiyeva and Höflich 2003). The salt-tolerant strain *Enterobacter* sp. NIASMVII increased plant biomass of wheat by producing IAA (Sorty et al. 2016). Plant-associated microbes use several mechanisms to improve stress tolerance including synthesis of osmoprotectants, exopolysaccharides, 1-aminocyclopropane-1-carboxylate (ACC) deaminase, and phytohormones and modulation of antioxidant enzymes (Berg et al. 2013; Mishra et al. 2016). For example, *Arthrobacter koreensis* associated with the halophyte shrub *Prosopis strombulifera* produced ABA, IAA, GA3, and jasmonic acid (Piccoli et al. 2011). *Pseudomonas* spp. and *Ochrobactrum* spp. strains produced IAA (Mishra et al. 2017). The symbiotic rhizobia that increased plant growth and development under stress have been found to produce auxins, cytokinins, and abscisic acids (Hayat et al. 2008). Moreover, the bacterial strains which colonize the plant root system were able to control cotton root disease. As reported earlier, salinated soils induce plant vulnerability to pathogens. However, the potential rhizosphere-colonizing bacteria *Pseudomonas* spp., *Pseudomonas putida*, *P. chlororaphis*, *Pseudomonas mendocina*, and *Pantoea agglomerans* isolated from cotton grown in salinated soil of Syrdarya province showed biocontrol potential. They were able to control root rot of cotton caused by *F. oxysporum* and reduced disease incidence by 19%, whereas control plants infested with *F. oxysporum* showed up to 50% disease incidence (Jabborova and Egamberdieva 2012). The bacterial inoculants *Pseudomonas alcaligenes* PsA15, *P. chlororaphis* TSAU13, and *P. extremorientalis* TSAU20 reduced significantly ($P < 0.05$) disease (up to 20%) over the *R. solani*-infected plants grown in saline soil. In general, mechanisms by which bacteria can promote plant growth and protect plants from phytopathogens include mobilization of nutrients (Lugtenberg and Kamilova 2009); production of phytohormones such as indole acetic acid (IAA), gibberellins (Cho et al. 2015), and hydrogen cyanide (HCN) (Compant et al. 2005); and synthesis of the enzymes such as 1-amino cyclopropane-1-carboxylate (ACC) deaminase (Glick et al. 2007).

11.4 Conclusion

In conclusion, salinity and drought negatively affect bacterial diversity. However, several microorganisms were able to survive under high salt concentrations and could contribute to recover fertility under hostile soil conditions. Soil microbes including free-living or rhizosphere-associated as well as endophytes adapted to salt and drought stress are diverse and positively affect soil biological activity and plant physiological processes under extreme environmental conditions. They play an essential role in nutrient cycles and nutrient acquisition, through their ability to fix atmospheric nitrogen, solubilize phosphate, or enhance decomposition of plant residues and furthermore improve plant stress tolerance. These stress-tolerant microbes have a great biotechnological potential to improve soil productivity and plant health of saline soils under arid conditions.

References

Abd_Allah EF, Alqarawi AA, Hashem A, Wirth S, Egamberdieva D (2017) Regulatory roles of 24-epibrassinolide in tolerance of *Acacia gerrardii* Benth to salt stress. Bioengineered 9(1):61–71

Ahmad P (2010) Growth and antioxidant responses in mustard (*Brassica juncea* L.) plants subjected to combined effect of gibberellic acid and salinity. Arch Agron Soil Sci 56(5):575–588

Ahmad P, Hashem A, Abd_Allah EF, Alqarawi AA, John R, Egamberdieva D, Gucel S (2015) Role of *Trichoderma harzianum* in mitigating NaCl stress in Indian mustard (*Brassica juncea* L) through antioxidative defense system. Front Plant Sci 6:868

Anjum MA, Saijad MR, Akhtar N, Qureshi MA, Iqbal A, Jami AR, Hasan M (2007) Response of cotton to plant growth promoting rhizobacteria (PGPR) inoculation under different levels of nitrogen. J Agric Res 45(2):135–143

Asaf S, Khan MA, Khan AL, Waqas M, Shahzad R, Kim AY, Kang SM, Lee IJ (2017) Bacterial endophytes from arid land plants regulate endogenous hormone content and promote growth in crop plants: an example of *Sphingomonas* sp. and *Serratia marcescens*. J Plant Interact 12(1):31–38

Belimov AA, Kojemiakov PA, Chuvarliyeva CV (1995) Interaction between barley and mixed cultures of nitrogen fixing and phosphate-solubilizing bacteria. Plant Soil 17:29–37

Berg G, Alavi M, Schmidt CS, Zachow C, Egamberdieva D, Kamilova F, Lugtenberg B (2013) Biocontrol and osmoprotection for plants under saline conditions. In: de Bruijn FJ (ed) Molecular microbial ecology of the rhizosphere. Wiley-Blackwell, Hoboken. https://doi.org/10.1002/9781118297674.ch55

Bertin C, Yang X, Weston LA (2003) The role of root exudates and allelochemicals in the rhizosphere. Plant Soil 256:67–83

Bharti N, Pandey SS, Barnawal D, Patel VK, Kalra A (2016) Plant growth promoting rhizobacteria *Dietzia natronolimnaea* modulates the expression of stress responsive genes providing protection of wheat from salinity stress. Sci Rep 6:34768

Bock E, Koops H-P, Harms H (1986) Nitrifying bacteria. In: Schlegel HG, Bowien B (eds) Autotrophic bacteria. Springer-Verlag, Berlin, pp 81–96

Brady NC, Well RR (2008) The nature and properties of soils. Pearson-Prentice Hall, Upper Saddle River, 990 pp

Bui EN (2013) Soil salinity: a neglected factor in plant ecology and biogeography. J Arid Environ 92:14–25

Butale SV, Raut AA, Sawant TB (2010) Application of moderately haloalkaliphilic nonsymbiotic diazotrophs of Lonar lake to saline soils. Int J Microbiol Res 2(2):1–4

Canfora L, Bacci G, Pinzari F, Lo Papa G, Dazzi C, Benedetti A (2014) Salinity and bacterial diversity: to what extent does the concentration of salt affect the bacterial community in a saline soil? PLoS One 9(9):e106662

Cardinale M, Grube M, Erlacher A, Quehenberger J, Berg G (2015) Bacterial networks and co-occurrence relationships in the lettuce root microbiota. Environ Microbiol 7(1):239–252

Cho ST, Chang HH, Egamberdieva D, Kamilova F, Lugtenberg B, Kuo CH (2015) Genome analysis of *Pseudomonas fluorescens* PCL1751: a rhizobacterium that controls root diseases and alleviates salt stress for its plant host. PLoS One 10:e0140231. https://doi.org/10.1371/journal.pone.0140231

Compant S, Duffy B, Nowak J, Cle'ment C, Barka EA (2005) Use of plant growth-promoting bacteria for biocontrol of plant diseases: principles, mechanisms of action, and future prospects. Appl Environ Microbiol 71:4951–4959

de Wit R, Mostajir B, Troussellier M, Do Chi T (2011) Environmental management and sustainable use of coastal lagoon ecosystems. In: Friedman AG (ed) Lagoons: biology, management and environmental impact. Nova Science Publishers, Inc, New York, pp 333–350

Egamberdieva D (2009) Alleviation of salt stress by plant growth regulators and IAA producing bacteria in wheat. Acta Physiol Plant 31:861–864

Egamberdieva D (2012) *Pseudomonas chlororaphis*: a salt tolerant bacterial inoculant for plant growth stimulation under saline soil conditions. Acta Physiol Plant 34:751–756

Egamberdieva D, Kucharova Z (2009) Selection for root colonising bacteria stimulating wheat growth in saline soils. Biol Fertil Soils 45:561–573

Egamberdieva D, Kamilova F, Validov S, Gafurova L, Kucharova Z, Lugtenberg B (2008) High incidence of plant growth-stimulating bacteria associated with the rhizosphere of wheat grown in salinated soil in Uzbekistan. Environ Microbiol 19:1–9

Egamberdieva D, Renella G, Wirth S, Islam R (2010a) Secondary salinity effects on soil microbial biomass. Biol Fertil Soils 46(5):445–449

Egamberdieva D, Berg G, Lindstrom K, Rasanen L (2010b) Root colonising *Pseudomonas* spp. improve growth and symbiosis performance of fodder galega (*Galega orientalis* LAM) grown in potting soil. Eur J Soil Biol 46(3–4):269–272

Egamberdieva D, Kucharova Z, Davranov K, Berg G, Makarova N, Azarova T, Chebotar V, Tikhonovich I, Kamilova F, Validov S, Lugtenberg B (2011) Bacteria able to control foot and root rot and to promote growth of cucumber in salinated soils. Biol Fertil Soils 47:197–205

Egamberdieva D, Berg G, Lindström K, Räsänen LA (2013) Alleviation of salt stress of symbiotic *Galega officinalis* L. (goat's rue) by co-inoculation of rhizobium with root colonising *Pseudomonas*. Plant Soil 369(1):453–465

Egamberdieva D, Li L, Wirth S, Lindström K, Räsänen LA (2016) Microbial cooperation in the rhizosphere improves liquorice growth under salt stress. Bioengineered 26:1–6

Egamberdieva D, Wirth S, Alqarawi AA, Abd_Allah EF, Hashem A (2017a) Phytohormones and beneficial microbes: essential components for plants to balance stress and fitness. Front Microbiol 8:2104. https://doi.org/10.3389/fmicb.2017.02104

Egamberdieva D, Wirth S, Shurigin V, Hashem A, Abd Allah EF (2017b) Endophytic bacteria improve plant growth, symbiotic performance of chickpea (*Cicer arietinum* L.) and induce suppression of root rot caused by *Fusarium solani* under salt stress. Front Microbiol 8:1887. https://doi.org/10.3389/fmicb.2017.01887

Egamberdieva D, Wirth S, Behrendt U, Parvaiz A, Berg G (2017c) Antimicrobial activity of medicinal plants correlates with the proportion of antagonistic endophytes. Front Microbiol 8:199. https://doi.org/10.3389/fmicb.2017.00199

Egamberdiyeva D (2005) Plant growth promoting rhizobacteria isolated from calcisol soil in a semiarid region of Uzbekistan: biochemical characterisation and effectiveness. Plant Nutr Soil Sci 168:91–99

Egamberdiyeva D, Hoflich G (2003) Influence of growth promoting bacteria on the growth of wheat at different soils and temperatures. Soil Biol Biochem 35:973–978

Egamberdiyeva D, Hoflich G (2004) Importance of plant growth promoting bacteria on growth and nutrient uptake of cotton and pea in semi-arid region Uzbekistan. J Arid Environ 56:293–301

Egamberdiyeva D, Gafurova L, Islam KR (2007) Salinity effects on irrigated soil chemical and biological properties in the syrdarya basin of Uzbekistan. In: Lal R, Sulaimanov M, Stewart B, Hansen D, Doraiswamy P (eds) Climate change and terrestrial C sequestration in Central Asia. Taylor-Francis, New York

El-Beltagy A, Madkour M (2012) Impact of climate change on arid lands agriculture. Agric Food Secur 1:3

FAO (2010) FAO land and plant nutrition management service. http://www.fao.org

Forchetti G, Masciarelli O, Izaguirre MJ, Alemano S, Alvarez D, Abdala G (2010) Endophytic bacteria improve seedling growth of sunflower under water stress, produce salicylic acid, and inhibit growth of pathogenic fungi. Curr Microbiol 61(6):485–493

Glick BR, Cheng Z, Czarny J, Duan J (2007) Promotion of plant growth by ACC deaminase-producing soil bacteria. Eur J Plant Pathol 119:329–339

Gougoulias C, Clark JM, Shaw LZ (2014) The role of soil microbes in the global carbon cycle: tracking the below-ground microbial processing of plant-derived carbon for manipulating carbon dynamics in agricultural systems. J Sci Food Agric 94(12):2362–2371

Hashem A, Abd_Allah EF, Alqarawi AA, Aldebasi A, Egamberdieva D (2015) Arbuscular mycorrhizal fungi enhances salinity tolerance of *Panicum turgidum* Forssk by altering photosynthetic and antioxidant pathways. J Plant Interact 10:230–242

Hashem A, Abd_Allah EF, Alqarawi A, Al-Huqail AA, Wirth S, Egamberdieva D (2016) The interaction between arbuscular mycorrhizal fungi and endophytic bacteria enhances plant growth of *Acacia gerrardii* under salt stress. Front Plant Sci 7:1089

Hashem A, Abd_Allah EF, Alqarawi AA, Soliman DFK, Wirth S, Egamberdieva D (2017) Regulation of secondary metabolites by inoculation with a Trichoderma harzianum in tomato (*Solanum lycopersicum* L.) under drought stress conditions. Integr Agric 16(8):1751–1757

Hayat S, Hasan SA, Fariduddin Q, Ahmad A (2008) Growth of tomato (*Lycopersicon esculentum*) in response to salicylic acid under water stress. J Plant Interact 3:297–304

Hidri R, Barea JM, Mahmoud OM, Abdelly C, Azc6n R (2016) Impact of microbial inoculation on biomass accumulation by *Sulla carnosa* provenances, and in regulating nutrition, physiological and antioxidant activities of this species under non-saline and saline conditions. J Plant Physiol 201:28–41

Jabborova D, Egamberdieva D (2012) Evaluation of root associated bacteria for control of cotton root rot caused by *Fusarium oxysporum* in salinated soils. J Inst Sci Technol 1(4):37–42

Jadhav GG, Salunkhe DS, Nerkar DP, Bhadekar RK (2010) Isolation and characterization of salt-tolerant nitrogen-fixing microorganisms from food. EurAsian J Biosci 4:33–40

Kapur M, Bhatia R, Pandey G, Pandey J, Paul D, Jain RK (2010) A case study for assessment of microbial community dynamics in genetically modified Bt cotton crop fields. Curr Microbiol 61:118–124

Kim MJ, Radhakrishnan R, Kang SM, You YH, Jeong EJ, Kim JG, Lee IJ (2017) Plant growth promoting effect of *Bacillus amyloliquefaciens* H-2-5 on crop plants and influence on physiological changes in soybean under soil salinity. Physiol Mol Biol Plants 23(3):571–580

Khan AL, Hamayun M, Kim Y-H, Kang S-M, Lee J-H, Lee I-J (2011) Gibberellins producing endophytic Aspergillus fumigatus sp. LH02 influenced endogenous phytohormonal levels, isoflavonoids production and plant growth in salinity stress. Process Biochem 46(2):440–447

Kouas S, Slanti T, Salah IB, Abdelly C (2010) Eco-physiological responses and symbiotic nitrogen fixation capacity of salt -exposed *Hedysarum carnosum* plants. Afr J Biotechnol 9(44):7462–7469

Lakhdar A, Rabhi M, Ghnaya T, Montemurro F, Jedidi N, Abdelly C (2009) Effectiveness of compost use in salt-affected soil. J Hazard Mater 171:29–37

Lauber CL, Hamady M, Knight R, Fierer N (2009) Pyrosequencing-based assessment of soil pH as a predictor of soil bacterial community structure at the continental scale. Appl Environ Microbiol 75:5111–5120

Lavania M, Nautiyal C (2013) Solubilization of tricalcium phosphate by temperature and salt tolerant Serratia marcescens NBRI1213 isolated from alkaline soils. Afr J Microbiol Res 7:4403–4413

Li X, Sun M, Zhang H, Xu N, Sun G (2016a) Use of mulberry–soybean intercropping in salt–alkali soil impacts the diversity of the soil bacterial community. Microb Biotechnol 9(3):293–304

Li Y, Li X, Liu Y, Wang ET, Ren C, Liu W, Xu H, Wu H, Jiang N, Li Y, Zhang X, Xie Z (2016b) Genetic diversity and community structure of rhizobia nodulating *Sesbania cannabina* in saline–alkaline soils. Syst Appl Microbiol 39(3):195–202

Li H, Lei P, Pang X, Li S, Xu H, Xu Z, Feng F (2017) Enhanced tolerance to salt stress in canola (*Brassica napus* L.) seedlings inoculated with the halotolerant *Enterobacter cloacae* HSNJ4. Appl Soil Ecol 119:26–34

Li-Hua C, Jin-Hai Z, Xiao-Hou S, Shan-Shan S, Zhi-Heng Y, Xin-Yu M, Ting-Ting C (2016) Effects of *Trichoderma harzianum* T83 on *Suaeda salsa* L. in coastal saline soil. Ecol Eng 91:58–64

Lijuan C, Changsheng L, Qi F, Yongping W, Hang Z, Yan Z, Yongjiu F, Huiya L (2017) Shifts in soil microbial metabolic activities and community structures along a salinity gradient of irrigation water in a typical arid region of China. Sci Total Environ 598:64–70

Lugtenberg BJJ, Kamilova F (2009) Plant growth-promoting rhizobacteria. Annu Rev Microbiol 63:541–556

Luo JY, Zhang S, Zhu XZ, Lu LM, Wang CY, Li CH, Cui JJ, Zhou ZG (2017) Effects of soil salinity on rhizosphere soil microbes in transgenic Bt cotton fields. J Integr Agric 16(7):1624–1633

Mahalingam R (2015) Consideration of combined stress: a crucial paradigm for improving multiple stress tolerance in plants. In: Mahalingam R (ed) Combined stresses in plants. Springer, Cham, pp 1–25

Marschner P, Crowley DE, Yang CH (2004) Development of specific rhizosphere bacterial communities in relation to plant species, nutrition and soil type. Plant Soil 261:199–208

Mavi MS, Sanderman J, Chittleborough DJ, Cox JW, Marschner P (2012) Sorption of dissolved organic matter in salt-affected soils: effect of salinity, sodicity and texture. Sci Total Environ 435–436:337–344

Mendes R, Garbeva P, Raaijmakers JM (2013) The rhizosphere microbiome: significance of plant beneficial, plant pathogenic, and human pathogenic microorganisms. FEMS Microbiol Rev 37:634–663

Mishra RR, Swain MR, Dangar TK, Thatoi H (2012) Diversity and seasonal fluctuation of predominant microbial communities in Bhitarkanika, a tropical mangrove ecosystem in India. Rev Biol Trop 60:909–924

Mishra BK, Meena KK, Dubey PN, Aishwath OP, Kant K, Sorty AM, Bitla U (2016) Influence on yield and quality of fennel (*Foeniculum vulgare* Mill.) grown under semi-arid saline soil, due to application of native phosphate solubilizing rhizobacteria isolates. Ecol Eng 97:327–333

Mishra SK, Khan MH, Misra S, Dixit KV, Khare P, Srivastava S, Chauhan PS (2017) Characterisation of *Pseudomonas* spp. and *Ochrobactrum* sp. isolated from volcanic soil. Antonie Van Leeuwenhoek 110:253–270

Moradi A, Tahmourespour A, Hoodaji M, Khorsandi F (2011) Effect of salinity on free living diazotroph and total bacterial populations of two saline soils. Afr J Microbiol Res 5(2):144–148

Nadeem SM, Ahmad M, Zahir ZA, Javaid A, Ashraf M (2014) The role of mycorrhizae and plant growth promoting rhizobacteria (PGPR) in improving crop productivity under stressful environments. Biotechnol Adv 32:429–448

Panosyan H, Hakobyan A, Birkeland NK, Trchounian A (2018) Bacilli community of saline-alkaline soils from the Ararat Plain (Armenia) assessed by molecular and culture-based methods. Syst Appl Microbiol 41:232. https://doi.org/10.1016/j.syapm.2017.12.002

Park J, Kim YS, Kim SG, Jung JH, Woo JC, Park CM (2011) Integration of auxin and salt signals by the NAC transcription factor NTM2 during seed germination in Arabidopsis. Plant Physiol 156:537–549

Park YG, Mun BG, Kang SM, Hussain A, Shahzad R, Seo CW, Kim AY, Lee SU, Oh KY, Lee DY, Lee IJ, Yun BW (2017) *Bacillus aryabhattai* SRB02 tolerates oxidative and nitrosative stress and promotes the growth of soybean by modulating the production of phytohormones. PLoS One 12(3):e0173203

Parray AP, Jan S, Kamili AN, Qadri RA, Ahmad ED (2016) Current perspectives on plant growth promoting rhizobacteria. Plant Growth Regul 35(3):877–902

Piccoli P, Travaglia C, Cohen A, Sosa L, Cornejo P, Masuelli R, Bottini R (2011) An endophytic bacterium isolated from roots of the halophyte *Prosopis strombulifera* produces ABA, IAA, gibberellins A1 and A3 and jasmonic acid in chemically-defined culture medium. Plant Growth Regul 64:207–210

Ragab R, Prudhomme C (2002) Climate change and water resources management in arid and semi-arid regions: prospective and challenges for the 21st century. Biosyst Eng 81(1):3–34

Rajput L, Imran A, Mubeen F, Hafeez FY (2013) Salt-tolerant PGPR strain *Planococcus rifietoensis* promotes the growth and yield of wheat (*Triticum aestivum* L.) cultivated in saline soil. Pak J Bot 45:1955–1962

Rangarajan S, Saleena LM, Nair S (2002) Diversity of *Pseudomonas* spp. isolated from rice rhizosphere populations grown along a salinity gradient. Microb Ecol 43:280–289

Renella G, Egamberdiyeva D, Landi L, Mench M, Nannipieri P (2006) Soil microbial activity and hydrolase activity during decomposition of model root exudates released by a model root surface in Cd-contaminated soils. Soil Biol Biochem 38:702–708

Sah S, Girija D, Deepa K, Nazeem PA, Firoz PM, Sunil E (2014) Diversity analysis of bacterial community in acid saline Pokkali soil of Kerala through metagenomic approach. J Trop Agric 52(1):123–130

Setia R, Marschner P, Baldock J, Chittleborough D (2010) Is CO2 evolution in saline soils affected by an osmotic effect and calcium carbonate? Biol Fertil Soils 46:781–792

Shahzad R, Waqas M, Khan AL, Asaf S, Khan MA, Kang SM, Yun BW, Lee IJ (2016) Seed-borne endophytic Bacillus amyloliquefaciens RWL-1 produces gibberellins and regulates endogenous phytohormones of *Oryza sativa*. Plant Physiol Biochem 106:236–243

Singh RP, Jha PN (2016) A halotolerant bacterium *Bacillus licheniformis* HSW-16 augments induced systemic tolerance to salt stress in wheat plant (*Triticum aestivum*). Front Plant Sci 7:1890

Singh RP, Jha P, Jha PN (2015) The plant-growth-promoting bacterium Klebsiella sp. SBP-8 confers induced systemic tolerance in wheat (Triticum aestivum) under salt stress. J Plant Physiol 184:57–67

Sorty AM, Meena KK, Choudhary K, Bitla UM, Minhas PS, Krishnani KK (2016) Effect of plant growth promoting bacteria associated with halophytic weed (*Psoralea corylifolia* L.) on germination and seedling growth of wheat under saline conditions. Appl Biochem Biotechnol 180:872–882

Turan M, Ekinci M, Yildirim E, Gneş A, Karagz K, Kotan R, Dursun A (2014) Plant growth-promoting rhizobacteria improved growth, nutrient, and hormone content of cabbage (*Brassica oleracea*) seedlings. Turk J Agric For 38:327–333

Upadhyay SK, Singh JS, Saxena AK, Singh DP (2012) Impact of PGPR inoculation on growth and antioxidant status of wheat under saline conditions. Plant Biol 14:605–611

Wallenstein MD, Hess AM, Lewis MR, Steltzer H, Ayres E (2010) Decomposition of aspen leaf litter results in unique metabolomes when decomposed under different tree species. Soil Biol Biochem 42:484–490

Yang H, Hu J, Long X, Liu Z, Rengel Z (2016) Salinity altered root distribution and increased diversity of bacterial communities in the rhizosphere soil of Jerusalem artichoke. Sci Rep 6:20687

Yuan BC, Li ZZ, Liu H, Gao M, Zhang YY (2007) Microbial biomass and activity in salt affected soils under arid conditions. Appl Soil Ecol 35:319–328

Halophilic *Actinobacteria* Biological Activity and Potential Applications

12

Osama Abdalla Abdelshafy Mohamad, Li Li, Jin-Biao Ma, Shaimaa Hatab, Bakhtiyor A. Rasulov, Zulpiya Musa, Yong-Hong Liu, and Wen-Jun Li

Abstract

Microbes belonging to the phylum *Actinobacteria* are prolific sources of antibiotics, bioactive compounds, and industrially relevant enzymes. There are tremendous diversity and novelty of *Actinobacteria*; the applications of halophilic *Actinobacteria* toward the industrially and medically important metabolites and

O. A. Abdelshafy Mohamad
Key Laboratory of Biogeography and Bioresource in Arid Land, Xinjiang Institute of Ecology and Geography, Chinese Academy of Sciences, Urumqi, China

Institute for Post Graduate Environmental Studies, Environmental Science Department, Arish University, North Sinai, Egypt

L. Li · J.-B. Ma · Y.-H. Liu
Key Laboratory of Biogeography and Bioresource in Arid Land, Xinjiang Institute of Ecology and Geography, Chinese Academy of Sciences, Urumqi, China

S. Hatab
Faculty of Agriculture and Environmental Science, Arish University, El-Arish, Egypt

B. A. Rasulov
Key Laboratory of Biogeography and Bioresource in Arid Land, Xinjiang Institute of Ecology and Geography, Chinese Academy of Sciences, Urumqi, China

Institute of Genetics and Plant Experimental Biology, Uzbekistan Academy of Sciences, Kybray District, Tashkent Province, Uzbekistan

Z. Musa
College of Life Science and Technology, Xinjiang University, Urumqi, China

W.-J. Li (✉)
Key Laboratory of Biogeography and Bioresource in Arid Land, Xinjiang Institute of Ecology and Geography, Chinese Academy of Sciences, Urumqi, China

State Key Laboratory of Biocontrol and Guangdong Provincial Key Laboratory of Plant Resources, School of Life Sciences, Sun Yat-Sen University, Guangzhou, China
e-mail: liwenjun3@mail.sysu.edu.cn

D. Egamberdieva et al. (eds.), *Extremophiles in Eurasian Ecosystems: Ecology, Diversity, and Applications*, Microorganisms for Sustainability 8,
https://doi.org/10.1007/978-981-13-0329-6_12

enzymes are gaining increasing attention by the scientific community. A large number of novel compounds and enzymes from halophilic actinomycetes have been isolated and characterized from various geographic regions around the world. In this chapter, we focus on the occurrence, characterization, and the specific metabolites concerning different industrial applications of the relevant biomolecules/bioactive compounds for agriculturally, pharmaceutically, and biotechnologically application are discussed. Halophilic actinomycetes may also serve as useful models for the production of essential metabolites and enzymes with respect to stress response.

Keywords
Halophilic actinomycetes · Occurrence · Metabolites · Enzymes · Applications

12.1 Introduction

Microbial biota is powerful; it can be found everywhere over a wide range of extreme conditions chemically or physically (salinity, temperature, pH, pressure, and nutrient conditions). Therefore, researchers have been interested in the mesmerizing organisms identified as extremophiles due to their ability to live in extreme environments. For each extreme environmental condition studied, there are a variety of organisms determined that not only can grow in these extreme conditions but that also often require those conditions for existence. Hypersaline environments are those which contain salt concentration in excess approximately 3.5% total dissolved salts, and they expressed typical examples of extreme microbial life, they are widely handing out on our planet due to their high salinity, and they are mainly shown by saline lakes and saline soils (Oren 2002; Ventosa 2006).

Halophilic microorganisms are the organism that survives in saturated salt environments; they are salt-loving organisms due to their capacity adapting themselves to balance the osmotic pressure of the environment (Enache and Kamekura 2010; Caton et al. 2009; Elshahed et al. 2004). Over the last decades, scientists have been intrigued to spend more efforts to characterize microorganisms from hypersaline environments due to the number of species exponentially growing included on different phylogenetic branches. Therefore, the world of the halophilic microorganisms is divided into two main physiological groups of halophiles in hypersaline area that defines various categories of halophilic microorganisms based on the salt concentration wherein they show optimal growth: extreme halophiles, which can grow optimally in media with 15–30% (2.5–5.2 M) NaCl, and moderate halophiles, which is growing optimally in media with 3–15% (0.5–2.5 M) NaCl (Kushner and Kamekura 1988).

Among the halophilic microorganisms, it is well known that *Actinobacteria* members were raised to the taxonomic rank of a phylum which is one of the primitive lineages in the prokaryotes (Ventura et al. 2007; Koch 2003); they are supposed to have derivative since 2.7 billion years ago (Battistuzzi and Hedges 2009). *Actinobacteria* are heterogeneous Gram-positive or Gram-variable, aerobic or anaerobic, motile or nonmotile, and spore or non-spore-forming microorganisms with a high G+C content >55 mol% (Garrity and Holt 2001; Ludwig and Klenk 2001; Baranasic et al. 2013).

Thus, *Actinobacteria* are an ecologically friendly group and considered as valuable resources for the development of novel biotechnological cycles and industrial applications. They are play a vital role in several biological processes such as pharmaceutically important bioactive compounds (antibiotics, anti-inflammatory compounds, antitumor agents, and enzyme inhibitors) and industrially such as proteases and amylases, biosurfactant production, biopolymers in oil recovery, poly-beta hydroxyalkanoate as biodegradable plastic, exopolysaccharide , plant growth promotion and bioremediation of contaminated hypersaline brines etc (Kanekar et al. 2012; Chen et al. 2015; Palaniyandi et al. 2013). On the other hand, *Actinobacteria* isolated from the extreme biosphere will be a rich source of novel natural products (Bull 2011; Berdy 2005; Olano et al. 2009). This chapter explores the halophilic *Actinobacteria* from different aspects. The focus will be on the exploitation of halophilic bacteria and its environmental importance, biotechnological applications, and biological survivability under salt stress conditions.

12.2 Where and How We Can Identify and Characterize Halophilic *Actinobacteria*

12.2.1 What Is *Actinobacteria*

Actinomycetes are the most valuable microorganisms for the production and synthesis of economically important therapeutic compounds and antibiotics. Actinomycete order is a group of unicellular organisms with fungal morphology such as special spores or conidia, which are considered as higher and filamentous bacteria. Gram-positive bacteria fall into two major phylogenetic divisions, "low GC" and "high GC." Gram-positive bacteria that belong to phylum *Actinobacteria* are divided into 6 classes, namely, *Actinobacteria*, *Acidimicrobiia*, *Coriobacteriia*, *Nitriliruptoria*, *Rubrobacteria*, and *Thermoleophilia*; 16 orders which are *Actinomycetales*, *Actinopolysporales*, *Bifidobacteriales*, *Catenulisporales*, *Corynebacteriales*, *Frankiales*, *Glycomycetales*, *Jiangellales*, *Kineosporiales*, *Micrococcales*, *Micromonosporales*, *Propionibacteriales*, *Pseudonocardiales*, *Streptomycetales*, *Streptosporangiales*, and *Incertae sedis*; subclass, *Actinobacteridae*; and order, *Actinomycetales*; they belong to 10 suborders, 30 families, and over 160 genera (Goodfellow et al. 2012). It has a high guanine-plus-cytosine (G+C) ratio of the DNA (>55mol%) (Baranasic et al. 2013).

12.2.2 Distribution of *Actinobacteria* in Nature

Actinomycetes adapted to high salt concentrations are found in quite a significant number of most ecological systems salt-loving microorganisms live in saline soils, lake sediments, aquatic habitat, salt lakes, brines, marine environment, alkaline habitats, plants and hypersaline regions are the main sources for novel halophilic actinomycetes due to their salt tolerance (Jones and Grant 1999; Duckworth et al. 1998; Sprusansky et al. 2005; George et al. 2012; Raja et al. 2010; Diraviyam et al. 2011). In fact, they were able to tolerate for extensive periods of time under high salt environment. Two hundred seventeen marine actinomycetes from sea surface micro-layer were isolated by Hakvåg et al. (2008). Moreover, Ghanem et al. (2000) and Salah et al. (2014) isolated halophytic actinomycetes from the marine sediments. In another study, 274 strains of actinomycetes were isolated by Tian et al. (2004), and 1755 actinomycete strains were isolated by Petrolini et al. (1995) from the plant's tissue. Al-Zarban et al. (2002b) isolated halophilic actinomycetes with optimum growth at 10% NaCl from marsh soil.

To explore halophilic microorgnisms world, many studies of microbial life at high salt concentrations reported from saltern ponds worldwide, Great Salt Lake, the Dead Sea (Al-Zarban et al. 2002b), saline lakes in Inner Mongolia (Pagaling et al. 2009), African soda lakes, deep-sea brines (Van der Wielen et al. 2005), Xinjiang salt lakes; Chinese salt mines; salterns in Goa India, Turkey, Spain, and Israel; South Siberian hypersaline lakes, thane Papke (Storrs, CT) in Spain and Algeria, hypersaline lake in Argentina and many others.

12.3 How We Can Make Morphological Observation of Actinomycetes

The preliminary differentiation could be done by morphological observation of the growth of actinomycetes referring to sporophore morphology when cultures became matured with spore mass, substrate mycelium and Pigments are convenient biomarkers types of halophiles. It is considered a stable and clearly defined feature for actinomycete classification (Williams et al. 1993). Therefore, based on different spore chain morphologies, it can be divided into seven main groups: rectus flexibilis (RF), retinaculum-apertum (RA), spira (S), monoverticillus (MV), monoverticillus-spira (MV-S), biverticillus (BIV), and biverticillus-spira (BIV-S) (Pridham and Lyons Jr 1961), whereas the spores could be divided into distinct ornamentation – smooth, spiny, rugose, knobby, hairy, warty, tuberculate, or verrucose (Vobis et al. 1997). In addition, every group was further classified into six "series" according to the color of spore mass: gray (light gray to mouse gray to brown gray to gray brown), white, olive buff (buff to tan to olive buff), yellow, blue (blue to blue green to green), and red (pink to red to lavender to lavender gray).

12.3.1 Identification of *Actinobacteria*

For actinomycete isolation, there are different cultivation methods such as integration of specific antibiotics, and chemicals were used in the composition of the cultivation medium to avoid the growth of fungi and bacteria (Amann et al. 1995). Some of the isolation media are not available commercially and must be prepared in the laboratory using colloidal chitin, soil extract, and plant extracts. In addition, humic acid and special vitamin agar medium carried out specific substrates for the isolation of actinomycetes. Therefore, certain interest must be paid in the identification of the rare halophilic actinomycetes. The molecular technique protocols for the amplification of specific genes such as 16S rRNA and rec A can be used in actinomycete classification and characterization; the phylogenetic relationship inference among microorganisms in order to characterize actinomycetes has been commonly accepted as a polyphasic taxonomic approach (Tian et al. 2007; Claridge and Campbell 2004).

The polymerase chain reaction (PCR)-based methods for the detection and identification of microbes are widely used (Tang and Persing 1999). The 16S rRNA gene sequence about 1550 base pairs (bp) long is composed of both variable and conserved regions. In particular, actinomycetes classified based on 16S rRNA gene to many species were recovered and characterized from hypersaline regions since 2000–2015 (Table 12.1).

Table 12.1 Halophilic and halotolerant actinomycetes isolated from hypersaline habitats since 2002–2015

Name of actinomycete isolate	Habitat	NaCl (%)	References
Nocardiopsis halotolerans	Salt soil, in desert area at Al-Khiran, southern Kuwait	0–15	Al-Zarban et al. (2002a)
Saccharomonospora halophila	Kuwait salt marsh soil	10	Al-Zarban et al. (2002a)
Saccharomonospora paurometabolica	Soil of Xinjiang Province, China	5–20	Li et al. (2003a)
Streptomonospora alba	Soil of Xinjiang Province, China	5–25	Li et al. (2003c)
Prauserella halophila, *Prauserella alba*	Soil of Xinjiang Province, China	0–25	Li et al. (2003b)
Nesterenkonia halotolerans, *Nesterenkonia xinjiangensis*	Soil of Xinjiang Province, China	0–25	Li et al. (2004a)
Nocardiopsis salina	Soil of Xinjiang Province, China	3–15	Li et al. (2004b)
Actinopolyspora sp.	Soil of Kuwait	–	Abbas (2006)
Microbispora sp.			
Amycolatopsis sp.			
Kocuria aegyptia	Desert soil, Egypt	1–5	Li et al. (2006c)
Nocardiopsis gilva	Soil of Xinjiang Province, Western China	5–18	Li et al. (2006b)
N. rhodophaea			
N. rosea			
N. chromatogenes			
N. baichengensis			
Nocardiopsis valliformis	Soil from alkaline lake near Buerjin (Xinjiang, China)		Yang et al. (2008)
Nesterenkonia halophila	Soil of Xinjiang Province, China	10	Li et al. (2008)
Streptomonospora halophila	Soil from Xinjiang Province, China	10	Cai et al. (2008)
Amycolatopsis marina	Sediment sample, South China Sea	0.5–12	Bian et al. (2009)

(continued)

Table 12.1 (continued)

Name of actinomycete isolate	Habitat	NaCl (%)	References
Marinactinospora thermotolerans	Sediment sample, the northern South China Sea	0–5	Tian et al. (2009)
Saccharomonospora saliphila	Muddy soil, Gulbarga, Karnataka Province, India	10	Syed et al. (2008)
Haloglycomyces albus	Soil from Xinjiang Province, Western China	8–12	Guan et al. (2009)
Kocuria halotolerans	Soil of Ganjiahu Suosuo Forest, Xinjiang, China	0–10	Tang et al. (2009a)
Spinactinospora alkalitolerans	Marine sediment, The Yellow Sea, China	3–8	Chang et al. (2011)
Streptomyces pharmamarensis	Marine sediment sample, Mediterranean Sea (Italy)	1–9	Carro et al. (2012)
Actinomadura sediminis	Sediments of Dugong Creek, Little Andaman, India	0–7	He et al. (2012)
Glycomyces halotolerans	Xinjiang Province, north-west China	4–5	Guan et al. (2011b)
Streptomyces oceani	Deep-sea sediment, South China Sea	2.5–12.5	Tian et al. (2012)
Gulosibacter chungangensis	Marine sediment near Oh Island, Republic of Korea	0–5	Park et al. (2012)
Actinopolyspora lacussalsi	Xinjiang Province, north-west China	12	Guan et al. (2013)
Actinopolyspora mzabensis	Soil samples, Ghardaïa, southern Algeria	–	Meklat et al. (2013)
Amycolatopsis cihanbeyliensis	Cihanbeyli Salt Mine, the Central Anatolia Region of Turkey	0–10	Tatar et al. (2013)
Glycomyces fuscus, *Glycomyces albus*	Xinjiang Province, North-west China		Han et al. (2014)
Amycolatopsis flava sp.	Sediment sample collected from the Dead Sea	1–30	Wei et al. (2015)
Prauserella isguenensis sp.	Soil samples from the arid region of Ahbas at Béni-Isguen (Mzab), located in the Algerian Sahara	–	Saker et al. (2015)
Streptimonospora salina sp.	Soil sample of the salt lake in Western China	–	Cui et al. (2001)
Streptomyces sodiphilus sp.	Mud sample of Chaka Salt Lake, Qinghai, China	–	Li et al. (2005)
Haloactinospora alba	Salt lake in Xinjiang Province, north-west China	–	Tang et al. (2008)
Saccharopolyspora lacisalsi	Soil of Lop Nur Salt Lake in Xinjiang, China	10	Guan et al. (2011a)
Saccharopolyspora qijiaojingensis	Salt lake in Xinjiang Province, north-west China	6–22	Tang et al. (2009b)
Actinomycete *Haloactinobacterium album*	Salt lake in Xinjiang Province, north-west China	1–15	Tang et al. (2010c)

(continued)

Table 12.1 (continued)

Name of actinomycete isolate	Habitat	NaCl (%)	References
Amycolatopsis halophila	Salt lake in Xinjiang Province, north-west China	–	Tang et al. (2010a)
Georgenia halophila	Salt lake in Xinjiang Province, north-west China	–	Tang et al. (2010b)
Myceligenerans halotolerans sp.	Salt lake in Xinjiang Province, north-west China	–	Wang et al. (2011)
Nocardioides aquaticus sp., *Friedmanniella lacustris* sp.	Waters samples of the hypersaline Ekho Lake located at Vestfold Hills, East Antarctica	–	Lawson et al. (2000)
Nesterenkonia lacusekhoensis	Hypersaline Ekho Lake (Vestfold Hills, East Antarctica)	–	Collins et al. (2002)
Actinopolyspora egyptensis sp.	Salty Lake Qaroun, Egypt	–	Hozzein and Goodfellow (2011)
Salinisphaera halophila	The brine of a salt well in Yunnan Province, China	14–19	Zhang et al. (2012a)

12.4 Industrially Important Enzyme Production by Halophilic Actinomycetes

12.4.1 Amylase Enzyme

The enzymatic hydrolysis α-amylases (E.C. 3.2.1.1, 1, 4-α-D-glucan glucanohydrolase) are one of the most vital industrial enzymes with a broad range of applications due to numerous advantages such as specificity of the reaction and stability of the final products (Kikani et al. 2010). At the moment, due to the growing demand for these enzymes in various industries, production of commercial amylase enzymes has been explored extensively, and these enzymes became about 30 % of the world's enzyme production (Elleuche et al. 2014; Raddadi et al. 2015; Chandrasekaran 1997).

α-Amylase enzymes stimulate the hydrolysis of internal a-1,4-O-glycosidic bonds in polysaccharides with the retention of a α-anomeric configuration in the products and are used broadly in numerous fields of biotechnology as well as pharmaceutical industry due to their cost-effective production techniques (Kadziola et al. 1998; Tonkova 2006; Machius et al. 1995; Pandey and Singh 2012). The α-amylase family can be basically divided into two main groups: the starch-hydrolyzing enzymes and the starch-transglycosylating enzymes.

Novel halophilic α-amylases have taken more interest due to their capability to remain stable in the presence of high salt concentrations, thermal, alkaline and organic solvent so as to be used in laundry detergent, food, fermentation (fuel alcohol production), textile, paper, and other industrial processes that require presence of high salt concentrations (Raja et al. 2010; Karan et al. 2012; Dalmaso et al. 2015; Chakraborty et al. 2012; Ratnakar 2013).

Peptone and maltose were noted as best nitrogen and carbon sources for maximum yield of amylase production. To prove the presence of amylolytic activity of bacteria could be screened by culturing them on starch agar medium. After incubation time, iodine solution added to the plates and the halo zone of clearance around the bacterial colony is a sign of starch hydrolysis (Sanchez-Porro et al. 2003; Cojoc et al. 2009; Rohban et al. 2009; Kumar and Khare 2012; Kumar et al. 2012b; Jayachandra et al. 2012; Neagu et al. 2014a). Many moderately and extremely halophilic actinomycetes isolated from different hyperfine environmental have been reported to produce strong amylolytic activity since to 2000–2013 (Table 12.2).

Table 12.2 Halophilic actinomycetes isolated from different hyperfine environments have been reported to produce strong amylolytic activity since 2002–2013

Name of actinomycete isolate	Habitat	References
Actinopolymorpha alba sp.	Hami Salt Lake, Xinjiang, China	Cao et al. (2009b)
Micromonospora sp.	Pudimadaka coast of Bay of Bengal, India	Haritha et al. (2010)
Streptomyces sp.		
Actinomycete sp.	Sediment samples of Machilipatnam	Ellaiah et al. (2002)
Actinomycete sp.	Kakinada coast, near Bay of Bengal, India	Ellaiah et al. (2004)
Actinomycete sp.	Tamil Nadu coast of the Bay of Bengal, India	Ramesh et al. (2006)
Streptomyces alboniger	Kodiyakarai coastal sediments, Bay of Bengal, India	Manivasagan et al. (2010)
Saccharopolyspora sp.	Sediment samples of west coast of India	Chakraborty et al. (2011)
Saccharopolyspora, Streptomyces, Actinopolyspora, Streptoverticillium, Microtetraspora, Actinokineospora, Nocardiopsis, Dactylosporangium	-	Meena et al. (2013)
Actinomycete sp.	Hypersaline soils in man-made solar salterns of Nellore district, India	Shameer and Babu (2013)
Halomonas meridiana	–	Coronado et al. (2000)
Halothermothrix orenii	–	Tan et al. (2008)
Streptomyces sp.	–	Chakraborty et al. (2009)
Chromohalobacter sp.	–	Prakash et al. (2009)
Streptomyces erumpens	Brick kiln soil nearby Bhubaneswar, India	Kar and Ray (2008)
Micrococcus sp.	Deep-sea sediments of the Southern Okinawa, China	Dang et al. (2009)
Nesterenkonia sp.	–	Dang et al. (2009)
Streptomyces aureofaciens	Shrimp pond, opposite to Vellar estuary	Poornima et al. (2008)

(continued)

Table 12.2 (continued)

Name of actinomycete isolate	Habitat	References
Nocardiopsis sp.	Deep-sea sediment of Prydz Bay, Antarctic	Zhang and Zeng (2008)
Streptomyces sp.	Sediments of Goa, Alibag	Chakraborty et al. (2009)
Saccharopolyspora sp.	Sediments, West Coast, India	Chakraborty et al. (2011)
Streptomyces sp.	Mushroom compost	Singh et al. (2012)
Streptomyces sp.	Sediments, West Coast of India	Chakraborty et al. (2012)
Streptomyces sp.	Coral reef sediments, Manoli Island in the Gulf of Mannar Biosphere Reserve, India	Sivakumar et al. (2012)
Streptomyces sp.	Antarctic vegetation samples from Progress Lake (East Antarctica)	Cotârleţ (2013)
Streptomyces rochei	Marine sediments, southeast coast of the Bay of Bengal, India	Acharyabhatta et al. (2013)
Streptomyces gulbargensis	–	Syed et al. (2009)
Nocardiopsis sp.	–	Liu et al. (2011)

12.4.2 Cellulose Enzyme

Cellulolytic enzymes (cellulases) are a group of glycosyl hydrolases classified into different families depending on their sequence homologies. Cellulase enzyme controls the enzymatic hydrolysis process which comprises three classes of soluble extracellular enzymes 1,4-β-endoglucanase ((E.C.3.2.1.4) (C x)), 1,4-β-exoglucanase (E.C.3.2.1.91), and β-D-glucoside glucohydrolase or cellobiase (E.C.3.2.1.21) (Aygan and Arıkan 2008; Wang et al. 2009; Karnchanatat et al. 2008; Bhat 2000; Yan and Wu 2013).

Cellulose enzymes are used for many purposes and known as industrially significant enzymes such as in the manufacturing of paper, textile fabric, agriculture, food additives, laundry industries, and chemical industries (Zhang et al. 2012b; Shokri and Adibkia 2013; Větrovský et al. 2014; Dorez et al. 2014). These enzymes are essential in hydrolyzing crystalline cellulose because of their processivity. In addition, hemicelluloses have a broad range of applications due to the biodegradability and nontoxicity which enable them to be used as dietary fiber (Dhingra et al. 2012; Dorez et al. 2014) and as a good coating over food stuff for their stabilization.

The capacity to degrade cellulose is found in many fungi, bacteria, and actinomycetes. Halophilic and halotolerant cellulases derived from *Actinobacteria* have been characterized and were reported in Table 12.3. Carboxymethyl cellulose (CMC) solid medium with 0.1% Congo red solution was used to screening the cellulose-producing bacteria. The clear halo zone around the colony is a sign of cellulolytic activity (Aunpad and Panbangred 2003; Rohban et al. 2009; Chen and Liu 2013).

Table 12.3 Halophilic actinomycetes isolated from different hyperfine environments have been reported to produce cellulases enzyme

No	Name of actinomycete isolate	Habitat	References
1	*Streptomyces* sp. BRC1 and *Streptomyces* sp. BRC2	Garden soil, Gujarat Vidyapith, Sadra, India	Chellapandi and Jani (2008)
2	*Streptomyces* sp. NIOT-VKKMA02	Bay of Bengal	Meena et al. (2013)
3	*Streptomyces* sp. NIOT-VKKMA26		
4	*Saccharopolyspora* sp. NIOT-VKKMA22		

12.4.3 Lipase Enzyme

Lipase enzymes are part of the family of hydrolases enzymes that act on the carboxylic ester bonds of triglycerides to diglycerides, monoglycerides, fatty acids, and glycerol; the reverse reaction in nonaqueous systems (Teo et al. 2003). Lipases are widely distributed throughout the microorganism. (1). Bacterial lipolytic enzymes are effective biocatalysts due to their broad substrate specificity and high chemo-, regio-, and stereoselectivity (de Lourdes Moreno et al. 2013; Houde et al. 2004; Jaeger and Eggert 2002; Park et al. 2009; Rodriguez et al. 2008; Snellman and Colwell 2004).

Thus, these enzymes are at present used in various fields of production of drugs in pharmaceutical, in the leather industries for the removal of subcutaneous fat, in the paper industry for the removal of impurities from raw cotton, and in dairy industry for the hydrolysis of milk fat and as detergent additives (Gomes and Steiner 2004; Sharma and Kanwar 2012; Hasan et al. 2006; Jaeger and Holliger 2010; Schmid et al. 2001; Schreck and Grunden 2014). However, industrial applications of lipases are not stable during the processes and loss of activity in the presence of the organic solvents. In this sense, the lipases enzyme isolated from extreme microorganisms represent an excellent alternative to the industrial processes (Pikuta et al. 2007). Therefore, isolation of salt-stable lipases from halophilic microorganism has been a growing attention these days (de Guzmân 2015). The availability of such enzymes would facilitate industrial processes, which need stability at high salt concentration and low water activity (Table 12.4).

12.4.4 Protease Enzyme

Proteases represent one of the most influential groups of enzymes and currently became the majority of worldwide enzyme sales due to the various applications in industry and biotechnological fields (Gohel and Singh 2012). Halophilic proteases have been commonly used in industry due to the stability and properties especially in laundry, detergents, and baking; brewing, cheese industry, leather industry, manufacturing of soy products, and tanning industry are the most common application of these enzymes (Li and Li 2009; Perez et al. 2012; Chand and Mishra 2003). In recent times, proteases enzymes in pharmaceutical industry and bioremediation process have attracted more attention. There has been considerable research on the halophilic

Table 12.4 Halophilic actinomycetes isolated from different hyperfine environments have been reported to produce lipase enzyme

No	Name of actinomycete isolate	Habitat	References
1	*Natronococcus* sp. strain TC6	–	Boutaiba et al. (2006)
2	*Streptomyces* sp. NIOT-VKKMA02	Bay of Bengal	Meena et al. (2013)
3	*Streptomyces* sp. NIOT-VKKMA26		
4	*Saccharopolyspora* sp. NIOT-VKKMA22		

Table 12.5 Halophilic actinomycetes producing protease enzymes since 2003–2014

No	Name of actinomycete isolate	References
1	*Pseudoalteromonas ruthenica*	Sanchez-Porro et al. (2003)
2	*Actinomycete* sp. MA1-1	Hayakawa et al. (2007a)
3	*Streptomyces* sp. LK3	Karthik et al. (2014)
4	*Pseudoalteromonas ruthenica* sp.	Chand and Mishra (2003)
5	*Chromohalobacter* sp. TVSP101	Vidyasagar et al. (2007)
6	*Nesterenkonia* sp.	Bakhtiar et al. (2005)
7	*Streptomyces clavuligerus* Mit-1	Thumar and Singh (2007)
8	*Nocardiopsis prasina* HA-4	Ningthoujam et al. (2009)
9	*Nocardiopsis alba* OK-5	Gohel and Singh (2012)
10	*Actinopolyspora* sp. VITSDK2	Suthindhiran et al. (2014)
11	*Actinopolyspora* sp. VITSDK2	Suthindhiran et al. (2014)
12	*Brachystreptospora xinjiangensis* OM-6	Gohel and Singh (2012)
13	*Streptomyces fungicidicus* MML1614	Ramesh et al. (2009)
14	*Nocardiopsis* sp. SD5	Saha et al. (2013)
15	*Streptomyces* sp. CW1	Kurzbaum et al. (2010)
16	*Micromonospora* sp.	Malviya et al. (2014)

proteases; many reports have been published on screening, production, and purification of proteases from microorganisms (Neagu et al. 2014b) (Sivaprakasam et al. 2011). Extracellular hydrolytic enzymes of bacteria could be screened by culturing them on a solid culture medium containing skim milk. After the suitable incubation time, clear halo zone around the colony is taken as evidence of proteolytic activity (Rohban et al. 2009). Recentily, there are a number of studies have been conducted to halophilic actinomycetes for the production of protease enzymes 2003–2014 (Table 12.5).

12.5 Role of Halophilic Actinomycetes as Potential Producer of Bioactive Compounds

12.5.1 Antibacterial

The antibacterial is an agent with various mechanisms that kill the bacteria or reduce the growth. Hence, they have drawn significant attention due to the urgent need to find novel antibacterial owing to the increase of bacterial resistance. The potential for antimicrobial resistance is one of the important considerations for physicians. Therefore, the discovery of novel antibiotics which have a powerful effect against resistant pathogenic bacteria is a major aspect of antibiotics research today. There are several studies focused on the isolation of new actinomycetes from different habitats in the context of the search for new sources (Sujatha et al. 2005). In general, the antibacterial activity of halophilic actinomycetes from marine environments is widely studied. On the other hand, the utilization of halophilic *Actinobacteria* as a source for the discovery of unique antibiotics is still at an early stage (DasSarma et al. 2010) (Table 12.6).

There are many examples of potential sources of secondary metabolites that can be used as a novel antibacterial compound isolated from halophilic actinomycete with potential use in the development of new pharmaceutical agents such as arenimycin and abyssomicin (Asolkar et al. 2010; Riedlinger et al. 2004). These compounds were classified as a new antibiotic based on the novel structure and the mode of action of this new compound and based on the inhibition of para-aminobenzoic acid biosynthesis resulting in the prohibition of the folic acid biosynthesis pathway (Abedin and Taha 2008; El Gamal 2010). There are many examples of halophilic actinomycetes that showed antibacterial activity since 2003–2014 (Table 12.6).

Table 12.6 Halophilic actinomycetes isolated from different hyperfine environments have been reported to produce unique antibiotics

Actinomycete isolate	Habitat	Pathogenic bacteria	Reference
Streptomyces viridiviolaceus	Lake Bardawil, Egypt	*C. michiganese*, *Staphylococcus* spp. *E. coli*, *Edwardsiella tarda*, and *P. solanacearum*	Rabeh et al. (2007)
Actinomycete sp.	Salt Lake Hami in Xinjiang, China	*Bacillus subtilis*	Cao et al. (2009a)
Actinomycete sp.	Salt lakes of Bay of Bengal, India	*P. aeruginosa*, *B. subtilis*, *S. epidermidis*, *E. coli*, and *C. albicans*	Ramesh and Mathivanan (2009)
Actinomycete sp.	Batim and Ribandar, Goa, India		Kamat and Kerkar (2011)
Saccharopolyspora	Salt pans, Arakkonam, Tamil Nadu, India	*K. pneumoniae*, *S. aureus*, *B. subtilis*, and *E. coli*	Suthindhiran and Kannabiran (2009a)
Streptomyces sp.	Ennore saltern, Tamil Nadu	*P. aeruginosa*, *E. coli*, *A. flavus*, and *A. fumigatus*	Lakshmipathy et al. (2010)
Streptoverticillium album	Kodiakarai, Vedaranyam, Nagapattinam, Tamil Nadu	*K. pneumoniae*, *S. aureus*, and *E. coli*	Gayathri et al. (2011)

(continued)

Table 12.6 (continued)

Actinomycete isolate	Habitat	Pathogenic bacteria	Reference
Streptomyces *Saccharomonospora*	Soil from salt pan regions of Cuddalore and Parangipettai (Porto-Novo)	*E. coli*, *K. pneumoniae*, *P. aeruginosa*, *S. aureus*, *Vibrio cholerae*, *S. typhi*, and *S. dysenteriae*	Dhanasekaran et al. (2005)
Streptomyces sp. VITSVK9	Marakkanam and Puducherry coast of the Bay of Bengal, India	*B. subtilis*, *E. coli*, *K. pneumoniae*, *A. niger*, *A. fumigatus*, and *C. albicans*	Saurav and Kannabiran (2010)
Nocardiopsis JAJ16	Salt pan soil	*B. subtilis*, *S. aureus*, *S. typhi*, *K. pneumoniae*, *Enterobacter* sp., and *P. aeruginosa*	
Streptomyces strain (JAJ06)	Salt pan soil collected at Tuticorin, India	*S. aureus*, *B. subtilis*, MRSA, *Enterobacter* sp., *K. pneumoniae*, *P. aeruginosa*, *S. typhi*, *C. albicans*	Arul Jose et al. (2011)
Actinomyces	Inland solar salterns at Sambhar Salt Lake	*S. aureus*, *B. subtilis*, *S. typhi*, *K. pneumoniae*, and *P. vulgaris*	Jose and Jebakumar (2014)
Streptomyces spp. BD21-2	Shallow water-sediment sample collected from Kailua Beach, Oahu, Hawaii	Against Gram-positive and Gram-negative bacteria	Schumacher et al. (2003)
Streptomyces sp. B6921	Coastal site of Mauritius (Indian Ocean)	*B. subtilis*, *S. aureus*, *E. coli*, and *S. viridochromogenes*	Maskey et al. (2003a)
Micromonosporaceae	Bismarck and Solomon Sea off the coast of Papua New Guinea	MDR Gram-positive bacteria	Magarvey et al. (2004)
Streptomyces B8005	Sediment of Laguna de Terminos at the Gulf of Mexico	*E. coli*, *S. viridochromogenes*, *S. aureus*, *C. albicans*, *C. vulgaris*, and *Mucor miehei*	Kock et al. (2005)
Streptomyces nodosus NPS007994	California, along the Pacific Coast of the United States	Drug-sensitive and drug-resistant Gram-positive reaction bacteria	Manam et al. (2005)
Streptomyces Merv8102	Platinum Coast on the Mediterranean Sea, north of Egypt	Gram-positive and Gram-negative bacteria	El-Gendy et al. (2008)
Streptomyces sp.	Seaside in Bigeum Island, southwest coast of South Korea	*B. subtilis*, *S. aureus*, *E. coli*, *S. typhi*, *A. niger*, *C. albicans*, and *S. cerevisiae*	Parthasarathi et al. (2010)

(continued)

Table 12.6 (continued)

Actinomycete isolate	Habitat	Pathogenic bacteria	Reference
Actinopolyspora AH1	Alibag coast, Maharashtra, India	*B. subtilis*, *S. aureus*, *S. epidermidis*	Kokare et al. (2004)
Streptomyces sp.	Water samples of the Asen fjord in the Trondheim fjord and Steinvikholmen, Norway	Gram-negative and Gram-positive bacteria	Hakvåg et al. (2008)
Streptomyces	Mangrove sediments in Zhangzhou, Fujian, China	*E. coli*, *S. aureus*, *B. subtilis*, *C. albicans*, and *R. solani*	Jiang et al. (2008)
Saccharomonospora			
Micromonospora			
Actinomadura			
Nocardiopsis			
Streptoverticillium album	Salt pans, Kodiakarai, Vedaranyam, Nagapattinam, Tamil Nadu, India	*S. aureus*, *K. pneumoniae*, and *E. coli*	Gayathri et al. (2011)
Nocardiopsis JAJ16	Salt pan soil	*Enterobacter* sp., *B. subtilis*, *P. aeruginosa*, *S. aureus*, MRSA, *K. pneumoniae*, *S. typhi*, *C. albicans*	Jose et al. (2010)
Verrucosispora AB-18-032	Japanese Sea	MDR bacteria and vancomycin-resistant *S. aureus*	Bister et al. (2004)
Streptomyces B4842	Laguna de Terminos, Gulf of Mexico	*S. aureus*, *B. subtilis*, *E. coli*, *C. albicans*	Kock et al. (2005)
Streptomyces chinaensis AUBN1/7	Bay of Bengal, India	*S. aureus*, *B. subtilis*, *B. pumilus*, *E. coli*, *P. aeruginosa*, *P. vulgaris*	Gorajana et al. (2007)
Streptomyces CNQ-418 marinopyrroles	La Jolla, California	MRSA	Hughes et al. (2008)
Marinispora	San Diego, South California	MRSA and vancomycin-resistant *E. faecium*	McArthur et al. (2008)
Nocardiopsis sp. VITSVK 5	Puducherry coast, South India	*E. coli*, *B. cereus*, *E. faecalis*, *K. pneumoniae*, *S. aureus*, *P. aeruginosa*, *Aspergillus* species	Vimal et al. (2009)
Streptomyces	Anyer West Coast, Java	*S. aureus*, *B. subtilis*, *E. coli*, *P. aeruginosa*, *C. albicans*	Sunaryanto et al. (2010)
Streptomyces, *Micromonospora*, *Actinopolyspora*, *Saccharopolyspora*	Puducherry coast, South India	*K. pneumoniae*, *B. subtilis*, *S. aureus*, *E. coli*	Suthindhiran and Kannabiran (2009b)

(continued)

Table 12.6 (continued)

Actinomycete isolate	Habitat	Pathogenic bacteria	Reference
Streptomyces, *Rhodococcus*	Pudimadaka coast, Andhra Pradesh, India	*S. aureus*, *P. aeruginosa*, *B. subtilis*, *B. cereus*, *E. coli*, *P. vulgaris*	Sivakumar et al. (2007)
Amycolatopsis alba	Visakhapatnam coast, Bay of Bengal, India	*A. formicans*, *B. subtilis*, *B. pumilus*, *S. aureus*, *E. coli*	Dasari et al. (2012)
Streptomyces coeruleorubidus	Visakhapatnam coast, Bay of Bengal, India	*B. subtilis*, *S. aureus*, *B. cereus*, *E. coli*, *P. vulgaris*, and *P. aeruginosa*	Kumar et al. (2012a)
Streptomyces, *Actinopolyspora*, and *Nocardia*	Dhanushkodi, Rathnapuram district, India	*P. aeruginosa*, *S. aureus*, *S. typhi*, *V. cholerae*	Devi et al. (2006)
Streptomyces rochei	Visakhapatnam coast, Bay of Bengal, India	*S. aureus*, *M. luteus*, *E. coli*, *A. hydrophylla*, *P. aeruginosa*, *P. fluorescens*, *V. alginolyticus*, *C. albicans*, *C. tropicana*	Reddy et al. (2011)
Streptomyces GA 22	Konkan Coast, Maharashtra, India	*S. aureus*, *P. vulgaris*, *B. subtilis*, and *E. coli*	Gulve and Deshmukh (2012)
Streptomyces alboniger	Vellar estuary, South India	*S. flexineri*, *B. subtilis*, *P. vulgaris*, *K. pneumoniae*, *V. cholerae*, and *S. aureus*	(Sahu et al. (2006))
S. vastus			
S. moderatus			
S. violaceus			
S. aureofaciens			
	Sundarbans, India	*S. aureus*, *A. protophormiae*, *B. subtilis*, *L. lactis*, *B. pumilus*, *K. pneumoniae*, *M. smegmatis*, *M. luteus*, *P. aeruginosa*, *S. marcescens*, *P. mirabilis*, and *E. coli*	Mitra et al. (2008)
Streptomyces sp. RM42, *Streptomyces* sp. RM17	Calicut mangrove, Kerala, India	*C. albicans*, *S. aureus*, *E. coli*, *S. typhi*, and *C. neoformans*	Remya and Vijayakumar (2008)
Streptomyces DPTD-5	Vellar estuary, South India	*C. tropicalis*, *S. aureus*, *E. coli*, *C. albicans*, *S. cerevisiae*, *Pseudomonas* sp., and *Bacillus* sp.	Dhanasekaran et al. (2009)
Actinomyces	Karanjal region, Sundarbans, India	*Plesiomonas*, *Hafnia* spp., *S. boydii*, *S. flexneri*, *S. sonnei*, *S. dysenteriae*, *V. cholerae*, *S. typhi*, *E. coli*	Arifuzzaman et al. (2010)
Nocardia			
Streptomyces			
Micromonospora			

(continued)

Table 12.6 (continued)

Actinomycete isolate	Habitat	Pathogenic bacteria	Reference
Actinomycetes PJS and BJS	Manakudi mangrove estuary, Arabian Sea, India	MRSA, *Enterobacter* sp., *S. typhi*, *B. subtilis*, *K. pneumoniae*, *P. vulgaris*, and *P. aeruginosa*	Jose et al. (2010)
S. neyagawaensis, *S. spheroides*, *A. aureocirculatus*, *S. albulus*, *S. antibioticus*, *S. mirabilis*, *S. umbrosus*	Muthupet mangrove ecosystem, Southeast India	*Pseudomonas* sp., *E. coli*, *Klebsiella* sp., and *Bacillus* sp.	Sathiyaseelan and Stella (2011)
Streptomyces sp. A107	Andaman and Nicobar Islands, India	*S. typhi*, *S. aureus*, *K. pneumoniae*, and *B. subtilis*	Baskaran et al. (2011)
Streptomyces sp.	Visakhapatnam, India	*E. coli*, *S. aureus*, *Bacillus subtilis*, *B. cereus*, *P. aeruginosa*, *P. vulgaris*, *S. cerevisiae*, *C. albicans*	Rao et al. (2012)
Pseudonocardia VUK-10	Nizampatnam, Andhra Pradesh, India	*S. aureus*, *Streptococcus mutans*, *B. subtilis*, *E. coli*, *E. faecalis*, *P. aeruginosa*, *C. albicans*	Mangamuri et al. (2012)
Streptomyces, *Nocardia*	Sharavathi estuary, Honnavar, Karnataka, India	*B. subtilis*, *S. aureus*, *S. typhi*, *E. coli*, *P. vulgaris*, and *P. aeruginosa*	Shobha and Vinutha (2014)

12.5.2 Antifungal

There are several fungicides used to control plant disease in agriculture, but the number of antifungal agents available in the market for controlling the fungal disease is still insufficient in comparison to antibacterial agents (Ambavane et al. 2014). A high demand of these chemical fungicide agents has generated a lot of environmental and health issues due to their toxicity; therefore, recently, the health-conscious society all over the world has made significant steps for farmers toward sustainable agriculture to detoxify the land by switching over to organic farming dispensing chemical fertilizers, pesticides, fungicides, and herbicides (Ambavane et al. 2014; Chavada et al. 2010). Therefore, the introduction of the microbial agents to control the plant diseases by eco-friendly fungicides referred as "biological control" is required. Thus, many types of research are conducted to the isolation of novel antifungals that are potentially effective against pathogenic fungi (Manivasagan et al. 2014).

Halophilic actinomycetes are a useful biological resource for the discovery of novel antifungal compound against pathogenic fungi to restrain the real risk to the future of mankind and environment. Nowadays, few studies recorded the isolation and characterization of antifungal agents from halophilic actinomycetes especially *Streptomyces*; they produce many viable antifungal compounds such as azalomycin F4a 2-ethylpentyl ester, bonactin, chandrananimycin, daryamides, and N-(2-hydroxyphenyl)-2-phenazinamine (NHP) (Subramani and Aalbersberg 2012; Maskey et al. 2003b; Gao et al. 2012). There are many examples of halophilic actinomycetes that showed antifungal activity since 2004–2016 (Table 12.7).

Table 12.7 A number of halophilic actinomycetes showed antifungal activity since 2004–2016

Name of actinomycete isolate	Habitat	Pathogenic fungi	Reference
Streptomyces species	Ennore saltern, Tamil Nadu, India	*A. flavus* and *A. fumigatus*	Gao et al. (2012)
Streptomyces spp. VITSVK9	Salt pans, Marakkanam and Puducherry coast Bay of Bengal, India	*A. niger*, *A. fumigatus*, *and C. albicans*	Saurav et al. (2013)
Nocardiopsis species JAJ16	Salt pan soil	*A. flavus* and *F. oxysporum*	Jose et al., (2010)
Actinopolyspora species AH1	Alibag coast, Maharashtra, India	*A. niger*, *A. fumigatus*, *A. flavus*, *Trichoderma*, and *Penicillium* species	Kokare et al. (2004)
Micromonosporaceae	Bismarck and the Solomon Sea, New Guinea	*C. neoformans* and *C. albicans*	Magarvey et al. (2004)
Streptomyces B8005	Laguna de Terminos, Gulf of Mexico	*C. albicans*, *C. vulgaris*, and *Mucor miehei*	Kock et al. (2005)
Streptomyces B4842	Laguna de Terminos, Gulf of Mexico	*C. albicans*	
Streptomyces, *Micromonospora*, *Nocardia*, *Streptoverticillium*, *Saccharopolyspora*	Andaman Coast, Bay of Bengal India	*C. albicans*	Peela et al (2005)
Streptomyces sp. BT-624	Andaman Coast, Bay of Bengal India	*C. albicans*	
Nocardiopsis sp. VITSVK	Puducherry coast, South India	*Aspergillus* species	Vimal et al. (2009)
Streptomyces	Anyer, West Coast, Java	*C. albicans* and *A. niger*	Sunaryanto et al. (2010)
Streptomyces hygroscopicus BDUS 49	Bigeum Island, Korea	*A. niger*, *C. albicans*, and *Saccharomyces cerevisiae*	Parthasarathi et al. (2012)
Streptomyces, *Actinopolyspora*, *Nocardia*	Dhanushkodi, Rathnapuram district, India	*Aspergillus* sp.	Devi et al. (2006)
Streptomyces	Asen fjord and Steinvikholmen islet, Norway	*C. albicans*	Hakvåg et al. (2008)

(continued)

Table 12.7 (continued)

Name of actinomycete isolate	Habitat	Pathogenic fungi	Reference
Streptomyces rochei	Visakhapatnam coast, Bay of Bengal, India	*V. alginolyticus*, *C. albicans*, and *C. tropicana*	Reddy et al. (2011)
Streptomyces roseolilacinus	Pichavaram, India	*C. albicans*	Sivakumar et al. (2007)
Streptomyces, *Micromonospora*, *Saccharomonospora*, *Actinomadura*, *Nocardiopsis*	Zhangzhou, Fujian, China	*C. albicans* and *R. solani*	Xiao et al. (2008)
Streptomyces sp.	Visakhapatnam, India	*C. albicans*, *A. niger*, and *A. flavus*	Rao et al. (2012)
Pseudonocardia VUK-10	Nizampatnam, Andhra Pradesh, India	*C. albicans*, *F. oxysporum*, and *A. niger*	Mangamuri et al. (2016)

12.5.3 Antiviral and Antitherapeutic

Although diseases compose a significant threat to the life of humans, until now, more than 30,000 viral or bacterial diseases have been discovered, and less than 30% of these diseases can be treated. New antiviral and therapeutic compound are urgently needed for pharmacological markets (Wright and Sutherland 2007). Natural products represent a major role in discovering new/novel medicine for the treatment of human diseases (Demain and Zhang 2005; Zhang 2005). Nevertheless, to date, antiviral and therapeutic compounds that have been isolated from natural products are still limited and few studies in this field in comparison to antibacterial and anti-fungi agents (Raveh et al. 2013).

In this regard, numerous investigations of bioactive compounds produced by microorganism especially halophilic actinomycetes from saline environments have been developed for controlling the human infections in the last few years (Lam 2006). Therefore, the promising antiviral and therapeutic compounds available commercially in markets are over 40 compounds (Yasuhara-Bell and Lu 2010; Abdel-Mageed et al. 2010). Among them are a few examples of some novel secondary metabolites during the period from 2003 to 2013 (Table 12.8)

12.6 Conclusions and Future Perspectives

Studies on halophilic actinomycetes are very limited, and the actinomycetes have been mentioned incidentally, on the microbial community in harsh conditions such as high salt. Further, only a little information is available on the halophilic actinomycetes with regard to their occurrence, distribution, commercially important enzymes

Table 12.8 A number of novel/new metabolites produced by halophilic actinomycetes during the period 2003–2013

Name of actinomycete isolate	Compound	Function	Reference
Streptomyces nitrosporeus	Benzastatin C	Antiviral	Abdel-Mageed et al. (2010)
Streptomyces kaviengensis	Antimycin A	Antiviral	Raveh et al. (2013)
Streptomyces sp.	Chinikomycins	Anticancer	Li et al. (2005)
Thermoactinomyces sp.	Mechercharmycins	Anticancer	Kanoh et al. (2005)
Salinispora tropica	Salinosporamide A	Anticancer	Prudhomme et al. (2008)
Salinispora arenicola	Saliniketal	Anticancer	Jensen et al. (2007)
Saccharomonospora sp.	Lodopyridone	Anticancer	Malet-Cascon et al. (2003)
Streptomyces sp.	1,8-Dihydroxy-2-ethyl-3-methylanthraquinone	Antitumor	Huang et al. (2006)
Streptomyces sp.	Caboxamycin	Anticancer	Hohmann et al. (2009)
Streptomyces sp.	Daryamides	Anticancer	Asolkar et al. (2006)
Actinomadura sp.	ZHD-0501	Anticancer	Han et al. (2005)
Streptomyces chinaensis	1-Hydroxy-1-norresistomycin	Anticancer	Gorajana et al. (2005)
Streptomyces sp.	3,6-Disubstituted indoles	Anticancer	Subramani and Aalbersberg (2012)
Streptomyces sp.	Caprolactones	Anticancer	Stritzke et al. (2004)
Marinispora	Marinomycins A-D	Anticancer	Kwon et al. (2006)
Salinispora arenicola	Arenicolides	Antitumor	Williams et al. (2007)
Streptomyces sp.	Aureolic acid	Antitumor	Lu et al. (2012)
Streptomyces aureoverticillatus	Aureoverticillactam	Antitumor	Mitchell et al. (2004)
Streptomyces sp.	Elaiomycins B and C	Antitumor	Subramani and Aalbersberg (2012)
Streptomyces sp.	Glyciapyrroles	Antitumor	Macherla et al. (2005)
Streptomyces lavendulae	Mitomycin C	Antitumor	Berdy (2005)
Streptomyces sp.	Staurosporine	Antitumor	Wu et al. (2006)
Streptomyces sp.	Piericidins	Antitumor	Hayakawa et al. (2007b)
Verrucosispora sp.	Proximicins	Antitumor	Fiedler et al. (2008)
Streptomyces sp.	Streptokordin	Antitumor	Jeong et al. (2006)
Streptoverticillium luteoverticillatum	Butenolides	Antitumor	Li et al. (2006a)

with various industrial applications of the relevant biomolecules/bioactive compounds for agriculturally, pharmaceutically, and biotechnologically application. It will be very interesting to study the mechanism of the stability properties of halophilic enzymes, which may lead to being significant novel biotechnological applications. In 2012, industrial enzymes consisted a global market of $4.5 billion, and this value increased to $4.8 billion in 2013. In addition, according to the BCC research, the enzyme market is expected to reach $7.1 billion by 2018. In respect to the growing demands for enzymes, identification of halophilic enzymes as adaptable agents against industrially harsh conditions seems to be an alternative approach. In addition to halophilic enzymes, stabilizing agents derived from halophiles have attracted extraordinary attention to several aspects of biotechnology. In conclusion, halophilic actinomycetes will be novel and useful host for the production of enzymes, chemicals, antibiotics, and biofuels in bulk with low cost.

References

Abbas IH (2006) A biological and biochemical studies of actinomycetes isolated from Kuwait saline soil-Kuwait. J Appl Sci Res 2(10):809–815

Abdel-Mageed WM, Milne BF, Wagner M, Schumacher M, Sandor P, Pathom-aree W, Goodfellow M, Bull AT, Horikoshi K, Ebel R (2010) Dermacozines, a new phenazine family from deep-sea dermacocci isolated from a Mariana Trench sediment. Org Biomol Chem 8(10):2352–2362

Abedin RM, Taha HM (2008) Antibacterial and antifungal activity of Cyanobacteria and green microalgae. Evaluation of medium components by ivaric-Burman design for antimicrobial activity of *Spirulina platensis*. Glob Hournal Biotechnol Biochem 3(1):22–31

Acharyabhatta A, Kandula SK, Terli R (2013) Taxonomy and polyphasic characterization of alkaline amylase producing marine actinomycete *Streptomyces rochei* BTSS 1001. Int J Microbiol 2013:1–8

Al-Zarban SS, Al-Musallam AA, Abbas I, Stackebrandt E, Kroppenstedt RM (2002a) *Saccharomonospora halophila* sp. nov., a novel halophilic actinomycete isolated from marsh soil in Kuwait. Int J Syst Evol Microbiol 52(2):555–558

Al-Zarban SS, Al-Musallam AA, Abbas IH, Fasasi YA (2002b) Noteworthy salt-loving actinomycetes from Kuwait. Kuwait S Sci Eng 29:99–109

Amann RI, Ludwig W, Schleifer K-H (1995) Phylogenetic identification and in situ detection of individual microbial cells without cultivation. Microbiol Rev 59(1):143–169

Ambavane V, Tokdar P, Parab R, Sreekumar E, Mahajan G, Mishra PD, D'Souza L, Ranadive P (2014) Caerulomycin A—an antifungal compound isolated from marine actinomycetes. Adv Microbiol 4(9):567–578

Arifuzzaman M, Khatun M, Rahman H (2010) Isolation and screening of actinomycetes from Sundarbans soil for antibacterial activity. Afr J Biotechnol 9(29):4615–4619

Arul Jose P, Satheeja Santhi V, Jebakumar SR (2011) Phylogenetic-affiliation, antimicrobial potential and PKS gene sequence analysis of moderately halophilic *Streptomyces* sp. inhabiting an Indian saltpan. J Basic Microbiol 51(4):348–356

Asolkar RN, Jensen PR, Kauffman CA, Fenical W (2006) Daryamides AC, weakly cytotoxic polyketides from a marine-derived actinomycete of the genus *Streptomyces* strain CNQ-085. J Nat Prod 69(12):1756–1759

Asolkar RN, Kirkland TN, Jensen PR, Fenical W (2010) Arenimycin, an antibiotic effective against rifampin-and methicillin-resistant *Staphylococcus aureus* from the marine actinomycete *Salinispora arenicola*. J Antibiot 63(1):37

Aunpad R, Panbangred W (2003) Cloning and characterization of the constitutively expressed chitinase C gene from a marine bacterium, *Salinivibrio costicola* strain 5SM-1. J Biosci Bioeng 96(6):529–536

Aygan A, Arıkan B (2008) A new halo-alkaliphilic, thermostable endoglucanase from moderately halophilic *Bacillus* sp. C14 isolated from Van Soda Lake. Int J Agric Biol 10(4):369–374

Bakhtiar S, Estiveira RJ, Hatti-Kaul R (2005) Substrate specificity of alkaline protease from alkaliphilic feather-degrading *Nesterenkonia* sp. AL20. Enzyme Microb Technol 37(5):534–540

Baranasic D, Gacesa R, Starcevic A, Zucko J, Blažič M, Horvat M, Gjuračić K, Fujs Š, Hranueli D, Kosec G (2013) Draft genome sequence of *Streptomyces rapamycinicus* strain NRRL 5491, the producer of the immunosuppressant rapamycin. Genome Announc 1(4):e00581–e00513

Baskaran R, Vijayakumar R, Mohan P (2011) Enrichment method for the isolation of bioactive actinomycetes from mangrove sediments of Andaman Islands, India. Malays J Microbiol 7(1):1–7

Battistuzzi FU, Hedges SB (2009) A major clade of prokaryotes with ancient adaptations to life on land. Mol Biol Evol 26(2):335–343

Berdy J (2005) Bioactive microbial metabolites. J Antibiot 58(1):1

Bhat M (2000) Cellulases and related enzymes in biotechnology. Biotechnol Advanc 18(5):355–383

Bian J, Li Y, Wang J, Song F-H, Liu M, Dai H-Q, Ren B, Gao H, Hu X, Liu Z-H (2009) *Amycolatopsis marina* sp. nov., an actinomycete isolated from an ocean sediment. Int J Syst Evol Microbiol 59(3):477–481

Bister B, Bischoff D, Ströbele M, Riedlinger J, Reicke A, Wolter F, Bull AT, Zähner H, Fiedler HP, Süssmuth RD (2004) Abyssomicin C—a polycyclic antibiotic from a marine *Verrucosispora* strain as an inhibitor of the p-Aminobenzoic acid/tetrahydrofolate biosynthesis pathway. Angew Chem Int Ed 43(19):2574–2576

Boutaiba S, Bhatnagar T, Hacene H, Mitchell D, Baratti J (2006) Preliminary characterisation of a lipolytic activity from an extremely halophilic archaeon, *Natronococcus* sp. J Mol Catal B: Enzym 41(1):21–26

Bull AT (2011) Actinobacteria of the extremobiosphere. In: Extremophiles handbook. Springer, Tokyo, pp 1203–1240

Cai M, Zhi X-Y, Tang S-K, Zhang Y-Q, Xu L-H, Li W-J (2008) *Streptomonospora halophila* sp. nov., a halophilic actinomycete isolated from a hypersaline soil. Int J Syst Evol Microbiol 58(7):1556–1560

Cao L, Yun W, Tang S, Zhang P, Mao P, Jing X, Wang C, Lou K (2009a) Biodiversity and enzyme screening of actinomycetes from Hami lake. Wei sheng wu xue bao=. Acta Microbiol Sin 49(3):287–293

Cao Y-R, Jiang Y, Wu J-Y, Xu L-H, Jiang C-L (2009b) *Actinopolymorpha alba* sp. nov., isolated from a rhizosphere soil. Int J Syst Evol Microbiol 59(9):2200–2203

Carro L, Zúñiga P, De la Calle F, Trujillo ME (2012) *Streptomyces pharmamarensis* sp. nov. isolated from a marine sediment. Int J Syst Evol Microbiol 62(5):1165–1170

Caton TM, Caton I, Witte LR, Schneegurt MA (2009) Archaeal diversity at the Great Salt Plains of Oklahoma described by cultivation and molecular analyses. Microb Ecol 58(3):519–528

Chakraborty S, Khopade A, Kokare C, Mahadik K, Chopade B (2009) Isolation and characterization of novel α-amylase from marine *Streptomyces* sp. D1. J Mol Catal B: Enzym 58(1):17–23

Chakraborty S, Khopade A, Biao R, Jian W, Liu X-Y, Mahadik K, Chopade B, Zhang L, Kokare C (2011) Characterization and stability studies on surfactant, detergent and oxidant stable α-amylase from marine haloalkaliphilic *Saccharopolyspora* sp. A9. J Mol Catal B: Enzym 68(1):52–58

Chakraborty S, Raut G, Khopade A, Mahadik K, Kokare C (2012) Study on calcium ion independent alpha-amylase from haloalkaliphilic marine *Streptomyces* strain A 3. Indian J Biotechnol 11(4):427–437

Chand S, Mishra P (2003) Research and application of microbial enzymes—India's contribution. In: Biotechnology in India II. Springer, Berlin, pp 95–124

Chandrasekaran M (1997) Industrial enzymes from marine microorganisms: the Indian scenario. J Mar Biotechnol 5:86–89

Chang X, Liu W, Zhang X-H (2011) *Spinactinospora alkalitolerans* gen. nov., sp. nov., an actinomycete isolated from marine sediment. Int J Syst Evol Microbiol 61(12):2805–2810

Chavada NB, Dave A, Patel B (2010) Study on diazotrophic and IAA producing bacteria isolated from desert soil submit an application for biofertilizer. Int J Appl Biol Pharm Technol 1(3):1067–1071

Chellapandi P, Jani HM (2008) Production of endoglucanase by the native strains of *Streptomyces* isolates in submerged fermentation. Braz J Microbiol 39(1):122–127

Chen WQ, Liu YY (2013) Isolation and identification of Halomonas sp. ZSCW-10: a moderately halophilic bacteria strain with cellulase activity. In: Advanced materials research. Trans Tech Publications, Zurich, pp 236–241
Chen M, Xu P, Zeng G, Yang C, Huang D, Zhang J (2015) Bioremediation of soils contaminated with polycyclic aromatic hydrocarbons, petroleum, pesticides, chlorophenols and heavy metals by composting: applications, microbes and future research needs. Biotechnol Advanc 33(6):745–755
Claridge G, Campbell I (2004) The biology of arid cryosols. In: Cryosols. Springer, Berlin, pp 573–580
Cojoc R, Merciu S, Popescu G, Dumitru L, Kamekura M, Enache M (2009) Extracellular hydrolytic enzymes of halophilic bacteria isolated from a subterranean rock salt crystal. Rom Biotechnol Lett 14:4658–4664
Collins MD, Lawson PA, Labrenz M, Tindall BJ, Weiss N, Hirsch P (2002) *Nesterenkonia lacusekhoensis* sp. nov., isolated from hypersaline Ekho Lake, East Antarctica, and emended description of the genus *Nesterenkonia*. Int J Syst Evol Microbiol 52(4):1145–1150
Coronado M-J, Vargas C, Hofemeister J, Ventosa A, Nieto JJ (2000) Production and biochemical characterization of an α-amylase from the moderate halophile *Halomonas meridiana*. FEMS Microbiol Lett 183(1):67–71
Cotârleţ M (2013) Medium optimization for the production of cold-active beta amylase by psychrotrophic *Streptomyces* MIUG 4 alga using response surface methodology. Microbiology 82(2):147–154
Cui X-L, Mao P-H, Zeng M, Li W-J, Zhang L-P, Xu L-H, Jiang C-L (2001) *Streptimonospora salina gen.* nov., sp. nov., a new member of the family *Nocardiopsaceae*. Int J Syst Evol Microbiol 51(2):357–363
Dalmaso GZL, Ferreira D, Vermelho AB (2015) Marine extremophiles: a source of hydrolases for biotechnological applications. Mar Drugs 13(4):1925–1965
Dang H, Zhu H, Wang J, Li T (2009) Extracellular hydrolytic enzyme screening of culturable heterotrophic bacteria from deep-sea sediments of the Southern Okinawa Trough. World J Microbiol Biotechnol 25(1):71–79
Dasari VRRK, Muthyala MKK, Nikku MY, Donthireddy SRR (2012) Novel Pyridinium compound from marine actinomycete, *Amycolatopsis alba var.* nov. DVR D4 showing antimicrobial and cytotoxic activities in vitro. Microbiol Res 167(6):346–351
de Guzmán MN (2015) Lipolytic enzyme production by halophilic/halotolerant microorganisms isolated from Laguna Verde, Bolivia. Revista Boliviana de Química 25(1):14–23
de Lourdes Moreno M, Pérez D, García MT, Mellado E (2013) Halophilic bacteria as a source of novel hydrolytic enzymes. Life 3(1):38–51
Demain AL, Zhang L (2005) Natural products and drug discovery. In: Natural products. Springer, Totowa, pp 3–29
Devi A, Jolitha AB, Ishii N (2006) Grape seed proanthocyanidin extract (GSPE) and antioxidant defense in the brain of adult rats. Medi Sci Monit 12(4):BR124–BR129
Dhanasekaran D, Rajakumar G, Sivamani P, Selvamani S, Panneerselvam A, Thajuddin N (2005) Screening of salt pans actinomycetes for antibacterial agents. Internet J Microbiol 1(2):1–8
Dhanasekaran D, Selvamani S, Panneerselvam A, Thajuddin N (2009) Isolation and characterization of actinomycetes in Vellar Estuary, Annagkoil, Tamil Nadu. Afr J Biotechnol 8(17):4159–4162
Dhingra D, Michael M, Rajput H, Patil R (2012) Dietary fibre in foods: a review. J Food Sci Technol 49(3):255–266
Diraviyam T, Radhakrishnan M, Balagurunathan R (2011) Antioxidant activity of melanin pigment from *Streptomyces* species D5 isolated from desert soil, Rajasthan, India. Drug Invent Today 3(3):12–13
Dorez G, Ferry L, Sonnier R, Taguet A, Lopez-Cuesta J-M (2014) Effect of cellulose, hemicellulose and lignin contents on pyrolysis and combustion of natural fibers. J Anal Appl Pyrolysis 107:323–331
Duckworth WC, Bennett RG, Hamel FG (1998) Insulin degradation: progress and potential 1. Endocr Rev 19(5):608–624
El Gamal AA (2010) Biological importance of marine algae. Saudi Pharm J 18(1):1–25
El-Gendy MM, Shaaban M, Shaaban KA, El-Bondkly AM, Laatsch H (2008) Essramycin: a first triazolopyrimidine antibiotic isolated from nature [dagger]. J Antibiot 61(3):149

Ellaiah P, Adinarayana K, Bhavani Y, Padmaja P, Srinivasulu B (2002) Optimization of process parameters for glucoamylase production under solid state fermentation by a newly isolated *Aspergillus* species. Process Biochem 38(4):615–620

Ellaiah P, Ramana T, Raju K, Sujatha P, Sankar AU (2004) Investigations on marine actinomycetes from bay of Bengal near Kakinada coast of Andhra Pradesh. Asian J Microbiol Biotechnol Environ Sci 6:53–56

Elleuche S, Schröder C, Sahm K, Antranikian G (2014) Extremozymes—biocatalysts with unique properties from extremophilic microorganisms. Curr opin Biotechnol 29:116–123

Elshahed MS, Najar FZ, Roe BA, Oren A, Dewers TA, Krumholz LR (2004) Survey of archaeal diversity reveals an abundance of halophilic Archaea in a low-salt, sulfide-and sulfur-rich spring. Appl Environ Microbiol 70(4):2230–2239

Enache M, Kamekura M (2010) Hydrolytic enzymes of halophilic microorganisms and their economic values. Rom J Biochem 47:47–59

Fiedler H-P, Bruntner C, Riedlinger J, Bull AT, Knutsen G, Goodfellow M, Jones A, Maldonado L, Pathom-Aree W, Beil W (2008) Proximicin A, B and C, novel aminofuran antibiotic and anticancer compounds isolated from marine strains of the *Actinomycete Verrucosispora* [dagger]. J Antibiot 61(3):158

Gao X, Lu Y, Xing Y, Ma Y, Lu J, Bao W, Wang Y, Xi T (2012) A novel anticancer and antifungus phenazine derivative from a marine actinomycete BM-17. Microbiol Res 167(10):616–622

Garrity GM, Holt JG (2001) The road map to the manual. In: Bergey's manual® of systematic bacteriology. Springer, New York, pp 119–166

Gayathri A, Madhanraj P, Panneerselvam A (2011) Diversity, antibacterial activity and molecular characterization of actinomycetes isolated from salt Pan Region of Kodiakarai, *Nagapattinam* DT. Asian J Pharm Technol 1(3):79–81

George M, Anjumol A, George G, Hatha AM (2012) Distribution and bioactive potential of soil actinomycetes from different ecological habitats. Afr J Microbiol Res 6(10):2265–2271

Ghanem NB, Sabry SA, El-Sherif ZM, Abu El-Ela GA (2000) Isolation and enumeration of marine actinomycetes from seawater and sediments in Alexandria. J Gen Appl Microbiol 46(3):105–111

Gohel SD, Singh SP (2012) Purification strategies, characteristics and thermodynamic analysis of a highly thermostable alkaline protease from a salt-tolerant alkaliphilic actinomycete, *Nocardiopsis alba* OK-5. J Chromatogr B 889:61–68

Gomes J, Steiner W (2004) The biocatalytic potential of extremophiles and extremozymes. Food Technol Biotechnol 42(4):223–235

Goodfellow M, Kämpfer P, Busse H-J, Trujillo ME, K-i S, Ludwig W, Whitman WB (2012) Bergey's manual® of systematic bacteriology: volume five the actinobacteria, part A. Springer, New York

Gorajana A, Kurada BV, Peela S, Jangam P, Vinjamuri S, Poluri E, Zeeck A (2005) 1-Hydroxy-1-norresistomycin, a new cytotoxic compound from a marine actinomycete, *Streptomyces chibaensis* AUBN1/7 [dagger]. J Antibiot 58(8):526

Gorajana A, Venkatesan M, Vinjamuri S, Kurada BV, Peela S, Jangam P, Poluri E, Zeeck A (2007) Resistoflavine, cytotoxic compound from a marine actinomycete, *Streptomyces chibaensis* AUBN 1/7. Microbiol Res 162(4):322–327

Guan T-W, Tang S-K, Wu J-Y, Zhi X-Y, Xu L-H, Zhang L-L, Li W-J (2009) *Haloglycomyces albus gen.* nov., sp. nov., a halophilic, filamentous actinomycete of the family *Glycomycetaceae*. Int J Syst Evol Microbiol 59(6):1297–1301

Guan T-W, Wu N, Xia Z-F, Ruan J-S, Zhang X-P, Huang Y, Zhang L-L (2011a) *Saccharopolyspora lacisalsi* sp. nov., a novel halophilic actinomycete isolated from a salt lake in Xinjiang, China. Extremophiles 15(3):373–378

Guan T-W, Xia Z-F, Xiao J, Wu N, Chen Z-J, Zhang L-L, Zhang X-P (2011b) *Glycomyces halotolerans* sp. nov., a novel actinomycete isolated from a hypersaline habitat in Xinjiang, China. Antonie Van Leeuwenhoek 100(1):137–143

Guan T-W, Wei B, Zhang Y, Xia Z-F, Che Z-M, Chen X-G, Zhang L-L (2013) *Actinopolysporalacussalsi* sp. nov., an extremely halophilic actinomycete isolated from a salt lake. Int J Syst Evol Microbiol 63(8):3009–3013

Gulve R, Deshmukh A (2012) Antimicrobial activity of the marine actinomycetes. Int Multidiscip Res J 2(3):16–22

Hakvåg S, Fjærvik E, Josefsen KD, Ian E, Ellingsen TE, Zotchev SB (2008) Characterization of *Streptomyces* spp. isolated from the sea surface microlayer in the Trondheim fjord, Norway. Mar Drugs 6(4):620–635

Han X-X, Cui C-B, Gu Q-Q, Zhu W-M, Liu H-B, Gu J-Y, Osada H (2005) ZHD-0501, a novel naturally occurring *staurosporine analog* from *Actinomadura* sp. 007. Tetrahedron Lett 46(36):6137–6140

Han X-X, Luo X-X, Zhang L-L (2014) *Glycomycesfuscus* sp. nov. and *Glycomycesalbus* sp. nov., actinomycetes isolated from a hypersaline habitat. Int J Syst Evol Microbiol 64(7):2437–2441

Haritha R, Siva Kumar K, Jagan Mohan Y, Ramana T (2010) Amylolytic and proteolytic Actinobacteria isolated from marine sediments of Bay of Bengal. Int J Microbiol Res 1(2):37–44

Hasan F, Shah AA, Hameed A (2006) Industrial applications of microbial lipases. Enzyme Microb Technol 39(2):235–251

Hayakawa Y, Shirasaki S, Kawasaki T, Matsuo Y, Adachi K, Shizuri Y (2007a) Structures of new cytotoxic antibiotics, piericidins C7 and C8. J Antibiot 60(3):201

Hayakawa Y, Shirasaki S, Shiba S, Kawasaki T, Matsuo Y, Adachi K, Shizuri Y (2007b) Piericidins C7 and C8, new cytotoxic antibiotics produced by a marine *Streptomyces* sp. Journal of Antibiot 60(3):196

He J, Xu Y, Sahu MK, Tian X-P, Nie G-X, Xie Q, Zhang S, Sivakumar K, Li W-J (2012) *Actinomadura sediminis* sp. nov., a marine actinomycete isolated from mangrove sediment. Int J Syst Evol Microbiol 62(5):1110–1116

Hohmann C, Schneider K, Bruntner C, Irran E, Nicholson G, Bull AT, Jones AL, Brown R, Stach JE, Goodfellow M (2009) Caboxamycin, a new antibiotic of the benzoxazole family produced by the deep-sea strain *Streptomyces* sp. NTK 937. J Antibiot 62(2):99–104

Houde A, Kademi A, Leblanc D (2004) Lipases and their industrial applications. Appl Biochem Biotechnol 118(1-3):155–170

Hozzein WN, Goodfellow M (2011) *Actinopolyspora egyptensis* sp. nov., a new halophilic actinomycete. Afr J Microbiol Res 5(2):100–105

Huang X, El-Sayed IH, Qian W, El-Sayed MA (2006) Cancer cell imaging and photothermal therapy in the near-infrared region by using gold nanorods. J Am Chem Soc 128(6):2115–2120

Hughes CC, Prieto-Davo A, Jensen PR, Fenical W (2008) The marinopyrroles, antibiotics of an unprecedented structure class from a marine *Streptomyces* sp. Organic Lett 10(4):629–631

Jaeger K-E, Eggert T (2002) Lipases for biotechnology. Curr Opin Biotechnol 13(4):390–397

Jaeger K-E, Holliger P (2010) Chemical biotechnology—a marriage of convenience and necessity. Curr Opin Biotechnol 21(6):711–712

Jayachandra S, Kumar A, Merley D, Sulochana M (2012) Isolation and characterization of extreme halophilic bacterium *Salinicoccus* sp. JAS4 producing extracellular hydrolytic enzymes. Recent Res Sci Technol 4(4):46–49

Jensen PR, Williams PG, Oh D-C, Zeigler L, Fenical W (2007) Species-specific secondary metabolite production in marine actinomycetes of the genus *Salinispora*. Appl Environ Microbiol 73(4):1146–1152

Jeong S-Y, Shin HJ, Kim TS, Lee H-S, S-k P, Kim HM (2006) Streptokordin, a new cytotoxic compound of the methylpyridine class from a marine-derived *Streptomyces* sp. KORDI-3238. J Antibiot 59(4):234

Jiang Y, Xu P, Lou K, Xu L H, Liu Z H (2008) Problems and countermeasure on development of pharmaceuticalsfrom actinobacteria resource. Microbiology 35: 272–274.

Jones B, Grant W (1999) Microbial diversity and ecology of the Soda Lakes of East Africa. In: Microbial biosystems: new frontiers: proceedings of the 8th international symposium for microbial ecology (2000). Atlantic Canada Society for Microbial Ecology, Halifax, pp 681–687

Jose PA, Jebakumar SRD (2014) Unexplored hypersaline habitats are sources of novel actinomycetes. Front Microbiol 5:242

Jose A, Satheeja Santhi V, Jebakumar Solomon R (2010) In vitro antimicrobial potential and growth characteristics of *Nocardiopsis* sp. JAJ16 isolated from crystallizer pond. Int J Curr Res 3:24–26

Kadziola A, Søgaard M, Svensson B, Haser R (1998) Molecular structure of a barley α-amylase-inhibitor complex: implications for starch binding and catalysis. J Mol Biol 278(1):205–217

Kamat T, Kerkar S (2011) Bacteria from Salt Pans: a potential resource of antibacterial metabolites. Rec Res Sci Technol 3(9):46–52

Kanekar P, Kanekar S, Kelkar A, Dhakephalkar P (2012) Halophiles–taxonomy, diversity, physiology and applications. In: Microorganisms in environmental management. Springer, Dordrecht, pp 1–34

Kanoh K, Matsuo Y, Adachi K, Imagawa H, Nishizawa M, Shizuri Y (2005) Mechercharmycins A and B, cytotoxic substances from marine-derived *Thermoactinomyces* sp. YM3-251. J Antibiot 58(4):289

Kar S, Ray R (2008) Partial characterization and optimization of extracellular thermostable Ca^{2+} inhibited α-amylase production by *Streptomyces erumpens* MTCC 7317. J Sci Ind Res 67:58–64

Karan R, Kumar S, Sinha R, Khare S (2012) Halophilic microorganisms as sources of novel enzymes. In: Microorganisms in sustainable agriculture and biotechnology. Springer, Dordrecht, pp 555–579

Karnchanatat A, Petsom A, Sangvanich P, Piapukiew J, Whalley AJ, Reynolds CD, Gadd GM, Sihanonth P (2008) A novel thermostable endoglucanase from the wood-decaying fungus *Daldinia eschscholzii* (Ehrenb.: Fr.) Rehm. Enzyme Microb Technol 42(5):404–413

Karthik L, Kumar G, Keswani T, Bhattacharyya A, Chandar SS, Rao KB (2014) Protease inhibitors from marine actinobacteria as a potential source for antimalarial compound. PloS one 9(3):e90972

Kikani B, Shukla R, Singh S (2010) Biocatalytic potential of thermophilic bacteria and actinomycetes. Curr Res Technol Educ Top Appl Microbiol Microbial Biotechnol 2:1000–1007

Koch AL (2003) Were Gram-positive rods the first bacteria? Trends Microbiol 11(4):166–170

Kock I, Maskey RP, Biabani MF, Helmke E, Laatsch H (2005) 1-Hydroxy-1-norresistomycin and resistoflavin methyl ether: new antibiotics from marine-derived *Streptomycetes* [dagger],[dagger][dagger]. J Antibiot 58(8):530

Kokare C, Mahadik K, Kadam S, Chopade B (2004) Isolation, characterization and antimicrobial activity of marine halophilic *Actinopolyspora* species AH1 from the west coast of India. Curr Sci 86(4):593–597

Kumar S, Khare S (2012) Purification and characterization of maltooligosaccharide-forming α-amylase from moderately halophilic Marinobacter sp. EMB8. Bioresour Technol 116:247–251

Kumar KS, Haritha R, Swathi A, Sirisha B, Ramana T (2012a) T axonomy and antimicrobial activity of *Streptomyces coeruleorubidus* sp. isolated from marine sediment. Res J Microbiol 7(3).171–181

Kumar S, Karan R, Kapoor S, Singh S, Khare S (2012b) Screening and isolation of halophilic bacteria producing industrially important enzymes. Braz J Microbiol 43(4):1595–1603

Kurzbaum E, Kirzhner F, Sela S, Zimmels Y, Armon R (2010) Efficiency of phenol biodegradation by planktonic *Pseudomonas pseudoalcaligenes* (a constructed wetland isolate) vs. root and gravel biofilm. Water Res 44(17):5021–5031

Kushner DJ, Kamekura M (1988) Physiology of halophilic eubacteria. In: Halophilic bacteria, vol 1. CRC Press, Boca Raton, pp 109–140

Kwon HC, Kauffman CA, Jensen PR, Fenical W (2006) Marinomycins AD, antitumor-antibiotics of a new structure class from a marine actinomycete of the recently discovered genus "*Marinispora*". J Am Chem Soc 128(5):1622–1632

Lakshmipathy TD, Prasad AA, Kannabiran K (2010) Production of biosurfactant and heavy metal resistance activity of *Streptomyces* sp. VITDDK3-a novel halo tolerant actinomycetes isolated from saltpan soil. Biol Res 4(2):108–115

Lam KS (2006) Discovery of novel metabolites from marine actinomycetes. Curr Opin Microbiol 9(3):245–251

Lawson PA, Collins MD, Schumann P, Tindall B, Hirsch P, Labrenz M (2000) New LL-diaminopimelic acid-containing actinomycetes from hypersaline, heliothermal and meromictic Antarctic Ekho Lake: *Nocardioides aquaticus* sp. nov. and Friedmannielly lacustris sp. nov. Syst Appl Microbiol 23(2):219–229

Li AN, Li DC (2009) Cloning, expression and characterization of the serine protease gene from *Chaetomium thermophilum*. J Appl Microbiol 106(2):369–380

Li W-J, Tang S-K, Stackebrandt E, Kroppenstedt RM, Schumann P, Xu L-H, Jiang C-L (2003a) *Saccharomonospora paurometabolica* sp. nov., a moderately halophilic actinomycete isolated from soil in China. Int J Syst Evol Microbiol 53(5):1591–1594

Li W-J, Xu P, Tang S-K, Xu L-H, Kroppenstedt RM, Stackebrandt E, Jiang C-L (2003b) *Prauserella halophila* sp. nov. and Prauserella alba sp. nov., moderately halophilic actinomycetes from saline soil. Int J Syst Evol Microbiol 53(5):1545–1549

Li W-J, Xu P, Zhang L-P, Tang S-K, Cui X-L, Mao P-H, Xu L-H, Schumann P, Stackebrandt E, Jiang C-L (2003c) *Streptomonospora alba* sp. nov., a novel halophilic actinomycete, and emended description of the genus *Streptomonospora* Cui et al. 2001. Int J Syst Evol Microbiol 53(5):1421–1425

Li W-J, Chen H-H, Zhang Y-Q, Schumann P, Stackebrandt E, Xu L-H, Jiang C-L (2004a) *Nesterenkonia halotolerans* sp. nov. and Nesterenkonia xinjiangensis sp. nov., actinobacteria from saline soils in the west of China. Int J Syst Evol Microbiol 54(3):837–841

Li W-J, Park D-J, Tang S-K, Wang D, Lee J-C, Xu L-H, Kim C-J, Jiang C-L (2004b) *Nocardiopsis salina* sp. nov., a novel halophilic actinomycete isolated from saline soil in China. Int J Syst Evol Microbiol 54(5):1805–1809

Li F, Maskey RP, Qin S, Sattler I, Fiebig HH, Maier A, Zeeck A, Laatsch H (2005) Chinikomycins A and B: isolation, structure elucidation, and biological activity of novel antibiotics from a marine *Streptomyces* sp. isolate M045#, 1. J Nat Prod 68(3):349–353

Li D-H, Zhu T-J, Liu H-B, Fang Y-C, Gu Q-Q, Zhu W-M (2006a) Four butenolides are novel cytotoxic compounds isolated from the marine-derived bacterium, *Streptoverticillium luteoverticillatum* 11014. Arch Pharm Res 29(8):624–626

Li W-J, Kroppenstedt RM, Wang D, Tang S-K, Lee J-C, Park D-J, Kim C-J, Xu L-H, Jiang C-L (2006b) Five novel species of the genus *Nocardiopsis* isolated from hypersaline soils and emended description of *Nocardiopsis* salina Li et al. 2004. Int J Syst Evol Microbiol 56(5):1089–1096

Li W-J, Zhang Y-Q, Schumann P, Chen H-H, Hozzein WN, Tian X-P, Xu L-H, Jiang C-L (2006c) *Kocuria aegyptia* sp. nov., a novel actinobacterium isolated from a saline, alkaline desert soil in Egypt. Int J Syst Evol Microbiol 56(4):733–737

Li W-J, Zhang Y-Q, Schumann P, Liu H-Y, Yu L-Y, Zhang Y-Q, Stackebrandt E, Xu L-H, Jiang C-L (2008) *Nesterenkonia halophila* sp. nov., a moderately halophilic, alkalitolerant actinobacterium isolated from a saline soil. Int J Syst Evol Microbiol 58(6):1359–1363

Liu J, Zhang Z, Liu Z, Zhu H, Dang H, Lu J, Cui Z (2011) Production of cold-adapted amylase by marine bacterium *Wangia* sp. C52: optimization, modeling, and partial characterization. Mar Biotechnol 13(5):837–844

Lu J, Ma Y, Liang J, Xing Y, Xi T, Lu Y (2012) Aureolic acids from a marine-derived *Streptomyces* sp. WBF16. Microbiol Res 167(10):590–595

Ludwig W, Klenk H-P (2001) Overview: a phylogenetic backbone and taxonomic framework for procaryotic systematics. In: Bergey's manual® of systematic bacteriology. Springer, New York, pp 49–65

Macherla VR, Liu J, Bellows C, Teisan S, Nicholson B, Lam KS, Potts BC (2005) Glaciapyrroles A, B, and C, pyrrolosesquiterpenes from a *Streptomyces* sp. isolated from an Alaskan marine sediment. J Nat Prod 68(5):780–783

Machius M, Wiegand G, Huber R (1995) Crystal structure of calcium-depleted *Bacillus licheniformis* α-amylase at 2.2 Å resolution. J Mol Biol 246(4):545–559

Magarvey NA, Keller JM, Bernan V, Dworkin M, Sherman DH (2004) Isolation and characterization of novel marine-derived actinomycete taxa rich in bioactive metabolites. Appl Environ Microbiol 70(12):7520–7529

Malet-Cascon L, Romero F, Espliego-Vazquez F, Grávalos D, Fernández-Puentes JL (2003) IB-00208, a new cytotoxic polycyclic xanthone produced by a marine-derived Actinomadura. I. Isolation of the strain, taxonomy and biological activites. J Antibiot 56(3):219–225

Malviya N, Yandigeri MS, Yadav AK, Solanki MK, Arora DK (2014) Isolation and characterization of novel alkali-halophilic actinomycetes from the Chilika brackish water lake, India. Ann Microbiol 64(4):1829–1838

Manam RR, Teisan S, White DJ, Nicholson B, Grodberg J, Neuteboom ST, Lam KS, Mosca DA, Lloyd GK, Potts BC (2005) Lajollamycin, a nitro-tetraene spiro-β-lactone-γ-lactam antibiotic from the marine actinomycete *Streptomyces n odosus*. J Nat Prod 68(2):240–243

Mangamuri UK, Muvva V, Poda S, Kamma S (2012) Isolation, identification and molecular characterization of rare actinomycetes from mangrove ecosystem of nizampatnam. Malays J Microbiol 8(2):83–91

Mangamuri UK, Vijayalakshmi M, Poda S, Manavathi B, Chitturi B, Yenamandra V (2016) Isolation and biological evaluation of N-(4-aminocyclooctyl)-3, 5-dinitrobenzamide, a new semisynthetic derivative from the Mangrove-associated actinomycete *Pseudonocardia endophytica* VUK-10. 3. Biotech 6(2):158

Manivasagan P, Gnanam S, Sivakumar K, Thangaradjou T, Vijayalakshmi S, Balasubramanian T (2010) Isolation, identification and characterization of multiple enzyme producing actinobacteria from sediment samples of Kodiyakarai coast, the Bay of Bengal. Afr J Microbiol Res 4(14):1550–1559

Manivasagan P, Kang K-H, Sivakumar K, Li-Chan EC, Oh H-M, Kim S-K (2014) Marine actinobacteria: an important source of bioactive natural products. Environ Toxicol Pharmacol 38(1):172–188

Maskey RP, Helmke E, Laatsch H (2003a) Himalomycin A and B: isolation and structure elucidation of new fridamycin type antibiotics from a marine *Streptomyces* isolate. J Antibiot 56(11):942–949

Maskey RP, Li FC, Qin S, Fiebig HH, Laatsch H (2003b) Chandrananimycins AC: production of novel anticancer antibiotics from a marine *Actinomadura* sp. isolate M048 by variation of medium composition and growth conditions. J Antibiot 56(7):622–629

McArthur KA, Mitchell SS, Tsueng G, Rheingold A, White DJ, Grodberg J, Lam KS, Potts BC (2008) Lynamicins A– E, chlorinated bisindole pyrrole antibiotics from a novel marine actinomycete. J Nat Prod 71(10):1732–1737

Meena B, Rajan LA, Vinithkumar NV, Kirubagaran R (2013) Novel marine actinobacteria from emerald Andaman & Nicobar Islands: a prospective source for industrial and pharmaceutical byproducts. BMC Microbiol 13(1):1

Meklat A, Bouras N, Zitouni A, Mathieu F, Lebrihi A, Schumann P, Spröer C, Klenk H-P, Sabaou N (2013) *Actinopolysporamzabensis* sp. nov., a halophilic actinomycete isolated from an Algerian Saharan soil. Int J Syst Evol Microbiol 63(10):3787–3792

Mitchell SS, Nicholson B, Teisan S, Lam KS, Potts BC (2004) *Aureoverticillactam*, a novel 22-atom macrocyclic lactam from the marine actinomycete Streptomyces aureoverticillatus. J Nat Prod 67(8):1400–1402

Mitra A, Santra SC, Mukherjee J (2008) Distribution of actinomycetes, their antagonistic behaviour and the physico-chemical characteristics of the world's largest tidal mangrove forest. Appl Microbiol Biotechnol 80(4):685–695

Neagu S, Enache M, Cojoc R (2014a) Extracellular hydrolytic activities of halophilic microorganisms isolated from Balta Albă salt lake. Roman Biotechnol Lett 19(1):8951–8958

Neagu S, Preda S, Anastasescu C, Zaharescu M, Enache M, Cojoc R (2014b) The functionalization of silica and titanate nanostructures with halotolerant proteases. Rom J Chem 2:97–103

Ningthoujam DS, Kshetri P, Sanasam S, Nimaichand S (2009) Screening, identification of best producers and optimization of extracellular proteases from moderately halophilic *alkalithermotolerant indigenous* actinomycetes. World Appl Sci J 7(7):907–916

Olano C, Méndez C, Salas JA (2009) Antitumor compounds from marine actinomycetes. Mar Drugs 7(2):210–248

Oren A (2002) Diversity of halophilic microorganisms: environments, phylogeny, physiology, and applications. J Ind Microbiol Biotechnol 28(1):56–63

Pagaling E, Wang H, Venables M, Wallace A, Grant WD, Cowan DA, Jones BE, Ma Y, Ventosa A, Heaphy S (2009) Microbial biogeography of six salt lakes in Inner Mongolia, China, and a salt lake in Argentina. Appl Environ Microbiol 75(18):5750–5760

Palaniyandi SA, Yang SH, Zhang L, Suh J-W (2013) Effects of actinobacteria on plant disease suppression and growth promotion. Appl Microbiol Biotechnol 97(22):9621–9636

Pandey S, Singh S (2012) Organic solvent tolerance of an α-amylase from haloalkaliphilic bacteria as a function of pH, temperature, and salt concentrations. Appl Biochem Biotechnol 166(7):1747–1757

Park JH, Ha HJ, Lee WK, Généreux-Vincent T, Kazlauskas RJ (2009) Molecular basis for the *Stereoselective Ammoniolysis* of N-alkyl aziridine-2-carboxylates catalyzed by candida antarctica Lipase B. Chembiochem 10(13):2213–2222

Park M-H, Traiwan J, Jung MY, Kim W (2012) *Gulosibacterchungangensis* sp. nov., an actinomycete isolated from a marine sediment, and emended description of the genus Gulosibacter. Int J Syst Evol Microbiol 62(5):1055–1060

Parthasarathi S, Kim C J, Kim P K, Sathya S, Manikandan M, Manikandan T, Karuppaiya B (2010)Taxonomic characterization and UV/VIS analysis of antagonistic marine actinomycete isolated from South WestCoast of South Korea. Int J Med Res, 1: 99-105

Parthasarathi S, Sathya S, Bupesh G, Samy RD, Mohan MR, Kumar GS, Manikandan M, Kim C, Balakrishnan K (2012) Isolation and characterization of antimicrobial compound from marine *Streptomyces hygroscopicus* BDUS 49. World J Fish Mar Sci 4(3):268–277

Peela S, Kurada VB, Terli R (2005) Studies on antagonistic marine actinomycetes from the Bay of Bengal. World J Microbiol Biotechnol 21(4):583–585

Perez D, Kovačić F, Wilhelm S, Jaeger K-E, García MT, Ventosa A, Mellado E (2012) Identification of amino acids involved in the hydrolytic activity of lipase LipBL from Marinobacter lipolyticus. Microbiology 158(8):2192–2203

Petrolini B, Quaroni S, Saracchi M, Sardi P, Fedeli L (1995) Actinoplanes strains isolated from roots: morphological investigations. Actinomycetes 6(2):29–37

Pikuta EV, Hoover RB, Tang J (2007) Microbial extremophiles at the limits of life. Crit Rev Microbiol 33(3):183–209

Poornima I, Brown SB, Bhashyam S, Parikh P, Bolukoglu H, Shannon RP (2008) Chronic glucagon-like peptide-1 infusion sustains left ventricular systolic function and prolongs survival in the spontaneously hypertensive, heart failure–prone rat. Circ Heart Fail 1(3):153–160

Prakash B, Vidyasagar M, Madhukumar M, Muralikrishna G, Sreeramulu K (2009) Production, purification, and characterization of two extremely halotolerant, thermostable, and alkali-stable α-amylases from *Chromohalobacter* sp. TVSP 101. Process Biochem 44(2):210–215

Pridham T, Lyons A Jr (1961) *Streptomyces albus* (Rossi-Doria) Waksman et Henrici: taxonomic study of strains labeled Streptomyces albus. J Bacteriol 81(3):431

Prudhomme J, McDaniel E, Ponts N, Bertani S, Fenical W, Jensen P, Le Roch K (2008) Marine actinomycetes: a new source of compounds against the human malaria parasite. PLoS One 3(6):e2335

Rabeh SA, Azab EA, Aly MM (2007) Studies on bacterioplankton and inhibitory strains of aquatic actinomycetes in Lake Bardawil, Egypt. World J Microbiol Biotechnol 23(2):167–176

Raddadi N, Cherif A, Daffonchio D, Neifar M, Fava F (2015) Biotechnological applications of extremophiles, extremozymes and extremolytes. Appl Microbiol Biotechnol 99(19):7907–7913

Raja S, Ganesan S, Sivakumar K, Thangaradjou T (2010) Screening of marine actinobacteria for amylase enzymes inhibitors. Indian J Microbiol 50(2):233–237

Ramesh S, Mathivanan N (2009) Screening of marine actinomycetes isolated from the Bay of Bengal, India for antimicrobial activity and industrial enzymes. World J Microbiol Biotechnol 25(12):2103–2111

Ramesh S, Jayaprakashvel M, Mathivanan N (2006) Microbial status in seawater and coastal sediments during pre-and post-tsunami periods in the Bay of Bengal, India. Mar Ecol 27(3):198–203

Ramesh S, Rajesh M, Mathivanan N (2009) Characterization of a thermostable alkaline protease produced by marine *Streptomyces fungicidicus* MML1614. Bioprocess Biosyst Eng 32(6):791–800

Rao VR, Rao DB, Rao TR (2012) Isolation and characterisation of antagonistic actinobacteria from mangrove soil. J Biochem Technol 3(4):361–365

Ratnakar D (2013) Use of halophile physiology and adaptations in various industrial applications. Res J Biotechnol 8:2

Raveh A, Delekta PC, Dobry CJ, Peng W, Schultz PJ, Blakely PK, Tai AW, Matainaho T, Irani DN, Sherman DH (2013) Discovery of potent broad spectrum antivirals derived from marine actinobacteria. PLoS One 8(12):e82318

Reddy N, Ramakrishna D, Raja Gopal S (2011) A morphological, physiological and biochemical studies of marine *Streptomyces rochei* (MTCC 10109) showing antagonistic activity against selective human pathogenic microorganisms. Asian J Biol Sci 4(1):1–14

Remya M, Vijayakumar R (2008) Isolation and characterization of marine antagonistic actinomycetes from west coast of India. Med Biol 15(1):13–19

Riedlinger J, Reicke A, Zähner H, Krismer B, Bull AT, Maldonado LA, Ward AC, Goodfellow M, Bister B, Bischoff D (2004) Abyssomicins, inhibitors of the para-aminobenzoic acid pathway produced by the marine *Verrucosispora* strain AB-18-032. J Antibiot 57(4):271–279

Rodriguez J, Mendoza L, Pezzotti F, Vanthuyne N, Leclaire J, Verger R, Buono G, Carriere F, Fotiadu F (2008) Novel chromatographic resolution of chiral diacylglycerols and analysis of the stereoselective hydrolysis of triacylglycerols by lipases. Analyt Biochem 375(2):196–208

Rohban R, Amoozegar MA, Ventosa A (2009) Screening and isolation of halophilic bacteria producing extracellular hydrolyses from Howz Soltan Lake, Iran. J Ind Microbiol Biotechnol 36(3):333–340

Saha S, Dhanasekaran D, Shanmugapriya S, Latha S (2013) *Nocardiopsis* sp. SD5: a potent feather degrading rare actinobacterium isolated from feather waste in Tamil Nadu, India. J Basic Microbiol 53(7):608–616

Sahu D, Lin S-Y, Huang J-L (2006) ZnO/Ag/ZnO multilayer films for the application of a very low resistance transparent electrode. Appl Surf Sci 252(20):7509–7514

Saker R, Bouras N, Meklat A, Zitouni A, Schumann P, Spröer C, Sabaou N, Klenk H-P (2015) *Prauserella isguenensis* sp. nov., a halophilic actinomycete isolated from desert soil. Int J Syst Evol Microbiol 65(5):1598–1603

Salah M, Gobalakrishnan R, Sivakumar K, Kannan L (2014) Actinobacterial research from the mangroves of India. J Sci 7(3):108–112

Sanchez-Porro C, Mellado E, Bertoldo C, Antranikian G, Ventosa A (2003) Screening and characterization of the protease CP1 produced by the moderately halophilic bacterium *Pseudoalteromonas* sp. strain CP76. Extremophiles 7(3):221–228

Sarma P, Coker JA, Huse V, DasSarma S (2010) Halophiles, industrial applications. In: Encyclopedia of industrial biotechnology. Wiley, New York

Sathiyaseelan K, Stella D (2011) Isolation, identification and antimicrobial activity of marine actinomycetes isolated from Parangipettai. Recent Res Sci Technol 3(9):74–77

Saurav K, Kannabiran K (2010) Diversity and optimization of process parameters for the growth of *Streptomyces* VITSVK9 spp isoled from Bay of Bengal, India. J Nat Environ Sci 1(2):56–65

Saurav K, Rajakumar G, Kannabiran K, Rahuman AA, Velayutham K, Elango G, Kamaraj C, Zahir AA (2013) Larvicidal activity of isolated compound 5-(2, 4-dimethylbenzyl) pyrrolidin-2-one from marine *Streptomyces* VITSVK5 sp. against *Rhipicephalus* (Boophilus) microplus, *Anopheles stephensi*, and *Culex tritaeniorhynchus*. Parasitol Res 112(1):215–226

Schmid A, Dordick J, Hauer B, Kiener A, Wubbolts M, Witholt B (2001) Industrial biocatalysis today and tomorrow. Nature 409(6817):258–268

Schreck SD, Grunden AM (2014) Biotechnological applications of halophilic lipases and thioesterases. Appl Microbiol Biotechnol 98(3):1011–1021

Schumacher RW, Talmage SC, Miller SA, Sarris KE, Davidson BS, Goldberg A (2003) Isolation and structure determination of an antimicrobial ester from a marine sediment-derived bacterium. J Nat Prod 66(9):1291–1293

Shameer S, Babu GP (2013) Paramageetham Ch. 2013. Extracellular enzymatic potential of halolakaliphiles from solar salterns of Nellore District AP India. Asian J Exp Biol Sci 4:302–305

Sharma C, Kanwar S (2012) Purification of a novel thermophilic lipase from B. *licheniformis* MTCC-10498. Int Res J Biol Sci 1:43–48

Shobha G, and Vinutha M (2014) Biological Synthesis of Copper Nanoparticles and its impact - a Review. Int. j.pharm. sci. Invent. 3(8): 28-38

Shokri J, Adibkia K (2013) Application of cellulose and cellulose derivatives in pharmaceutical industries. Cellulose—medical, pharmaceutical and electronic applications. Intech 3:47–66

Singh R, Kapoor V, Kumar V (2012) Production of thermostable, Ca+ 2-independent, maltose producing α-amylase by *Streptomyces* sp. MSC702 (MTCC 10772) in submerged fermentation using agro-residues as sole carbon source. Ann Microbiol 62(3):1003–1012

Sivakumar K, Sahu MK, Thangaradjou T, Kannan L (2007) Research on marine actinobacteria in India. Indian J Microbiol 47(3):186–196

Sivakumar K, Karuppiah V, Sethubathi G, Thangaradjou T, Kannan L (2012) Response surface methodology for the optimization of α-amylase production by *Streptomyces* sp. ML12 using agricultural byproducts. Biologia 67(1):32–40

Sivaprakasam S, Dhandapani B, Mahadevan S (2011) Optimization studies on production of a salt-tolerant protease from *Pseudomonas aeruginosa* strain BC1 and its application on tannery saline wastewater treatment. Braz J Microbiol 42(4):1506–1515

Snellman EA, Colwell RR (2004) Acinetobacter lipases: molecular biology, biochemical properties and biotechnological potential. J Ind Microbiol Biotechnol 31(9):391–400

Sprusansky O, Stirrett K, Skinner D, Denoya C, Westpheling J (2005) The bkdR gene of *Streptomyces coelicolor* is required for morphogenesis and antibiotic production and encodes a transcriptional regulator of a branched-chain amino acid dehydrogenase complex. J Bacteriol 187(2):664–671

Stritzke K, Schulz S, Laatsch H, Helmke E, Beil W (2004) Novel caprolactones from a marine *streptomycete*. J Nat Prod 67(3):395–401

Subramani R, Aalbersberg W (2012) Marine actinomycetes: an ongoing source of novel bioactive metabolites. Microbiol Res 167(10):571–580

Sujatha P, Raju KB, Ramana T (2005) Studies on a new marine *streptomycete* BT-408 producing polyketide antibiotic SBR-22 effective against methicillin resistant *Staphylococcus aureus*. Microbiol Res 160(2):119–126

Sunaryanto R, Marwoto B, Irawadi TT, Mas'ud ZA, Hartoto L (2010) Antibiotic compound from marine Actinomycetes (*Streptomyces* sp. A11): isolation and structure elucidaton. Indones J Chem 10(2):219–225

Suthindhiran K, Kannabiran K (2009a) Cytotoxic and antimicrobial potential of actinomycete species *Saccharopolyspora salina* VITSDK4 isolated from the Bay of Bengal Coast of India. A J Infect Dis 5(2):90–98

Suthindhiran K, Kannabiran K (2009b) Hemolytic activity of *Streptomyces* VITSDK1 spp. isolated from marine sediments in Southern India. Journal de Mycologie Médicale/J Med Mycol 19(2):77–86

Suthindhiran K, Jayasri MA, Dipali D, Prasar A (2014) Screening and characterization of protease producing actinomycetes from marine saltern. J Basic Microbiol 54(10):1098–1109

Syed DG, Tang S-K, Cai M, Zhi X-Y, Agasar D, Lee J-C, Kim C-J, Jiang C-L, Xu L-H, Li W-J (2008) *Saccharomonospora saliphila* sp. nov., a halophilic actinomycete from an Indian soil. Int J Syst Evol Microbiol 58(3):570–573

Syed DG, Agasar D, Pandey A (2009) Production and partial purification of α-amylase from a novel isolate *Streptomyces gulbargensis*. J Ind Microbiol Biotechnol 36(2):189–194

Tan T-C, Mijts BN, Swaminathan K, Patel BK, Divne C (2008) Crystal structure of the polyextremophilic α-amylase AmyB from Halothermothrix orenii: details of a productive enzyme–substrate complex and an N domain with a role in binding raw starch. J Mol Biol 378(4):852–870

Tang Y, Persing D (1999) Molecular detection and identification of microorganisms. In: Manual of clinical microbiology, 7th edn. American Society for Microbiology, Washington, DC, pp 215–244

Tang S-K, Tian X-P, Zhi X-Y, Cai M, Wu J-Y, Yang L-L, Xu L-H, Li W-J (2008) *Haloactinospora alba* gen. nov., sp. nov., a halophilic filamentous actinomycete of the family Nocardiopsaceae. Int J Syst Evol Microbiol 58(9):2075–2080

Tang S-K, Wang Y, Lou K, Mao P-H, Xu L-H, Jiang C-L, Kim C-J, Li W-J (2009a) *Kocuria halotolerans* sp. nov., an actinobacterium isolated from a saline soil in China. Int J Syst Evol Microbiol 59(6):1316–1320

Tang S-K, Wang Y, Wu J-Y, Cao L-L, Lou K, Xu L-H, Jiang C-L, Li W-J (2009b) *Saccharopolyspora qijiaojingensis* sp. nov., a halophilic actinomycete isolated from a salt lake. Int J Syst Evol Microbiol 59(9):2166–2170

Tang S-K, Wang Y, Guan T-W, Lee J-C, Kim C-J, Li W-J (2010a) *Amycolatopsis halophila* sp. nov., a halophilic actinomycete isolated from a salt lake. Int J Syst Evol Microbiol 60(5):1073–1078

Tang S-K, Wang Y, Lee J-C, Lou K, Park D-J, Kim C-J, Li W-J (2010b) *Georgenia halophila* sp. nov., a halophilic actinobacterium isolated from a salt lake. Int J Syst Evol Microbiol 60(6):1317–1421

Tang S-K, Zhi X-Y, Wang Y, Wu J-Y, Lee J-C, Kim C-J, Lou K, Xu L-H, Li W-J (2010c) *Haloactinobacterium album* gen. nov., sp. nov., a halophilic actinobacterium, and proposal of Ruaniaceae fam. nov. Int J Syst Evol Microbiol 60(9):2113–2119

Tatar D, Sazak A, Guven K, Cetin D, Sahin N (2013) *Amycolatopsiscihanbeyliensis* sp. nov., a halotolerant actinomycete isolated from a salt mine. Int J Syst Evol Microbiol 63(10):3739–3743

Teo JW, Zhang L-H, Poh CL (2003) Cloning and characterization of a novel lipase from *Vibrio harveyi* strain AP6. Gene 312:181–188

Thumar JT, Singh SP (2007) Secretion of an alkaline protease from a salt-tolerant and alkaliphilic, *Streptomyces clavuligerus* strain MIT-1. Braz J Microbiol 38(4):766–772

Tian X, Cao L, Tan H, Zeng Q, Jia Y, Han W, Zhou S (2004) Study on the communities of endophytic fungi and endophytic actinomycetes from rice and their antipathogenic activities in vitro. World J Microbiol Biotechnol 20(3):303–309

Tian X, Cao L, Tan H, Han W, Chen M, Liu Y, Zhou S (2007) Diversity of cultivated and uncultivated actinobacterial endophytes in the stems and roots of rice. Microb Ecol 53(4):700–707

Tian X-P, Tang S-K, Dong J-D, Zhang Y-Q, Xu L-H, Zhang S, Li W-J (2009) *Marinactinospora thermotolerans* gen. nov., sp. nov., a marine actinomycete isolated from a sediment in the northern South China Sea. Int J Syst Evol Microbiol 59(5):948–952

Tian X-P, Xu Y, Zhang J, Li J, Chen Z, Kim C-J, Li W-J, Zhang C-S, Zhang S (2012) *Streptomyces oceani* sp. nov., a new obligate marine actinomycete isolated from a deep-sea sample of seep authigenic carbonate nodule in South China Sea. Antonie Van Leeuwenhoek 102(2):335–343

Tonkova A (2006) Microbial starch converting enzymes of the α-amylase family. Microbial Biotechnol Hortic 1:421–472

Van der Wielen P, Bolhuis H, Borin S, Daffonchio D, Corselli C, Giuliano L, D'Auria G, de Lange G, Huebner A, Varnavas S (2005) Biodeep scientific party (2005). The enigma of prokaryotic life in deep hypersaline anoxic basins. Science 307:121–123

Ventosa A (2006) Unusual micro-organisms from unusual habitats: hypersaline environments. In: Symposia-society for general microbiology, vol 1999. Cambridge University Press, Cambridge, p 223

Ventura M, Canchaya C, Tauch A, Chandra G, Fitzgerald GF, Chater KF, van Sinderen D (2007) Genomics of Actinobacteria: tracing the evolutionary history of an ancient phylum. Microbiol Mol Biol Rev 71(3):495–548

Větrovský T, Steffen KT, Baldrian P (2014) Potential of cometabolic transformation of polysaccharides and lignin in lignocellulose by soil Actinobacteria. PloS one 9(2):e89108

Vidyasagar M, Prakash S, Jayalakshmi S, Sreeramulu K (2007) Optimization of culture conditions for the production of halothermophilic protease from halophilic bacterium *Chromohalobacter* sp. TVSP101. World J Microbiol Biotechnol 23(5):655–662

Vimal V, Rajan BM, Kannabiran K (2009) Antimicrobial activity of marine actinomycete, *Nocardiopsis* sp. VITSVK 5 (FJ973467). Asian J Med Sci 1(2):57–63

Vobis G, Miyadoh S, Hamada M, Hotta K, Kudo T, Seino T (1997) Morphology of actinomycetes. In: Atlas of Actinomycetes. Asakura Publishing, Tokio, pp 180–191

Wang C-Y, Hsieh Y-R, Ng C-C, Chan H, Lin H-T, Tzeng W-S, Shyu Y-T (2009) Purification and characterization of a novel halostable cellulase from *Salinivibrio* sp. strain NTU-05. Enzyme Microb Technol 44(6):373–379

Wang Y, Tang S-K, Li Z, Lou K, Mao P-H, Jin X, Klenk H-P, Zhang L-X, Li W-J (2011) *Myceligenerans halotolerans* sp. nov., an actinomycete isolated from a salt lake, and emended description of the genus *Myceligenerans*. Int J Syst Evol Microbiol 61(4):974–978

Wei X, Jiang Y, Chen X, Jiang Y, Lai H (2015) *Amycolatopsis flava* sp. nov., a halophilic actinomycete isolated from Dead Sea. Antonie Van Leeuwenhoek 108(4):879–885

Williams S, Locci R, Beswick A, Kurtböke D, Kuznetsov V, Le Monnier F, Long P, Maycroft K, Palma R, Petrolini B (1993) Detection and identification of novel actinomycetes. Res Microbiol 144(8):653–656

Williams PG, Miller ED, Asolkar RN, Jensen PR, Fenical W (2007) Arenicolides AC, 26-membered ring macrolides from the marine actinomycete *Salinispora arenicola*. J Org Chem 72(14):5025–5034

Wright GD, Sutherland AD (2007) New strategies for combating multidrug-resistant bacteria. Trends Mol Med 13(6):260–267

Wu SJ, Fotso S, Li F, Qin S, Kelter G, Fiebig HH, Laatsch H (2006) N-Carboxamido-staurosporine and Selina-4 (14), 7 (11)-diene-8, 9-diol, New Metabolites from a Marine *Streptomyces* sp.[dagger]. J Antibiot 59(6):331

Xiao J, Xu J, Xie S, Zhang X, Yu Z, Xu J (2008) Isolation of mangrove actinomycetes and their antagonistic activities. Chin J Appl Environ Biol 14:244–248

Yan S, Wu G (2013) Secretory pathway of cellulase: a mini-review. Biotechnol Biofuels 6(1):1

Yang R, Zhang L-P, Guo L-G, Shi N, Lu Z, Zhang X (2008) *Nocardiopsis valliformis* sp. nov., an alkaliphilic actinomycete isolated from alkali lake soil in China. Int J Syst Evol Microbiol 58(7):1542–1546

Yasuhara-Bell J, Lu Y (2010) Marine compounds and their antiviral activities. Antivir Res 86(3):231–240

Zhang L (2005) Integrated approaches for discovering novel drugs from microbial natural products. In: Natural products. Springer, Totowa, pp 33–55

Zhang J-W, Zeng R-Y (2008) Purification and characterization of a cold-adapted α-amylase produced by *Nocardiopsis* sp. 7326 isolated from Prydz Bay, Antarctic. Mar Biotechnol 10(1):75–82

Zhang Y-J, Tang S-K, Shi R, Klenk H-P, Chen C, Yang L-L, Zhou Y, Li W-J (2012a) *Salinisphaera halophila* sp. nov., a moderately halophilic bacterium isolated from brine of a salt well. Int J Syst Evol Microbiol 62(9):2174–2179

Zhang Y-J, Zhou Y, Ja M, Shi R, Chun-Yu W-X, Yang L-L, Tang S-K, Li W-J (2012b) *Virgibacillus albus* sp. nov., a novel moderately halophilic bacterium isolated from Lop Nur salt lake in Xinjiang province, China. Antonie Van Leeuwenhoek 102(4):553–560

13 Microbial Diversity in Asian Deserts: Distribution, Biotechnological Importance, and Environmental Impacts

Nimaichand Salam, Zi-Wen Yang, Mipeshwaree Devi Asem, Wael N. Hozzein, and Wen-Jun Li

Abstract

Desert is the driest and largest terrestrial biomes constituting about 35% of Earth's surface. Apart from being the most arid zone, this environment is subjected to different stresses including geochemical and physical stresses. Despite the limitations, diverse and unique groups of microorganisms are able to sustain life in this dryland. It is exemplified by the fact that more than 60 novel bacterial taxa have been isolated over the past decade from deserts located in Asian countries. This chapter reviewed the microbial diversity in the deserts located in Asia with special emphasis on its distribution, adaptation, and biotechnological importance.

N. Salam · W.-J. Li (✉)
Key Laboratory of Biogeography and Bioresource in Arid Land, Xinjiang Institute of Ecology and Geography, Chinese Academy of Sciences, Ürümqi, People's Republic of China

State Key Laboratory of Biocontrol and Guangdong Provincial Key Laboratory of Plant Resources, School of Life Sciences, Sun Yat-Sen University, Guangzhou, People's Republic of China
e-mail: liwenjun3@mail.sysu.edu.cn

Z.-W. Yang · M. D. Asem
State Key Laboratory of Biocontrol and Guangdong Provincial Key Laboratory of Plant Resources, School of Life Sciences, Sun Yat-Sen University, Guangzhou, People's Republic of China

W. N. Hozzein
Zoology Department, College of Science, King Saud University, Riyadh, Saudi Arabia

Botany and Microbiology Department, Faculty of Science, Beni-Suef University, Beni-Suef, Egypt

D. Egamberdieva et al. (eds.), *Extremophiles in Eurasian Ecosystems: Ecology, Diversity, and Applications*, Microorganisms for Sustainability 8,
https://doi.org/10.1007/978-981-13-0329-6_13

Keywords
Asian deserts · *Cyanobacteria* · *Proteobacteria* · *Actinobacteria* · Microbial diversity

13.1 Introduction

Desert, by definition, is a barren area of land where little precipitation occurs. This region is often characterized by extreme physicochemical conditions such as extreme desiccation, high salinity, intense radiation, and oligotrophic nutrient availability, thereby making condition inimical for life. In fact, it is due to these reasons that US National Aeronautics and Space Administration (NASA) considered the environment of Atacama Desert to be closest to the Martian environment that is available on Earth and adopted the Viking mission in 1975 (Opfell and Zebel 1967; Bull et al. 2016). Despite the failure for detection of life during earlier microbiological work from the Viking mission, an important finding in the form of consumption of organic material was detected, thereby confirming the presence of oxidants in Atacama Desert soils (Bull et al. 2016). In other words, desert environments were beyond the threshold of life and can be considered as a source for extremophiles (Stetter 1999). Likewise, if we consider the several hypotheses for the origin of life, it is valid to assume that life begins under extraordinary circumstances and under unusual environments, and deserts provide the right environment for search of extremophilic microorganisms.

Extremophiles are adapted to grow optimally at or near the extreme ranges of temperature, pH, pressure, and/or salinity (Cavicchioli 2002; Bull 2011). Relatively little is known about the genetic diversity underlying the functional processes of microorganisms under extreme environment especially deserts. Understanding the extremophiles, therefore, will provide a link between the microbial community structure and their function in nature (Stevenson et al. 2015). Taking into consideration the implacable role these organisms play in biogeochemical processes of extreme environments, they are even considered as "ecosystem engineers" (Jones et al. 1994).

Studies on extremophiles have helped us in the development of many applications in the field of biotechnology. In fact, the impact of biotechnology on our lives is inescapable (Coker 2016). It is because every category of these microbes has unique characteristics that can be harnessed for use in biotechnological industries (Tango and Islam 2002). For instance, they are an effective and environmentally friendly means for bioremediation of hydrocarbon contaminants in oil fields (Al-Mailem et al. 2010a). Several biopolymers including biosurfactants, exopolysaccharides, and bioplastics have been developed from halophilic microorganisms (Tango and Islam 2002). Extremozymes have been found to be a more efficient and cost-effective means to replace currently used mesophilic enzymes in industries (Coker 2016).

In this chapter, we will be discussing the distribution, adaptation, and biotechnological importance of a specific group of microorganisms sustaining life under

Fig. 13.1 Distribution of major deserts in Asian continent

salinity stress (halophilic) and temperature stress (thermophilic). This group of microorganisms survives desiccation and resumes life on subsequent wetting thereby playing a major role in the control of dryland systems (Pointing and Belnap 2012).

13.2 Deserts in Asia

Being one of the most extensive terrestrial biomes, the number of deserts distributed on the Earth's surface is extensive. Identified on the basis of their size, the world's two largest deserts are the Antarctic desert (~14 million sq. km) and Arctic desert (13 million sq. km), both cold or polar deserts. The largest among the hot deserts is the Sahara desert in Africa which extends over an area of 9.4 million sq. km. On the basis of aridity level, Atacama Desert located in part of Chile and Peru is considered to be one of the most arid (or hyper-arid) dry lands (McKay et al. 2003; Navarro-Gonzalez et al. 2003; Drees et al. 2006; López et al. 2016).

The deserts in Asian countries are confined to small sizes. The largest of these deserts is the Arabian Desert which covers an area of 2.3 million sq. km. and extending over the countries of Iraq, Jordan, Kuwait, Oman, Qatar, Saudi Arabia, United Arab Emirates, and Yemen (Fig. 13.1). The largest among the cold deserts, and the second largest one in size, is the Gobi Desert located in part of Mongolia and China. A list of the major deserts in the Asian countries (area greater than or equal to 50,000 sq. km.) is listed in Table 13.1. Many of the other deserts are of smaller size and therefore not included in the list (for more details about the deserts in the world, see: https://en.wikipedia.org/wiki/List_of_deserts).

Table 13.1 List of deserts[a] by area distributed in Asian countries

Rank	Name	Type	Area (sq. km)	Location
1	Arabian Desert	Subtropical	2,330,000	Western Asia
2	Gobi Desert	Cold	1,000,000	Central Asia
3	Syrian Desert	Subtropical	520,000	Western Asia
4	Karakum Desert	Cold	350,000	Turkmenistan
5	Kyzylkum Desert	Cold	300,000	Central Asia
6	Taklamakan Desert	Cold	270,000	China
7	Thar Desert	Subtropical	200,000	South Asia
8	Dasht-e Margo	Subtropical	150,000	Afghanistan
9	Registan Desert	Subtropical	146,000	Afghanistan
10	Ordos Desert (Kubuqi + Maowusu Deserts)	–	90,650	China
11	Dasht-e Kavir	Subtropical	77,000	Iran
12	Dasht-e Lut	Subtropical	52,000	Iran
13	Lop Desert	–	50,000	China

Source: Wikipedia
[a]Area > 50,000 sq. km

13.3 Desert and Microbes

Microbial colonization in deserts is largely defined by its ability to adapt to environmental stresses such as low moisture availability, low or high temperatures, and ultraviolet (UV) irradiation. Moisture in deserts is maintained by limited precipitation in the form of rainfall and fog-derived moisture (Warren-Rhodes et al. 2006; Azúa-Bustos et al. 2011). However the availability of water is largely influenced by the composition of biological soil crusts and the presence of diaphanous rocks which determined the level of precipitation and evaporation (Cockell and Stokes 2004; Rajeev et al. 2013; Hagemann et al. 2015). Unlike moisture, temperature is at its extreme end in desert environments. While temperature in hot deserts may exceed 60 °C, it may drop to less than −20 °C in cold deserts (Warren-Rhodes et al. 2007; Tracy et al. 2010). Another important factor characterizing desert is the incident solar radiation of the spectrum below 400 nm wavelength. Two major spectra, viz., ultraviolet A (315–400 nm) and ultraviolet B (280–315 nm), can penetrate the atmosphere and reach the Earth's surface. Quantity of these radiations reaching Earth's surface is however small but enough to cause detrimental effect on biomolecules and thereby life in general (Jeffrey et al. 1996; Gao and Garcia-Pichel 2011).

In view of the different extremes contributing dryland ecosystems, sustainability of microbial life under arid environments is limited to stable soil structures such as the biological soil crusts and stony desert pavements (Warren-Rhodes et al. 2007; Pointing and Belnap 2012). Biological soil crusts are considered as ecosystem engineers due to their pivotal role in soil stabilization (Belnap et al. 2003), carbon fixation, and nitrogen recycling (Elbert et al. 2012). These delicate structures are the dominant functional vegetation units and thus function to serve as the food source for other organisms inhabiting the desert ecosystems (Grube et al. 2009). Physical disturbances and alteration in temperature or precipitation can however cause irreparable damage to this delicate structure (Belnap and Gillette 1998; Belnap et al. 2004; Kidron et al. 2008, 2012; Kuske et al. 2012). The stony desert pavements, on the other hand, are composed of surface soils covered by gravels and bedrock debris (Friedmann and Galun 1974; Golubic et al. 1981). This habitat is primarily occupied by the photoautotrophic cyanobacterial communities, mainly of the genus *Chroococcidiopsis*, often in association with other filamentous cyanobacterial and heterotrophic taxa (Grilli Caiola et al. 1993; Billi et al. 2000; Pointing et al. 2007). Frequently, their distribution is limited by the level of aridity. It has been found that the abundance of hypolithic *Cyanobacteria* drops from 28 to <0.1% over a gradient of low to high arid core of Atacama Desert, while the molecular diversity declines threefold as compared to the less arid region (Warren-Rhodes et al. 2006). The major microorganisms surviving the extreme conditions of deserts are, therefore, the ones that can sustain desiccation, tolerate extreme temperature, and develop mechanisms for resisting ultraviolet radiation (Potts 1994; Billi et al. 2000).

Recent discoveries have uncovered certain mechanisms on how these microorganisms adapt the hyper-arid conditions. In the simplest mode, these organisms gain protection from solar radiation and also receive increase moisture by inhabiting diaphanous rocks and minerals, e.g., quartz, granite, gypsum, halite, and sandstone (Friedmann et al. 1967; Cockell and Stokes 2004; Cockell et al. 2005). Some microorganisms, however, produce certain types of secondary metabolites under radiation stress which act as ultraviolet "sunscreen" (Gao and Garcia-Pichel 2011). These compounds include scytonemin, mycosporines, and naphthalene-based melanins (Soule et al. 2007; Gao and Garcia-Pichel 2011). This protective mechanism remains functional even during long periods of dormancy that are typically endured by poikilohydric life forms (Garcia-Pichel et al. 1992; Böhm et al. 1995; Geng et al. 2008; Balskus and Walsh 2010; Gonzalez et al. 2010). Formation of biofilms during water stress is another means for enduring desiccation tolerance in desert's microbes (Flemming et al. 2016). Biofilms are formed by microbial communities that are embedded in a self-produced matrix of extracellular polymeric substances. The presence of these extracellular polymeric substances, which is a direct effect of salt stress (Liu and Buskey 2000; Abdullahi et al. 2006), provides several advantages, including water absorption and retention, soil adhesion, reduced evaporation, and nutrient capture (Foster 1981; Lynch and Bragg 1985; Grilli Caiola et al. 1993; Mazor et al. 1996; Philippis and Vincenzini 1998; de los Rios et al. 2004; Warren-Rhodes et al. 2007). Interestingly, in an interdune sabkha (salt flats) in the Rub'al Khali (the Empty Quarter), United Arab Emirates, an unusual inverted saline microbial mat community was observed (McKay et al. 2016). In this microbial mat (endoevaporite mat), the salinity gradient is inverted as compared to most salt flat communities with the salt crust at the uppermost layer, followed by a thin layer of halophilic bacteria and layers of green photosynthetic organisms, thereby protecting the photosynthetically based microbial ecosystem from the arid environment.

13.4 Microbial Community Assemblage in Asian Deserts

Studies of viable microbiota in deserts have been assumed earlier to be either low or very low, partly by the limitation of culturing techniques and partly by the several geographical and environmental constraints (Cary et al. 2010). However numerous developments especially in the application of molecular methods have advanced the analysis of arid or hyper-arid environments. For instance, many of the recent inferences on microbial diversity of Asian deserts have been from the analysis of biological soil crusts using high-throughput sequencing technology. Analyses of biological desert crusts which are relatively common in arid deserts have indicated the presence of halotolerant, thermotolerant, and UV-resistant *Cyanobacteria* (*Cyanothece*, *Chroococcidiopsis*, *Dactylococcopsis*, *Euhalothece*, *Haloleptolyngbya*, and *Halomicronema*) (Abed et al. 2015). These cyanobacterial strains are tactically important for desert's life in that they form flatbed on the soil crusts by binding the sand through the formation of extracellular polymeric substances (Friedmann 1980; Wong et al. 2010). While this group maintained the structural platform for other microbial cultures to sustain life on the extremes, energy sources are primarily provided by another class of *Cyanobacteria* affiliated to heterocytous type such as *Nostoc*, *Scytonema*, *Brasilonema*, and *Petalonema* (Abed et al. 2010). Microbial diversity can therefore be discussed on the basis of the different geographical pattern of desert.

13.4.1 Cyanobacterial Mats

Abed et al. (2015) investigated the bacterial diversity of hypersaline cyanobacterial mats from Wadi Muqshin, located inland near the Empty Quarter desert (Rub'al Khali) – the central portion of the Arabian Desert by pyrosequencing analysis. Apart from *Cyanobacteria*, many of the OTUs identified were affiliated to the phyla *Proteobacteria*, *Bacteroidetes*, *Clostridia*, and *Chloroflexi*. While *Proteobacteria* (especially members of the class *Alpha-*, *Gamma-*, and *Deltaproteobacteria*) made up for 13–32% of the total sequences, the phylum *Bacteroidetes* constituted 9–22% of the total OTUs. Unlike the above phyla, much of the *Chloroflexi* sequences belonged to uncultured families including uncultured *Anaerolineaceae*, uncultured *Caldilineaceae*, and Candidatus *Chlorothrix*, indicating the presence of untapped microbial diversity. A lower proportion of *Verrucomicrobia*, *Acidobacteria*, *Actinobacteria*, *Chlorobi*, *Firmicutes*, and *Deferribacteres* were also detected. In cyanobacterial mats of the Arabian Gulf coast of Saudi Arabia, oil-degrading bacteria belonging to *Beta-*, *Gamma-*, and *Deltaproteobacteria*, *Cytophaga-Flavobacterium-Bacteroides* group, and *Spirochetes* were detected (Abed et al. 2006).

Two different groups had simultaneously analyzed the ecological factors influencing the distribution of *Cyanobacteria* along environmental gradients in hot and cold deserts of western China (Pointing et al. 2007; Warren-Rhodes et al. 2007). Their research findings indicated that moisture is an important determinant for bacterial diversity in arid environments, while other factors including substrate availability, temperature, or rainfall are of lesser importance. In Thar Desert (India), the structure of community composition was found to be more homogeneous in non-sandy, crusted, and vegetated soils than in sandy, non-crusted, and barren soils indicating that incidence and colonization of desert soils by *Cyanobacteria* are dependent on agro-ecological conditions (Bhatnagar et al. 2008). The finding of Li and co-workers (2013) however indicated that the selection and growth of bacterial communities were dependent on the salinity conditions of the desert crusts. In their study, they found that cyanobacteria of the order *Oscillatoriales* predominate in low saline crusts, while other phototrophs such as diatoms were the main microbial group responsible for photosynthesis in high saline crusts. In addition, the higher salt content in crusts stimulates the growth of *Deinococcus-Thermus*, *Bacteroidetes*, and some members of the subdivision of *Proteobacteria*.

13.4.2 Desert Sands

Culture-dependent methods of isolation in hypersaline environments in Kuwait led to the discovery of the bacterial genera *Halomonas*, *Chromohalobacter*, *Marinobacter*, *Exiguobacterium*, *Stenotrophomonas*, *Pseudomonas*, *Salinivibrio*, and *Bacillus* and the haloarcheal genera *Haloferax* and *Halobacterium* (Al-Mailem et al. 2014). However a different microbial composition was detected by the culture-independent methods which include the genera *Ochrobactrum*, *Stenotrophomonas*, *Rhodococcus*, *Halomicrobium* (all bacterial phylotypes), *Halorussus*, *Halomicrobium*, and *Haloorientalis* (archaeal phylotypes) (Al-Mailem et al. 2014). In a similar study by Dashti et al. (2015), the isolation of *Agrobacterium*, *Sphingomonas*, and *Pseudomonas* from oily desert soil was reported, while *Nesiobacter*, *Nitratireductor*, *Acinetobacter*, *Alcanivorax*, *Arthrobacter*, *Marinobacter*, *Pseudoalteromonas*, *Vibrio*, *Diatzia*, *Mycobacterium*, and *Microbacterium* were isolated from the Arabian/Persian Gulf water body.

In a very recent study, our group has also tried to explore the microbial reserves of desert samples from Riyadh (Saudi Arabia). The culture-based isolation technique involved a series of different permutation and combination of culturing methods; thereby maximizing the chance for isolation of novel undiscovered bacteria (Yang et al. unpublished data). During our assessment, a wide variety of bacterial strains were isolated, many of which were earlier thought to be present in smaller proportion. These microbes are represented in a phylogenetic dendrogram (Fig. 13.2) generated with neighbor-joining algorithm (Saitou and Nei 1987). Majority of these strains were affiliated to the phylum *Actinobacteria*, followed by the phylum *Proteobacteria*.

A separate study on an interdune sabkha in the Rub'al Khali (the Empty Quarter), United Arab Emirates, indicated high abundance of *Bacteroidetes* and *Proteobacteria* in the top and middle layers of endoevaporite mat, while higher proportions of *Proteobacteria* and *Cyanobacteria* in the bottom and sediment layers of the mat were found (McKay et al. 2016). Another finding based on pyrosequencing analysis of surface sand samples from Taklamakan and Gobi deserts

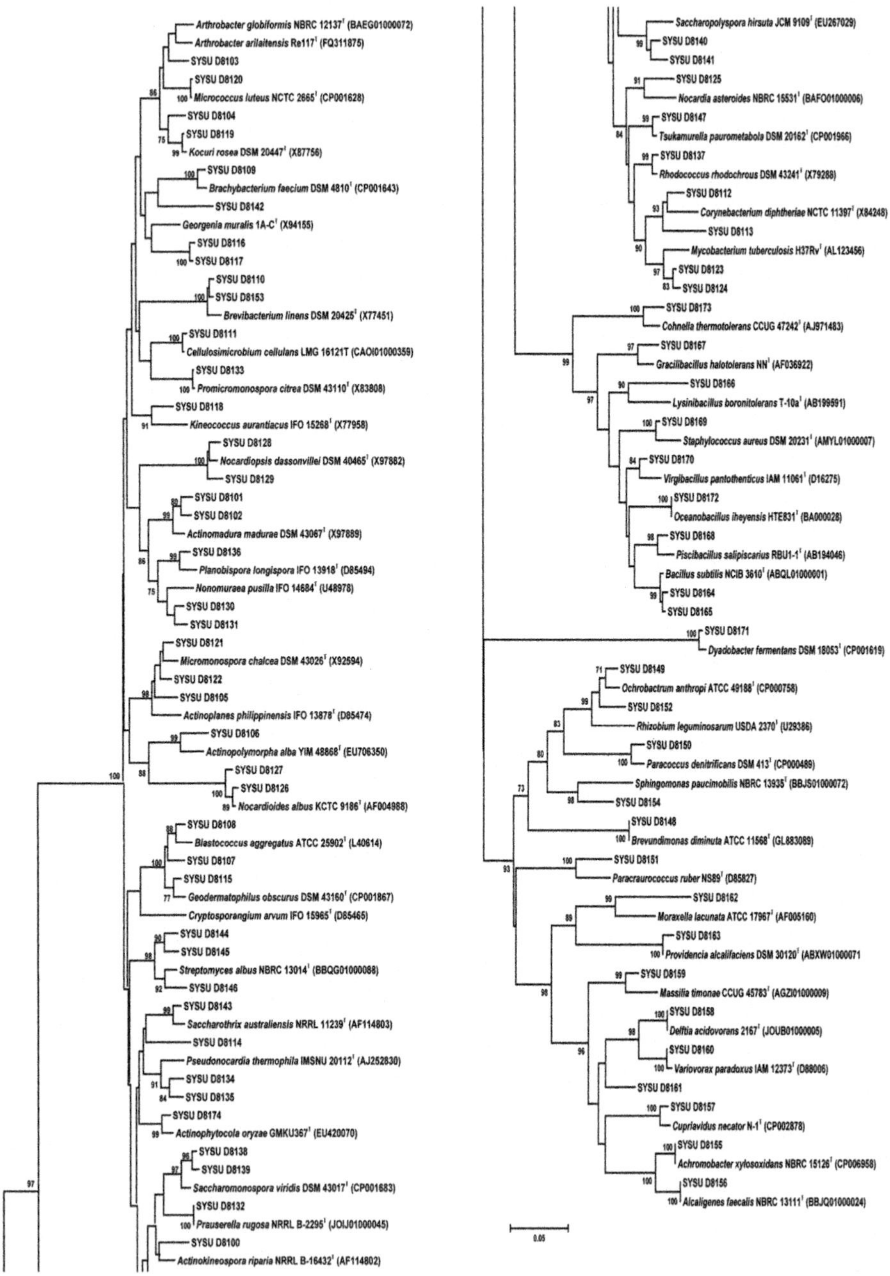

Fig. 13.2 Neighbor-joining phylogenetic dendrogram indicating the affiliation of bacterial strains isolated from a desert sample in Riyadh, Saudi Arabia

indicated the presence of large bacterial diversity in harsh desert environments, with C/N ratio playing a possible role in determining bacterial richness (An et al. 2013). The 4088 OTUs determined during the study were grouped into 102 families belonging to 15 phyla, the most abundant being the phyla *Firmicutes* (genera *Bacillus* and *Planomicrobium*), *Proteobacteria* (genera *Pseudomonas*, *Acinetobacter*, *Massilia*, *Lysobacter*, *Herbaspirillum*, *Devosia*, *Paracoccus*, *Sphingomonas*, *Novosphingobium*, and *Comamonas*), *Bacteroidetes* (genera *Effluvibacter*, *Adhaeribacter*, *Flavisolibacter*, *Pedobacter*, *Pontibacter*, and *Salinimicrobium*), and *Actinobacteria* (*Arthrobacter*, *Nocardioides*, *Blastococcus*, and *Marmoricola*).

13.4.3 Desert Halophyte

In addition to the microbial community within desert crusts, halophytes and their associated microbes are equally important in structuring the desert ecology. In a recent study, Zhao et al. (2016) characterized 219 nodule isolates from an endangered evergreen legume, *Ammopiptanthus mongolicus*, which is widely distributed in deserts of Northwest China. These isolates represented nine genospecies of the genera *Ensifer*, *Neorhizobium*, *Agrobacterium*, *Pararhizobium*, and *Rhizobium* which are known as nitrogen-fixing microorganisms. Analysis of the culturable bacterial community associated with roots of the perennial grass *Lasiurus sindicus* of Thar Desert indicated the presence of Gram-negative diazotrophs such as *Azospirillum* and *Rhizobium* (Chowdhury et al. 2007). Apart from the Gram-negative diazotrophs, the presence of Gram-positive bacteria such as *Staphylococcus*, *Bacillus*, *Micrococcus*, and *Microbacterium* and Gram-negative bacteria such as *Agrobacterium*, *Inquilinus*, *Ralstonia*, *Variovorax*, *Bordetella*, *Pseudomonas*, and *Stenotrophomonas* was also reported.

Due to the rise in global temperature and glacial recession in higher altitudes, vascular plants have been observed to be migrated upward (Chen et al. 2011). This effect has however been determined to the presence of endophytic microbial communities of the order *Sphingomonadales* and *Sphigobacteriales* in the roots of the vascular plants (Angel et al. 2016). These endophytic microorganisms along with some early colonizers, especially *Cyanobacteria*, help in maturing the nascent soil into soil crusts that are capable of supporting plant growth (Schmidt et al. 2008).

Despite being harsh ecosystems with temperature, salinity, and radiation stresses, it is clear from the above findings that the microbial resources in desert environments are unique and diverse. A clearer picture of the potential of these extremophiles is provided when these cultures are available in pure forms. Recent efforts have resulted in the isolation of many novel strains of bacteria from deserts. A list of these bacterial strains isolated from deserts located in Asian countries is provided in Table 13.2.

Table 13.2 List of novel microorganisms isolated from desert samples located in Asian countries

Sl. no.	Organism	Taxa level	Source location	Characteristics	References
1	*Actinoalloteichus spitiensis*	Novel species	Indian Himalayas	–	Singla et al. (2005)
2	*Actinophytocola gilvus*	Novel species	Tengger Desert	–	Sun et al. (2014)
3	*Aeromicrobium halotolerans*	Novel species	Turpan desert	Halotolerant	Yan et al. (2016)
4	*Agrococcus lahaulensis*	Novel species	Indian Himalayas	Halotolerant	Mayilraj et al. (2006a)
5	*Alcaligenes endophyticus*[a]	Novel species	Takeermohuer desert	Halotolerant	Lu et al. (2017)
6	*Altererythrobacter soli*	Novel species	Tengger Desert	Alkalitolerant	Zhao et al. (2017)
7	*Altererythrobacter xinjiangensis*	Novel species	Xinjiang	Alkalitolerant	Xue et al. (2012)
8	*Arthrobacter deserti*	Novel species	Turpan desert	Halotolerant	Hu et al. (2016)
9	*Arthrobacter liuii*	Novel species	Xinjiang	–	Yu et al. (2015)
10	*Bacillus deserti*	Novel species	Xinjiang	–	Zhang et al. (2011)
11	*Bacillus gobiensis*	Novel species	Gobi Desert	–	Liu et al. (2016)
12	*Caenispirillum deserti*	Novel species	Kutch	Alkalitolerant, halotolerant	Divyasree et al. (2015)
13	*Cesiribacter roseus*	Novel species	Xinjiang	Alkalitolerant, halotolerant	Liu et al. (2012)
14	*Corynebacterium deserti*	Novel species	Western China	–	Zhou et al. (2012)
15	*Deinococcus gobiensis*	Novel species	Gobi Desert	γ- and UV radiation resistant	Yuan et al. (2009)
16	*Deinococcus xinjiangensis*	Novel species	Xinjiang	UV radiation resistant	Peng et al. (2009a)
17	*Delftia deserti*	Novel species	Turpan desert	–	Li et al. (2015)
18	*Desertibacter roseus*	Novel genus	Taklamakan Desert	γ radiation resistant, alkalitolerant	Liu et al. (2011a)
19	*Dietzia kunjamensis*	Novel species	Indian Himalayas	Alkalitolerant	Mayilraj et al. (2006b)
20	*Dyadobacter alkalitolerans*	Novel species	Xinjiang	Alkalitolerant	Tang et al. (2009)

(continued)

Table 13.2 (continued)

Sl. no.	Organism	Taxa level	Source location	Characteristics	References
21	*Falsirhodobacter deserti*	Novel species	Xinjiang	Halotolerant	Wang et al. (2015)
22	*Gemmatimonas phototrophica*	Novel species	Gobi Desert	Facultative photoheterotrophic	Zeng et al. (2015)
23	*Hymenobacter deserti*	Novel species	Xinjiang	Alkalitolerant	Zhang et al. (2009)
24	*Hymenobacter xinjiangensis*	Novel species	Xinjiang	γ-radiation resistant	Zhang et al. (2007)
25.	*Jiangella gansuensis*	Novel genus	Gansu	–	Song et al. (2005)
26	*Kineococcus xinjiangensis*	Novel species	Xinjiang	–	Liu et al. (2009b)
27	*Kocuria himachalensis*	Novel species	Indian Himalayas	–	Mayilraj et al. (2006c)
28	*Kribbella deserti*	Novel species	Hangjin Banner	–	Sun et al. (2017)
29	*Kurtzmanomyces shapotouensis*	Novel species	Tengger Desert	–	Zhang et al. (2013a)
30	*Lysobacter xinjiangensis*	Novel species	Xinjiang	Alkalitolerant	Liu et al. (2011b)
31	*Mesorhizobium gobiense*[a]	Novel species	Xinjiang	–	Han et al. (2008)
32	*Mesorhizobium tarimense*[a]	Novel species	Xinjiang	–	Han et al. (2008)
33	*Microbacterium radiodurans*	Novel species	Gobi Desert	UV radiation resistant, alkalitolerant, halotolerant	Zhang et al. (2010)
34	*Microvirga pakistanensis*	Novel species	Cholistan	–	Amin et al. (2016)
35	*Mycetocola manganoxydans*	Novel species	Taklamakan Desert	Alkalitolerant	Luo et al. (2012)
36	*Natronobacillus azotifigens*	Novel genus	Libyan Desert	Obligate alkaliphile, halophilic	Sorokin et al. (2008)
37	*Nesterenkonia populi*[a]	Novel species	Taklamakan Desert	Alkaliphilic, moderately halophilic	Liu et al. (2015a)
38	*Nesterenkonia rhizosphaerae*	Novel species	Fukang	Alkaliphilic, halotolerant	Wang et al. (2014)
39	*Nocardioides deserti*	Novel species	Taklamakan Desert	Alkalitolerant, halotolerant	Tuo et al. (2015)
40	*Ornithinicoccus halotolerans*	Novel species	Xinjiang	Alkaliphilic, halotolerant	Zhang et al. (2016)

(continued)

Table 13.2 (continued)

Sl. no.	Organism	Taxa level	Source location	Characteristics	References
41	*Ornithinimicrobium kibberense*	Novel species	Indian Himalayas	Halotolerant	Mayilraj et al. (2006d)
42	*Paenibacillus gansuensis*	Novel species	Gansu	–	Lim et al. (2006)
43	*Paenibacillus harenae*	Novel species	Gansu	–	Jeon et al. (2009)
44	*Paenibacillus tarimensis*	Novel species	Xinjiang	–	Wang et al. (2008)
45	*Pedobacter xinjiangensis*	Novel species	Xinjiang	–	Tang et al. (2010)
46.	*Pelagibacterium lixinzhangensis*	Novel species	Xinjiang	Alkalitolerant, moderately halophilic	Yang and Sun (2016)
47.	*Planobacterium taklimakanense*	Novel genus	Taklamakan Desert	Alkalitolerant	Peng et al. (2009b)
48.	*Planococcus stackebrandtii*	Novel species	Indian Himalayas	Alkalitolerant, halotolerant	Mayilraj et al. (2005)
49.	*Pontibacter akesuensis*	Novel species	Akesu	Alkalitolerant	Zhou et al. (2007)
50.	*Pontibacter deserti*	Novel species	Kutch	–	Subhash et al. (2014)
51.	*Pontibacter diazotrophicus*	Novel species	Taklamakan Desert	Diazotroph, halotolerant	Xu et al. (2014)
52.	*Pontibacter korlensis*	Novel species	Xinjiang	Alkalitolerant, halotolerant	Zhang et al. (2008)
53.	*Pontibacter ruber*	Novel species	Kutch	–	Subhash et al. (2014)
54.	*Pontibacter soli*	Novel species	Xinjiang	Alkalitolerant	Dai et al. (2014)
55.	*Pontibacter yuliensis*	Novel species	Taklamakan Desert	Halotolerant	Cao et al. (2014)
56.	*Prauserella endophytica*[a]	Novel species	Taklamakan Desert	Alkalitolerant, halotolerant	Liu et al. (2015b)
57.	*Prauserella shujinwangii*	Novel species	Xinjiang	Halotolerant	Liu et al. (2014)
58.	*Pseudomonas duriflava*	Novel species	Taklamakan Desert	–	Liu et al. (2008)
59.	*Pseudomonas xinjiangensis*	Novel species	Xinjiang	Alkalitolerant, halotolerant	Liu et al. (2009c)
60.	*Rhizobium tianshanense*[a]	Novel species	Xinjiang	–	Chen et al. (1995)
61.	*Rhizobium yanglingense*[a]	Novel species	Northwest China	–	Tan et al. (2001)
62.	*Rhodococcus kroppenstedtii*	Novel species	Indian Himalayas	Halotolerant	Mayilraj et al. (2006e)

(continued)

Table 13.2 (continued)

Sl. no.	Organism	Taxa level	Source location	Characteristics	References
63.	*Saccharibacillus deserti*	Novel species	Kubuqi Desert	–	Sun et al. (2016)
64.	*Saccharibacillus kuerlensis*	Novel species	Xinjiang	–	Yang et al. (2009)
65.	*Shinella curvata*	Novel species	Kuwait	–	Subhash and Lee (2016a)
66.	*Skermanella rosea*	Novel species	Kuwait	–	Subhash and Lee (2016b)
67.	*Skermanella rubra*	Novel species	Xinjiang	Halotolerant	Zhang et al. (2015)
68.	*Skermanella xinjiangensis*	Novel species	Xinjiang	–	An et al. (2009)
69.	*Sphingobacterium deserti*	Novel species	Western Desert, China	Moderately alkaliphilic	Teng et al. (2015)
70.	*Sphingobacterium gobiense*	Novel species	Gobi Desert	Alkalitolerant	Zhao et al. (2014)
71.	*Sphingomonas xinjiangensis*	Novel species	Xinjiang	–	An et al. (2011)
72.	*Spirosoma soli*	Novel species	Kubuqi Desert	–	Yang et al. (2016)
73.	*Streptomyces fukangensis*	Novel species	Xinjiang	Moderately alkaliphilic, halotolerant	Zhang et al. (2013b)
74.	*Tenggerimyces mesophilus*	Novel genus	Tengger Desert	Alkalitolerant	Sun et al. (2015)
75.	*Yuhushiella deserti*	Novel genus	Xinjiang desert	Alkaliphilic, halotolerant	Mao et al. (2011)

[a]Isolated as endophyte from desert's halophyte

13.5 Environmental Significance

Biological soil crusts play a crucial role in ecological succession in arid regions. Gases, nutrients and water are held in the uppermost few centimeters of intact crust soil, whereby most of the biological activity that is found in desert soils occurs in this top layer. This layer may yet prove to be the largest carbon sinks in desert areas.

Oil contamination is a major phenomenon being observed in oil-rich desert countries. Salinity and temperature are important key environmental parameters that influence the degradation process of petroleum compounds. On one hand, presence of hydrocarbonoclastic microflora in hypersaline areas indicates effective potential for oil mineralization therein (Al-Mailem et al. 2014), while on the other hand, consumption of crude oil by these halophilic microorganisms highlights the self-cleaning potential of hypersaline area from oil contamination (Al-Mailem et al. 2010a, b, 2012; Dashti et al. 2015). Many of these bacteria are positive for nitrogenase activity, consume crude oil, and therefore provide a cost-effective, environmentally friendly bioremediation of hydrocarbon contaminants (Dashti et al. 2015).

Another new perspective for utilization of desert microorganisms is restoration of plant growth under salt stress (Nabti et al. 2015). While majority of the seed germination step to maturity in plant life including phytohormone synthesis and regulation, normal root and shoot development, nutrient uptake, etc. is abolished by high salinity (François et al. 1986), diazotrophs and associated microorganisms benefit plant growth by restoration of essential activities. Some halotolerant plant growth-promoting bacteria are able to colonize plants and produce various antimicrobial metabolites against pathogenic fungi and bacteria (Nabti et al. 2015).

13.6 Environmental Impacts on Microbial Diversity

Biological soil crusts are very important to ensure proper structuring and functioning of desert ecosystems (Belnap 2003; Belnap and Eldridge 2001; Eldridge and Greene 1994). As species distribution in arid desert ecosystem has direct correlation with the environmental factors, it is necessary to preserve the soil crusts (Ding et al. 2013; El-Ghani 1998; Fierer et al. 2010). However increase global warming and intensified human activities have pose serious problems on structure of crusts (Belnap and Eldridge 2001), thereby causing severe negative impacts on the movements of nutrients and transfer of energy between soil and diazotrophs or atmosphere (Pointing and Belnap 2012). Disturbance of soil crust and desert pavement enhances desertification and increases evapotranspiration resulting in sand drift or desert aerosol which can be a mean for major intercontinental trajectories for desert dust (Pease et al. 1998; Al-Awadhi 2005). Desert dust not only affects the microbial diversity but also the health of desert animals.

It may also be noted that different environmental constraints select the highly adapted and tolerant genotypes among the microbial community. In an experiment conducted by Aded and co-workers to check the variability on distribution of cyanobacterial mats across different tidal zones, it was determined that frequent alteration of air exposure and inundation promoted the growth of contiguous pinnacle mats on well-drained elevations in middle tidal zones and severe dryness in the higher tidal zones (Abed et al. 2008). These *Cyanobacteria* lie dormant during most of the year but photosynthesize immediately when seasonal rains fall (Downing et al. 2015; Richer et al. 2015). These cyanobacteria produce a variety of toxins (Cox et al. 2009; Metcalf et al. 2012, 2015), which persist in the environment even after removal of the crust itself (Richer et al. 2012). The potential large reservoir of cyanobacterial toxins may over the year give rise to exposure to particulate matter in desert dust thereby resulting in loss of soil structure and also posing risks to human and animal life (Richer et al. 2015).

Another factor influencing the variability in soil bacterial community abundance and diversity in deserts is crude oil contamination, which is most commonly observed in various oil fields (Al-Mailem et al. 2014; Gerdes et al. 2005). Under oil contamination stress, the overall abundance of soil bacteria, archaea, and fungi decreased to 10%, 40%, and 80% of those in the pristine soil, respectively (Liang et al. 2009). Along with it, the level of several functional genes in the families *pgl*, *rbcL*, *nifH*, and *nor* and those encoding cellulase, laccase, chitinase, urease, and key enzymes in metabolizing organic compounds was significantly decreased with oil contamination (Liang et al. 2011; Yang et al. 2014), resulting in an overall bacterial community shift (Khamehchiyan et al. 2007; Liu et al. 2009a; Liao et al. 2015).

13.7 Conclusion

Cumulative analyses of the microbial diversity in Asian deserts give a clear indication that microbial resources in deserts are unique and diverse. They are largely untapped for industrial applications, apart from the fact that hydrocarbonoclastic microorganisms can be applied for phytoremediation of oil contaminants. Since most aspects of living systems are based on the variability and complexity of organisms that constitute the biodiversity of a given geographical region, it is necessary that special measures are taken for biodiversity conservation especially of arid lands for continuous and sustainable life of all living communities.

References

Abdullahi AS, Underwood GJC, Gretz MR (2006) Extracellular matrix assembly in diatoms (*Bacillariophyceae*). V. Environmental effects of polysaccharide synthesis in the model diatom *Phaeodactylum tricornutum*. J Phycol 42:363–378

Abed RMM, Al-Thukair A, de Beer D (2006) Bacterial diversity of a cyanobacterial mat degrading petroleum compounds at elevated salinities and temperatures. FEMS Microbiol Ecol 57:290–301

Abed RMM, Kohls K, Schoon R, Scherf AK, Schacht M, Palinska KA, Al-Hassani H, Hamza W, Rullkötter J, Golubic S (2008) Lipid biomarkers, pigments and cyanobacterial diversity of microbial mats across intertidal flats of the arid coast of the Arabian gulf (Abu Dhabi, UAE). FEMS Microbiol Ecol 65:449–462

Abed RMM, Al Kharusi S, Schramm A, Robinson MD (2010) Bacterial diversity, pigments and nitrogen fixation of biological desert crusts from the Sultanate of Oman. FEMS Microbiol Ecol 72:418–428

Abed RMM, Klempová T, Gajdoš P, Čertík M (2015) Bacterial diversity and fatty acid composition of hypersaline cyanobacterial mats from an inland desert Wadi. J Arid Environ 115:81–89

Al-Awadhi J (2005) Dust fallout and characteristics in Kuwait: a case study. Kuwait J Sci Eng 32:135–152

Al-Mailem DM, Sorkhoh NA, Al-Awadhi H, Eliyas M, Radwan SS (2010a) Biodegradation of crude oil and pure hydrocarbons by extreme halophilic archaea from hypersaline coasts of the Arabian Gulf. Extremophiles 14:321–328

Al-Mailem DM, Sorkhoh NA, Marafic M, Al-Awadhi H, Eliyas M, Radwan SS (2010b) Oil phytoremediation potential of hypersaline coasts of the Arabian Gulf using rhizosphere technology. Bioresour Technol 101:5786–5792

Al-Mailem DM, Eliyas M, Radwan SS (2012) Enhanced haloarchaeal oil removal in hypersaline environments via organic nitrogen fertilization and illumination. Extremophiles 16:751–758

Al-Mailem D, Eliyas M, Khanafer M, Radwan S (2014) Culture-dependent and culture-independent analysis of hydrocarbonoclastic microorganisms indigenous to hypersaline environments in Kuwait. Microb Ecol 67:857–865

Amin A, Ahmed I, Habib N, Abbas S, Hasan F, Xiao M, Hozzein WN, Li WJ (2016) *Microvirga pakistanensis* sp. nov., a novel bacterium isolated from desert soil of Cholistan, Pakistan. Arch Microbiol 198:933–939

An H, Zhang L, Tang Y, Luo X, Sun T, Li Y, Wang Y, Dai J, Fang C (2009) *Skermanella xinjiangensis* sp. nov., isolated from the desert of Xinjiang, China. Int J Syst Evol Microbiol 59:1531–1534

An H, Xu M, Dai J, Wang Y, Cai F, Qi H, Peng F, Fang C (2011) *Sphingomonas xinjiangensis* sp. nov., isolated from desert sand. Int J Syst Evol Microbiol 61:1865–1869

An S, Couteau C, Luo F, Neveu J, DuBow MS (2013) Bacterial diversity of surface sand samples from the Gobi and Taklamakan deserts. Microb Ecol 66:850–860

Angel R, Conrad R, Dvorsky M, Kopecky M, Kotilínek M, Hiiesalu I, Schweigruber F, Doležal J (2016) The root-associated microbial community of the world's highest growing vascular plants. Microb Ecol 72:394–406

Azúa-Bustos A, González-Silva C, Mancilla RA, Salas L, Gómez-Silva B, McKay CP, Vicuña R (2011) Hypolithic cyanobacteria supported mainly by fog in the coastal range of the Atacama Desert. Microb Ecol 61:568–581

Balskus EP, Walsh CT (2010) The genetic and molecular basis for sunscreen biosynthesis in cyanobacteria. Science 329:1653–1656

Belnap J (2003) The world at your feet: desert biological soil crust. Front Ecol Environ 1:181–189

Belnap J, Eldridge DJ (2001) Discovery and recovery of biological soil crust. In: Belnap J, Lange OL (eds) Biological soil crusts: structure, function, and management. Springer, Berlin, pp 363–384

Belnap J, Gillette DA (1998) Vulnerability of desert biological soil crusts to wind erosion: the influences of crust development, soil texture and disturbances. J Arid Environ 39:133–142

Belnap J, Büdel B, Lange OL (2003) Biological soil crusts: characteristics and distribution. In: Belnap J, Lange OL (eds) Biological soil crusts: structure, function, and management. Ecological studies, vol 150, 2nd edn. Springer, Berlin, pp 3–30

Belnap J, Phillips SL, Miller ME (2004) Response of desert biological soil crusts to alterations in precipitation frequency. Oecologia 141:306–316

Bhatnagar A, Makandar MB, Garg MK, Bhatnagar M (2008) Community structure and diversity of cyanobacteria and green algae in the soils of Thar Desert (India). J Arid Environ 72:73–83

Billi D, Friedmann EI, Hofer K, Grilli-Caiola M, Ocampo-Friedmann R (2000) Ionizing-radiation resistance in the desiccation-tolerant cyanobacterium *Chroococcidiopsis*. Appl Environ Microbiol 66:1489–1492

Böhm GA, Pfleiderer W, Böger P, Scherer S (1995) Structure of a novel oligosaccharide-mycosporine amino acid ultraviolet A/B sunscreen pigment from the terrestrial cyanobacterium *Nostoc commune*. J Biol Chem 270:8536–8539

Bull AT (2011) Actinobacteria of the extremobiosphere. In: Horikoshi K (ed) Extremophile handbook. Springer, Tokyo, pp 1203–1240

Bull AT, Asenjo JA, Goodfellow M, Gómez-Silva B (2016) The Atacama Desert: technical resources and the growing importance of novel microbial diversity. Annu Rev Microbiol 70:215–234

Cao H, Nie Y, Zeng XC, Xu L, He L, He Z, Luo X, Wu R (2014) *Pontibacter yuliensis* sp. nov., isolated from soil. Int J Syst Evol Microbiol 64:968–972

Cary SC, McDonald IR, Barrett JE, Cowan DA (2010) On the rocks: the microbiology of Antarctic dry valley soils. Nat Rev Microbiol 8:129–138

Cavicchioli R (2002) Extremophiles and the search for extraterrestrial life. Astrobiology 2:281–292

Chen W, Wang E, Wang S, Li Y, Chen X, Li Y (1995) Characteristics of *Rhizobium tianshanense* sp. nov., a moderately and slowly growing root nodule bacterium isolated from an arid saline environment in Xinjiang, People's Republic of China. Int J Syst Bacteriol 45:153–159

Chen IC, Hill JK, Ohlemüller R, Roy DB, Thomas CD (2011) Rapid range shifts of species associated with high levels of climate warming. Science 333:1024–1026

Chowdhury SP, Schmid M, Hartmann A, Tripathi AK (2007) Identification of diazotrophs in the culturable bacterial community associated with roots of *Lasiurus sindicus*, a perennial grass of Thar Desert, India. Microb Ecol 54:82–90

Cockell C, Stokes D (2004) Widespread colonization by polar hypoliths. Nature 431:414

Cockell C, Schuerger A, Billi D, Friedmann EI, Panitz C (2005) Effects of a simulate martian flux on the cyanobacterium, *Chroococcidiopsis* sp. 029. Astrobiology 5:127–140

Coker JA (2016) Extremophiles and biotechnology: current uses and prospects. F1000 Res 5: F1000 Faculty Rev-396. https://doi.org/10.12688/f1000reserach.7432.1

Cox PA, Richer R, Metcalf JS, Banack SA, Codd GA, Bradley WG (2009) Cyanobacteria and BMAA exposure from desert dust: a possible link to sporadic ALS among Gulf War veterans. Amytroph Lateral Scler 10:109–117

Dai J, Xu M, Peng F, Jiang F, Chen X, Wang Z, Fang C (2014) *Pontibacter soli* sp. nov., isolated from the soil of a *Populus* rhizosphere in Xinjiang, China. Antonie Van Leeuwenhoek 105:65–72

Dashti N, Ali N, Eliyas M, Khanafer M, Sorkhoh NA, Radwan SS (2015) Most hydrocarbonoclastic bacteria in the total environment are diazotrophic, which highlights their value in the bioremediation of hydrocarbon contaminants. Microbes Environ 30:70–75

De los Rios A, Wierzchos J, Sancho L, Ascaso C (2004) Exploring the physiological state of continental Antarctic endolithic microorganisms by microscopy. FEMS Microbiol Ecol 50:143–152

Ding G, Piceno YM, Heuer H, Weinert N (2013) Changes of soil bacterial diversity as a consequence of agricultural land use in a semi-arid ecosystem. PLoS One 8:e59497. https://doi.org/10.1371/journal.pone.0059497

Divyasree B, Lakshmi KV, Bharti D, Sasikala C, Ramana CV (2015) *Caenispirillum deserti* sp. nov., a spheroplast-forming bacterium isolated from a salt desert. Int J Syst Evol Microbiol 65:3119–3124

Downing TG, Phelan RR, Downing S (2015) A potential physiological role for cyanotoxins in cyanobacteria of arid environments. J Arid Environ 112:147–151

Drees KP, Neilson JW, Betancourt JL, Quade J, Henderson DA, Pryor BM, Maier RM (2006) Bacterial community structure in the hyperarid core of the Atacama Desert, Chile. Appl Environ Microbiol 72:7902–7908

Elbert W, Weber B, Burrows S, Steinkamp J, Büdel B, Andreae MO, Pöschi U (2012) Contribution of cryptogamic covers to the global cycles of carbon and nitrogen. Nat Geosci 5:459–462

Eldridge DJ, Greene RSB (1994) Microbiotic soil crusts: a view of their roles in soil and ecological processes in the rangelands of Australia. Aus J Soil Res 32:389–415

El-Ghani MMA (1998) Environmental correlates of species distribution in arid desert ecosystems of eastern Egypt. J Arid Environ 38:297–313

Fierer N, Nemergut D, Knight R, Craine JM (2010) Changes through time: integrating microorganisms into the study of succession. Res Microbiol 161:635–642

Flemming HC, Wingender J, Szewzyk U, Steinberg P, Rice SA, Kjelleberg S (2016) Biofilms: an emergent form of bacterial life. Nat Rev Microbiol 14:563–575

Foster R (1981) Polysaccharides in soil fabrics. Science 214:665–667

François LE, Maas EV, Donovan TJ, Young VL (1986) Effect of salinity on grain yield and quality, vegetative growth, and germination of semi-dwarf and durum wheat. Agron J 78:1053–1058

Friedmann EI (1980) Endolithic microbial life in hot and cold deserts. Orig Life 10:223–235

Friedmann EI, Galun M (1974) Desert algae, lichens and fungi. In: Brown G (ed) Desert biology. Academic Press, New York, pp 165–212

Friedmann EI, Lipkin Y, Ocampo-Paus R (1967) Desert algae of the Negev (Isreal). Phycologia 6: 185–196

Gao Q, Garcia-Pichel F (2011) Microbial ultraviolet sunscreens. Nat Rev Microbiol 9:791–802

Garcia-Pichel F, Sherry ND, Castenholz RW (1992) Evidence for an ultraviolet sunscreen role of the extracellular pigment scytonemin in the terrestrial cyanobacterium *Chlorogloeopsis* sp. Photochem Photobiol 56:17–23

Geng J, Tang W, Wan X, Zhou Q, Wang XJ, Shen P, Lei TC, Chen XD (2008) Photoprotection of bacterial-derived melanin against ultraviolet A-induced cell death and its potential application as an active sunscreen. J Euro Acad Dermatol Venereol 22:852–858
Gerdes B, Brinkmeyer R, Dieckmann G, Helmke E (2005) Influence of crude oil on changes of bacterial communities in Arctic sea-ice. FEMS Microbiol Ecol 53:129–139
Golubic S, Friedmann EI, Schneider J (1981) The lithobiontic ecological niche, with special reference to microorganisms. J Sediment Petrol 51:475–478
Gonzalez S, Philips N, Gilaberte Y (2010) Photoprotection: update in UV-filter molecules, the "new wave" of sunscreens. G Ital Dermatol Venereol 145:515–523
Grilli Caiola M, Ocampo-Friedmann R, Friedmann IE (1993) Cytology of long-term desiccation in the desert cyanobacterium *Chroococcidiopsis* (Chroococcales). Phycologia 32:315–322
Grube M, Cardinale M, de Castro JV, Müller H, Berg G (2009) Species-specific structural and functional diversity of bacterial communities in lichen symbiosis. ISME J 3:1105–1115
Hagemann M, Henneberg M, Felde VJMNL, Drahorad SL, Berkowicz SM, Felix-Henningsen P, Kaplan A (2015) Cyanobacterial diversity in biological soil crusts along a precipitation gradient, Northwest Negev Desert, Isreal. Microb Ecol 70:218–230
Han TX, Han LL, Wu LJ, Chen WF, Sui XH, Gu JG, Wang ET, Chen WX (2008) *Mesorhizobium gobiense* sp. nov. and *Mesorhizobium tarimense* sp. nov., isolated from wild legumes growing in desert soils of Xinjiang, China. Int J Syst Evol Microbiol 58:2610–2618
Hu QW, Chu X, Xiao M, Li CT, Yan ZF, Hozzein WN, Kim CJ, Zhi XY, Li WJ (2016) *Arthrobacter deserti* sp. nov., isolated from a desert soil sample. Int J Syst Evol Microbiol 66:2035–2040
Jeffrey WH, Aas P, Lyons MM, Coffin RB, Pledger RJ, Mitchell DL (1996) Ambient solar radiation-induced photodamage in marine bacterioplankton. Photochem Photobiol 64:419–427
Jeon CO, Lim JM, Lee SS, Chung BS, Park DJ, Xu LH, Jiang CL, Kim CJ (2009) *Paenibacillus harenae* sp. nov., isolated from desert sand in China. Int J Syst Evol Microbiol 59:13–17
Jones CG, Lawton JH, Shachak M (1994) Organisms as ecosystem engineers. Oikos 69:373–386
Khamehchiyan M, Hossein Charkhabi A, Tajik M (2007) Effects of crude oil contamination on geotechnical properties of clayey and sandy soils. Eng Geol 89:220–229
Kidron GJ, Vonshak A, Abeliovich A (2008) Recovery rates of microbiotic crusts within a dune ecosystem in the Negev desert. Geomorphology 100:444–452
Kidron GJ, Barinova S, Vonshak A (2012) The effects of heavy winter rains and rare summer rains on biological soil crusts in the Negeve Desert. Catena 95:6–11
Kuske CR, Yeager CM, Johnson S, Ticknor LO, Belnap J (2012) Response and resilience of soil biocrust bacterial communities to chronic physical disturbance in shrub lands. ISME J 6:886–897
Li K, Liu R, Zhang H, Yun J (2013) The diversity and abundance of bacteria and oxygenic phototrophs in saline biological desert crusts in Xinjiang, Northwest China. Microb Ecol 66:40–48
Li CT, Yan ZF, Chu X, Hussain F, Xian WD, Yunus Z, Hozzein WN, Abaydulla G, Li WJ (2015) *Delftia deserti* sp. nov., isolated from a desert soil sample. Antonie Van Leeuwenhoek 107:1445–1450
Liang Y, Li G, Van Nostrand JD, He Z, Wu L, Deng Y, Zhang X, Zhou J (2009) Microarray-based analysis of microbial functional diversity along an oil contamination gradient in oil field. FEMS Microbiol Ecol 70:324–333
Liang Y, Van Nostrand JD, Deng Y, He Z, Wu L, Zhang X, Li G, Zhou J (2011) Functional gene diversity of soil microbial communities from five oil-contaminated fields in China. ISME J 5:403–413
Liao J, Wang J, Huang Y (2015) Bacterial community features are shaped by geographic location, physicochemical properties and oil contamination of soil in main oil fields of China. Microb Ecol 70:380–389
Lim JM, Jeon CO, Lee JC, Xu LH, Jiang CL, Kim CJ (2006) *Paenibacillus gansuensis* sp. nov., isolated from desert soil of Gansu Province in China. Int J Syst Evol Microbiol 56:2131–2134

Liu H, Buskey EJ (2000) Hypersalinity enhances the production of extracellular polymeric substance (EPS) in the Texas brown tide alga, *Aureoumbra lagunensis* (*Pelagophyceae*). J Phycol 36:71–77

Liu R, Liu H, Feng H, Wang X, Zhang CX, Zhang KY, Lai R (2008) *Pseudomonas duriflava* sp. nov., isolated from a desert soil. Int J Syst Evol Microbiol 58:1404–1408

Liu YJ, Chen YP, Jin PK, Wang XC (2009a) Bacterial communities in a crude oil gathering and transferring system (China). Anaerobe 15:214–218

Liu M, Peng F, Wang Y, Zhang K, Chen G, Fang C (2009b) *Kineococcus xinjiangensis* sp. nov., isolated from desert sand. Int J Syst Evol Microbiol 59:1090–1093

Liu M, Luo X, Zhang L, Dai J, Wang Y, Tang Y, Li J, Sun T, Fang C (2009c) *Pseudomonas xinjiangensis* sp. nov., a moderately thermotolerant bacterium isolated from desert sand. Int J Syst Evol Microbiol 59:1286–1289

Liu M, Dai J, Liu Y, Cai F, Wang Y, Rahman E, Fang C (2011a) *Desertibacter roseus* gen. nov., sp. nov., a gamma radiation-resistant bacterium in the family *Rhodospirillaceae*, isolated from desert sand. Int J Syst Evol Microbiol 61:1109–1113

Liu M, Liu Y, Wang Y, Luo X, Dai J, Fang C (2011b) *Lysobacter xinjiangensis* sp. nov., a moderately thermotolerant and alkalitolerant bacterium isolated from a gamma-irradiated sand soil sample. Int J Syst Evol Microbiol 61:433–437

Liu M, Qi H, Luo X, Dai J, Peng F, Fang C (2012) *Cesiribacter roseus* sp. nov., a pink-pigmented bacterium isolated from desert sand. Int J Syst Evol Microbiol 62:96–99

Liu M, Zhang L, Ren B, Yang N, Yu X, Wang J, Ding L, Liu X, Liu Z, Goodfellow M, Zhang L (2014) *Prauserella shujinwangii* sp. nov., from a desert environment. Int J Syst Evol Microbiol 64:3833–3837

Liu JM, Tuo L, Habden X, Guo L, Jiang ZK, Liu XF, Chen L, Zhang YQ, Sun CH (2015a) *Nesterenkonia populi* sp. nov., an actinobacterium isolated from *Populus euphratica*. Int J Syst Evol Microbiol 65:1474–1479

Liu JM, Habden X, Guo L, Tuo L, Jiang ZK, Liu SW, Liu XF, Chen L, Li RF, Zhang YQ, Sun CH (2015b) *Prauserella endophytica* sp. nov., an endophytic actinobacterium isolated from *Tamarix taklamakanensis*. Antonie Van Leeuwenhoek 107:1401–1409

Liu B, Liu GH, Cetin S, Schumann P, Pan ZZ, Chen QQ (2016) *Bacillus gobiensis* sp. nov., isolated from a soil sample. Int J Syst Evol Microbiol 66:379–384

López RP, Squeo FA, Armas C, Kelt DA, Gutiérrez JR (2016) Enhanced facilitation at the extreme end of the aridity gradient in the Atacama Desert: a community-level approach. Ecology 97:1593–1604

Lu CY, Li YQ, Tian Y, Han MX, Narsing Rao MP, Li YR, Zhu ZN, Wei DQ, An DD, Li WJ (2017) *Alcaligenes endophyticus* sp. nov., isolated from roots of *Ammodendron bifollium*. Int J Syst Evol Microbiol 67:939–943

Luo X, Wang J, Zeng XC, Wang Y, Zhou L, Nie Y, Dai J, Fang C (2012) *Mycetocola manganoxydans* sp. nov., an actinobacterium isolated from the Taklamakan desert. Int J Syst Evol Microbiol 62:2967–2970

Lynch J, Bragg E (1985) Microorganisms and soil aggregate stability. Adv Soil Sci 2:133–171

Mao J, Wang J, Dai HQ, Zhang ZD, Tang QY, Ren B, Yang N, Goodfellow M, Zhang LX, Liu ZH (2011) *Yuhushiella deserti* gen. nov., sp. nov., a new member of the suborder *Pseudonocardineae*. Int J Syst Evol Microbiol 61:621–630

Mayilraj S, Prasad GS, Suresh K, Saini HS, Shivaji S, Chakrabarti T (2005) *Planococcus stackebrandtii* sp. nov., isolated from a cold desert of the Himalayas, India. Int J Syst Evol Microbiol 55:91–94

Mayilraj S, Suresh K, Schumann P, Kroppenstedt RM, Saini HS (2006a) *Agrococcus lahaulensis* sp. nov., isolated from a cold desert of the Indian Himalayas. Int J Syst Evol Microbiol 56:1807–1810

Mayilraj S, Suresh K, Kroppenstedt RM, Saini HS (2006b) *Dietzia kunjamensis* sp. nov., isolated from the Indian Himalayas. Int J Syst Evol Microbiol 56:1667–1671

Mayilraj S, Kroppenstedt RM, Suresh K, Saini HS (2006c) *Kocuria himachalensis* sp. nov., an actinobacterium isolated from the Indian Himalayas. Int J Syst Evol Microbiol 56:1971–1975

Mayilraj S, Saha P, Suresh K, Saini HS (2006d) *Ornithinimicrobium kibberense* sp. nov., isolated from the Indian Himalayas. Int J Syst Evol Microbiol 56:1657–1661

Mayilraj S, Krishnamurthi S, Saha P, Saini HS (2006e) *Rhodococcus kroppenstedtii* sp. nov., a novel actinobacterium isolated from a cold desert of the Himalayas, India. Int J Syst Evol Microbiol 56:979–982

Mazor G, Kidron G, Vonshak A, Abeliovich A (1996) The role of cyanobacterial exopolysaccharides in structuring desert microbial crusts. FEMS Microbiol Ecol 21:121–130

McKay CP, Freedman EI, Gómez-Silva B, Cáceres-Villanueva L, Anderson DT, Landheim R (2003) Temperature and moisture conditions for life in the extreme arid region of the Atacama Desert: four years of observation including the El Niño of 1997–1998. Astrobiology 3:393–406

McKay CP, Rask JC, Detweller AM, Bebout BM, Everroad RC, Lee JZ, Chanton JP, Mayer MH, Caraballo AAL, Kapili B, Al-Awar M, Al-Farraj A (2016) An unusual inverted saline microbial mat community in an interdune sabkha in the Rub'al Khali (the Empty Quarter), United Arab Emirates. PLoS One 11:e0150342. https://doi.org/10.1371/journal.pone.0150342

Metcalf JS, Richer R, Cox PA, Codd GA (2012) Cyanotoxins in desert environments may present a risk to human health. Sci Total Environ 421–422:118–123

Metcalf JS, Banack SA, Richer R, Cox PA (2015) Neurotoxic amino acids and their isomers in desert environments. J Arid Environ 112:140–144

Nabti E, Schmid M, Hartmann A (2015) Application of halotolerant bacteria to restore plant growth under salt stress. In: Maheshwari DK, Saraf M (eds) Halophiles. Sustainable development and biodiversity, vol 6. Springer, Cham, pp 235–259

Navarro-Gonzalez R, Rainey F, Molina P, Bagaley DR, Hollen BJ, de la Rosa J, Small AM, Quinn RC, Grunthaner FJ, Cáceres L, Gómez-Silva B, McKay CP (2003) Mars-like soils in the Atacama Desert, Chile and the dry limit of microbial life. Science 302:1018–1021

Opfell JB, Zebal GP (1967) Ecological patterns of micro-organisms in desert soil. Life Sci Space Res 5:187–203

Pease PP, Tchakerian VP, Tindale NW (1998) Aerosols over the Arabian Sea: geochemistry and source areas for aeolian desert dust. J Arid Environ 39:477–496

Peng F, Zhang L, Luo X, Dai J, An H, Tang Y, Fang C (2009a) *Deinococcus xinjiangensis* sp. nov., isolated from desert soil. Int J Syst Evol Microbiol 59:709–713

Peng F, Liu M, Zhang L, Dai J, Luo X, An H, Fang C (2009b) *Planobacterium taklimakanense* gen. nov., sp. nov., a member of the family *Flavobacteriaceae* that exhibits swimming motility, isolated from desert soil. Int J Syst Evol Microbiol 59:1672–1678

Philippis R, Vincenzini M (1998) Exocellular polysaccharides from cyanobacteria and their possible application. FEMS Microbiol Rev 22:151–175

Pointing SB, Belnap J (2012) Microbial colonization and controls in dryland systems. Nat Rev Microbiol 10:551–562

Pointing S, Warren-Rhodes KA, Lacap D, Rhodes K, McKay CP (2007) Hypolithic community shifts occur as a result of liquid water availability along environmental gradients in China's hot and cold hyperarid deserts. Environ Microbiol 9:414–424

Potts M (1994) Desiccation tolerance of prokaryotes. Microbiol Mol Biol Rev 58:755–805

Rajeev L, da Roche UN, Klitgord N, Luning EG, Fortney J, Axen SD, Shih PM, Bouskil NJ, Bowen BP, Kerfeld CA, Garcia-Pichel F, Brodie EL, Northen TR, Mukhopadhyay A (2013) Dynamic cyanobacterial response to hydration and dehydration in a desert biological soil crust. ISME J 7:2178–2191

Richer R, Anchassi D, El-Assaad I, El-Matbouly M, Ali F, Makki I, Metcalf JS (2012) Variation in the coverage of biological soil crusts in the State of Qatar. J Arid Environ 78:187–190

Richer R, Banack SA, Metcalf JS, Cox PA (2015) The persistence of cyanobacterial toxins in desert soils. J Arid Environ 112:134–139

Saitou N, Nei M (1987) The neighbour-joining method: a new method for reconstructing phylogenetic trees. Mol Biol Evol 4:406–425

Schmidt SK, Reed SC, Nemergut DR, Grandy AS, Cleveland CC, Weintraub MN, Hill AW, Costello EK, Meyer AF, Neff JC, Martin AM (2008) The earliest stages of ecosystem succession in high-elevation (5000 metres above sea level), recently deglaciated soils. Proc Biol Sci 275:2793–2802

Singla AK, Mayilraj S, Kudo T, Krishnamurthi S, Prasad GS, Vohra RM (2005) *Actinoalloteichus spitiensis* sp. nov., a novel actinobacterium isolated from a cold desert of the Indian Himalayas. Int J Syst Evol Microbiol 55:2561–2564

Song L, Li WJ, Wang QL, Chen GZ, Zhang YS, Xu LH (2005) *Jiangella gansuensis* gen. nov., sp. nov., a novel actinomycete from a desert soil in north-west China. Int J Syst Evol Microbiol 55:881–884

Sorokin ID, Zadorina EV, Kravchenko IK, Boulygina ES, Tourova TP, Sorokin DY (2008) *Natronobacillus azotifigens* gen. nov., sp. nov., an anaerobic diazotrophic haloalkaliphile from soda-rich habitats. Extremophiles 12:819–827

Soule T, Stout V, Swingley WD, Meeks JC, Garcia-Pichel F (2007) Molecular genetics and genomic analysis of scytonemin biosynthesis in *Nostoc punctiforme* ATCC 29133. J Bacteriol 189:4465–4472

Stetter KO (1999) Extremophiles and their adaptation to hot environments. FEBS Lett 452:22–25

Stevenson A, Burkhardt J, Cockell CS, Cray JA, Dijksterhuis J, Fox-Powell M, Kee TP, Kminek G, McGenity TJ, Timmis KN, Timson DJ, Voytek MA, Westall F, Yakimov MM, Hallsworth JE (2015) Multiplication of microbes below 0.690 water activity: implications for terrestrial and extraterrestrial life. Environ Microbiol 17:257–277

Subhash Y, Lee SS (2016a) *Shinella curvata* sp. nov., isolated from hydrocarbon contaminated desert sands. Int J Syst Evol Microbiol 66:3929–3934

Subhash Y, Lee SS (2016b) *Skermanella rosea* sp. nov., isolated from hydrocarbon contaminated desert sands. Int J Syst Evol Microbiol 66:3951–3956

Subhash Y, Sasikala C, Ramana CV (2014) *Pontibacter ruber* sp. nov. and *Pontibacter deserti* sp. nov., isolated from desert. Int J Syst Evol Microbiol 64:1006–1011

Sun HM, Zhang T, Yu LY, Lu XX, Mou XZ, Zhang YQ (2014) *Actinophytocola gilvus* sp. nov., isolated from desert soil crusts, and emended description of the genus *Actinophytocola* Indananda et al. 2010. Int J Syst Evol Microbiol 64:3120–3125

Sun HM, Zhang T, Wei YZ, Liu HY, Yu LY, Zhang YQ (2015) *Tenggerimyces mesophilus* gen. nov., sp. nov., a member of the family *Nocardioidaceae*. Int J Syst Evol Microbiol 65:3359–3364

Sun JQ, Wang XY, Wang LJ, Xu L, Liu M, Wu XL (2016) *Saccharibacillus deserti* sp. nov., isolated from desert soil. Int J Syst Evol Microbiol 66:623–627

Sun JQ, Xu L, Guo Y, Li WL, Shao ZQ, Yang YL, Wu XL (2017) *Kribella deserti* sp. nov., isolated from rhizosphere soil of *Ammopiptanthus mongolicus*. Int J Syst Evol Microbiol 67:692–696

Tan ZY, Kan FL, Peng GX, Wang ET, Reinhold-Hurek B, Chen WX (2001) *Rhizobium yanglingense* sp. nov., isolated from arid and semi-arid regions in China. Int J Syst Evol Microbiol 51:909–914

Tang Y, Dai J, Zhang L, Mo Z, Wang Y, Li Y, Ji S, Fang C, Zheng C (2009) *Dyadobacter alkalitolerans* sp. nov., isolated from desert sand. Int J Syst Evol Microbiol 59:60–64

Tang Y, Wang Y, Ji S, Zhang K, Dai J, Zhang L, Peng F, Fang C (2010) *Pedobacter xinjiangensis* sp. nov., from desert, Xinjiang. J Microbiol Biotechnol 20:397–402

Tango MSA, Islam MR (2002) Potential of extremophiles for biotechnology and petroleum application. Energy Sources 24:543–559

Teng C, Zhou Z, Molnár I, Li X, Tang R, Chen M, Wang L, Su S, Zhang W, Lin M (2015) Whole-genome optical mapping and finished genome sequence of *Sphingobacterium deserti* sp. nov., a new species isolated from the Western Desert of China. PLoS One 10:e0122254. https://doi.org/10.1371/journal.pone.0122254

Tracy CR, Streten-Joyce C, Dalton R, Nussear KE, Gibb KS, Christian KA (2010) Microclimate and limits to photosynthesis in a diverse community of hypolithic cyanobacteria in northern Australia. Environ Microbiol 12:592–607

Tuo L, Dong YP, Habden X, Liu JM, Guo L, Liu XF, Chen L, Jiang ZK, Liu SW, Zhang YB, Zhang YQ, Sun CH (2015) *Nocardioides deserti* sp. nov., an actinobacterium isolated from desert soil. Int J Syst Evol Microbiol 65:1604–1610

Wang M, Yang M, Zhou G, Luo X, Zhang L, Tang Y, Fang C (2008) *Paenibacillus tarimensis* sp. nov., isolated from sand in Xinjiang, China. Int J Syst Evol Microbiol 58:2081–2085

Wang HF, Zhang YG, Chen JY, Hozzein WN, Li L, Wadaan MA, Zhang YM, Li WJ (2014) *Nesterenkonia rhizosphaerae* sp. nov., an alkaliphilic actinobacterium isolated from rhizophere soil in a saline-alkaline desert. Int J Syst Evol Microbiol 64:4021–4026

Wang L, Zhou Z, Wu G, Chen M, Lin M, Zhang W, Chen W (2015) *Falsirhodobacter deserti* sp. nov., isolated from sandy soil. Int J Syst Evol Microbiol 65:650–655

Warren-Rhodes KA, Rhodes KL, Pointing SB, Ewing SA, Lacap DC, Gómez-Silva B, Amundson R, Imre Friedmann E, McKay CP (2006) Hypolithic cyanobacteria, dry limit of photosynthesis and microbial ecology in the hyperarid Atacama Desert. Microb Ecol 52:389–398

Warren-Rhodes KA, Rhodes KL, Boyle LN, Pointing SB, Chen Y, Liu S, Zhou P, McKay CP (2007) Cyanobacterial ecology across environmental gradients and spatial scales in China's hot and cold deserts. FEMS Microbiol Ecol 61:470–482

Wong FKY, Lau MCY, Lacap DC, Aitchison JC, Cowan DA, Pointing SB (2010) Endolithic microbial colonization of limestone in a high-altitude arid environment. Microb Ecol 59:689–699

Xu L, Zeng XC, Nie Y, Luo Z, Zhou E, Zhou L, Pan Y, Li W (2014) *Pontibacter diazotrophicus* sp. nov., a novel nitrogen-fixing bacterium of the family *Cytophagaceae*. PLoS One 9:e92294. https://doi.org/10.1371/journal.pone.0092294

Xue X, Zhang K, Cai F, Dai J, Wang Y, Rahman E, Peng F, Fang C (2012) *Altererythrobacter xinjiangensis* sp. nov., isolated from desert sand, and emended description of the genus *Altererythrobacter*. Int J Syst Evol Microbiol 62:28–32

Yan ZF, Lin P, Chu X, Kook M, Li CT, Yi TH (2016) *Aeromicrobium halotolerans* sp. nov., isolated from desert soil sample. Arch Microbiol 198:423–427

Yang N, Sun C (2016) *Pelagibacterium lixinzhangensis* sp. nov., a novel member of the genus *Pelagibacterium*. Curr Microbiol 72:551–556

Yang SY, Liu H, Liu R, Zhang KY, Lai R (2009) *Saccharibacillus kuerlensis* sp. nov., isolated from a desert soil. Int J Syst Evol Microbiol 59:953–957

Yang YY, Wang J, Liao JQ, Huang Y, Xie SG (2014) Distribution of naphthalene dioxygenase genes in crude oil-contaminated soils. Microb Ecol 68:785–793

Yang SS, Tang K, Zhang JX, Wang XJ, Wang BX, Feng YF, Li H (2016) *Spirosoma soli* sp. nov., isolated from biological soil crusts. Int J Syst Evol Microbiol 66:5568–5574

Yu XY, Zhang L, Ren B, Yang N, Liu M, Liu XT, Zhang LX, Ding LX (2015) *Arthrobacter liuii* sp. nov., resuscitated from Xinjiang desert soil. Int J Syst Evol Microbiol 65:896–901

Yuan M, Zhang W, Dai S, Wu J, Wang Y, Tao T, Chen M, Lin M (2009) *Deinococcus gobiensis* sp. nov., an extremely radiation-resistant bacterium. Int J Syst Evol Microbiol 59:1513–1517

Zeng Y, Selyanin V, Lukeš M, Dean J, Kaftan D, Feng F, Koblížek M (2015) Characterization of the microaerophilic, bacteriochlorophyll a-containing bacterium *Gemmatimonas phototrophica* sp. nov., and emended descriptions of the genus *Gemmatimonas* and *Gemmatimonas aurantiaca*. Int J Syst Evol Microbiol 65:2410–2419

Zhang Q, Liu C, Tang Y, Zhou G, Shen P, Fang C, Yokota A (2007) *Hymenobacter xinjiangensis* sp. nov., a radiation-resistant bacterium isolated from the desert of Xinjiang, China. Int J Syst Evol Microbiol 57:1752–1756

Zhang L, Zhang Q, Luo X, Tang Y, Dai J, Li Y, Wang Y, Chen G, Fang C (2008) *Pontibacter korlensis* sp. nov., isolated from the desert of Xinjiang, China. Int J Syst Evol Microbiol 58:1210–1214

Zhang L, Dai J, Tang Y, Luo X, Wang Y, An H, Fang C, Zhang C (2009) *Hymenobacter deserti* sp. nov., isolated from the desert of Xinjiang, China. Int J Syst Evol Microbiol 59:77–82

Zhang W, Zhu HH, Yuan M, Yao Q, Tang R, Lin M, Yang SZ, Li ZK, Chen M (2010) *Microbacterium radiodurans* sp. nov., a UV radiation-resistant bacterium isolated from soil. Int J Syst Evol Microbiol 60:2665–2670

Zhang L, Wu GL, Wang Y, Dai J, Fang CX (2011) *Bacillus deserti* sp. nov., a novel bacterium isolated from the desert of Xinjiang, China. Antonie Van Leeuwenhoek 99:221–229

Zhang T, Jia RL, Zhang YQ, Liu HY, Su J, Yu LY (2013a) *Kurtzmanomyces shapotouensis* sp. nov., an anamorphic basidiomycetous yeast isolated from desert soil crusts. Int J Syst Evol Microbiol 63:3892–3895

Zhang YG, Wang HF, Liu Q, Hozzein WN, Wadaan MA, Cheng J, Chen YJ, Zhang YM, Li WJ (2013b) *Streptomyces fukangensis* sp. nov., a novel alkaliphilic actinomycete isolated from a saline-alkaline soil. Antonie Van Leeuwenhoek 104:1227–1233

Zhang ZY, Gao XH, Zhang YJ, Jia M, Lu XJ, Ma YC, Tian F, Xie Q, Tang SK (2015) *Skermanella rubra* sp. nov., a bacterium isolated from the desert of Xinjiang, China. Antonie Van Leeuwenhoek 108:627–632

Zhang YG, Wang HF, Yang LL, Guo JW, Xiao M, Huang MJ, Hozzein WN, Li WJ (2016) *Ornithinicoccus halotolerans* sp. nov., and emended description of the genus *Ornithinicoccus*. Int J Syst Evol Microbiol 66:1894–1899

Zhao P, Zhou Z, Chen M, Lin W, Zhang W, Wei G (2014) *Sphingobacterium gobiense* sp. nov., isolated from soil of the Gobi Desert. Int J Syst Evol Microbiol 64:3931–3935

Zhao Q, Li HR, Han QQ, He AL, Nie CY, Wang SM, Zhang JL (2017) *Altererythrobacter soli* sp. nov., isolated from desert sand sample. Int J Syst Evol Microbiol 67:454–459

Zhao L, Wang X, Huo H, Yuan G, Sun Y, Zhang D, Cao Y, Xu L, Wei G (2016) Phylogenetic diversity of *Ammopiptanthus* rhizobia and distribution of rhizobia associated with *Ammopiptanthus mongolicus* in diverse regions of Northwest China. Microb Ecol 72:231–239

Zhou Y, Wang X, Liu H, Zhang KY, Zhang YQ, Lai R, Li WJ (2007) *Pontibacter akesuensis* sp. nov., isolated from a desert soil in China. Int J Syst Evol Microbiol 57:321–325

Zhou Z, Yuan M, Tang R, Chen M, Lin M, Zhang W (2012) *Corynebacterium deserti* sp. nov., isolated from desert sand. Int J Syst Evol Microbiol 62:791–794

Trichoderma from Extreme Environments: Physiology, Diversity, and Antagonistic Activity

14

Laith Khalil Tawfeeq Al-Ani

Abstract

The fungus *Trichoderma* is spreading throughout different climate zones. Therefore, this enhances the chance to get some isolates having the ability to confront poor conditions. Several extreme conditions affect *Trichoderma*. In this chapter I focus on important parameters that have large effects on growth, bioactivity, and antagonism as biological control agents. On the basis of these effects, some parameters are appropriate for every strain of *Trichoderma*: main factors such as temperature, pH, nutrient substrate, and water potential, and minor factors such as light and humidity. The temperature parameter is the first main factor that is suggested here to be responsible for alteration in *Trichoderma* life phases and bioactivity. *Trichoderma* has shown a high tolerance for temperature (range 0–50 °C). Most *Trichoderma* spp. showed high efficacy at moderate temperatures. *Trichoderma* spp. can tolerate pH from 2.0 to 13, but more *Trichoderma* tend toward acidic media. Nutrient substrate, water potential, light, and humidity were effective factors related to one or two activities of *Trichoderma*. However, parameters are very important in determining the efficacy of *Trichoderma* for use in controlling plant pathogens. Therefore, we can consider four points to confront these weaknesses of some *Trichoderma*-derived biopesticides and biofertilizers to control plant pathogens.

Keywords

Trichoderma · Extreme soils · Diversity · Biology · Plant

L. K. T. Al-Ani (✉)
Department of Plant Protection, College of Agriculture, University of Baghdad, Baghdad, Iraq

School of Biology Science, Universiti Sains Malaysia, Minden, Pulau Pinang, Malaysia

D. Egamberdieva et al. (eds.), *Extremophiles in Eurasian Ecosystems: Ecology, Diversity, and Applications*, Microorganisms for Sustainability 8,
https://doi.org/10.1007/978-981-13-0329-6_14

14.1 Introduction

Trichoderma can spread throughout a wide range of ecological niches. It is ubiquitous in soil and the rhizosphere. The environment for this genus is very attractive because of its ability to attack and compete within different habitats. *Trichoderma* spp. have unique properties that help them grow at high densities in any habitat (Chet et al. 1997). The efficacy of *Trichoderma* activity is depending on an asexual cycle. Indeed, in a habitat under optimal conditions, *Trichoderma* spp. produce enzymes, secondary metabolites, and proteins to compete and that are useful for getting nutrients to grow and disperse through the asexual cycle. Hjeljord et al. (2000) suggested that the growth and antagonism of *Trichoderma* against plant pathogens are decreasing because of poor nutrient levels. The enzymes and secondary metabolites of *Trichoderma* are used in different fields of study. Mastouri et al. (2010, 2012) showed that *Trichoderma* has a high ability to improve the resistance of plants against abiotic and biotic stresses. Therefore, many bioproducts (e.g., biopesticides and biofertilizers) with *Trichoderma* formulations are used to control plant pathogens and to enhance plant growth, as well as, Trichoderma is having the ability to enhance for the tolerance the hard condition such as Salinity, extreme temperature, and water stress (Balestrini et al. 2017). In addition, *Trichoderma* produces several enzymes with high activity that are helpful in biotechnology and remove waste from the environment. Conidial suspension are used to control plant pathogens and other activity by *Trichoderma*. Carreras-Villaseñor et al. (2012) mention of the role of conidia in an "asexual cycle"; it has different beneficial functions including in the biocontrol of plant pathogens and in the industry. Such as textile, medical/pharmaceutical, and animal feed by utilizing the compounds and enzymes that produced of *Trichoderma*.

Interestingly, *Trichoderma* suspensions are used in different industries, such as agriculture. A conidial suspension of *T. harzianum* (1×10^{-7} spores/ml) can control *Sclerotinia sclerotiorum* (Zhang et al. 2016). A high density (10^{-10}) of *T. harzianum* and *T. viride* have efficacy to control *Meloidogyne javanica* on tomato (Al-Hazmi and TariqJaveed 2016). The growth of peas improved after seeds were treated with 10^6 spores from two *Trichoderma* strains (T4 and N47) (Naseby et al. 2000). Conidia of *Trichoderma* are used in various activities because they spread quickly and germinate easily. Extreme environments are, however, certainly affecting the efficacy of *Trichoderma* in biocontrol of the life cycle of plant pathogens. In general, conidia are available in normal environments but not under difficult conditions. *Trichoderma* spp. are sensitive to changes in the environment (Carreras-Villaseñor et al. 2012).

This sensitivity is leading to change in attributes such as stop in the growth of hyphae for *Trichoderma* as the status of adaptation to confront the hard environment. These hyphae start the differentiation to specific structure by thickening the wall and create the resting spores as a tool for survival. Molecular mechanisms are responsible for adaptation and response to diverse cues from the environment (Bahn et al. 2007). Extreme conditions induce the fungus to produce resting or dormant spores such as thick-walled chlamydospores. However, some species of *Trichoderma* produce microsclerotia (Jackson et al. 2017). Difficult conditions and biological rhythms have a role in inducing the formation of chlamydospores and in the sexual cycle. Chlamydospores of *Trichoderma* are beneficial not only for survival and dispersal but also for export as a biological control agent (BCA), as mentioned by Mishra et al. (2012). Indeed, changes from useful gene expression to other expressions produce certain proteins and enzymes as a way to protect the survival of the fungal thallus . Interactions and sensing between fungi and the environment happen at the molecular level in response to environmental cues (Bahn et al. 2007). Some interesting factors affect the physiological activities of *Trichoderma*. These factors place high stress on the success of *Trichoderma* in confronting plant pathogens. Also, *Trichoderma* is a very interesting agent in different fields and is used within industry and production as a biopesticide and biofertilizer. Conditions of an extreme environment—major factors such as pH, temperature, nutrient substrate, and water potential, and some minor factors such as light and humidity—affect the success of *Trichoderma* products. Some parameters such as pH, Carbon content, and carbon : nitrogen (C:N) ratio were affected Trichoderma including the growth, sporulation, and the time of spore production (Agosin et al. 1997).

On the other hand, three points are important to discuss here in order to explain the effect of extreme environments on *Trichoderma* used as a BCA. For example, *Trichoderma* biopesticides and biofertilizers may be not beneficial for use in fields. Many farmers think *Trichoderma* products have low efficacy in controlling plant pathogens and enhancing plant growth. Also, many researchers think isolates or strains of *Trichoderma* are not suitable for use as antagonists or to improve plant growth in fields. Therefore, we must determine the parameters of extreme environments to provide some information that may be of benefit in improving *Trichoderma*-derived bioproducts. However, extreme environments include many factors that affect physiological activities and diversity, and that antagonize *Trichoderma*. This chapter shows the relation between extreme environmental conditions and the life cycle of *Trichoderma*.

14.2 Growth of *Trichoderma*

Determining optimal and extreme conditions is very helpful in determining the ability of *Trichoderma* to grow in different habitats. *Trichoderma* species can grow within a specific temperature range. Its growth comprises germination of spores and mycelia, and sporulation, which allow the fungus to spread. This study is useful for understanding the utilization of *Trichoderma* in many actions as mentioned previously such as biopesticides, biofertilizer, and industry. Therefore, it must invoke the role of extreme conditions affecting the very interesting part of a life cycle for *Trichoderma*, as following.

14.2.1 Germination

Fungal growth begins with conidia (spore) germination and mycelial growth. In general, the best and fastest growth of conidia and mycelial mass occurs under optimal circumstances. These conditions are limited to a particular range, which is different among *Trichoderma* species. Conidia and hyphae are exposed to different conditions including water availability, temperature, and pH. These factors are important for determining the rate and density of growth, rate of hyphal extension, and tube length; water availability and temperature are most important and most effective (Hjeljord and Tronsmo 2003). Danielson and Davey (1973) mention the large role of temperature and pH in the growth of seven species of *Trichoderma*, such as *T. pseudokoningii* and *T. saturnisporum*; they showed no growth at low and medium temperatures, and growth was very effective in extremely acidic or alkaline conditions. Gervais et al. (1988) suggested that the main factor affecting germination of *Trichoderma* was water potential. And Jackson et al. (1991) mentioned pH, temperature, and water potential as three factors affecting germination of *Trichoderma*. The duration of radiation did not affect conidial germination or growth of *Trichoderma* (Wibowo 1999). However, Hjeljord and Tronsmo (2003) indicated that the nutrient substrate is an effective factor in the germination of *Trichoderma* conidia; therefore, some of the conidia population fails to initiate germination on nutrient-poor substrates and in dilute inocula. According to the temperature factor, two levels of the conidia germination for *Trichoderma* such as low tolerance range and high tolerance range are noticed. Germination can occur within two temperature ranges: a low-tolerance range (<20 °C) and a high-tolerance range (≥20 °C).

T. harzianum, *T. longibrachiatum*, and *T. viride* grow in the low-tolerance range at 12–20 °C (they cannot grow at high temperatures) and at water potential between −0.7 and −2.8 MPa (Magan 1988). Some strains of *Trichoderma* are, however, able to tolerate the extreme environments. Cold-tolerant *Trichoderma* strains such as *T. aureoviride*, *T. harzianum*, and *T. viride* isolated from a forest at Asotthalom in southern Hungary grew at a low temperature (5 °C) (Antal et al. 2000).

T. viride grew at 5 °C but did not grow at 40 °C; it grew at pH ranging between 4.6 and 6.8; and the water potential decreased over the range of −0.7 to −14.0 MPa, but germination or growth occurred at −14.0 MPa (Jackson et al. 1991). Conidial germination and growth of *T. harzianum* occur within a pH range of 5–9; conidial germination prefers a temperature between 20 and 30 °C and was inhibited at 10 °C and 40 °C; mycelia grew within a temperature range of 10–30 °C, but their growth was inhibited at 40 °C (Wibowo 1999). *T. koningii* can grow at high temperatures (5–29 °C), in soil containing 10–80% moisture (water holding capacity), and a pH 5.8 (Wakelin et al. 1999). *T. koningii* growth increased after the ammonium (NH^{+4}-N) was added, which affected the acidity, but was suppressed when nitrate (NO_3^-) was added (Wakelin et al. 1999). Upon initiation of germination, the conidia of *Trichoderma* are more sensitive to desiccation after only 2 h when incubated on a nutrient-rich substrate at 23 °C (Hjeljord and Tronsmo 2003).

The thermophilic strain of *T. reesei* (RL-P31) grows quickly at 37 °C but does not grow at 28 °C (Sharma 1992). *T. harzianum*, *T. viride*, and *T. koningii* grew at temperatures between 9 and 35 °C and at pH within the range of 4–12, but the best growth occurred at 24 °C and pH 5.5 (Ghildiyal and Pandy 2008). Some *Trichoderma* species—*T. harzianum*, *T. viride*, *T. asperellum*, *T. koningii*, *T. atroviride*, *T. longibrachiatum*, and *T. virens*—were able to grow at temperatures of 25–30 °C and at pH values between 5.5 and 7.5 (Singh et al. 2014). Two strains of *Trichoderma*—*T. viride* (Td50) and *T. pseudokoningii* (Td85)—grow between 25 °C and 30 °C (and grow very slowly at 15 °C) and favor a pH from 4.5 to 5.5 (Petrisor et al. 2016). *T. asperelloides* IBLF 908 is able to grow at 12–37 °C, but maximum growth occurred at 27 °C (Domingues et al. 2016). *T. asperellum* can grow at 50 °C (Montoya-Gonzalez et al. 2016). Indeed, *T. polysporum* strains from Norway (a polar region) grew at temperatures between 0 °C and 28 °C, with higher growth at 20 °C (Kamo et al. 2016). *T. harzianum*, *T. viride*, *T. asperellum*, and *T. hamatum* showed favorable growth at pH ranging from 4.6 to 7.6, but the species grew best at different temperatures: *T. harzianum* and *T. viride* grew at temperatures between 25 °C and 40 °C, and *T. asperellum* and *T. hamatum* preferred 25–35 °C (Zehra et al. 2017). Isolates of *T. harzianum*, *T. viride*, and *T. koningii* could tolerate high reductions in temperature and grew under conditions between 4 and 42 °C and at a pH of 3–13 (Sharma et al. 2013). Finally, the spores and mycelia of *T. harzianum* strain T22 could germinate at 25 °C and at a water potential between −0.03 and −0.50 MPa (Innocenti et al. 2015). On the other hand, the concentration of salt in the habitation of Trichoderma is affecting the germination. Zehra et al. (2017) found that 1000 μM NaCl (salt) affects Trichoderma species, including T. harzianum, T. viride, T. asperellum, and T. hamatum.

14.2.2 Sporulation

Many different environmental factors affect sporulation of *Trichoderma* spp.: Nutrient substrate, temperature, humidity, light, and pH are very important factors in this context (Galun and Gressel 1966; Wibowo 1999; Berrocal-Tito et al. 1999; Jayaswal et al. 2003; Casas-Flores et al. 2004). The sporulation of *T. harzianum* was enhanced when receiving 24 h of light and at temperatures between 10 and 30 °C, but it did not produce conidia at 40 °C. In addition, acidity increases sporulation but alkaloids greatly affect it (Wibowo 1999). Blue light (400–480 nm) induces synchronous sporulation (Casas-Flores et al. 2004).

T. stromaticum can sporulate at temperatures from 20 to 25 °C and at 100% humidity, but it cannot sporulate at 75% humidity (Sanogo et al. 2002). *T. viride* produces maximum conidia at pH of 4.5–5.5 and a temperature of 20–37 °C, but this production is inhibited at temperatures below 20 °C, and is very poor with carbon sources (rhamnose, sorbitol, and pyruvic acid) (Jayaswal et al. 2003). *T. harzianum* was produced at 30 °C in a medium containing 30 g/L glucose and a carbon-to-nitrogen ratio of 24 (Said 2007). *T. hamatum* and *T. asperellum* preferred temperatures between 25 and 35 °C for sporulation, but *T. harzianum* and *T. viride* preferred 25–40 °C; the best sporulation for these four species of *Trichoderma* occurred in the pH range of 4.6–7.6 (Zehra et al. 2017). It is striking that *Trichoderma* can sporulate in extreme environments. Sharma et al. (2013) note that sporulation was induced at 0 °C in three isolates of *Trichoderma*—*T. harzianum*, *T. viride*, and *T. koningii*.

14.3 Bioactivity of *Trichoderma*

This genus has a high ability to attack and kill other fungi. This bioactivity is a part of the *Trichoderma* life cycle; production changes in accordance with alterations in the environment.

14.3.1 Production of Enzymes

Trichoderma secretes several important enzymes that are used for survival and to compete with and attack other organisms. *Trichoderma* can produce enzymes within appropriate temperature and pH conditions, and on appropriate nutrient substrates. Water potential and pH affect the production of enzymes by *Trichoderma* species (Kredics et al. 2004). Different conditions discriminate between isolates and species of *Trichoderma*.

At high temperatures, *T. reesei* strain RL-P37, cultivated at 37 °C in medium containing lactose, hypersecreted the xylanase enzyme (Suh et al. 1988). *T. viride* SL-1 produced cellulase at temperatures ranging from 30 to 50 °C (Tao et al. 1997). Temperature can change the cellulase enzyme in the subsequent stages. The activity of cellulase from *T. reesei* was not affected until the temperature reached 37 °C; enzyme activity decreases at temperatures from 37 to 50 °C, an no activity occurred at temperatures above 70 °C (Andreaus et al. 1999). *T. harzianum* 1073D3 produced the xylanase enzyme, with high activity at 60 °C and pH 5 in medium containing 1% xylan (Isil and Nilufer 2005). *T. lignorum* (Tode) Harz produced cellulolytic enzymes on banana waste at an optimal temperature of 45 °C and at a pH of 5.6–5.8 (Baig 2005). *Trichoderma* sp. produced cellulase at an optimal temperature (45 °C) and pH (6.5), on a nutrient substrate with an appropriate carbon-to-nitrogen ratio, such as cellulase (municipal solid waste residue), peptone, and yeast extract (Gautam et al. 2011). Cellulase was produced by *T. reesei* strain HY07 cultivated on a nutrient substrate (1.5% ammonium sulfate) at 30 °C (Guoweia et al. 2011). *T. harzianum* and *T. viride* produced the active enzyme chitosanase at pH 5.0; *T. koningii* and *T. polysporum* produced this enzyme at pH 5.5, and a temperature between 40 and 50 °C did not affect chitosanase activity (Da Silva et al. 2012). The highest production of glucose through secretion of the cellulase enzyme by *T. reesei* occurred at 30 °C and pH 4.5 (Silas et al. 2017). *T. asperellum* was producing the cell wall-degrading enzymes (CWDEs) and be high activity at 36°C (Qiu et al. 2017).

At moderate temperatures, *Trichoderma* strains such as *T. aureoviride* T122, *T. harzianum* T66 and T334, and *T. viride* T114 and T228 produce different extracellular enzymes, including of β-glucosidase, cellobiohydrolase, and β-xylosidase; at 25 °C, this production is related to two factors: water potential and pH (Kredics et al. 2004). Some isolates of *Trichoderma* produce enzymes at low temperatures (extreme environments). *T. aureoviride*, *T. harzianum*, and *T. viride* were cold-resistance strains and produced high levels of various extracellular enzymes that are active highly at 5 °C: chitinases, proteases, and β-glucosidases (Antal et al. 2000).

14.3.2 Production of Secondary Metabolites

Trichoderma produces secondary metabolites (volatile and nonvolatile compounds) within particular environments. Extreme environments already affect the capacity of *Trichoderma* to produce these compounds. Temperature and pH affect the efficacy of *Trichoderma*. A new isolate of *T. harzianum*, SQR-T037, was highly efficacious in producing volatile and nonvolatile compounds at 30 °C and pH 6, but very few compounds were produced under extreme conditions (Raza et al. 2013). Tronsmo and Dennis (1978) mention the high ability of some *Trichoderma* isolates to produce nonvolatile antibiotics at low temperatures; others, however, produce at high temperatures. Mukherjee and Raghu (1997) suggested that the fungitoxic metabolites are produced with higher concentrations of *Trichoderma* at higher temperatures. Also, six new peptaibol compounds, asperelines A–F (1–6), were produced by *T. asperellum* isolated from Penguin Island in the Antarctic (Ren et al. 2009). Four isolates of *Trichoderma*—*T. parareesei* T26, *T. koningii* TR102, and *T. harzianum* Tveg1 and TL5—produced 30 possible antifungal compounds at 28–30 °C under 12 h of darkness and 12 h of light (Al-Ani 2017).

14.4 Diversity of *Trichoderma*

The genus *Trichoderma* is widely diverse worldwide. The varying diversity of *Trichoderma* is dependent on climate and soil traits. *Trichoderma* spp. are predominant in all climate zones and are free-living organisms that grow in soil, root, and foliar environments (Harman et al. 2004). The four main climate zones globally are tropical, subtropical, temperate, and polar. Extreme environments are found in all climate zones that comprise characteristics such as lack of water potential (drought), high or low temperatures, high elevation, and extremely high pH. Lupo et al. (2002) classified the genus *Trichoderma* as mesophilic organisms. Kredics et al. (2003) mention that most *Trichoderma* isolates are mesophiles.

T. harzianum, *T. virens*, *T. spirale*, *T. koningii*, *T. atroviride*, *T. asperellum*, *T. reesei*, *T. viride*, *T. hamatum*, and *T. ghanense* were isolated from Taiwan and Western Indonesia in Southeast Asia (the tropical zone) (Kubicek et al. 2003). *T. koningii* growth is restricted in eastern North America and Europe, but other species, such as *T. koningiopsis*, *T. caribbaeum* var. *aequatoriale*, *T. ovalisporum*,

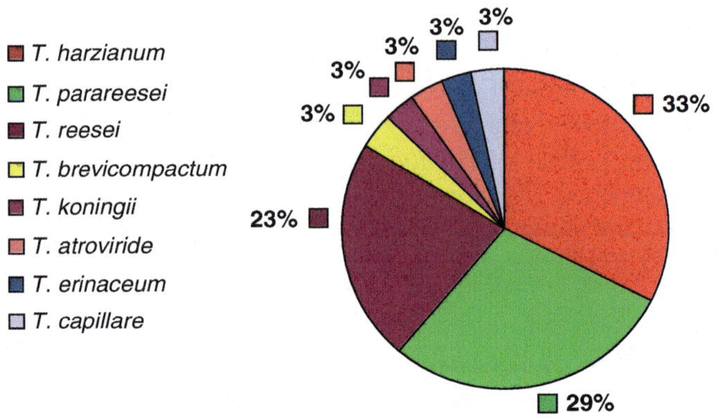

Fig. 14.1 Percentages of several species of *Trichoderma* (*T. harzianum*, *T. reesei*, *T. parareesei*, *T. brevicompactum*, *T. koningii*, *T. atroviride*, *T. erinaceum*, and *T. capillare*)

T. ovalisporum, and *T. stilbohypoxyli*, grow in tropical areas (Samuels et al. 2006). Many isolates of *Trichoderma* species—*T. harzianum*, *T. hamatum*, *T. asperelloides*, and *T. spirale*—were detected in the rhizosphere in the coffee-growing region in the highlands of Ethiopia and are more endemic in tropical regions such as Africa (Mulaw et al. 2010). *T. reesei* and *T. parareesei* are widespread throughout the pantropical region (Druzhinina et al. 2010). many strains of *Trichoderma* were identified in the neotropical region of Mexico, including *T. asperellum*, *T. brevicompactum*, *T. harzianum*, *T. koningiopsis*, *T. longibrachiatum*, *T. pleuroticola*, *T. reesei*, *T. spirale*, and *T. virens* (Torres-De la Cruz et al. 2015). Most species of *Trichoderma*, such as *T. harzianum*, *T. reesei*, and *T. parareesei*, show diverse isolates from Pulau Penang, Malaysia (Al-Ani 2017) (Fig. 14.1).

Three strains of *Trichoderma*—one strain of *T. harzianum* and two of *T. asperellum*—were isolated from a subtropical desert (Montoya-Gonzalez et al. 2016). *T. asperellum*, *T. virens*, *T. harzianum*, *T. sinensis*, *T. citrinoviride*, *T. longibrachiatum*, *T. koningii*, *T. atroviride*, *T. viride*, *T. velutinum*, and *T. cerinum* were isolated from subtropical and temperate zones in northern, southern, and eastern China; almost half of those species were *T. harzianum* (Zhang et al. 2005). *T. viride* and *T. harzianum* were found on Mount Moosilauke in New Hampshire in the United States (temperate zone), where they grew under cold temperatures in winter and moderate temperatures in summer, ranging from −10 to 25 °C. Several isolates of *T. harzianum* (21 strains), *T. rossicum* (13 strains), *T. cerinum* (4 strains), *T. hamatum* (2 strains), and *T. atroviride* and *T. koningii* (1 strain each) were identified in southeast Austria (Wuczkowski et al. 2003). *T. harzianum*, *T. koningii*, *T. longibrachiatum*, *T. viride*, and *T. citrinoviride* were collected from decaying wood in Poland (Błaszczyk et al. 2011). Many species of *Trichoderma* were isolated from three mountains in different regions of Poland, including *T. atroviride*, *T. citrinoviride*, *T. cremeum*, *T. gamsii*, *T. harzianum*, *T. koningii*, *T. koningiopsis*, *T. longibrachiatum*, *T. longipile*, *Trichoderma* sp. (*Hypocrea parapilulifera*), *T. viride*, and *T. viridescens*;

one species, *T. viride*, comprised 53% of the isolates (Błaszczyk et al. 2016). *T. orientale*, *T. spirale*, *T. tomentosum*, *T. albolutescens*, and *T. asperelloides* were found to be the first recorded *Trichoderma* species in Korea (Jang et al. 2017).

Six strains of *T. polysporum* were isolated from arctic wetlands in Norway (polar zone) (Yamazaki et al. 2011). *T. asperellum* was isolated from sediment on Penguin Island in the Antarctic (Ren et al. 2009). The pH factor is affecting the diversity of Trichoderma in some time. *Trichoderma* did not grow, or grew only little, at pH below 2.0 or above 6.0 (Kredics et al. 2003). For example, soil pH might be affected on the distribution of T. harzianum under pH 6.2 and over pH 7.9 (Eastburn and Butler 1988). In addition, the distribution of T. koningii was related with a soil pH (Muniappan and Muthukumar 2014).

14.5 *Trichoderma* as an Antagonistic BCA

Trichoderma is used widely as a BCA against many plant pathogens. The *Trichoderma* fungus is important worldwide as an alternative to chemical pesticides. *Trichoderma* spp. are used as BCAs (biopesticides) and are safe for the ecosystem. For successful biocontrol of plant pathogens, the isolate, strain, or species must be selected appropriately based on the ecology in the location of use. Therefore, environmental conditions are an influential factor in the antagonism of *Trichoderma* through use as a biopesticide. pH, temperature, and water potential affect it's biocontrol status against plant pathogens. Mukherjee and Raghu (1997) mention that temperature is the critical factor that influences BCAs. In addition, Kredics et al. (2004) presented the importance of water potential and pH in antagonism.

In low-temperature environments, *Trichoderma* increasingly suppressed plant pathogens such as *Gaeumannomyces graminis* var. *tritici*; the suppression for this pathogen can be highly in acidic soils through the addition of ammonium sulfate at 15 °C (Simon et al. 1988). Three cold strains of such as *T. aureoviride*, *T. viride*, and *T. harzianum* used as BCAs were actively antagonistic and showed high interactions that produced appressoria at different temperatures (5 °C, 10 °C, and 20 °C) (Antal et al. 2000). By producing the antibiotics, an arctic strain of *T. polysporum* could control *Pythium iwayamai*, which causes snow rot (Kamo et al. 2016).

In moderate temperatures, *Trichoderma* spp. were antagonistic against the pathogen *Sclerotium rolfsii* in dual cultures at temperatures ranging from 25 to 30 °C, but *Trichoderma* spp. do not suppress *S. rolfsii* at temperatures above 30 °C (Mukherjee and Raghu 1997). Water potential and pH affected the mycoparasitism of some species of *Trichoderma* (*T. harzianum*, *T. viride*, and *T. aureoviride*) at 25 °C (Kredics et al. 2004). At 25 °C, the ability of *T. harzianum* to antagonize *Verticillium dahliae* was reduced in high-salinity soils (Regragui and Lahlou 2005). *T. harzianum* strain T22 was very antagonistic against *Fusarium oxysporum* f. sp. *lactucae* strain 365.07; *T. harzianum* caused this *Fusarium* to wilt at 25 °C and at high extremes of water potential (−0.03 and −0.50 MPa) (Innocenti et al. 2015). *T. harzianum* LU698 has more influence on *Sclerotinia sclerotiorum* and reduced the viability of sclerotia under water potential values of −0.1 and −0.3 MPa, but *T. asperellum* LU697 was affected at water potentials of −0.01 and −1.5 MPa and at 25 °C (Jones et al. 2016). Al-Ani (2017) isolated 32 different strains of *Trichoderma* from regions in northern and in the middle of Malaysia and found that more of these isolates were highly antagonistic against *F. oxysporum* f. sp. *cubense* Tropical Race 4 that was isolated from the same region.

In high-temperature regions, strains of *T. harzianum*, *T. viride*, *T. hamatum*, *T. pseudokoningii*, *T. koningii*, and *T. longibrachiatum* could control *Macrophomina phaseolina* and showed maximal inhibition at 35 °C, but *T. pseudokoningii* was inhibitory at a temperature of 40 °C (Malathi and Doraisamy 2003). *T. harzianum* Th2 inhibited the growth of *F. oxysporum* f. sp. *ciceri* to a minimal level (10–12%) in sandy clay at 35 °C and water potential of −0.3 MPa (Inam-Ul-Haq et al. 2009). In addition, 14 *Trichoderma* species isolated from soils in a desert in Algeria were very antagonistic against three plant pathogens. *T. harzianum* 8.4, *T. asperellum* 12-2, and *T. asperellum* BP60 were isolated from sandy soils in a desert but only the *T. asperellum* BP60 isolate was active at temperatures below 50 °C (Montoya-Gonzalez et al. 2016). This isolate was able to control *Setophoma terrestris*, which causes pink root rot on green onions, under extreme temperatures and produced siderophores and chitinases (Montoya-Gonzalez et al. 2016). *Trichoderma* isolates had highly antagonistic activity against *F. oxysporum* f. sp. *cubense* Tropical Race 4 at temperatures ranging from 28 to 32 °C (Al-Ani et al. 2013; Al-Ani 2017; Al-Ani and Albaayit 2018).

14.6 Conclusion

Trichoderma has an amazing ability to survival in extreme environments, but this survival depends on the species and environmental factors. Indeed, temperature has more of an impact on *Trichoderma* than other factors such as pH, water potential, and nutrient substrate. The optimal temperature for all physiological actions of *Trichoderma* (e.g., germination of conidia and hyphae, sporulation, production of active enzymes, and antagonism) lies within the temperate range. Temperatures in the range of 0–50 °C can be considered the main extreme factor affecting *Trichoderma* . The second most important factor that affects *Trichoderma* is pH. Species of the *Trichoderma* genus can grow at high or low pH values (2.0–13), but the level of growth differs from one species to another. The *Trichoderma* population is increasing in acidic soils that contain the ammonium sulfate.

The third factor affecting *Trichoderma* growth is the nutrient substrate, which can be efficacious but is not as important as temperature and pH. The nutrient substrate is important for inducing the growth and antagonism of *Trichoderma*. The fourth factor with an effect on *Trichoderma* growth and activity is water potential. This factor is also important in determining the diversity and distribution of *Trichoderma* in soil. The lack of water, and pH, have potential effects on the growth and mycoparasitic action of *Trichoderma*. Light and humidity are important factors in the sporulation and dispersal of *Trichoderma*, as well as, Highland is affecting the diversity of Trichoderma. These major and minor factors are the main parameters for estimating the effects of *Trichoderma* strains against plant pathogens.

Four points can be considered to clarify how to confront the decrease in efficiency of *Trichoderma* products used against plant pathogens and to enhance plant growth in fields. First, the appropriate parameters must be determined before being used. Second, *Trichoderma* must be isolated from the same climate zone or an area nearest the plant pathogens. Isolates of *Trichoderma* from the same area as the plant pathogens or plants will have higher efficacy when used. This may guarantee success in the biocontrol of plant pathogens, enhancement of plant growth, and use in other activities. Third, the genome of *Trichoderma* is responsible for controlling the cells upon confrontation of difficult conditions that affect germination, dispersal, and survival. Therefore, mutations in *Trichoderma* can improve the traits necessary to tolerate poor conditions. The fourth and final point is the importance of *Trichoderma* to resist unfavorable conditions in order to be beneficial in controlling phytopathogens; this helps the plants by enhancing their capability to resist the stress of a difficult environment. The high tolerance of *Trichoderma* strains to extreme environments is useful when applying it to the different crops and in helping growth and improving and increasing production within several climate zones.

References

Agosin E, Volpe D, MunÄoz G, San Martin R, Crawford A (1997) Effect of culture conditions on spore shelf life of the biocontrol agent *Trichoderma harzianum*. World J Microbiol Biotechnol 13:225–232. https://doi.org/10.1023/A:1018502217083

Al-Ani LKT (2017) Potential of utilizing biological and chemical agents in the control of Fusarium wilt of banana. PhD School of Biology Science, Universiti Sains Malaysia Pulau, Pinang, Malaysia, p 259

Al-Ani LKT, Albaayit SFA (2018) Antagonistic of some Trichoderma against Fusarium oxysporum sp. f. cubense tropical race 4 (FocTR4). In: International Conference on Research in Education & Science, ICRES April 28 – May 1, Marmaris, Turkey, p 271

Al-Ani LKT, Salleh B, Ghazali AHA (2013) Biocontrol of Fusarium wilt of banana by *Trichoderma* spp. 8th PPSKH colloquium Pust Pengajian Sains Kajihayat/School of Biological Sciences USM June 5–6

Al-Hazmi AS, TariqJaveed M (2016) Effects of different inoculum densities of *Trichoderma harzianum* and *Trichoderma viride* against *Meloidogyne javanica* on tomato. Saudi J Biol Sci 23:288–292. https://doi.org/10.1016/j.sjbs.2015.04.007

Andreaus J, Azevedo H, Cavaco-Paulo A (1999) Effects of temperature on the cellulose binding ability of cellulase enzymes. J Mol Catal B Enzym 7:233–239. https://doi.org/10.1016/S1381-1177(99)00032-6

Antal Z, Manczinger L, Szakaacs G, Tengerdy RP, Ferenczy L (2000) Colony growth, *in vitro* antagonism and secretion of extracellular enzymes in cold-tolerant strains *Trichoderma* species. Mycol Res 104:545–549. https://doi.org/10.1017/S0953756299001653

Bahn YS, Xue C, Idnurm A, Rutherford JC, Heitman J, Cardenas ME (2007) Sensing the environment: lessons from fungi. Nat Rev Microbiol 5:57–69. https://doi.org/10.1038/nrmicro1578

Baig MMV (2005) Cellulolytic enzymes of Trichoderma lignorum produced on banana agro-waste: optimisation of culture medium and conditions. J Sci Ind Res 64:57–60. http://hdl.handle.net/123456789/4995

Balestrini R, Chitarra W, Fotopoulos V, Ruocco M (2017) Potential role of beneficial soil microorganisms in plant tolerance to abiotic stress factors. In: Lukac M, Grenni P, Gamboni M (eds) Soil biological communities and ecosystem resilience (Sustainability in plant and crop protection). Springer, Cham, pp 191–207. https://doi.org/10.1007/978-3-319-63336-7_12

Berrocal-Tito G, Sametz-Baron L, Eichenberg K, Horwitz BA, Herrera-Estrella A (1999) Rapid blue light regulation of a *Trichoderma harzianum* photolyase gene. J Biol Chem 274:14288–14294. https://doi.org/10.1074/jbc.274.20.14288

Błaszczyk L, Popiel D, Chełkowski J, Koczyk G, Samuels GJ, Sobieralski K, Siwulski M (2011) Species diversity of Trichoderma in Poland. J Appl Genet 52:233–243. https://doi.org/10.1007/s13353-011-0039-z

Błaszczyk L, Strakowska J, Chełkowski J, Gąbka-Buszek A, Kaczmarek J (2016) *Trichoderma* species occurring on wood with decay symptoms in mountain forests in Central Europe: genetic and enzymatic characterization. J Appl Genet 57:397–407. https://doi.org/10.1007/s13353-015-0326-1

Carreras-Villaseñor N, Sánchez-Arreguín JA, Herrera-Estrella AH (2012) *Trichoderma*: sensing the environment for survival and dispersal. Microbiology 158:3–16. https://doi.org/10.1099/mic.0.052688-0

Casas-Flores S, Rios-Momberg M, Bibbins M, Ponce-Noyola P, Herrera-Estrella A (2004) BLR-1 and BLR-2, key regulatory elements of photoconidiation and mycelial growth in *Trichoderma atroviride*. Microbiology 150:3561–3569. https://doi.org/10.1099/mic.0.27346-0

Chet I, Inbar J, Hadar I (1997) Fungal antagonists and mycoparasites. In: Wicklow DT, Söderström B (eds) The Mycota IV: environmental and microbial relationships. Springer, Berlin, pp 165–184 ISBN-13: 978-3540580058

Da Silva LCA, Honorato TL, Cavalcante RS, Franco TT, Rodrigues S (2012) Effect of pH and temperature on enzyme activity of Chitosanase produced under solid stated fermentation by *Trichoderma* spp. Indian J Microbiol 52(1):60–65. https://doi.org/10.1007/s12088-011-0196-0

Danielson RM, Davey CB (1973) Non nutritional factors affecting the growth of *Trichoderma* in culture. Soil Biol Biochem 5:495–504. https://doi.org/10.1016/0038-0717(73)90039-4

Domingues MVPF, Moura KE, Salomão D, Elias LM, Patricio FRA (2016) Effect of temperature on mycelial growth of *Trichoderma*, *Sclerotinia minor* and *S. sclerotiorum*, as well as on mycoparasitism. Summa Phytopathol 42(3):222–227. https://doi.org/10.1590/0100-5405/2146

Druzhinina IS, Komoń-Zelazowska M, Atanasova L, Seidl V, Kubicek CP (2010) Evolution and ecophysiology of the industrial producer *Hypocrea jecorina* (Anamorph *Trichoderma reesei*) and a new sympatric agamospecies related to it. PLoS One 5:e9191. https://doi.org/10.1371/journal.pone.0009191

Eastburn DM, Butler EE (1988) Microhabitat characterization of *Trichoderma harzianum* in natural soil: evaluation of factors affecting distribution. Soil Biol Biochem 20:547–553. https://doi.org/10.1016/0038-0717(88)90071-5

Galun E, Gressel J (1966) Morphogenesis in *Trichoderma*: suppression of photoinduction by 5-fluorouracil. Science (New York) 151:696–698. https://doi.org/10.1126/science.151.3711.696

Gautam SP, Bundela PS, Pandey AK, Khan J, Awasthi MK, Sarsaiya S (2011) Optimization for the production of cellulase enzyme from municipal solid waste residue by two novel cellulolytic fungi. Biotechnol Res Int 2011:1: 1–1: 8. https://doi.org/10.4061/2011/810425

Gervais P, Fasquel J-P, Molin P (1988) Water relations of fungal spore germination. Appl Microbiol Biotechnol 29(6):586–592. https://doi.org/10.1007/BF00260989

Ghildiyal G, Pandy A (2008) Isolation of cold tolerant antifungal strains of *Trichoderma* sp. from glacial sites of Indian Himalayan region. Res J Microbiol 3(8):559–564. https://doi.org/10.3923/jm.2008.559.564

Guoweia S, Man H, Shikai W, He C (2011) Effect of some factors on production of cellulase by *Trichoderma reesei* HY07. Procedia Environ Sci 8:357–361. https://doi.org/10.1016/j.proenv.2011.10.056

Harman GE, Howell CR, Viterbo A, Chet I, Lorito M (2004) *Trichoderma* species – opportunistic, avirulent plant symbionts. Nat Rev Microbiol 2:43–56. https://doi.org/10.1038/nrmicro797

Hjeljord LG, Tronsmo A (2003) Effect of germination initiation on competitive capacity of *Trichoderma atroviride* P1 conidia. Phytopathology 93:1593–1598. https://doi.org/10.1094/PHYTO.2003.93.12.1593

Hjeljord LG, Stensvand A, Tronsmo A (2000) Effect of temperature and nutrient stress on the capacity of commercial *Trichoderma* products to control *Botrytis cinerea* and *Mucor piriformis* in greenhouse strawberries. Biol Control 19:149–160. https://doi.org/10.1006/bcon.2000.0859

Inam-Ul-Haq M, Javed N, Ahsan Khan M, Jaskani MJ, Khan MM, Khan HU, Irshad G, Gowen SR (2009) Role of temperature, moisture and *Trichoderma* species on the survival of *Fusarium oxysporum ciceri* in the rainfed areas of Pakistan. Pak J Bot 41(4):1965–1974

Innocenti G, Roberti R, Piattoni F (2015) Biocontrol ability of *Trichoderma harzianum* strain T22 against fusarium wilt disease on water-stressed lettuce plants. BioControl 60(4):573–581. https://doi.org/10.1007/s10526-015-9662-7

Isil S, Nilufer A (2005) Investigation of factors affecting xylanase activity from *Trichoderma harzianum* 1073 D3. Braz Arch Biol Technol 48(2):187–193. https://doi.org/10.1590/S1516-89132005000200004

Jackson AM, Whipps JM, Lynch JM (1991) Effects of temperature, pH and water potential on growth of four fungi with disease biocontrol potential. World J Microbiol Biotechnol 7(4):494–501. https://doi.org/10.1007/BF00303376

Jackson MA, Mascarin GM, Kobori NN (2017) *Trichoderma* microsclerotia and methods of making. US patent US9642372B2

Jang S, Jang Y, Kim C-W, Lee H, Hong J-H, Heo YM, Lee YM, Lee DW, Lee HB, Kim J-J (2017) Five new records of soil-derived *Trichoderma* in Korea: *T. albolutescens*, *T. asperelloides*, *T. orientale*, *T. spirale*, and *T. tomentosum*. Mycobiology 45(1):1–8. https://doi.org/10.5941/MYCO.2017.45.1.1

Jayaswal RK, Singh R, Lee YS (2003) Influence of physiological and environmental factors on growth and sporulation of an antagonistic strain of *Trichoderma viride* RSR 7. Mycobiology 31(1):36–41. https://doi.org/10.4489/MYCO.2003.31.1.036

Jones EE, Bienkowski DA, Stewart A (2016) The importance of water potential range tolerance as a limiting factor on *Trichoderma* spp. biocontrol of *Sclerotinia sclerotiorum*. Ann Appl Biol 168:41–51. https://doi.org/10.1111/aab.12240

Kamo M, Tojo M, Yamazaki Y, Itabashi T, Takeda H, Wakana D, Hosoe T (2016) Isolation of growth inhibitors of the snow rot pathogen *Pythium iwayamai* from an arctic strain of *Trichoderma polysporum*. J Antibiot 69:451–455. https://doi.org/10.1038/ja.2015.130

Kredics L, Antal Z, Manczinger L, Szekeres A, Kevei F, Nagy E (2003) Influence of environmental parameters on Trichoderma strains with biocontrol potential. Food Technol Biotechnol 47:37–42

Kredics L, Manczinger L, Antal Z, Pénzes Z, Szekeres A, Kevei F, Nagy E (2004) *In vitro* water activity and pH dependence of mycelial growth and extracellular enzyme activities of *Trichoderma* strains with biocontrol potential. J Appl Microbiol 96:491–498. https://doi.org/10.1111/j.1365-2672.2004.02167.x

Kubicek CP, Bissett J, Druzhinina IS, Kullnig-Gradinger CM, Szakacs G (2003) Genetic and metabolic diversity of *Trichoderma*: a case study on south east Asian isolates. Fungal Genet Biol 38:310–319. https://doi.org/10.1016/S1087-1845(02)00583-2

Lupo S, Dupont J, Bettucci L (2002) Ecophysiology and saprophytic ability of *Trichoderma* spp. Cryptogam Mycol 23:71–80

Magan N (1988) Effects of water potential and temperature on spore germination and germ-tube growth *in vitro* and on straw leaf sheaths. Trans Br Mycol Soc 90(1):97–107. https://doi.org/10.1016/S0007-1536(88)80185-2

Malathi P, Doraisamy S (2003) Effect of temperature on growth and antagonistic activity of *Trichoderma* spp. against *Macrophomina phaseolina*. J Biol Control 17(2):153–159. https://doi.org/10.18311/jbc/2003/3973

Mastouri F, Bjorkman T, Harman GE (2010) Seed treatment with *Trichoderma harzianum* alleviates biotic, abiotic, and physiological stresses in germinating seeds and seedlings. Phytopathology 100:1213–1221. https://doi.org/10.1094/PHYTO-03-10-0091

Mastouri F, Bjorkman T, Harman GE (2012) *Trichoderma harzianum* enhances antioxidant defense of tomato seedlings and resistance to water deficit. Mol Plant-Microbe Interact 25:1264–1271. https://doi.org/10.1094/MPMI-09-11-0240

Mishra DS, Prajapati CR, Gupta AK, Sharma SD (2012) Relative bio-efficacy and shelf-life of mycelial fragments, conidia and chlamydospores of *Trichoderma harzianum*. Vegetos Int J Plant Res 25(1):225–232

Montoya-Gonzalez AH, Quijano-Vicente G, Morales-Maza A, Ortiz-Uribe N, Hernandez-Martinez R (2016) Isolation of Trichoderma Spp. from desert soil, biocontrol potential evaluation and liquid culture production of conidia using agricultural fertilizers. J Fertil Pestic 7:163. https://doi.org/10.4172/jbfbp.1000163

Mukherjee PK, Raghu K (1997) Effect of temperature on antagonistic and biocontrol potential of shape *Trichoderma* sp. on *Sclerotium rolfsii*. Mycopathologia 139(3):151–155. https://doi.org/10.1023/A:1006868009184

Mulaw TB, Kubicek CP, Druzhinina IS (2010) The rhizosphere of Coffea Arabica in its native highland forests of Ethiopia provides a niche for a distinguished diversity of *Trichoderma*. Diversity 2:527–549. https://doi.org/10.3390/d2040527

Muniappan V, Muthukumar T (2014) Influence of crop species and edaphic factors on the distribution and abundance of Trichoderma in Alfisol soils of southern India. Acta Bot Croat 73(1):37–50

Naseby DC, Pascual JA, Lynch JM (2000) Effect of biocontrol strains of *Trichoderma* on plant growth, *Pythium ultimum* populations, soil microbial communities and soil enzyme activities. J Appl Microbiol 88:161–169. https://doi.org/10.1046/j.1365-2672.2000.00939.x

Petrisor C, Paica A, Constantinescu F (2016) Influence of abiotic factors on *in vitro* growth of *Trichoderma* strains. Proc Romanian Acad Ser B 18(1):11–14

Qiu Z, Wu X, Zhang J, Huang C (2017) High temperature enhances the ability of *Trichoderma asperellum* to infect *Pleurotus ostreatus* mycelia. PLoS One 12(10):e0187055. https://doi.org/10.1371/journal.pone.0187055

Raza W, Faheem M, Yousaf S, Rajer FU, Yamin M (2013) Volatile and non-volatile antifungal compounds produced by *Trichoderma harzianum* SQR-T037 suppressed the growth of *Fusarium oxysporum* f sp. *niveum*. Sci Lett 1:21–24

Regragui A, Lahlou H (2005) Effect of salinity on *in vitro Trichoderma harzianum* antagonism against *Verticillium dahliae*. Pak J Biol Sci 8(6):872–876. https://doi.org/10.1080/03235408.2017.1357360

Ren J, Xue C, Tian L, Xu M, Chen J, Deng Z, Proksch P, Lin W (2009) Asperelines A–F, peptaibols from the marine-derived fungus *Trichoderma asperellum*. J Nat Prod 72:1036–1044. https://doi.org/10.1021/np900190w

Said SD (2007) Spore production of biocontrol agent *Trichoderma harzianum*: effect of C/N ratio and glucose concentration. Jurnal Rekayasa Kimia dan Lingkungan 6(1):35–40

Samuels GJ, Dodd SL, Lu BS, Petrini O, Schroers HJ, Druzhinina IS (2006) The *Trichoderma koningii* aggregate species. Stud Mycol 56:67–133. https://doi.org/10.3114/sim.2006.56.03

Sanogo S, Pomella A, Hebbar PK, Bailey B, Costa JCB, Samuels GJ, Lumsden RD (2002) Production and germination of conidia of *Trichoderma stromaticum*, a mycoparasite of *Crinipellis perniciosa* on cacao. Phytopathology 92:1032–1037. https://doi.org/10.1094/PHYTO.2002.92.10.1032

Sharma KK (1992) Effects of temperature and ethanol on maturation and germination of *Trichoderma reesei*. Master thesis, Lehigh University USA, p 104

Sharma PK, Gothalwal R, Tiwari RKS (2013) Isolation of cold tolerant antifungal strains of *Trichoderma* sp. from northern hilly zones of Chhattisgarh. Int J Plant Prot 6(2):236–240

Silas K, Bitrus HK, Wadinda JM, Zubairu A (2017) Effects of temperature and pH on *Trichoderma reseei* cellulase activity in glucose production from water melon peel. Int J Sci Res Pub 7(3):253–259

Simon A, Sivasithamparam K, Macnish GC (1988) Effect of application of soil nitrogenous fertilizers and lime on biological suppression of *Gaeumannomyces graminis* var. tritici. Trans Br Mycol Soc 91:287–294. https://doi.org/10.1016/S0007-1536(88)80217-1

Singh A, Shahid M, Srivastava M, Pandey S, Sharma A, Kumar V (2014) Optimal physical parameters for growth of *Trichoderma* species at varying pH, temperature and agitation. Virol Mycol 3(1):127. https://doi.org/10.4172/2161-0517.1000127

Suh DH, Becker TC, Sands JA, Montenecourt BS (1988) Effects of temperature on xylanase secretion by *Trichoderma reesei*. Biotechnol Bioeng 32:821–825. https://doi.org/10.1002/bit.260320614

Tao S, Beihui L, Deming L, Zuohu L (1997) Effect of elevated temperature on *Trichoderma viride* SL-1 in solid state fermentations. Biotechnol Lett 19(2):171–174. https://doi.org/10.1023/A:1018372616637

Torres-De la Cruz M, Ortiz-García CF, Bautista-Muñoz C, Ramírez-Pool JA, Ávalos-Contreras N, Cappello-García S, De la Cruz-Pérez A (2015) Diversidad de *Trichoderma* en el agroecosistema cacao del estado de Tabasco, México. Revista Mexicana de Biodiversidad 86(4):947–961. https://doi.org/10.1016/j.rmb.2015.07.012

Tronsmo A, Dennis C (1978) Effect of temperature on antagonistic properties of *Trichoderma* species. Trans Br Mycol Soc 71(3):469–474. https://doi.org/10.1016/S0007-1536(78)80075-8

Wakelin SA, Sivasithamparam K, Cole ALJ, Skipp RA (1999) Saprophytic growth in soil of a strain of *Trichoderma koningii*. N Z J Agric Res 42:337–345. https://doi.org/10.1080/00288233.1999.9513383

Wibowo A (1999) The effect of environmental factors on conidial germination, sporulation and growth of *Trichoderma harzianum in vitro*. Jurnal Perlindungan Tanaman Indonesia 5(2):108–113. https://doi.org/10.22146/jpti.12759

Wuczkowski M, Druzhinina I, Gherbawy Y, Klug B, Prillinger HJ, Kubicek CP (2003) Species pattern and genetic diversity of Trichoderma in a mid-European, primeval floodplain-forest. Microbiol Res 158(2):125–133. https://doi.org/10.1078/0944-5013-00193

Yamazaki Y, Tojo M, Hoshino T, Kida K, Sakamoto T, Ihara H, Yumoto I, Tronsmo AM, Kanda H (2011) Characterization of *Trichoderma polysporum* from Spitsbergen, Svalbard archipelago, Norway, with species identity, pathogenicity to moss, and polygalacturonase activity. Fungal Ecol 4:15–21. https://doi.org/10.1016/j.funeco.2010.06.002

Zehra A, Dubey MK, Meena M, Upadhyay RS (2017) Effect of different environmental conditions on growth and sporulation of some *Trichoderma* species. J Environ Biol 38:197–203. https://doi.org/10.22438/jeb/38/2/MS-251

Zhang CL, Druzhinina IS, Kubicek CP, Xu T (2005) Biodiversity of *Trichoderma* in China: evidence for a north to south difference of species distribution in East Asia. FEMS Microbiol Lett 251:251–257. https://doi.org/10.1016/j.femsle.2005.08.034

Zhang F, Ge H, Zhang F, Guo N, Wang Y, Chen L, Ji X, Li C (2016) Biocontrol potential of *Trichoderma harzianum* isolate T-aloe against *Sclerotinia sclerotiorum* in soybean. Plant Physiol Biochem 100:64–74. https://doi.org/10.1016/j.plaphy.2015.12.017

15 Exopolysaccharide-Producing Microorganisms from Extreme Areas: Chemistry and Application

Annarita Poli, Paola Di Donato, Giuseppina Tommonaro, Gennaro Roberto Abbamondi, Ilaria Finore, and Barbara Nicolaus

15.1 Introduction

EPS-producing microorganisms have been isolated from different natural sources of both aquatic and terrestrial environments, like freshwater, marine water, wastewater, soils, biofilms and also extreme niches such as hot springs, cold environments, hypersaline and halophilic environments, salt lakes and salterns (Maugeri et al. 2002; Nichols et al. 2005a; Mata et al. 2006; Poli et al. 2007; Satpute et al. 2010; Poli et al. 2010; Andersson et al. 2011; Nicolaus et al. 2016).

Extreme environments, generally characterized by atypical temperatures, pH, pressure, salinity, toxicity and radiation levels, are inhabited by various microorganisms specifically adapted to these particular conditions.

These extreme environments have been identified as an important source of bacteria, archaea, algae and fungi with interesting applications, and the organisms, living there, have developed different strategies to cope with adverse living conditions, and the production of EPSs is a frequent survival strategy (Nicolaus et al. 2004; Nichols et al. 2005a). For example, bacteria living in extreme marine environments such as those found in the cold waters of polar regions, in ocean trenches or in deep-sea hydrothermal vents often use EPSs as an efficient protective barrier (Nichols et al. 2005a).

A. Poli · G. Tommonaro · G. R. Abbamondi · I. Finore · B. Nicolaus (✉)
Consiglio Nazionale delle Ricerche (CNR) -Institute of Biomolecular Chemistry (ICB), Pozzuoli, NA, Italy
e-mail: bnicolaus@icb.cnr.it

P. Di Donato
Consiglio Nazionale delle Ricerche (CNR) -Institute of Biomolecular Chemistry (ICB), Pozzuoli, NA, Italy

Department of Science and Technology, University of Naples Parthenope – Centro Direzionale, Naples, Italy

D. Egamberdieva et al. (eds.), *Extremophiles in Eurasian Ecosystems: Ecology, Diversity, and Applications*, Microorganisms for Sustainability 8,
https://doi.org/10.1007/978-981-13-0329-6_15

The protection conferred by EPSs in these hostile environments is achieved by the formation of biofilms to withstand high pressure and/or temperature or by decreasing the freezing point of water in the vicinity of the bacteria (Nichols et al. 2005a). Similar strategies are used by thermophilic bacteria found in terrestrial habitats (Lin et al. 2011).

Polysaccharides produced by microbes can be generally classified by their biological functions into intracellular storage polysaccharides (glycogen), capsular polysaccharides which are closely linked to the cell surface (e.g. K30 O-Antigen) and extracellular bacterial polysaccharides (e.g. xanthan, sphingan, alginate, cellulose, etc.) that are important for biofilm formation and pathogenicity. This article will focus on the latter, also termed EPS, which are secreted to the surrounding environment and therefore can be efficiently harvested from cell-free culture supernatant in a continuous and cost-effective manufacturing process (Schmid et al. 2015).

The capability to synthesize exopolysaccharides has been observed for microorganisms belonging to both Archaea and *Bacteria* domains and in all kind of extremophilic microorganisms, by means of thermophiles, halophiles, psychrophiles, acidophiles, anaerobes and so on (Poli et al. 2007; VanFossen et al. 2008; Michel et al. 2009; Radchenkova et al. 2013; Casillo et al. 2017; Poli et al. 2017).

Due to their many interesting physicochemical and rheological properties, these biopolymers possess novel functionality that is generally superior to petrochemical-derived polymers in aspects that embrace biodegradability and environmental and human compatibility. Consequently, biopolymers of extremophiles are widely used in foods, cosmetics, pharmaceutical products, textiles, detergents, adhesives, oil recovery from wells, brewing and waste treatment processes (Poli et al. 2009).

15.2 Ecological and Physiological Roles

The EPS production process resulted to be a physiological mechanism for some microbial genera, such as *Xanthomonas*, *Leuconostoc*, *Pseudomonas* and *Alcaligenes*, which synthesized xanthan, dextran, gellan and curdlan (Finore et al. 2014), but also a response to biotic and abiotic stress factors (Donot et al. 2012).

These biomolecules carried out an ecological role, allowing the bacteria to proliferate in stressful environmental conditions; by means of high or low values of temperature, high salt concentration and extreme of pHs; and in the presence of more stress factors simultaneously (Nicolaus et al. 2010; Finore et al. 2015; Poli et al. 2017).

Microorganisms synthesized and released out of the cell polymers for their survival. Therefore, the role of energy reservoir and defensive agent has been attributed to the EPS; in addition their production can influence the cell functioning, the osmotic regulation and the symbiosis and sustain the microorganism in all vital function, from the adaptation to the cell reproduction (Steinbüchel 2001; Vijayendra and Shamala 2014).

The EPS production process necessitated a conspicuous energy expenditure for the microorganism, up to 70% of carbon investment. Evidently, the advantage coming from the EPS production was much more higher with respect to their survival (Wolfaardt et al. 1999).

Hot niches hosted a wide variety of prokaryotic microorganisms. They represented an interesting source of many bioactive compounds, including exopolysaccharides. Thermophilic microorganisms proliferated in a wide range of temperature, from 122 °C of hyperthermophile *Methanopyrus* species (Takai et al. 2008) up to 50–60 °C of thermotolerant microorganisms. Thermophiles producing EPS have been isolated from both *Bacteria* (*Aeribacillus*, *Bacillus*, *Brevibacillus*, *Geobacillus*, *Thermotoga* and *Thermus*) and Archaea (*Sulfolobus* and *Thermococcus*) domains (Kambourova et al. 2016). The exopolymers surrounded the microbial cells by contributing to their survival: (a) the roles of protection against predators, (b) the energy and carbon source reservoir and (c) the regular nutrient uptake even in environments wherein they would tend to be dispersed. In particular, marine thermophiles, isolated from deep-sea hydrothermal vents, showed ability to grow in the presence of metal ions and toxic substances; this capability was derived from the presence of exopolysaccharides bound with high-affinity cations and trace metals (Loaëc et al. 1997).

Cold environments are distributed all over the world and are characterized by a low nutritive substance diffusion; psychrophiles and psychrotrophs are microbes that thrive in these places, and need or tolerate low temperature values, respectively. Their capability to proliferate in freezing niches is related to different cellular mechanisms, from membrane lipid compositions to the cold-stable RNA conformation up to exopolysaccharide synthesis (Poli et al. 2017). The high amount of polyhydroxyl groups of EPS decreased both the freezing point of water and the ice nucleation temperature (De Maayer et al. 2014). The EPSs assumed a gelatinous aspect in nature, playing a cryoprotection role, because they modified the immediate surrounding environment of the cell (McLean 1918; Ewert and Deming 2013).

Abundant amount of exopolysaccharides have been found both in Antarctic and Arctic marine bacteria and in all cold environment (Poli et al. 2017). These polymers altered the chemical parameters around the microbial organisms, contributing to the adhesion of cells to surfaces with water and nutrient sequester, improving their uptake. In addition, the EPS can preserve the extracellular enzymes against the freezing temperatures, avoiding their denaturation (D'Amico et al. 2006). The EPSs protected the cells from viral attacks and influence the osmosis (Deming and Young 2017).

The obligately marine and psychrophilic γ-proteobacterium, *Colwellia psychrerythraea* strain 34H, is reported as an EPS-producing bacterium. The production of EPS did not change over growth-permissive temperatures of ~10 to −4 °C, but from −8 to −14 °C when samples froze, EPS production rose dramatically. Moreover, in salinity tests at 10‰–100‰ (and −1 and 5 °C), EPS production also increased at the freshest salinity tested, and the strain 34H recovered best from deep-freezing to −80 °C if first supplemented with a preparation of its own EPS, rather than other cryoprotectants like glycerol. These results suggested that the EPS represented a survival strategy of microorganisms in a harsh environment and an interesting compound with potentially properties for biotechnological application (Marx et al. 2009). In a following paper, the detailed molecular primary and secondary structures of capsular polysaccharide from *C. psychrerythraea* 34H cells were reported. The polysaccharide consisted of a tetrasaccharidic repeating unit containing two amino sugars and two uronic acids bearing threonine as substituent. The structural features of this EPS resemble those present in antifreeze proteins and glycoproteins. These results suggested a possible correlation between the capsule structure and the ability of *C. psychrerythraea* to colonize subfreezing marine environments and, more, confirmed the potential properties of this polymer (Carillo et al. 2015).

In literature have been reported many examples of halophilic microorganisms able to synthesize exopolysaccharides and this property has been linked to a specific regulation role in the presence of salts. The polymers around the microbial cell attenuated the physical stress due to the salinity. Many halophilic microorganisms possessed exopolysaccharides around the cell, for protecting membrane integrity (DasSarma and DasSarma 2001; Poli et al. 2010; Qurashi and Sabri 2012; Oren 2013). Halophilic Archaea producing EPS are *Haloarcula*, *Halococcus*, *Haloferax* and *Natronococcus* (Nicolaus et al. 2010). Also halophilic *Bacteria* are good producers of EPSs, for example, *Halomonas maura* produced mauran, an exopolysaccharide deeply investigated and with a wide commercial use (Arias et al. 2003).

15.3 Microbial Exopolysaccharides' Isolation, Purification and Structure Definition

Exopolysaccharides are produced as exocellular polymers that generally account for about 40% to 95% of the extracellular polymeric substances (Flemming and Wingender 2001). Exopolysaccharides can be dispersed in the biofilm matrix surrounding the cell, or they can be found as a discrete layer enveloping the cell: usually the polysaccharides belonging to the cell envelope of the bacteria are also referred to as capsular polysaccharides (CPSs) and lipopolysaccharides (LPSs), the latter being present only in Gram-negative bacteria. In general, the term EPS indicates the extracellular polysaccharide molecules that are not tightly bound to the cell surface but sloughed off to form slime, although the release of polysaccharides from the cell surface is not an absolute criterion to distinguish EPSs from the other carbohydrate capsular components (Roberts 1996).

Isolation of EPSs is a challenging task since these polymers are found embedded in a complex matrix also containing proteins and other biomolecules, i.e. the biofilm matrix. Therefore, the quantitative recovery of an EPS is very difficult to achieve because usually a fraction can remain bound to the cell and because the sample can be contaminated from intracellular materials released during isolation procedures after cell disruption. There is no single isolation and purification protocol generally efficient for EPS recovery; indeed the isolation procedure can change depending on the microbial source of EPSs.

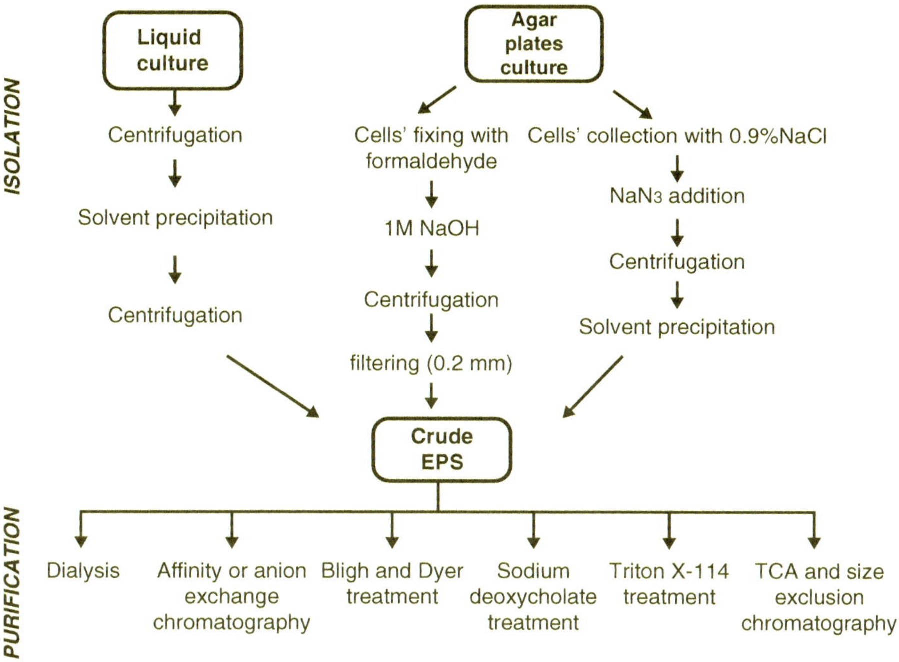

Fig. 15.1 General scheme of water-soluble EPSs' isolation and purification. (Partially adapted from Di Donato et al. (2016))

Isolation of water-soluble EPSs is usually implemented by cold ethanol precipitation, although also other solvents can be used, for example, acetone, isopropyl alcohol or methanol. The protocols to be adopted strongly depends on the bacterial growth method: indeed, in the case of static mode (seeding on agar plates), a previous step of cell's fixing or removal is required in order to avoid EPS contamination from endocellular molecules; then washing with an alkali solution and centrifugation will afford a crude EPS sample (Bales et al. 2013). Another strategy is represented by washing with NaCl solution and adding a bacteriostatic agent in order to preserve cell integrity; after cell removal by centrifugation, addition of solvent will precipitate the EPS for further purification. If a liquid culture has been implemented, the solvent precipitation has to be forerun by centrifugation to remove intact cells (Fig. 15.1) (Di Donato et al. 2016).

Dialysis is the classical method of EPSs' purification, although in order to remove contaminants, for example, LPSs, other methods can be used including chromatography; dissolution in 0.01 M EDTA followed by extraction with chloroform/methanol (Bligh and Dyer treatment); suspension in 0.05 M Tris HCl and addition of sodium deoxycholate followed by acidification with acetic acid (20%) and centrifugation to remove LPSs; dissolution in water or 0.01 M EDTA in the presence of Triton X-114, followed by addition of NaCl 2% w/v and cold ethanol to precipitate the purified EPS (Du et al. 2017); or finally trichloroacetic acid (TCA) addition to remove nucleic acids followed by ethanol precipitation and then gel filtration chromatography for EPS final purification (Bales et al. 2013).

In the case of water-insoluble EPSs, for example, cellulose, isolation is carried out in harsh conditions, such as treatment with acetic acid and nitric acid at 95 °C or with NaOH at 80 °C, followed by washing with distilled water and neutralization acetic acid, thus achieving a purified polysaccharide (Rangaswamy et al. 2015).

The complete structural definition of the EPSs is carried out by means of chemical, analytical and spectral techniques. First of all, the gross chemical composition of a purified EPS is assessed by determination of the total carbohydrate content (DuBois et al. 1956), of the total protein content (Bradford 1976), of the nucleic acids and of the uronic acids (Spanò et al. 2013). The monosaccharide composition and the determination of linkage positions are carried out by hydrolysis of the polymer followed by liquid or gas chromatography. The hydrolysis is usually carried out by treatment in trifluoroacetic acid (TFA) at 110–120 °C followed by analysis of the resulting mixture by means of TLC or of high-performance anion-exchange chromatography with pulsed amperometric detection (HPAE-PAD). The hydrolysed polymer can also be subjected to derivatization (by per-acetylation or silylation treatments), and in these cases, the identification of monomer sugars is carried out by means of gas chromatography (GC) analysis. The determination of linkage positions of sugars in the EPS is accomplished by methylation analysis (MA), i.e. treatment with methyl iodide followed by acidic hydrolysis, reduction and acetylation/silylation: the so-obtained volatile alditol acetates or methylsilanes are then identified by GC-MS. The chemical analysis is completed by determination of functional and substituting groups that is commonly implemented by means of Fourier-transform infrared spectroscopy (FTIR) or nuclear magnetic resonance (NMR) (Mishra and Jha 2013). The spectral NMR and FTIR techniques are useful to confirm the chemical composition of the EPSs (determination of the number and the type of monomer sugar residues identified by chemical degradation), but they also allow to gain other fundamental information like the anomeric configurations of the monosaccharides and their sequence in the polymer backbone.

NMR spectroscopy is a useful tool for the determination of EPSs' backbone composition and conformation; indeed thanks to 1D and 2D ^{1}H- and ^{13}C-NMR techniques, coupled with the use of relevant databases such as Carb-Bank, SUGABASE or CASPER, it is possible to estimate the number of sugar residues present in an EPS and their mode of linking. In particular, the 2D heteronuclear techniques like HSQC, HMQC or HMBC are useful for the determination of the anomeric configuration of monomer sugars, the homonuclear TOCSY or DQF-COSY is useful for the identification of the single monosaccharides, and finally NOESY and HMBC techniques allow the determination of glycosidic linkages sequence along the polysaccharide backbone (Duus et al. 2000). FTIR is also a valuable tool in the structural definition of EPSs since it allows to recognize the presence of the peculiar functional groups characterizing either monosaccharides or probable substituents in the polymer backbone (Wiercigroch et al. 2017). Molecular weight's determination is another important issue to be addressed for a complete structural characterization of an EPS. Several classical techniques are available for such a study, for example, light scattering, analytical ultracentrifugation, viscosity determination and size-exclusion chromatography (SEC). More recently some innovative techniques like high-performance size-exclusion chromatography (HPSEC) coupled with refractive index (RI) detection or multi-angle laser light scatter (MALLS) detection have gained increasing attention. Such techniques require smaller quantities of samples, compared to the other methods, and enable faster analyses: in particular RI is a useful tool for the determination of molecular weight distribution, while MALLS detector allows to evaluate the absolute molecular weight with higher accuracy (Gómez-Ordóñez et al. 2012).

15.4 Examples of Polysaccharide-Producing Extremophilic Microorganisms

The demand of biomolecules is growing quickly because of their advantageous application in a wide variety of market segments (e.g. biotechnology, biomedicine, cosmetics, pharmaceutical industry, food processing, etc.).

Polysaccharides represent one of the more interesting classes of biomolecules for biotechnological application, due to their wide range of functional properties which make them able to form gels, films and membranes. In particular, carbohydrate polymers from natural sources have the significant advantage to be biocompatible, biodegradable, bioadhesive and nontoxic.

Moreover, the research of polysaccharides from microbial origin is very interesting, in particular those produced by extremophilic microorganisms. Extremophiles are able to thrive in a wide variety of harsh habitats because of their capability to counterbalance extreme physical or chemical parameters by means of different strategies. One of these mechanisms is the synthesis of special biomolecules with unique proprieties. Extreme biomolecules have the important advantage of resisting and be effective even in the harsh environmental conditions in which the extremophilic microorganisms live (such as extreme temperature, pH, salt concentration and hydrostatic pressure). These parameters are very close to those of biotechnological processes; therefore this kind of biomolecules can be considered an important source of special compounds for industrial application. In addition, the unique properties of these substances make them possible to carry biotechnological processes at high temperatures or high saline concentrations; thus the risk of contamination is reduced to a minimum (Raddadi et al. 2015).

In this paragraph we report the different polysaccharides isolated from extremophilic microorganisms from Eurasia of which the chemical characterization has been completely or partially performed.

Thermophilic bacteria are the largest group of polysaccharides extracted from extremophilic microorganisms. In Table 15.1 are reported some examples.

Concerning psychrophiles, most of polysaccharide-producing microorganisms have been isolated from Antarctic regions. In literature, there is actually only one example of polysaccharide isolated from psychrophilic bacteria from Eurasia. Ko

Table 15.1 Examples of polysaccharides from thermophilic bacteria

Producer	Sampling site	Chemical composition	EPS/yields	Reference(s)
Thermotoga maritima Cocultivation with *Methanococcus jannaschii*	Geothermal-heated marine sediment at Vulcano, Italy	Glc/Rib/Man (1:0.06:0.03)	5% of the dry mass	Rinker and Kelly (2000), Johnson et al. (2005), and VanFossen et al. (2008)
Bacillus licheniformis B3-15	Shallow marine hot spring, Vulcano, Italy	Man/Glc (1:0.3)	165 mg/l	Maugeri et al. (2002) and Arena et al. (2006)
Bacillus licheniformis T14	Shallow hydrothermal vent, Panarea Island, Italy	Fru/Fuc/Glc/GalN/Man (1.0:0.75:0.28:tr:tr)	Maximum EPS yield of 366 mg/l with sucrose as C-source	Spanò et al. (2013) and Gugliandolo et al. (2014)
Geobacillus sp. 4001	Shallow hydrothermal vents and marine hot springs, Flegrean Region, Italy	Man/Glc/Gal/ManN (1:0.1:tr:tr)	Maximum EPS yield of 60 μg/l with trehalose as C-source	Nicolaus et al. (2002, 2003, 2004)
Geobacillus sp. 4004	Shallow hydrothermal vents and marine hot springs, Flegrean Region, Italy	Gal/Man/GlcN/Ara (1:0.8:0.4:0.2)	50 μg/l	Nicolaus et al. (2002, 2004)
Geobacillus thermodenitrificans B3-72	Shallow hydrothermal vent, Vulcano island, Italy	Man/Glc (0.3:1)	Maximum EPS yield of 70 mg/l with glucose/sucrose as C-sources	Nicolaus et al. (2000) and Arena et al. (2009)
Geobacillus tepidamans V264	Hot spring, Velingrad, Bulgaria	Glc/Gal/Fuc/Fru (1:0.07:0.04:0.02)	Maximum EPS yield of 111.4 mg/l with maltose as C-source	Kambourova et al. (2009)
Aeribacillus pallidus 418	Hot spring, Rupi basin, Bulgaria	EPS1: Man/Glc/GalN/GlcN/Gal/Rib(1:0.16:0.09:0.08:0.07:0.04) EPS2: Man/Gal/Glc/GalN/GlcN/Rib/Ara (1:0.5:0.46:0.35:0.24:0.16:0.14)	130 mg/l	Radchenkova et al. (2013, 2014)
Brevibacillus thermoruber strain 423	Hot spring, Blagoevgrad region, Bulgaria	Glc/Gal/GalN/Man/Man (1:0.3:0.25:0.16:0.04)	Maximum EPS yield of 863 mg/l with maltose as C-source	Yasar Yildiz et al. (2014)
Rhodothermus marinus DSM 4252^T	Submarine hot spring, Iceland	Main components: Xyl, Ara and Glc	8.8 mg/g cell dry weight	Sardari et al. (2017)
Rhodothermus marinus MAT 493	Submarine hot spring, Iceland	Main components: Glc, Ara, Xyl and Man	13.7 mg/g cell dry weight	Sardari et al. (2017)

Monosaccharide codes: D-arabinose (Ara), D-fructose (Fru), D-fucose (Fuc), D-galactose (Gal), D-galactosamine (GalN), D-glucose (Glc), D-glucosamine (GlcN), D-mannose (Man), D-mannosamine (ManN), D-ribose (Rib), D-xylose (Xyl)

et al. (2000) isolated an extracellular polysaccharide (molecular mass over 2×10^6 Da) from the marine isolate *Hahella chejuensis* (Ko et al. 2000). The monosaccharidic composition of the carbohydrate polymer was partially characterized and consisted of galactose, glucose, xylose and ribose.

The isolation of polysaccharides from halophilic bacteria is reported for only six strains, from which six carbohydrate polymers have been chemically characterized in total. The examples of polysaccharides from *Halophiles*, together with the one isolated from the psychrophile *Hahella chejuensis*, are reported in Table 15.2.

The last group is represented by microorganisms belonging to the Archaea domain. In literature eight polysaccharide-producing Archaea have been chemically characterized (Table 15.3). It is interesting to notice that glucose and mannose are almost always present, often as main monosaccharides.

The research on polysaccharides from Eurasian extremophilic microorganisms has mainly developed over the last 20 years. In fact, analysing all the tables, it is possible to notice that most of the studies are dated after 2000. The growing interest in this topic has essentially two reasons: first, it is important to deeply investigate the physiological mechanisms at the basis of the polysaccharide production, in order to better understand the ecological role of these biomolecules in extremophiles; in addition, the unique proprieties of extremophilic carbohydrate polymers make them highly attractive to biotechnological industry. These subjects will be deeply investigated and discussed in the following paragraphs.

15.5 Application and Biological Activities

In an extreme environment, the synthesis of exopolysaccharides (EPSs) is in response to adaption to prohibitive conditions. Therefore, these biomolecules produced by extremophiles showed unique features for adapting to extreme conditions. EPS-producing microorganisms, in particular those from extreme habitats, have become the natural source of polysaccharides of growing interest for their bioactivities and physicochemical properties; therefore they represent very promising compounds for biotechnological applications. Herein, we report the most representative examples of EPSs isolated from extremophiles having great potential in application in numerous industrial sectors such as tissue engineering, drug delivery and cosmetic (Table 15.4).

15.5.1 Halophiles

Among halophiles microorganisms, *Halomonas* represents the most common genera producing EPSs. Mauran is a highly polyanionic sulphated exopolysaccharide produced by a moderately halophilic bacterium *Halomonas maura*. For its unique physicochemical properties, mauran has been successfully employed in the nanoparticle synthesis and application for sustained drug delivery, cancer chemotherapy and bioimaging and for its antioxidant defence mechanism along with

Table 15.2 Examples of polysaccharides from psychrophilic and halophilic bacteria

Producer	Sampling site	Chemical composition	EPS/yields	Reference(s)
Psychrophiles				
Hahella chejuensis	Marado, Jeju Island, Republic of Korea	Gal/Glc/Xyl/Rib	9.23 g/l	Ko et al. (2000) and Lee et al. (2001)
Halophiles				
Halomonas smyrnensis	Çamaltı Saltern area, Sasalı, İzmir province, Aegean Region of Turkey	Levan. Repeating unit of β-(2,6)-d-fructofuranosyl residues	Maximum EPS yield of 1.844 g/l with sucrose as C-source	Poli et al. (2009)
Halomonas anticariensis strain FP35	Fuente de Piedra saline wetland, province of Málaga, southern Spain	Man/GalA/Glc (1:0.82:0.33)	296.5 mg/l	Mata et al. (2006)
Halomonas anticariensis strain FP36	Fuente de Piedra saline wetland, province of Málaga, southern Spain	Man/GalA/Glc (1:0.92:0.4)	499.5 mg/l	Mata et al. (2006)
Halomonas ventosae strain Al12	Saline soils in Jaén, southeastern Spain	Man/Glc/Gal (1:0.43:0.25)	283.5 mg/l	Mata et al. (2006)
Halomonas ventosae strain Al16	Saline soils in Jaén, southeastern Spain	Man/Glc/Gal (1:0.42:0.22)	289.5 mg/l	Mata et al. (2006)
Halomonas eurihalina strain H212	Saline soil in Alicante, Southern Spain	High sulphate content and significant amounts of uronic acid	1.6 g/l	Béjar et al. (1998)

Monosaccharide codes: D-fructose (Fru), D-galactose (Gal), D-galacturonic acid (GalA), D-glucose (Glc), D-mannose (Man), D-ribose (Rib), D-xylose (Xyl)

Table 15.3 Examples of polysaccharides from Archaea

Producer	Sampling site	Chemical composition	EPS/yields	Reference(s)
Archaea				
Thermococcus litoralis	Shallow marine thermal spring, Naples, Italy	Mannan: Man (only monosaccharidic constituent)		Rinker and Kelly (1996, 2000)
Sulfolobus solfataricus MT4	Hot acidic spring, Agnano, Italy	Glc/Man/GlcN/Gal (1:0.8:0.15:0.11)	Exopolysaccharide yield of 8.4 mg/l (fermentor culture)	Nicolaus et al. (1993)
Sulfolobus solfataricus MT3	Hot acidic spring, Agnano, Italy	Glc/Man/GlcN/Gal (1:0.8:0.54:0.61)	Exopolysaccharide yield of 7.0 mg/l (fermentor culture)	Nicolaus et al. (1993)
Sulfolobus tokodaii	Beppu hot springs, Japan	Man/Glc/Gal/GlcNAc		Koerdt et al. (2010)
Haloarcula hispanica ATCC33960	Solar saltern, Alicante, Spain	Acidic exopolysaccharide: Man/Gal/Glc (1:0.77:0.02)	EPS yield of 30 mg/l (in AS-168 medium)	Lü et al. (2017)
Haloterrigena turkmenica	Sulphate saline soil in Turkmenistan (Central Asia)	Sulphated heteropolysaccharide: Glc/GlcN/GlcA/Gal/GalN (1:0.65:0.24:0.22:0.02)	Maximum EPS yield of 206.8 mg/l in HTR complex medium +1% Glc (w/v)	Squillaci et al. (2016)
Haloferax gibbonsii	Saltern, Spain	Man/Glc/Gal/Rha (0.6:0.3:1:0.3)		Paramonov et al. (1998)
Haloferax mediterranei	Salt ponds, Alicante, Spain	Man/Glc/Gal/amino sugars/uronic acids, Man as major component	Maximum EPS yield of 2.6 g/l	Antón et al. (1988) and Parolis et al. (1996)

Monosaccharide codes: D-galactose (Gal), D-galactosamine (GalN), D-glucose (Glc), D-glucuronic acid (GlcA), D-glucosamine (GlcN), N-acetyl-D-glucosamine (GlcNAc), D-mannose (Man), D-rhamnose (Rha)

Table 15.4 Examples of the most relevant EPSs from extremophiles with potential application in different industrial sectors

EPS-producing microorganisms	EPS *(as reported in literature)*	Biological activities/application	Reference(s)
Halophiles			
Halomonas maura	Mauran	High viscosity and pseudoplasticity, metal-binding capacity; antioxidant, haemocompatibility/water treatment, drug delivery, cancer chemotherapy, bioimaging, nanotechnology	Bouchotroch et al. (2001), Arias et al. (2003), and Raveendran et al. (2013a, b)
Halomonas eurihalina H96	EPS H96	Gelificant ability/biodetoxification and water treatment	Béjar et al. (1998)
Halomonas ventosae strains Al12^{T} and Al16	/	Emulsifying activity/surfactant	Mata et al. (2006)
Halomonas anticariensis strains FP35^{T} and FP36	/	Emulsifying activity/surfactant	Mata et al. (2006)
Halomonas smyrnensis strain AAD6^{T}	Levan	Anti-cytotoxic, high biocompatibility/ cosmetic, food and medical sectors	Poli et al. (2009), Sam et al. (2011), and Sezer et al. (2011)
Salipiger mucosus	/	Emulsifying activity/surfactant	Llamas et al. (2010)
Thermophiles			
Alteromonas infernus	EPS GY785/GY785 DROS (after depolymerization and sulphation)	Activation of normal human serum (NHS), bone and skin regeneration/ drug delivery, tissue engineering	Raguénès et al. (1997a, b), Zanchetta et al. (2003a, b), and Poli et al. (2017)
Vibrio diabolicus	EPS HE800/HE800 DROS (after depolymerization and sulphation)	Activation of normal human serum (NHS), bone and skin regeneration/ drug delivery, tissue engineering	Raguénès et al. (1997a, b), Zanchetta et al. (2003a, b), and Poli et al. (2017)
Bacillus licheniformis strain B3-15	EPS2-B3-15	Immunomodulatory, antiviral/drug delivery	Nicolaus et al. (2000), Maugeri et al. (2002), Arena et al. (2006), Spanò et al. (2013), and Gugliandolo et al. (2015)
Geobacillus thermodenitrificans strain B3-72	EPS2-B3-72	Immunomodulatory, antiviral/drug delivery	Nicolaus et al. (2000), Maugeri et al. (2002), Arena et al. (2006), Spanò et al. (2013), and Gugliandolo et al. (2015)

Bacillus licheniformis strain T14	EPS1-T14	Immunomodulatory, antiviral/drug delivery	Nicolaus et al. (2000), Maugeri et al. (2002), Arena et al. (2006), Spanò et al. (2013), and Gugliandolo et al. (2015)
Thermus aquaticus YT-1	EPS TA-1	Immunomodulatory	(Lin et al. 2011)
Geobacillus tepidamans V264	/	Anti-cytotoxic/drug delivery	Kambourova et al. (2009)
Aeribacillus pallidus 418	/	Emulsifying activity/surfactant	Radchenkova et al. (2013, 2014)
Brevibacillus thermoruber strain 423	EPS-FT; EPS-P	High biocompatibility/tissue engineering, drug delivery	Nwodo et al. (2012) and Yasar Yildiz et al. (2014)
Psychrophiles			
Colwellia psychrerythraea strain 34H	/	Cryoprotective/food, pharmaceutical and cosmetic industries	Carillo et al. (2015)
Pseudomonas sp. ID1	/	Cryoprotective/food, pharmaceutical and cosmetic industries	Carrión et al. (2015)
Pseudoalteromonas sp. SM9913	EPS SM9913	Good flocculating agent, good adsorptive effect/biodetoxification and water treatment	Qin et al. (2007) and Li et al. (2008)
Archaea			
Haloterrigena turkmenica	/	Emulsifying, antioxidant/food, pharmaceutical and cosmetic industries	Squillaci et al. (2016)

haemocompatibility under in vitro conditions using L929 (mouse fibroblast cell line) and mice liver homogenate (Bouchotroch et al. 2001; Arias et al. 2003; Raveendran et al. 2013a, b). Other halophilic EPS producers belonging to the genus *Halomonas* are *H. eurihalina*, *H. ventosae* and *H. anticariensis*. Nineteen strains belonging to *H. eurihalina* were studied for their ability to produce EPS in two different culture media. Results showed that the chemical composition of the polysaccharides was affected by the strain and by the culture medium. All EPS exhibited an unusually high sulphate content. Moreover, the EPS from strain H96 contained significant amounts of uronic acid. EPS from strain H96, cultivated in defined NH medium (minimal medium), showed an interesting rheological feature reaching a viscosity value of 30,000 cP at pH 3.0. This gelificant ability, probably due to its high uronic acid content, is attractive for industrial application, for example, in biodetoxification and water treatment (Béjar et al. 1998).

H. ventosae strains Al12^T and Al16 produced polymers showing a molecular mass of about 50 kDa, and their main components were glucose, mannose and galactose. Moreover, they exhibited emulsifying activity on several hydrophobic substrates. *H. anticariensis* strains FP35^T and FP36 also excreted polymers having a molecular mass of about 20 and 46 kDa, respectively, and were composed mainly of glucose, mannose and galacturonic acid. All EPSs produced solutions of low viscosity and pseudoplastic features. Furthermore, they also exhibited a high affinity for binding cations and incorporated considerable quantities of sulphates, just as do those produced by *H. maura* and *H. eurihalina*, which is very uncommon in bacterial polysaccharides, but represents an advantageous feature for biotechnological application. Both bacteria formed biofilms both in polystyrene wells and borosilicate test tubes. In particular, *H. ventosae* strain Al16 gave the best results in biofilm formation assays, possibly due to the high emulsifying activity of its polysaccharide (Mata et al. 2006).

Halomonas smyrnensis strain AAD6^T, isolated from soil samples taken from Çamaltı Saltern area in Turkey, was found to produce high levels of levan (Poli et al. 2009, 2013). This EPS did not affect cellular viability and proliferation in two different cellular systems tested, osteoblasts and murine macrophages, demonstrating its high biocompatibility. Besides, it displayed a protective effect against the toxic activity of avarol implying its additional use as an anti-cytotoxic agent. The potential applications of levan as an industrial gum, a blood plasma extender, a sweetener, an emulsifier, a formulation aid, a stabilizer, a thickener, a surface-finishing agent, an encapsulating agent and a carrier for flavour and fragrances are known (Shih et al. 2005; Beine et al. 2008). Then, *Halomonas* sp. AAD6 represented an alternative cheap source of levan polymer when grown on defined media hypothesizing its larger employment in industrial application being a non-pathogenic microorganism (Sam et al. 2011; Sezer et al. 2011).

A species of halophilic, EPS-producing bacterium belonging to the *Alphaproteobacteria*, is the type strain (A3^T) of *Salipiger mucosus*, isolated on the Mediterranean seaboard. The EPS produced by *S. mucosus* was able to emulsify high percentages of pure hydrocarbons (tetradecane, octane, kerosene, xylene and crude oil) more than other chemical surfactants used in comparison. This ability could be ascribed to the presence of acetyl groups which render the EPS somewhat hydrophobic. Furthermore, the EPS was also able to bind cations and to incorporate high quantities of sulphates, which represent a very unusual feature in bacterial polysaccharides (Llamas et al. 2010).

15.5.2 Thermophiles

EPS producers were also found among thermophiles isolated from different thermophilic habitats. Remarkable antiviral and immunomodulatory activities against herpes simplex virus type 2 (HSV-2) were showed by EPSs produced by *Bacillus licheniformis* strain B3-15, *Geobacillus thermodenitrificans* strain B3-72 and *B. licheniformis* strain T14, three thermophilic and thermotolerant bacilli isolated from Aeolian Islands shallow vents. All EPSs were not cytotoxic towards peripheral blood mononuclear cells (PBMC) at the concentration of 300 $\mu g \cdot mL^{-1}$. They were able to interfere HSV-2 replication in PBMC. This ability, expressed as logarithm, was higher for EPS2-B3-15 (0.82) compared with EPS2-B3-72 (0.49) and EPS1-T14 (0.63). Further investigations showed a correlation between the antiviral effect of EPSs and the immune response involved in the controlling viral replication. Indeed, EPS treatment caused high production of Th1 cytokines (IFN-γ, IFN-α, TNF-α, IL-12 and IL-18) by PBMC, which means the inhibition of viral replication by induction of antiviral state in neighbouring cells (i.e. IFNs) or the destruction of virus-infected cells (i.e. TNF-α and IL-18). EPS2-B3-15 exhibited the best antiviral potential compared with the other EPSs assayed (Nicolaus et al. 2000; Maugeri et al. 2002; Arena et al. 2006; Spanò et al. 2013; Gugliandolo et al. 2015; Marino-Merlo et al. 2017).

The extracellular polysaccharide TA-1 secreted by the thermophilic bacteria *Thermus aquaticus* YT-1 also showed immunomodulatory activity by stimulation of macrophage cells to produce the cytokines TNF-α and IL-6, which increases the immune response. The presence of D-galactofuranose residues in the EPS TA-1 could be probably responsible for observed immunoregulatory activity through Toll-like receptor 2 within macrophages, the first line of host defence against bacterial infection (Lin et al. 2011).

Geobacillus tepidamans V264, a thermophilic bacteria isolated from Velingrad hot spring, Bulgaria, secreted an extracellular polysaccharide exhibiting an anti-cytotoxic activity evaluated by means of brine shrimp test, towards avarol, a natural toxic sesquiterpene hydroquinone isolated from *Dysidea avara* sponge (Tommonaro et al. 2015). The biopolymer increased the value of LD_{50} of avarol, more than 12-fold, from 0.18 $\mu g\ mL^{-1}$ up to 2.24 $\mu g\ mL^{-1}$. The activity exerted by EPS could be related to the adhesion of toxic compounds to the surface of the polysaccharide. Hence, this EPS could be used in pharmacy as anti-cytotoxic drugs (Kambourova et al. 2009).

From southwest of Bulgaria, in Rupi Basin hot springs, the strain *Aeribacillus pallidus* 418 producing an exopolysaccharide was isolated. The EPS exhibited good emulsifying properties, which could be improved using mixtures with other biopolymers. In particular, the mixture of EPS from *A. pallidus* 418 with xanthan showed the best synergy in terms of stability of emulsion. Both these properties (good emulsifying properties and the enhanced synergistic activity) of EPS represented valuable features for its industrial exploration (Radchenkova et al. 2013, 2014).

In the same region of Bulgaria, from a hot spring close to the village Gradeshnitsa, Blagoevgrad region, a thermophilic microorganism, which belonged to the phylum *Firmicutes* and closely related with other strains from the species *Brevibacillus thermoruber*, *B. thermoruber* strain 423, was isolated. Its colonies exhibited high mucoidity, and it was a high-level exopolysaccharide (EPS)-producing thermophile. Chemical studies showed that the EPS was a heteropolymer composed of glucose as prevailing monomer unit. At first, it was purified in two fractions, as a flow through column (EPS-FT) and peak of salt elution (EPS-P), and next assayed for its biocompatibility with the monkey kidney fibroblast cell line Cos-7, considering that biocompatibility is one of key factors for medical applications. Results showed a no pathogenicity of the pure EPS fractions on cellular line used together with their high biocompatibility, and then this study suggested their potential use in biomedical applications, such as scaffolds or matrices in tissue engineering, drug delivery and wound dressing (Nwodo et al. 2012; Yasar Yildiz et al. 2014).

From deep-sea hydrothermal vent located in the Gulf of California, two EPS-producing bacteria have been isolated, *Vibrio diabolicus* and *Alteromonas infernus*. The EPS GY785 produced by *A. infernus* was a branched, sulphated polysaccharide and showed a high molecular weight (up to 10^6 Da), while EPS HE800 produced by *V. diabolicus* was a linear glycosaminoglycan and showed a molecular mass of about 8×10^5 Da. Both EPSs exhibited very interesting biological activity after depolymerization and, next, sulphation of the hydroxyl groups present on the low molecular weight (LMW) EPSs. The over-sulphated EPSs, named HE800 DROS and GY785 DROS, interacted with C1q protein of the complement pathway system by activation of normal human serum (NHS) incubated with various amounts of GY785 DR or HE800 DR, to restore the haemolytic activity of serum deficient in complement protein C1q. However, EPS HE800 already showed very interesting biological properties in regard to bone and skin regeneration (Raguénès et al. 1997a, b; Zanchetta et al. 2003a, b; Courtois et al. 2014; Poli et al. 2017).

15.5.3 Psychrophiles

Cold-adapted microorganisms (psychrophiles and psychrotolerant) are widespread in terrestrial environments and marine ecosystem. Despite the large number of psychrophilic microorganisms reported in literature, few of them are described as EPS-producing microorganisms. Psychrophilic γ-proteobacterium *Pseudomonas* sp. ID1 is a cold-adapted bacterium isolated from a marine sediment sample collected from South Shetland Islands (Antarctica). This microorganism produced an EPS mainly composed of glucose, galactose and fucose and had a molecular mass higher than 2×10^6 Da. This biopolymer exhibited emulsifying activity against different food and cosmetic oils much higher than commercial gums (xanthan gum and Arabic gum), cryoprotective activity, pseudoplastic flow behaviour, low thixotropy and yield stress. All these properties of EPS of *Pseudomonas* sp. ID1 suggested its significant cryoprotection role for the strain and make it a promising alternative to commercial polysaccharides as emulsifier and cryoprotectant agent for food, pharmaceutical and cosmetic industries (Carrión et al. 2015).

Psychrotolerant bacterium *Pseudoalteromonas* sp. SM9913 secreted large quantities of EPSs. The yield of EPS increased as the temperature decreased in the tested range, indicating that the EPS production of strain SM9913 had cold adaptation. Under optimal growth conditions (15 °C, 52 h), the yield of EPS reached 5.25 g l21 (dry weight), which was higher than that reported for the EPSs produced by other psychrotolerant microorganisms (Nichols et al. 2005b). Structural analysis of EPS SM9913 showed that it consisted mainly of glucose, with arabinose, xylose and a minor peak for mannose. This biopolymer enhanced the thermostability of protease MCP-01 (the main protease secreted by strain SM9913) at 40 °C, by preventing its autolysis. In the presence of EPS (1% w/v), the protease activity of MCP-01 showed no evident change after 150 min incubation. In contrast, the protease activity in the absence of EPS was quickly lost, with 90% of the activity lost after 135 min incubation at 40 °C. In addition, the flocculation experiment showed that the EPS could make colloidal and suspended particles in solution conglomerate, suggesting that the EPS was a very good flocculating agent and had a good adsorptive effect. Therefore, it might play an important role for strain SM9913 in enriching nutrient particles in the deep-sea environment (Qin et al. 2007; Li et al. 2008).

15.5.4 Archaea

The most valuable example of biotechnologically interesting EPS produced by an Archaea is that reported by Squillaci et al. (2016). In that paper the isolation and the chemical characterization of the EPS secreted by *Haloterrigena turkmenica* together with its applicative properties are discussed. The microorganism produced the EPS mainly in the middle exponential growth phase and reached the maximal production (20.68 mg EPS per 100 ml of culture medium) in the stationary phase. Results obtained by means of anion-exchange chromatography and SEC-TDA Viscotek indicated that the EPS was composed of two main fractions of 801.7 and 206.0 kDa. It was a sulphated heteropolysaccharide containing glucose, galactose, glucosamine, galactosamine and glucuronic acid. EPS exhibited interesting emulsifying activity towards *n*-hexane while was capable of producing stable emulsions with vegetable oils. EPSs supplied with emulsifying ability could be employed in the food industry as emulsifier and stabilizer agents (Duboc and Mollet 2001). Moreover, EPS displayed also a moderate antioxidant activity evaluated by means of DPPH, FRAP and TAC assays. In DPPH assay, at a concentration of 10 mg/ml, the radical scavenging activity of the EPS was 68.2% ± 1.1 with IC50 value of 6.03 mg/ml, whereas hyaluronic acid (standard used) did not show scavenging capacity. In TAC and FRAP assays also, the EPS showed the ability to react with both Mo^{6+} and Fe^{3+}

ions showing a linear dose-dependent antioxidant activity. All these features make the EPS produced by *H. turkmenica* a possible candidate for wide applications in several industrial sectors (Squillaci et al. 2016).

15.6 Biosynthesis of EPSs and Genetic Strategy for Their Hyperproduction

Bacteria exopolysaccharides are synthesized via different biosynthesis pathways, and the genes responsible for the synthesis are often clustered within the genome. The knowledges related to EPS biosynthetic processes and the genetic regulation are essential to produce tailor-made biopolymers. Intracellular synthesis of homo- and heteropolysaccharides is a complex process that proceeds via intracellular assembly of sugar nucleotide precursors. In EPS biosynthesis, different enzymes and regulatory molecules are involved in several metabolic pathways. It begins with the entry of the sugars in the cell, which are catabolized by periplasmic oxidation or intracellular phosphorylation: sugars that do not take part in the central metabolic pathways act as a raw material for EPS manufacture (Freitas et al. 2011). The intracellular EPS-synthetic machinery requires charged and energy-rich precursor monosaccharides: these letters are in the form of nucleotide diphosphate/monophosphate sugars (NDP-/NMP-sugar). This is a crucial step of biosynthesis: sugars often are in the form of sugar-1P and rarely in the form of sugar-2P or sugar- 6P and serve as activated primary residues (Madhuri and Prabhakar 2014). Some intermediates like fructose-6P or glucose-1P, in majority of cases, lead to synthesis of uridine diphosphate-N-acetyl glucosamine (UDP-GlcNAc), uridine diphosphate-N-acetyl galactosamine (UDP- GalNAc) and dTDP-rhamnose, precursor molecules for EPS synthesis (Boels et al. 2001). In subsequent step, phosphoglucomutase enzyme catalyses the conversion of sugar-6P to sugar-1P, and UDP-glucose pyrophosphorylase and dTDP-glucose pyrophosphorylase catalyse the conversion of sugar-1P to UDP-glucose and dTDP-glucose. The conversion of UDP-GlcNAc to UDP-GalNAc in *Lactobacillus rhamnosus* has been found to be catalysed by UDP-N-acetylglucosamine 4-epimerase (Boels et al. 2001). At present four general mechanisms are known in bacteria: (a) the Wzx-/Wzy-dependent pathway, (b) the ATP-binding cassette (ABC) transporter-dependent pathway, (c) the synthase-dependent pathway and (d) the extracellular synthesis by the use of a single sucrase protein. In the first three biosynthesis pathways, the precursor molecules, which are necessary for the stepwise elongation of the polymer strands, are realized inside the cell, while for the extracellular production, the polymer strand is elongated by direct addition of monosaccharides obtained by cleavage of di- or trisaccharides (Fig. 15.2).

(a) In the Wzx-/Wzy-dependent pathway the repeating units, linked to an undecaprenol diphosphate anchor (C55) at level of the inner membrane, are assembled by different glycosyltransferases (GTs) and translocated across the cytoplasmic membrane by a Wzx protein (flippase protein). The Wzy polymerase and the polysaccharide co-polymerase (PCP) protein are responsible for polymerization

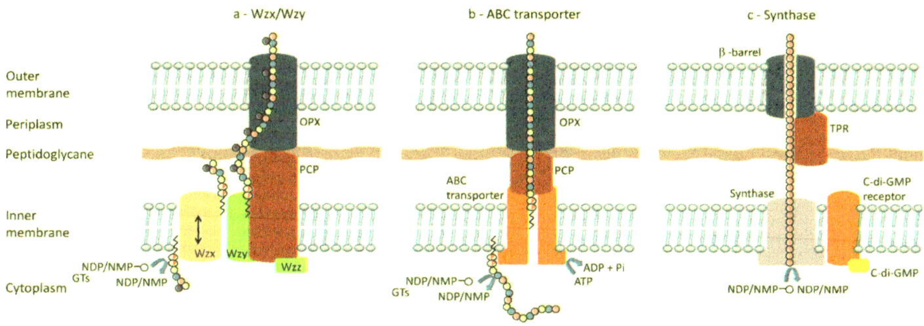

Fig. 15.2 Schematic representation of EPS biosynthesis pathways

process. The polysaccharide export (OPX) protein families and the PCP carry the transport across the membranes. (b) In the ABC transporter-dependent pathway, the EPS chain is assembled on a lipid carrier situated in the inner leaflet of the inner membrane before the transportation across the inner membrane by an ABC transporter. The EPS is then exported through the periplasm and across the outer membrane by the OPX and the PCP families of proteins. (c) In the synthase-dependent pathway, a complete polymer chain is polymerized and secreted across the inner membrane by inner-membrane synthase proteins. The activity of the polysaccharide synthase is post-translationally regulated by an inner-membrane c-di-GMP receptor. The EPS is then exported across the outer membrane by TRP-containing protein and an integral outer-membrane beta-barrel.

In the first mechanism, the repeating units, which are linked to an undecaprenol diphosphate anchor (C55) at level of the inner membrane, are assembled by different glycosyltransferases (GTs) and translocated across the cytoplasmic membrane by a Wzx protein, also known as flippase. Then, at level of the periplasmic space, the repeating units are polymerized by the Wzy protein before they will be exported to the cell surface (Islam and Lam 2014). Subsequently, the transport of polymers from the periplasm to the cell surface is due to the polysaccharide co-polymerase (PCP) and the outer-membrane polysaccharide export (OPX; formerly OMA) protein families (Cuthbertson et al. 2009; Morona et al. 2009). EPSs assembled by the Wzx/Wzy pathway are heteropolymers (e.g. xanthan) and possess different sugar patterns, up to four or five types of sugars in their chemical structure. All strains using this pathway carry the genes for the flippase (Wzx) and the polymerase (Wzy) within their extracellular polysaccharide operons.

In the synthase-dependent pathway, complete polymer strands are secreted across the membranes and the cell wall: this pathway is independent of a flippase for the translocation of the repeating units. Both polymerization and translocation processes are performed by a single synthase protein. In the case of alginate and cellulose, for example, the synthase protein is a single subunit of an envelope-spanning multiprotein complex (Rehm 2010; Whitney and Howell 2013). This pathway is often utilized for the synthesis of homopolymers: this is the case of curdlan biosynthesis, for example, in which only β-(1–3)-linked glucose is present or in the case of bacterial cellulose, consisting only of β-(1–4)-linked glucose units.

In the extracellular synthesis pathway, the biosynthesis of polymers, such as dextran or levan, occurs via GTs, which are secreted and covalently linked to the cell surface. In this case the responsible enzymes involved in the process transfer the activated precursor monosaccharides from substrate to growing polysaccharide that assembles in a final structure by the formation of linkage pattern and degree of branching. The Wzx-/Wzy-dependent pathway is responsible for the biosynthesis of diutan, gellan, welan, xanthan and colanic acid also known as the M antigen (an EPS with no commercial application but is of high interest due to pathogenicity in enterobacteria studies). Alginate, cellulose, curdlan and hyaluronic acid are examples of EPSs in which the synthase-dependent mechanism is responsible for their synthesis; dextran, levan and mutan are the common examples that require dextran sucrase and levan sucrase as enzymes and sucrose as a substrate, respectively (extracellular pathway) (Boels et al. 2001).

Bacterial polysaccharides have interesting and unique properties for industrial applications and are used as emulsifiers, viscosifiers, stabilizers or gelling agents. Due to these valuable properties, several studies were performed to genetically engineer the producing organisms in order to improve the yield of production or to generate new polysaccharide variants. Putative targets for engineering are the molecular weight, addition of substituents, composition and sequence of sugar components. Recently, intensive research focused on mechanisms underlying EPS biosynthesis pathways, genome sequencing, protein structure analysis and new bioinformatics tools aid to understand the principles of EPS formation. Engineering strategies can be subdivided into different categories. One is an increased volumetric productivity: these studies were mostly aiming at increasing sugar nucleotide precursors to enhance the carbon flux towards the final polymer, and the genes of precursor biosynthesis were overexpressed (Schmid et al. 2015). Overexpression of genes involved in the EPS assembly such as GTs, Wzx and Wzy, both as single genes and whole cluster, resulted in enhanced yields and precursor conversion rates, while in other cases, it had a negative effect presumably due to distorting the multidomain protein complex involved in polymerization and secretion (van Kranenburg et al. 1999). Another approach is to increase the EPS productivity by increasing transcription of the operons, which encode the EPS biosynthesis proteins. Single-gene knockouts were also described to enhance yield and alter EPS chemical structure, as shown in *Azotobacter vinelandii* (Gaytán et al. 2012). However, the strategy to enhance productivity based on genetic engineering might be interesting for EPS with reduced viscosifying properties, for example, due to lower molecular weight. The optimization of manufacturing process parameters might be more promising than engineering EPS biosynthesis for many established industrial EPS producers. The highest titres of highly viscous EPS such as xanthan are around 30–50 g/L and represent the current maximum amount, which is manageable by existing bioprocess units (Hublik 2012).

Another strategy of engineering EPS biosynthesis is to alter the molecular structure and therefore the chemical characteristics and behaviour of the final biopolymer. These modifications can be based on deleting substituents or monomeric sugars from the side chain or binding new substituents: in these cases a change of the ratio of decoration, such as the degree of acetylation and pyruvylation, occurred (Donati and Paoletti 2009). The degree of acetylation and pyruvylation has opposite effects on viscosity. A high degree of pyruvylation resulted in higher viscosity, whereas the presence of more acetyl groups decreased viscosity of EPS. This finding is a general rule for polysaccharides and can be used in tailoring the EPS viscosity.

Other engineering approaches with respect to the production of xanthan variants included the length of the side chain. A truncated tetramer xanthan version, obtained by deletion of the terminal mannose via inactivation of the glycosyltransferase (GT GumI), resulted in a much lower viscosity. The further removal of the glucuronic acid from the side chain by inactivation of GT GumK resulted in enhanced viscosity compared to the wild type (Schmid et al. 2015). The molecular weight of xanthan was synthetically adjusted by controlling the expression level of the Wzy polymerase Gum E (Galván et al. 2013), while for alginate a similar effect was observed by an overexpression of alginate polymerase *alg8/alg44* in *Azotobacter vinelandii* producing a high molecular weight alginate type (Díaz-Barrera et al. 2012).

In some EPS, overexpression or mutation of genes involved in the polymerization/degradation process (e.g. synthase, Wzy, PCP/lyases, glucosidases) represented another possibility to change the rheological properties of the polymers (Rehm 2010; Galván et al. 2013).

As concerning the haloarchaeal EPSs, although the chemical structures have been solved, little is known about their biosynthesis. The EPS from *Haloferax mediterranei* ATCC 33500 was identified to be a heteropolysaccharide containing mannose as the major component. The repeating unit of EPS in *H. mediterranei* contains one mannosyl and two N-acetyl-glucosaminuronyl moieties, and one N-acetyl-glucosaminuronyl group is modified by a sulfonic group. Based on the complete genome sequence of *H. mediterranei*, a gene cluster involved in EPS biosynthesis in *H. mediterranei* was identified. Deletion of the gene cluster eliminated EPS synthesis. The mutant strain deficient of EPS biosynthesis showed a remarkable decrease in viscosity and foaming propensity of culture broth and increase in content of dissolved oxygen and enhanced the production of polyhydroxyalkanoate (Zhao et al. 2013). Lü et al. (2017) purified an acidic exopolysaccharide from an extremely halophilic archaeon *Haloarcula hispanica* ATCC 33960, which mainly composed of mannose and galactose with a small amount of glucose in a molar ratio of 55.9:43.2:0.9. The authors reported the identification of two glycosyltransferase genes (HAH_1662 and HAH_1667), responsible for the synthesis of EPS. Deletion of either HAH_1662 or HAH_1667 led to loss of the EPS production. In addition, the mutants of *Haloarcula hispanica* displayed a different cell surface morphology, retarded growth in low salty environment, an increased adhesion and swimming ability, suggesting that its biosynthesis might act as an adaptable mechanism to protect the cells against harsh environments.

In conclusion, as emerged from this overview, the genes involved in the different biosynthesis pathways encode various types of GTs, polymerizing and branching enzymes, but also enzymes responsible for addition of substituents or modifications of sugar chain. However, several steps are currently understood, and sometimes the differences between the pathways become less defined. Genome or plasmids contain the genes encoding these enzymes in most of the EPS-producing bacteria (Rehm 2010). The clustering of several GTs and polymerizing as well as secreting enzymes facilitate the identification of EPS operons. Even if many gene clusters involved in the EPS biosynthesis have been known for several years, the function and mode of action of most of the genes and proteins are not completely clarified.

Moreover, the identification of novel EPS clusters by next-generation sequencing approaches will enhance our understanding of EPS synthesis pathway variation and modification. By using different tools, such as bioinformatics, structural information of proteins and EPSs, it will enable the implementation of synthetic biology approaches for tailoring microbial EPS. The insights in Wzx and Wzy topology and mechanism might open up the opportunity for incorporation of desired sugars or sugar derivatives resulting in modified EPS structures with hitherto unknown material properties (Rehm 2015). Recently, an innovative bi-enzymatic process was reported, stating that from sucrose, the production of short-chain fructooligosaccharides and oligolevans was obtained. This system was based on an immobilized levansucrase and an endo-inulinase, resulting in a highly efficient synthesis system with a yield of more than 65% and a productivity of 96 g/L/h (Tian et al. 2014). The utilization and combination of several carbohydrate-modifying enzymes create the potential for industrial production of different low molecular weight oligo- or polysaccharides with applications as food additives (prebiotics) or in medicine.

15.7 Conclusion and Future Perspectives

Microbial EPSs are ubiquitous in the extreme environments where they are crucial for microbial survival. Most of the functions attributed to EPSs are related to a protective role, which are dependent on the ecological niches in which the producer microorganisms live. They could support the microbial communities to tolerate extremes of temperature, salt concentration and nutrient supply, building an interface between the bacterial cell and its surrounding environment. Several EPSs produced by microorganisms from extreme habitats show biotechnological promise. By examining their structure and chemical/physical characteristics, it is possible to gain insight into their commercial application; they are employed in several fields ranging from food-processing to pharmaceutical industries, through to the bioremediation ability of polluted areas. Considering that most of extreme ecosystems and, therefore, the respective microbial communities are still unexplored, it is reasonable to hypothesize that the isolation of new microorganisms, together with their biomolecules, in particular exopolysaccharides, will provide interesting perspectives for new industrial processes in several fields.

Acknowledgements The manuscript was partially supported by the funding of Programma Nazionale di Ricerca in Antartide- PNRA16_00020 – A1 and PNRA16_00274-A1.

References

Andersson S, Dalhammar G, Kuttuva Rajarao G (2011) Influence of microbial interactions and EPS/polysaccharide composition on nutrient removal activity in biofilms formed by strains found in wastewater treatment systems. Microbiol Res 166(6):449–457. https://doi.org/10.1016/j.micres.2010.08.005

Antón J, Meseguer I, Rodríguez-Valera F (1988) Production of an extracellular polysaccharide by *Haloferax mediterranei*. Appl Environ Microbiol 54(10):2381–2386

Arena A, Gugliandolo C, Stassi G, Pavone B, Iannello D, Bisignano G, Maugeri TL (2009) An exopolysaccharide produced by *Geobacillus thermodenitrificans* strain B3-72: antiviral activity on immunocompetent cells. Immunol Lett 123(2):132–137. https://doi.org/10.1016/j.imlet.2009.03.001

Arena A, Maugeri TL, Pavone B, Iannello D, Gugliandolo C, Bisignano G (2006) Antiviral and immunoregulatory effect of a novel exopolysaccharide from a marine thermotolerant *Bacillus licheniformis*. Int Immunopharmacol 6(1):8–13. https://doi.org/10.1016/j.intimp.2005.07.004

Arias S, del Moral A, Ferrer MR, Tallon R, Quesada E, Béjar V (2003) Mauran, an exopolysaccharide produced by the halophilic bacterium *Halomonas maura*, with a novel composition and interesting properties for biotechnology. Extremophiles 7(4):319–326. https://doi.org/10.1007/s00792-003-0325-8

Bales PM, Renke EM, May SL, Shen Y, Nelson DC (2013) Purification and characterization of biofilm-associated EPS exopolysaccharides from ESKAPE organisms and other pathogens. PLoS One 8(6):e67950. https://doi.org/10.1371/journal.pone.0067950

Beine R, Moraru R, Nimtz M, Na'amnieh S, Pawlowski A, Buchholz K, Seibel J (2008) Synthesis of novel fructooligosaccharides by substrate and enzyme engineering. J Biotechnol 138(1):33–41. https://doi.org/10.1016/j.jbiotec.2008.07.1998

Béjar V, Llamas I, Calvo C, Quesada E (1998) Characterization of exopolysaccharides produced by 19 halophilic strains of the species Halomonas eurihalina. J Biotechnol 61(2):135–141. https://doi.org/10.1016/S0168-1656(98)00024-8

Boels IC, Rv K, Hugenholtz J, Kleerebezem M, de Vos WM (2001) Sugar catabolism and its impact on the biosynthesis and engineering of exopolysaccharide production in lactic acid bacteria. Int Dairy J 11(9):723–732. https://doi.org/10.1016/S0958-6946(01)00116-9

Bouchotroch S, Quesada E, del Moral A, Llamas I, Béjar V (2001) *Halomonas maura* sp. nov., a novel moderately halophilic, exopolysaccharide-producing bacterium. Int J Syst Evol Microbiol 51(5):1625–1632. https://doi.org/10.1099/00207713-51-5-1625

Bradford MM (1976) A rapid and sensitive method for the quantification of microgram quantities of protein utilizing the principle of protein-dye binding. Anal Biochem 72(1–2):248–254. https://doi.org/10.1016/0003-2697(76)90527-3

Carillo S, Casillo A, Pieretti G, Parrilli E, Sannino F, Bayer-Giraldi M, Cosconati S, Novellino E, Ewert M, Deming JW, Lanzetta R, Marino G, Parrilli M, Randazzo A, Tutino ML, Corsaro MM (2015) A unique capsular polysaccharide structure from the psychrophilic marine bacterium *Colwellia psychrerythraea* 34H that mimics antifreeze (Glyco)proteins. J Am Chem Soc 137(1):179–189. https://doi.org/10.1021/ja5075954

Carrión O, Delgado L, Mercade E (2015) New emulsifying and cryoprotective exopolysaccharide from Antarctic *Pseudomonas* sp. ID1. Carbohydr Polym 117:1028–1034. https://doi.org/10.1016/j.carbpol.2014.08.060

Casillo A, Parrilli E, Sannino F, Mitchell DE, Gibson MI, Marino G, Lanzetta R, Parrilli M, Cosconati S, Novellino E, Randazzo A, Tutino ML, Corsaro MM (2017) Structure-activity relationship of the exopolysaccharide from a psychrophilic bacterium: a strategy for cryoprotection. Carbohydr Polym 156:364–371. https://doi.org/10.1016/j.carbpol.2016.09.037

Courtois A, Berthou C, Guézennec J, Boisset C, Bordron A (2014) Exopolysaccharides isolated from hydrothermal vent bacteria can modulate the complement system. PLoS One 9(4):e94965. https://doi.org/10.1371/journal.pone.0094965

Cuthbertson L, Mainprize IL, Naismith JH, Whitfield C (2009) Pivotal roles of the outer membrane polysaccharide export and polysaccharide copolymerase protein families in export of extracellular polysaccharides in gram-negative bacteria. Microbiol Mol Biol Rev 73(1):155–177. https://doi.org/10.1128/mmbr.00024-08

D'Amico S, Collins T, Marx JC, Feller G, Gerday C, Gerday C (2006) Psychrophilic microorganisms: challenges for life. EMBO Rep 7(4):385

DasSarmaS,DasSarmaP(2001)Halophiles.In:eLS.Wiley.https://doi.org/10.1002/9780470015902.a0000394.pub4

De Maayer P, Anderson D, Cary C, Cowan DA (2014) Some like it cold: understanding the survival strategies of psychrophiles. EMBO Rep 15(5):508–517. https://doi.org/10.1002/embr.201338170

Deming JW, Young JN (2017) The role of exopolysaccharides in microbial adaptation to cold habitats. In: Margesin R (ed) Psychrophiles: from biodiversity to biotechnology. Springer International Publishing, Cham, pp 259–284. https://doi.org/10.1007/978-3-319-57057-0_12

Di Donato P, Poli A, Taurisano V, Abbamondi G, Nicolaus B, Tommonaro G (2016) Recent advances in the study of marine microbial biofilm: from the involvement of quorum sensing in its production up to biotechnological application of the polysaccharide fractions. J Mar Sci Eng 4(2):34

Díaz-Barrera A, Soto E, Altamirano C (2012) Alginate production and alg8 gene expression by *Azotobacter vinelandii* in continuous cultures. J Ind Microbiol Biotechnol 39(4):613–621. https://doi.org/10.1007/s10295-011-1055-z

Donati I, Paoletti S (2009) Material properties of alginates. In: Rehm BHA (ed) Alginates: biology and applications. Springer, Berlin/Heidelberg, pp 1–53. https://doi.org/10.1007/978-3-540-92679-5_1

Donot F, Fontana A, Baccou JC, Schorr-Galindo S (2012) Microbial exopolysaccharides: main examples of synthesis, excretion, genetics and extraction. Carbohydr Polym 87(2):951–962. https://doi.org/10.1016/j.carbpol.2011.08.083

Du AG-L, Zykwinska A, Sinquin C, Ratiskol J, Weiss P, Vinatier C, Guicheux J, Delbarre-Ladrat C, Colliec-Jouault S (2017) Purification of the exopolysaccharide produced by *Alteromonas infernus*: identification of endotoxins and effective process to remove them. Appl Microbiol Biotechnol 101(17):6597–6606. https://doi.org/10.1007/s00253-017-8364-8

Duboc P, Mollet B (2001) Applications of exopolysaccharides in the dairy industry. Int Dairy J 11(9):759–768. https://doi.org/10.1016/S0958-6946(01)00119-4

DuBois M, Gilles KA, Hamilton JK, Rebers PA, Smith F (1956) Colorimetric method for determination of sugars and related substances. Anal Chem 28(3):350–356. https://doi.org/10.1021/ac60111a017

Duus JØ, Gotfredsen CH, Bock K (2000) Carbohydrate structural determination by NMR spectroscopy: modern methods and limitations. Chem Rev 100(12):4589–4614. https://doi.org/10.1021/cr990302n

Ewert M, Deming JW (2013) Sea ice microorganisms: environmental constraints and extracellular responses. Biology 2(2):603–628. https://doi.org/10.3390/biology2020603

Finore I, Di Donato P, Mastascusa V, Nicolaus B, Poli A (2014) Fermentation technologies for the Optimization of marine microbial exopolysaccharide production. Mar Drugs 12(5):3005–3024. https://doi.org/10.3390/md12053005

Finore I, Lama L, Poli A, Di Donato P, Nicolaus B (2015) Biotechnology implications of extremophiles as life pioneers and wellspring of valuable biomolecules. In: Kalia VC (ed) Microbial factories: biodiversity, biopolymers, bioactive molecules, vol 2. Springer India, New Delhi, pp 193–216. https://doi.org/10.1007/978-81-322-2595-9_13

Flemming H-C, Wingender J (2001) Relevance of microbial extracellular polymeric substances (EPSs) – part I: structural and ecological aspects. Water Sci Technol 43(6):1–8

Freitas F, Alves VD, Reis MAM (2011) Advances in bacterial exopolysaccharides: from production to biotechnological applications. Trends Biotechnol 29(8):388–398. https://doi.org/10.1016/j.tibtech.2011.03.008

Galván EM, Ielmini MV, Patel YN, Bianco MI, Franceschini EA, Schneider JC, Ielpi L (2013) Xanthan chain length is modulated by increasing the availability of the polysaccharide copolymerase protein GumC and the outer membrane polysaccharide export protein GumB. Glycobiology 23(2):259–272. https://doi.org/10.1093/glycob/cws146

Gaytán I, Peña C, Núñez C, Córdova MS, Espín G, Galindo E (2012) *Azotobacter vinelandii* lacking the Na+-NQR activity: a potential source for producing alginates with improved properties and at high yield. World J Microbiol Biotechnol 28(8):2731–2740. https://doi.org/10.1007/s11274-012-1084-4

Gómez-Ordóñez E, Jiménez-Escrig A, Rupérez P (2012) Molecular weight distribution of polysaccharides from edible seaweeds by high-performance size-exclusion chromatography (HPSEC). Talanta 93(Supplement C):153–159. https://doi.org/10.1016/j.talanta.2012.01.067

Gugliandolo C, Spanò A, Lentini V, Arena A, Maugeri TL (2014) Antiviral and immunomodulatory effects of a novel bacterial exopolysaccharide of shallow marine vent origin. J Appl Microbiol 116(4):1028–1034. https://doi.org/10.1111/jam.12422

Gugliandolo C, Spanò A, Maugeri T, Poli A, Arena A, Nicolaus B (2015) Role of bacterial exopolysaccharides as agents in counteracting immune disorders induced by. Herpes Virus Microorganisms 3(3):464

Hublik G (2012) 10.11 – Xanthan A2 – Matyjaszewski, Krzysztof. In: Möller M (ed) Polymer science: a comprehensive reference. Elsevier, Amsterdam, pp 221–229. https://doi.org/10.1016/B978-0-444-53349-4.00262-4

Islam ST, Lam JS (2014) Synthesis of bacterial polysaccharides via the Wzx/Wzy-dependent pathway. Can J Microbiol 60(11):697–716. https://doi.org/10.1139/cjm-2014-0595

Johnson MR, Montero CI, Conners SB, Shockley KR, Bridger SL, Kelly RM (2005) Population density-dependent regulation of exopolysaccharide formation in the hyperthermophilic bacterium *Thermotoga maritima*. Mol Microbiol 55(3):664–674. https://doi.org/10.1111/j.1365-2958.2004.04419.x

Kambourova M, Mandeva R, Dimova D, Poli A, Nicolaus B, Tommonaro G (2009) Production and characterization of a microbial glucan, synthesized by *Geobacillus tepidamans* V264 isolated from Bulgarian hot spring. Carbohydr Polym 77(2):338–343. https://doi.org/10.1016/j.carbpol.2009.01.004

Kambourova M, Radchenkova N, Tomova I, Bojadjieva I (2016) Thermophiles as a promising source of exopolysaccharides with interesting properties. In: Rampelotto PH (ed) Biotechnology of extremophiles: advances and challenges. Springer International Publishing, Cham, pp 117–139. https://doi.org/10.1007/978-3-319-13521-2_4

Ko S-H, Lee HS, Park SH, Lee HK (2000) Optimal conditions for the production of exopolysaccharide by marine microorganism *Hahella chejuensis*. Biotechnol Bioprocess Eng 5(3):181. https://doi.org/10.1007/bf02936591

Koerdt A, Gödeke J, Berger J, Thormann KM, Albers S-V (2010) Crenarchaeal biofilm formation under extreme conditions. PLoS One 5(11):e14104. https://doi.org/10.1371/journal.pone.0014104

Lee HK, Chun J, Moon EY, Ko SH, Lee DS, Lee HS, Bae KS (2001) *Hahella chejuensis* gen. nov., sp. nov., an extracellular-polysaccharide-producing marine bacterium. Int J Syst Evol Microbiol 51(2):661–666. https://doi.org/10.1099/00207713-51-2-661

Li WW, Zhou WZ, Zhang YZ, Wang J, Zhu XB (2008) Flocculation behavior and mechanism of an exopolysaccharide from the deep-sea psychrophilic bacterium Pseudoalteromonas sp. SM9913. Bioresour Technol 99(15):6893–6899. https://doi.org/10.1016/j.biortech.2008.01.050

Lin M-H, Yang Y-L, Chen Y-P, Hua K-F, Lu C-P, Sheu F, Lin G-H, Tsay S-S, Liang S-M, Wu S-H (2011) A novel exopolysaccharide from the biofilm of *Thermus aquaticus* YT-1 induces the immune response through toll-like receptor 2. J Biol Chem 286(20):17736–17745. https://doi.org/10.1074/jbc.M110.200113

Llamas I, Mata JA, Tallon R, Bressollier P, Urdaci MC, Quesada E, Béjar V (2010) Characterization of the exopolysaccharide produced by *Salipiger mucosus* A3(T), a halophilic species belonging to the *Alphaproteobacteria*, isolated on the Spanish Mediterranean seaboard. Mar Drugs 8(8):2240–2251. https://doi.org/10.3390/md8082240

Loaëc M, Olier R, Guezennec J (1997) Uptake of lead, cadmium and zinc by a novel bacterial exopolysaccharide. Water Res 31(5):1171–1179. https://doi.org/10.1016/S0043-1354(96)00375-2

Lü Y, Lu H, Wang S, Han J, Xiang H, Jin C (2017) An acidic exopolysaccharide from Haloarcula hispanica ATCC33960 and two genes responsible for its synthesis. Archaea 2017:5842958. https://doi.org/10.1155/2017/5842958

Madhuri KV, Prabhakar KV (2014) Microbial exopolysaccharides: biosynthesis and potential applications. Orient J Chem 30(3):1401–1410. https://doi.org/10.13005/ojc/300362

Marino-Merlo F, Papaianni E, Maugeri TL, Zammuto V, Spanò A, Nicolaus B, Poli A, Di Donato P, Mosca C, Mastino A, Gugliandolo C (2017) Anti-herpes simplex virus 1 and immunomodulatory activities of a poly-γ- glutamic acid from Bacillus horneckiae strain APA of shallow vent origin. Appl Microbiol Biotechnol. https://doi.org/10.1007/s00253-017-8472-5

Marx JG, Carpenter SD, Deming JW (2009) Production of cryoprotectant extracellular polysaccharide substances (EPS) by the marine psychrophilic bacterium Colwellia psychrerythraea strain 34H under extreme conditions This article is one of a selection of papers in the special issue on polar and alpine microbiology. Can J Microbiol 55(1):63–72. https://doi.org/10.1139/w08-130

Mata JA, Béjar V, Llamas I, Arias S, Bressollier P, Tallon R, Urdaci MC, Quesada E (2006) Exopolysaccharides produced by the recently described halophilic bacteria *Halomonas ventosae* and *Halomonas anticariensis*. Res Microbiol 157(9):827–835. https://doi.org/10.1016/j.resmic.2006.06.004

Maugeri TL, Gugliandolo C, Caccamo D, Panico A, Lama L, Gambacorta A, Nicolaus B (2002) A halophilic thermotolerant *Bacillus* isolated from a marine hot spring able to produce a new exopolysaccharide. Biotechnol Lett 24(7):515–519. https://doi.org/10.1023/a:1014891431233

McLean AL (1918) Bacteria of ice and snow in Antarctica. Nature 102:35–39. https://doi.org/10.1038/102035a0

Michel C, Bény C, Delorme F, Poirier L, Spolaore P, Morin D, d'Hugues P (2009) New protocol for the rapid quantification of exopolysaccharides in continuous culture systems of acidophilic bioleaching bacteria. Appl Microbiol Biotechnol 82(2):371. https://doi.org/10.1007/s00253-008-1824-4

Mishra A, Jha B (2013) Microbial exopolysaccharides. In: Rosenberg E, EF DL, Lory S, Stackebrandt E, Thompson F (eds) The prokaryotes: applied bacteriology and biotechnology. Springer, Berlin/Heidelberg, pp 179–192. https://doi.org/10.1007/978-3-642-31331-8_25

Morona R, Purins L, Tocilj A, Matte A, Cygler M (2009) Sequence-structure relationships in polysaccharide co-polymerase (PCP) proteins. Trends Biochem Sci 34(2):78–84. https://doi.org/10.1016/j.tibs.2008.11.001

Nichols CAM, Guezennec J, Bowman JP (2005a) Bacterial exopolysaccharides from extreme marine environments with special consideration of the southern ocean, sea ice, and deep-sea hydrothermal vents: a review. Mar Biotechnol 7(4):253–271. https://doi.org/10.1007/s10126-004-5118-2

Nichols CM, Bowman JP, Guezennec J (2005b) Effects of incubation temperature on growth and production of exopolysaccharides by an antarctic sea ice bacterium grown in batch culture. Appl Environ Microbiol 71(7):3519–3523. https://doi.org/10.1128/aem.71.7.3519-3523.2005

Nicolaus B, Kambourova M, Oner ET (2010) Exopolysaccharides from extremophiles: from fundamentals to biotechnology. Environ Technol 31(10):1145–1158. https://doi.org/10.1080/09593330903552094

Nicolaus B, Lama L, Panico A, Moriello VS, Romano I, Gambacorta A (2002) Production and characterization of exopolysaccharides excreted by thermophilic bacteria from shallow, marine hydrothermal vents of Flegrean Ares (Italy). Syst Appl Microbiol 25(3):319–325. https://doi.org/10.1078/0723-2020-00128

Nicolaus B, Manca MC, Ramano I, Lama L (1993) Production of an exopolysaccharide from two thermophilic archaea belonging to the genus *Sulfolobus*. FEMS Microbiol Lett 109(2–3):203–206. https://doi.org/10.1111/j.1574-6968.1993.tb06168.x

Nicolaus B, Moriello VS, Lama L, Poli A, Gambacorta A (2004) Polysaccharides from extremophilic microorganisms. Orig Life Evol Biosph 34(1):159–169. https://doi.org/10.1023/B:ORIG.0000009837.37412.d3

Nicolaus B, Moriello V, Maugeri T, Gugliandolo C, Gambacorta A (2003) *Bacilli* from shallow Mediterranean marine vents producers of exopolysaccharides. In: Recent Res Dev Microbiol 7:197–208. Research Signpost 37/661 (2), Fort P.O., Trivandrum, India

Nicolaus B, Panico A, Manca MC, Lama L, Gambacorta A, Maugeri T, Gugliandolo C, Caccamo D (2000) A thermophilic *Bacillus* isolated from an Eolian shallow hydrothermal vent able to produce exopolysaccharides. Syst Appl Microbiol 23(3):426–432. https://doi.org/10.1016/S0723-2020(00)80074-0

Nicolaus B, Poli A, Donato P, Romano I, Laezza G, Gioiello A, Ulgiati S, Fratianni F, Nazzaro F, Orlando P, Dumontet S (2016) Pb^{2+} effects on growth, lipids, and protein and DNA profiles of the thermophilic bacterium *Thermus thermophilus*. Microorganisms 4(4(4)):pii: E45. https://doi.org/10.3390/microorganisms4040045

Nwodo U, Green E, Okoh A (2012) Bacterial exopolysaccharides: functionality and prospects. Int J Mol Sci 13(11):14002

Oren A (2013) Life at high salt concentrations, intracellular KCl concentrations, and acidic proteomes. Front Microbiol 4:315. https://doi.org/10.3389/fmicb.2013.00315

Paramonov NA, Parolis LAS, Parolis H, Boán IF, Antón J, Rodrıguez-Valera F (1998) The structure of the exocellular polysaccharide produced by the Archaeon *Haloferax gibbonsii* (ATCC 33959). Carbohydr Res 309(1):89–94. https://doi.org/10.1016/S0008-6215(98)00102-5

Parolis H, Parolis LAS, Boán IF, Rodríguez-Valera F, Widmalm G, Manca MC, Jansson P-E, Sutherland IW (1996) The structure of the exopolysaccharide produced by the halophilic Archaeon *Haloferax mediterranei* strain R4 (ATCC 33500). Carbohydr Res 295(Supplement C):147–156. https://doi.org/10.1016/S0008-6215(96)90134-2

Poli A, Anzelmo G, Nicolaus B (2010) Bacterial exopolysaccharides from extreme marine habitats: production, characterization and biological activities. Mar Drugs 8(6):1779

Poli A, Esposito E, Orlando P, Lama L, Giordano A, de Appolonia F, Nicolaus B, Gambacorta A (2007) *Halomonas alkaliantarctica* sp. nov., isolated from saline lake Cape Russell in Antarctica, an alkalophilic moderately halophilic, exopolysaccharide-producing bacterium. Syst Appl Microbiol 30(1):31–38. https://doi.org/10.1016/j.syapm.2006.03.003

Poli A, Finore I, Romano I, Gioiello A, Lama L, Nicolaus B (2017) Microbial diversity in extreme marine habitats and their biomolecules. Microorganisms 5(2):25

Poli A, Kazak H, Gürleyendağ B, Tommonaro G, Pieretti G, Öner ET, Nicolaus B (2009) High level synthesis of Levan by a novel *Halomonas* species growing on defined media. Carbohydr Polym 78(4):651–657. https://doi.org/10.1016/j.carbpol.2009.05.031

Poli A, Nicolaus B, Denizci AA, Yavuzturk B, Kazan D (2013) *Halomonas smyrnensis* sp. nov., a moderately halophilic, exopolysaccharide-producing bacterium. Int J Syst Evol Microbiol 63(1):10–18. https://doi.org/10.1099/ijs.0.037036-0

Qin G, Zhu L, Chen X, Wang PG, Zhang Y (2007) Structural characterization and ecological roles of a novel exopolysaccharide from the deep-sea psychrotolerant bacterium *Pseudoalteromonas* sp. SM9913 Microbiology 153(5):1566–1572. https://doi.org/10.1099/mic.0.2006/003327-0

Qurashi AW, Sabri AN (2012) Bacterial exopolysaccharide and biofilm formation stimulate chickpea growth and soil aggregation under salt stress. Braz J Microbiol 43(3):1183–1191. https://doi.org/10.1590/s1517-83822012000300046

Radchenkova N, Vassilev S, Martinov M, Kuncheva M, Panchev I, Vlaev S, Kambourova M (2014) Optimization of the aeration and agitation speed of *Aeribacillus palidus* 418 exopolysaccharide production and the emulsifying properties of the product. Process Biochem 49(4):576–582. https://doi.org/10.1016/j.procbio.2014.01.010

Radchenkova N, Vassilev S, Panchev I, Anzelmo G, Tomova I, Nicolaus B, Kuncheva M, Petrov K, Kambourova M (2013) Production and properties of two novel exopolysaccharides synthesized by a thermophilic bacterium *Aeribacillus pallidus* 418. Appl Biochem Biotechnol 171(1):31–43. https://doi.org/10.1007/s12010-013-0348-2

Raddadi N, Cherif A, Daffonchio D, Neifar M, Fava F (2015) Biotechnological applications of extremophiles, extremozymes and extremolytes. Appl Microbiol Biotechnol 99(19):7907–7913. https://doi.org/10.1007/s00253-015-6874-9

Raguénès G, Christen R, Guezennec J, Pignet P, Barbier G (1997a) *Vibrio diabolicus* sp. nov., a new polysaccharide-secreting organism isolated from a Deep-Sea hydrothermal vent polychaete annelid, *Alvinella pompejana*. Int J Syst Evol Microbiol 47(4):989–995. https://doi.org/10.1099/00207713-47-4-989

Raguénès GHC, Peres A, Ruimy R, Pignet P, Christen R, Loaec M, Rougeaux H, Barbier G, Guezennec JG (1997b) *Alteromonas infernus* sp. nov., a new polysaccharide-producing bacterium isolated from a deep-sea hydrothermal vent. J Appl Microbiol 82(4):422–430. https://doi.org/10.1046/j.1365-2672.1997.00125.x

Rangaswamy BE, Vanitha KP, Hungund BS (2015) Microbial cellulose production from bacteria isolated from rotten fruit. Int J Polym Sci 2015:8. https://doi.org/10.1155/2015/280784

Raveendran S, Palaninathan V, Chauhan N, Sakamoto Y, Yoshida Y, Maekawa T, Mohanan PV, Kumar DS (2013a) In vitro evaluation of antioxidant defense mechanism and hemocompatibility of mauran. Carbohydr Polym 98(1):108–115. https://doi.org/10.1016/j.carbpol.2013.04.050

Raveendran S, Poulose AC, Yoshida Y, Maekawa T, Kumar DS (2013b) Bacterial exopolysaccharide based nanoparticles for sustained drug delivery, cancer chemotherapy and bioimaging. Carbohydr Polym 91(1):22–32. https://doi.org/10.1016/j.carbpol.2012.07.079

Rehm BHA (2010) Bacterial polymers: biosynthesis, modifications and applications. Nat Rev Microbiol 8(8):578–592

Rehm BHA (2015) Synthetic biology towards the synthesis of custom-made polysaccharides. Microb Biotechnol 8(1):19–20. https://doi.org/10.1111/1751-7915.12241

Rinker KD, Kelly RM (1996) Growth physiology of the hyperthermophilic archaeon *Thermococcus litoralis*: development of a sulfur-free defined medium, characterization of an exopolysaccharide, and evidence of biofilm formation. Appl Environ Microbiol 62(12):4478–4485

Rinker KD, Kelly RM (2000) Effect of carbon and nitrogen sources on growth dynamics and exopolysaccharide production for the hyperthermophilic archaeon *Thermococcus litoralis* and bacterium *Thermotoga maritima*. Biotechnol Bioeng 69(5):537–547. https://doi.org/10.1002/1097-0290(20000905)69:5<537::aid-bit8>3.0.co;2-7

Roberts IS (1996) The biochemistry and genetics of capsular polysaccharide production in bacteria. Annu Rev Microbiol 50(1):285–315. https://doi.org/10.1146/annurev.micro.50.1.285

Sam S, Kucukasik F, Yenigun O, Nicolaus B, Oner ET, Yukselen MA (2011) Flocculating performances of exopolysaccharides produced by a halophilic bacterial strain cultivated on agro-industrial waste. Bioresour Technol 102(2):1788–1794. https://doi.org/10.1016/j.biortech.2010.09.020

Sardari RRR, Kulcinskaja E, Ron EYC, Björnsdóttir S, Friðjónsson ÓH, Hreggviðsson GÓ, Karlsson EN (2017) Evaluation of the production of exopolysaccharides by two strains of the thermophilic bacterium *Rhodothermus marinus*. Carbohydr Polym 156(Supplement C):1–8. https://doi.org/10.1016/j.carbpol.2016.08.062

Satpute SK, Banat IM, Dhakephalkar PK, Banpurkar AG, Chopade BA (2010) Biosurfactants, bioemulsifiers and exopolysaccharides from marine microorganisms. Biotechnol Adv 28(4):436–450. https://doi.org/10.1016/j.biotechadv.2010.02.006

Schmid J, Sieber V, Rehm B (2015) Bacterial exopolysaccharides: biosynthesis pathways and engineering strategies. Front Microbiol 6:496. https://doi.org/10.3389/fmicb.2015.00496

Sezer AD, Kazak H, Öner ET, Akbuğa J (2011) Levan-based nanocarrier system for peptide and protein drug delivery: optimization and influence of experimental parameters on the nanoparticle characteristics. Carbohydr Polym 84(1):358–363. https://doi.org/10.1016/j.carbpol.2010.11.046

Shih I-L, Yu Y-T, Shieh C-J, Hsieh C-Y (2005) Selective production and characterization of Levan by *Bacillus subtilis* (Natto) Takahashi. J Agric Food Chem 53(21):8211–8215. https://doi.org/10.1021/jf058084o

Spanò A, Gugliandolo C, Lentini V, Maugeri TL, Anzelmo G, Poli A, Nicolaus B (2013) A novel EPS-producing strain of *Bacillus licheniformis* isolated from a shallow vent off Panarea Island (Italy). Curr Microbiol 67(1):21–29. https://doi.org/10.1007/s00284-013-0327-4

Squillaci G, Finamore R, Diana P, Restaino OF, Schiraldi C, Arbucci S, Ionata E, La Cara F, Morana A (2016) Production and properties of an exopolysaccharide synthesized by the extreme halophilic archaeon *Haloterrigena turkmenica*. Appl Microbiol Biotechnol 100(2):613–623. https://doi.org/10.1007/s00253-015-6991-5

Steinbüchel A (2001) Perspectives for biotechnological production and utilization of biopolymers: metabolic engineering of polyhydroxyalkanoate biosynthesis pathways as a successful example. Macromol Biosci 1(1):1–24. https://doi.org/10.1002/1616-5195(200101)1:1<1::aid-mabi1>3.0.co;2-b

Takai K, Nakamura K, Toki T, Tsunogai U, Miyazaki M, Miyazaki J, Hirayama H, Nakagawa S, Nunoura T, Horikoshi K (2008) Cell proliferation at 122°C and isotopically heavy CH(4) production by a hyperthermophilic methanogen under high-pressure cultivation. Proc Natl Acad Sci U S A 105(31):10949–10954. https://doi.org/10.1073/pnas.0712334105

Tian F, Karboune S, Hill A (2014) Synthesis of fructooligosaccharides and oligolevans by the combined use of levansucrase and endo-inulinase in one-step bi-enzymatic system. Innovative Food Sci Emerg Technol 22(Supplement C):230–238. https://doi.org/10.1016/j.ifset.2013.12.004

Tommonaro G, Iodice C, Abd El-Hady FK, Guerriero G, Pejin B (2015) The Mediterranean sponge *Dysidea avara* as a 40 year inspiration of marine natural product chemists. J Biodivers Endanger Species S1:001. https://doi.org/10.4172/2332-2543.10.4172/2332-2543.S1.001

van Kranenburg R, Boels IC, Kleerebezem M, de Vos WM (1999) Genetics and engineering of microbial exopolysaccharides for food: approaches for the production of existing and novel polysaccharides. Curr Opin Biotechnol 10(5):498–504. https://doi.org/10.1016/S0958-1669(99)00017-8

VanFossen AL, Lewis DL, Nichols JD, Kelly RM (2008) Polysaccharide degradation and synthesis by extremely thermophilic anaerobes. Ann N Y Acad Sci 1125(1):322–337. https://doi.org/10.1196/annals.1419.017

Vijayendra SVN, Shamala TR (2014) Film forming microbial biopolymers for commercial applications—a review. Crit Rev Biotechnol 34(4):338–357. https://doi.org/10.3109/07388551.2013.798254

Whitney JC, Howell PL (2013) Synthase-dependent exopolysaccharide secretion in gram-negative bacteria. Trends Microbiol 21(2):63–72. https://doi.org/10.1016/j.tim.2012.10.001

Wiercigroch E, Szafraniec E, Czamara K, Pacia MZ, Majzner K, Kochan K, Kaczor A, Baranska M, Malek K (2017) Raman and infrared spectroscopy of carbohydrates: a review. Spectrochim Acta A Mol Biomol Spectrosc 185(Supplement C):317–335. https://doi.org/10.1016/j.saa.2017.05.045

Wolfaardt GM, Lawrence JR, Korber DR (1999) Function of EPS. In: Wingender J, Neu TR, Flemming H-C (eds) Microbial extracellular polymeric substances: characterization, structure and function. Springer Berlin/Heidelberg, pp 171–200. https://doi.org/10.1007/978-3-642-60147-7_10

Yasar Yildiz S, Anzelmo G, Ozer T, Radchenkova N, Genc S, Di Donato P, Nicolaus B, Toksoy Oner E, Kambourova M (2014) Brevibacillus themoruber: a promising microbial cell factory for exopolysaccharide production. J Appl Microbiol 116(2):314–324. https://doi.org/10.1111/jam.12362

Zanchetta P, Lagarde N, Guezennec J (2003a) A new bone-healing material: a hyaluronic acid-like bacterial exopolysaccharide. Calcif Tissue Int 72(1):74–79. https://doi.org/10.1007/s00223-001-2091-x

Zanchetta P, Lagarde N, Guezennec J (2003b) Systemic effects on bone healing of a new hyaluronic acid-like bacterial exopolysaccharide. Calcif Tissue Int 73(3):232–236. https://doi.org/10.1007/s00223-002-2081-7

Zhao D, Cai L, Wu J, Li M, Liu H, Han J, Zhou J, Xiang H (2013) Improving polyhydroxyalkanoate production by knocking out the genes involved in exopolysaccharide biosynthesis in *Haloferax mediterranei*. Appl Microbiol Biotechnol 97(7):3027–3036. https://doi.org/10.1007/s00253-012-4415-3

16 Why Settle for Mediocre, When Extremophiles Exist?

Shivanshi Vashist and Rohit Sharma

Abstract

The ever-increasing uses of microorganisms and enzymes in the food, medical, pharmaceutical, detergent, leather, and textile industries has triggered a great amount of research in "extreme" enzymology. In areas of research that are based on solving environmental problems, by methods such as bioremediation, considerable attention has been paid to enzymes/microorganisms that can survive in extreme environments. Such entities include thermostable and organic solvent-tolerating microorganisms/enzymes. The study of enzymes (such as amylases, proteases, lipases, and nitrilases) that can tolerate high organic solvent concentrations has revolutionized the way science and industry work together and evolve. Organic solvent-rich environments provide an edge with respect to enzyme behavior and applications as compared with aqueous environments. These behavioral attributes in organic solvent-rich environments include thermal stability, a positive shift in the thermodynamic equilibrium, simple removal of solvent from the system, and enhanced enantio-recognition and stereo-stability. Non-aqueous biocatalysis is a key area of research that has led us in various directions through the exploration of the stated properties of such enzymes. The applications of non-aqueous biocatalysis include the biocatalytic synthesis of cardiovascular drugs and anti-inflammatory agents, the resolution of racemic acids and alcohols, and fatty acid ester synthesis.

S. Vashist · R. Sharma (✉)
Department of Microbial Biotechnology, Panjab University, Chandigarh, India

Cluster Innovation Centre in Biotechnology, Panjab University, Chandigarh, India

BioNEST- Panjab University, Chandigarh, India
e-mail: rohit28@pu.ac.in

D. Egamberdieva et al. (eds.), *Extremophiles in Eurasian Ecosystems: Ecology, Diversity, and Applications*, Microorganisms for Sustainability 8,
https://doi.org/10.1007/978-981-13-0329-6_16

This chapter narrates the journey of these extremists—these solvent-tolerant microorganisms/enzymes—from the initial need for their identification to their multifarious applications in solving environmental, industrial, and biotechnological issues.

Keywords
Organic solvent tolerance · Biocatalysis · Organic synthesis · Bioremediation

16.1 Introduction

This chapter outlines the present and forecasted applications of extreme environment-tolerating microorganisms and their enzymes in the field of biotechnology, particularly in the production and processing of chemicals in textiles, biocatalytic synthesis, and agriculture. Microbial biotechnology can offer environmental benefits, process efficiency, product quality, and economic benefits in the production of high value commercials, along with bio-eco-friendly waste management. Microbial biotechnology/engineering assists in the invention/discovery of sustainable technologies that offer a number of process and market benefits.

The sustainable production of existing and novel products is a major advantage of microbial biotechnology, and this technology also offers reduced dependence on nonrenewable fuels and other resources, improving the economics of production. Since the 1970s, biotechnology, particularly enzymology, has substantially affected healthcare and pharmaceuticals; food and agriculture; environmental protection; and the production of various materials and chemicals. "White biotechnology" was primarily dependent on aqueous enzymology until enzymes that could withstand organic and biphasic environments were explored. Organic solvent-rich environments, which are lethal for most organisms/enzymes, enhance hydrolytic activity in a few useful extremophiles. The study of enzymes that can tolerate high organic solvent concentrations has revolutionized the way science and industry work together and evolve. Organic solvent-rich environments provide an edge with respect to enzyme behavior and applications as compared with aqueous environments. These behavioral attributes in organic solvent-rich environments include thermal stability, a positive shift in thermodynamic equilibrium, simple removal of solvent from the system, and enhanced enantio-recognition and stereo-stability. Non-aqueous biocatalysis is a key area of research that has led us in various directions through the exploration of the stated properties of such enzymes. The applications of non-aqueous biocatalysis include the biocatalytic synthesis of cardiovascular drugs and anti-inflammatory agents, the resolution of racemic acids and alcohols, and fatty acid ester synthesis (Martinkova et al. 2017).

Interest in exploring the benign applications of enzymes was promoted by the successful production of nicotinic acid (Lonza, Guangzhou, China) and (*R*)-(−)-mandelic acid (Mitsubishi Rayon, Tokyo, Japan; BASF, Ludwigshafen, Germany) and this has motivated multiple studies in the field to take the findings from the laboratory to scaled-up products/processes in the market.

16.2 Adequate Process Design

Traditionally, primary screening for microorganism/s producing extreme environment-tolerating enzyme/s was done in a selective enrichment medium. But it takes more than screening to take the product/enzyme to the market. In modern chemistry, a primary established mainstream discipline, various non-organic catalysts are used, but the industry is now embracing green catalysis (also known as biotransformation). This is the biochemical conversion/alteration of a substance by the use of biocatalysts, with a compound being transformed into a relevant form via the action of biological agents. The biological agents can be plants, plant products, microorganisms, or microbial products.

With an effect on the global economy, enzymes such as nitrilases now have proven prominence in the field of green chemistry, with an advantageous ratio of waste produced and product obtained (Gong et al. 2012). The major requirement is to develop/discover shorter/smarter/faster alternatives to the conventional methods to produce relevant products. Therefore, research is now directed toward green energy.

16.3 Biological Conversion of Hazardous Compounds

The hazardous nature of compounds in the environment such as nitriles and cyanide has led to the exploration of nitrile-hydrolyzing enzymes by industries and by enzymologists, who are investigating biotransformation and the widespread applications of biologically transformed hazardous product/s. Various studies have reported the increased research in the area (e.g., Ludmila and Křen 2010; Gong et al. 2012). There are multiple approaches for the conversion of an undesirable product into a desirable one. With chemical (acid-catalyzed or base-catalyzed) hydrolysis posing environmental and economic threats, biological processes have gained importance and present a better solution.

16.4 Industrial Tenability

The term ‘industrial tenability’ relates to the achievement of sustainability with regard to production, processing, and economic landmarks. Bioprocessing and biotechnology engineering, as a fourth discipline, along with genetic engineering, protein engineering, and metabolic engineering, is required for the commercial production of biotechnology products and their delivery. The industrial chemical method of production and commercialization relies excessively on nonrenewable energy and resources and on environmentally damaging production processes that can be unsafe. These processes produce toxic products and waste and products that are not readily recyclable and degradable after their useful life. Further, the industrial chemical method involves excessive regional concentration of production so that the social benefits of production are limited.

16.5 Advantages of Organic Solvents Over Aqueous Media for Biotransformation

Apart from their action in eliminating microbial contamination, which is an underrated advantage, the organic solvent-tolerating enzymes that assist in efficient biotransformation present a number of other advantages. These advantages, compared with processes in aqueous media, include better solubility of the substrate and/or product, the shifting of thermodynamic equilibrium toward the synthesis of the product, simpler removal of solvent (most organic solvents have lower boiling points than that of water), and a reduction in the water-dependent side of the reaction (hydrolysis of acid anhydrates). Furthermore, the organic solvents offer better enzyme thermal stability, since aqueous media inactivate enzymes at higher temperatures.

A new study has presented ionic liquids as an alternative to organic solvents, with these liquids overcoming the problem of the volatile nature of organic solvents. These designer solvents are environmentally friendly and have a defined set of properties. The properties and aspects of ionic liquids in catalytic organic transformation are well discussed by Vekariya (2017).

As well as the class of ionic liquids, there is another important and promising solvent class: Bronsted acidic ionic liquids (BAILS). This newly studied, versatile, and fast-evolving category is employed in various essential organic reactions, such as hydrogenation and dehydrogenation, oxidation, transesterification, esterification, and alkylation. The great potential of BAILs in various reactions has led to what has been called a new era in acid-catalyzed transformation (Vafaeezadeh and Alinezhad 2016).

With particular focus on biomass conversion processes, alkyl phenols have also managed to enter the organic solvent race (Jérôme et al. 2017). β-Sitosterol, cholesterol, and campesterol are some of the steroids that can be converted into industrially important compounds by *Mycobacterium* sp. (de Carvalho 2017). The introduction of an organic phase, in contrast to water, limits the possible slow solubility of steroids. The employment of an organic solvent instead of aqueous media in the biocatalysis of β-sitosterol, 4 androstadiene-3, 17 (AD), and 1,4 androstadiene-3, 17 (ADD) leads to better chances of exploring potentially useful enzymes (de Carvalho 2017).

Better yields, such as that seen in the increased production of ethyl lactate through enzyme esterification in green solvents, have involved the use of heterologous *Rhizopus oryzae* and *Candida rugosa*, which retain high activity in organic solvents such as chloroform (Koutinasa et al. 2018). Another cost-effective method was the immobilization of a lipase from *C. rugosa*; its reusable stability was confirmed, and its esterification activity was retained for up to 60% of its maximum activity after five reuse cycles in solvents such as cyclohexane (V_{max} = 26.4 mMol/min) (Lidija et al. 2017).

Reports on organic solvents suggest that the appropriate optimization of the type of enzyme, e.g., lipase, and the use of an appropriate reaction medium lead to an enantiomerically pure product (Sikora et al. 2017). Lipo amino acids were synthesized using enzymes from *Pseudomonas stutzeri* (lipase) and *Bacillus subtilis* (protease) and the selectivity and synthesis were dependent upon the reaction conditions (Bernal et al. 2018). Low-cost mass cultivation of organic solvent-tolerant enzymes in *Pseudomonas sp.* BCNU 106 in toluene supplemented with glycerol can be employed in biotransformation and biodegradation (Choi et al. 2017).

16.6 Applications of Organic Solvent-Tolerant Enzymes

Basic chemicals or commodity chemicals include the products of industries that are generally involved in processing applications (e.g., pulp and paper, oil refining, metal recovery), as well as the raw materials used for producing other basic chemicals, specialty chemicals, and consumer products, including manufactured goods.

Specialty chemicals are derived from basic chemicals, but are more technologically advanced and are used in lower volumes than the basic chemicals. Examples of specialty chemicals include adhesives and sealants, catalysts, coatings, and plastic additives. Specialty chemicals command higher profit margins and their demand is less cyclic than that of basic chemicals. Specialty chemicals have a higher value-added component because they are not easily duplicated by other producers or are protected from competition by patents. Consumer-care products, including soaps, detergents, bleaches, laundry aids, hair care products, skin care products, and fragrances, are one of the oldest segments of the chemical industry. Other products include pharmaceuticals, products for crop protection, and products of modern biotechnology.

C. cylindracea lipase, a thermostable lipase with potential for use in the oleochemical industry for soap production, has been studied and reported on. Processes such as hydrolysis and glycolysis are catalyzed by this enzyme. Similarly, lipases from bacteria such as *Pseudomonas thermomyces* sp. have been used in detergent making (Choudhary and Bhunia 2015).

Other microorganisms, especially fungi such as *Aspergillus niger*, *C. antarctica*, *C. rugosa, C. viscosum, Mucor miehei, P. fluorescens, P. cepacia, P. lypolyticum, and Thermomyces lanuginosus* have been well studied in the production of biodiesel (Choudhary and Bhunia 2015). A novel organo solvent-tolerant esterase, from *Monascus purpureus* strain M7, retained 99%–110% relative activity (minimum 20%) in hydrophilic organic solvents such as methanol and ethanol (Kang et al. 2017). As well as bacteria and fungi, microalgae are potential candidates for use in molecular- to industrial-scale biocatalysis (Miazek et al. 2017). In the past decade there has been an exponential increase in microbiologists' interest in studying the potential of microalgae for biocatalysis; this can be ascertained by the huge number of reports in the field, covering diverse microalgae (Bayat et al. 2015; Hunt et al. 2010). The target product range is wide, ranging from lipids to pigments. The microalgae studied tolerate high concentrations of organic solvents such as ethylene glycol, benzene, xylene, acetaldehyde, chloroform, waste organic solvents, and ionic liquids. The microalgae studied include *Chlorella minutissima, Chlamydomonas reinhardtii, Chlorella sorokiniana, Euglena gracilis, Botryococcus braunii*, and *Dunaliella tertiolecta*; the growth of these was promoted by methanol (Miazek et al. 2017). Active aggregates of an organic solvent-tolerant lipase from *Marinobacter sp.* EMB5 have also been reported (Hemamalini and Khare 2016); this lipase from this halophilic bacteria is stable for long incubation periods in organic solvents. These studies are usually linked with studies of the bioremediation of undesired and/or toxic material.

The applications of the extreme microorganisms noted above, and their enzymes and products, are well established, as evidenced by diverse reports from all over the globe (Fernandes et al. 2003; Li et al. 1998). There are also reports and discussions on the mechanisms underlying the causes/effects of the microorganisms' actions. Manefield et al. (2017) discuss mechanisms such as efflux pumps in bacterial resistance to antimicrobial compounds, in terms of organic solvent-tolerance. Other mechanisms of microorganisms' resistance to antimicrobial compounds include biofilm formation, motility, and the formation of endospores.

Pan-genome studies of *P. putida,* a microbe generally recognized as organic solvent-tolerant, have revealed 30% of genes belong to *Pseudomonas.* A highly organic solvent-tolerant *Pseudomonas* strain, dot-t1e, has also been identified (Molina-Santiago and Udaondo 2017). With biofilm formation as one of the "favorite" mechanisms of organic solvent-tolerant microorganisms, biomass quantification of *P. taiwanensis* VLB120ΔC biofilm was done in the presence of n-butanol; this study showed a robust organism capable of tolerating and adapting to increased concentrations of reactants and products that can be toxic to microorganisms (Halan 2017). In another study, an ethylene glycol-tolerant lignolytic ascomycete strain, *Pseudo Cochliobolus verruculosa* NFCCI3818, was investigated for its utility in waste management (Nikama et al. 2017).

In biotechnology, biocatalysis and metabolic engineering are the two fast-evolving fields that have the potential to replace and drive transformation in the conventional chemical industry (Martinkova et al. 2017). Genetic engineering and molecular biology techniques have been used to obtain many modified enzymes with enhanced properties compared with their natural counterparts. Some well established biotechnology products include bioethanol, L-glutamic acid (MSG), citric acid, L-lysine, lactic acid, food-processing enzymes, vitamin C, gluconic acid, antibiotics, feed enzymes, xanthan, L-threonine, L-dihydroxyphenylalanine, 6-aminopenicillanic acid, nicotinamide, D-p-hydroxyphenylglycine, vitamin F, 7-aminocephalosporanic acid, aspartame, L-methionine, dextran, vitamin B12, and provitamin D2.

16.7 Enzymatic Processes

Enzymes are being increasingly used in the chemical industry as catalysts for numerous reactions. The global microbial identification market alone is estimated to reach 1194 million $US by 2019 (de Carvalho 2017). Millions of years of evolution have provided enzymes with an unparalleled capacity for facilitating life reactions in ways that are sustainable. Compared with conventional chemical catalysis, enzyme catalysis is highly specific and it functions under temperatures, pressures, and pH conditions that are compatible with life. Unlike many processes in conventional synthetic chemistry, enzymes require nontoxic and noncorrosive conditions.

About 75% of enzyme use by value is accounted for by the detergent, food, and starch-processing industries. These industries mostly use hydrolytic enzymes such as proteases, amylases, lipases, and cellulases. Specialty enzymes account for around 10% of the enzyme market and they are finding numerous analytical uses, as well as increasing uses in the development of new drugs and medical diagnostics. Modern biotechnology has contributed to more than 60% of commercialized products and/or enzymes such as Novozymes' Cellic®, Shire's Velaglucerasealfa VPRIV™, and Taliglucerasealfa Elelyso™ (Li et al. 2012).

Some industrial enzymes and their various substrates include proteases-proteins, carbohydrases-carbohydrates, lipases-fats and oils, amylases-polysaccharides, cellulases-cellulose, pectinases-pectin, and nitrilases-nitriles. The reactions include proteolysis, hydrolysis of carbohydrates to sugars, hydrolysis of fats to fatty acids and glycerol, hydrolysis of pectin, hydrolysis of cellulose, hydrolysis of starch to sugars, and hydrolysis of hazardous nitriles to high-value commercial products (Martinkova et al. 2017).

The industries that primarily require/explore such enzymes include the detergent, food, pharmaceutical, synthetic food, feed, pulp and paper, sugar, and textile industries. Analytical applications include the development of enzymes for the production, degradation, and biotransformation of chemicals, foods and feeds, agricultural produce, and textiles. For example, isomalto-oligosaccharides, produced using glucosyl transferases, are used to suppress tooth decay and prevent baked goods from becoming stale; cellulases, which synergistically break down cellulose, are used because of their potential for providing fuel, food, and other chemicals from widely available cellulose. Enzymes such as amylases and proteases are added to animal feed to improve digestibility by supplementing the animals' own enzymes. The addition of enzymes such as beta-glucanases and arabino-xylanase to feed cereals breaks down non-starch polysaccharide anti-nutritional factors, aiding the digestion and absorption of nutrients.

16.8 Pharmaceuticals: Exploring Biotransformation

Pharmaceutics, chiral intermediates, enantiomers, and precursors are some of the terms used by the pharmaceutical industry today to describe their products. Many pharmaceutical companies adopt chemical methods for the synthesis of chiral intermediate, enantiomer, and precursor compounds such as α-hydroxy acids, α-hydroxyl amides, α- and β-amino acids, and mono/di acids. These chemical methods, apart from being environmentally unfriendly, raise global issues when employed on a large scale. They also lead to reduced overall product yields because of the formation of nonspecific and unwanted by-products. The process is expensive owing to the addition of chemical substances for enhancing enantiomeric selectivity; in contrast, enantiomeric selectivity is naturally provided by some microorganisms. Some shrewder manufacturers in the field employ smart microorganisms that show "extremophilicity" in more than one aspect, be it organo-solvent tolerance, thermophilicity, halophilicity, or alkalophilicity.

These microorganisms present the most desirable trait in the industry; that is, enantio-selectivity and/or enantio-retentivity. These traits reduce the downstream processing cost, bringing down the overall cost of the process by selecting and/or retaining the wanted enantiomer. These factors help the industry not only in regard to reducing costs, but also in regard to overcoming the regulatory pressures that every pharmaceutical company faces today. Justifying the regulatory pressure, the final product must contain the active pharmacological enantiomer of the desired compound and not the racemic mixture. The complex key intermediates can be synthesized in an environmentally friendly, cost-effective manner; as noted previously, such synthesis has been exemplified by the successful production of nicotinic acid (Lonza) and (*R*)-(−)-mandelic acid (Mitsubishi Rayon; BASF). This success has motivated multiple studies in the field to take the laboratory findings to a scaled-up product/process in the market on a large scale (Yamada and Kobayashi 2014).

This section presents some microorganisms recently used in studies across the globe; the enzymes from these organisms include lipases, nitrilases, nitrile hydrolases, amidases, and laccases. Different *Aspergillus* spp. secrete lipases that show multiple characteristics of extremophilicity, including thermostability, organic solvent tolerance, enantio-selectivity, and pH stability (Contesini et al. 2016).

Of note, Li et al. (2017) have reported the use of structure-guided saturation mutagenesis to produce high-quality mutant libraries. Also, other authors have discussed examples of stereoselectivity in enzymes overcoming the distinct traditional limitations of the processes (Maksimova et al. 2017; Mazmouza et al. 2018). The asymmetrical synthesis of chiral intermediates has now reached a point of resolution as a result of these studies. Gurung et al. (2013) have reported that lipases from *Candida rugosa* synthesize lovastatin, a drug that lowers serum cholesterol level. Lipase from *Serratia marcescens* has been reported to asymmetrically hydrolyse trans-3-phenylglycidic acid ester, the key intermediate in the synthesis of diltiazem hydrochloride (Matsumae et al. 1993; Singh and Banerjee 2005).

Sun et al. (2018) have described reductases, oxidases, hydrolases, lyases, isomerases, and transaminases in relation to expression of enzyme activity, specificity, thermostability, and solvent-tolerance. For example, reduction of 4-oxo-4-[3-(trifluoromethyl)-5,6-dihydro-[1,2,4] triazolo [4,3-a] pyrazin-7(8H)-yl]-1-(2,4,5-trifluorophenyl)butan-2-one (OTPP) by *Pseudomonas pseudoalcaligenes* XW-40 (Wei et al. 2016). The challenging nature of the enzymatic synthesis of complex natural compounds, such as by smart single-step conversion followed by cascade reactions, has been highlighted in a study by Classen and Pietruszka (2017). Another class of enzymes, the nitrilases, have proven to be valuable for their potential use in the biotransformation of various hazardous nitrile compounds to useful intermediates and corresponding carboxylic acid, for example, acrylonitrile to acrylic acid, etc. (Sharma and Vashist 2017). One of the most inspiring success stories in this regard is the biosynthesis of acrylamide using nitrilase on a commercial scale. Some nitrilases have also been successfully applied to practical production in food industries, chemical manufacturing, pharmaceutical processes, wastewater treatment, and textile industries.

From the production of (R)-mandelic acid from (R,S)-mandelonitrile through *Aspergillus niger* reported by Vesela et al. (2015) and the production of (R)-mandelic acid through *Burkholderia cenocepacia* reported by Ni et al. (2013) and Wang et al. (2015a), there is evidence that establishes nitrilases as promising biocatalysts. Other examples include the production of (R)-o-chloromandelic acid from (R,S)-o-chloro-mandelonitrile, using *Burkholderia cenocepacia* (nitrilase mutant I113M/Y199G) (Wang et al. 2015b), the production of (R)-phenylglycine from (R,S)-2-amino-2-phenylacetonitrile, using *Sphingomonas wittichii* (Qiu et al. 2014a, b), and the production of β-alanine from 3-aminopropionitrile, using *Bradyrhizobium japonicum* (Han et al. 2015). Further examples include the production of 1-cyanocyclohexylacetic acid, (s)-2-cyano-2-methylpentanoic acid, and iminodiacetic acid from 1-cyanocyclohexylacetonitrile (Xue et al. 2015), 2-methyl-2-propylmalononitrile (Yoshida et al. 2013), and iminodiacetonitrile (Cai et al. 2014; Liu et al. 2013), respectively, using the microorganisms *Acidovorax facilis* (nitrilase mutant F168V), *Rhodococcus rhodochrous*, *Arthrobacter aurescens*, and *Acidovorax facilis* (nitrilase mutant F168V/L201N/S192F).

Apart from these examples, glucose isomerases have been employed in the pharmaceutical industry to convert aldoses and ketoses from *Streptomyces rubiginosus*, and their crystal structures have been elucidated by Eun Bae et al. (2017). With potential applications in l-ribose production, Tseng et al. (2017) studied the overproduction and characterization of a recombinant l-ribose isomerase from *Actinotalea fermentans* ATCC 43279.

Martínez et al. (2017) showed that oxidoreductases were potential candidates for use in biotransformation, reporting reactions such as 1 naphthol, 2,5-hydroxyvitamin D3 drug metabolism catalyzed by peroxygenases, copper oxidases, and laccases, hence elucidating the characteristics of peroxidases from fungi, including Basidiomycetes, along with their limitations. Truppo (2017), in a study that showed a dramatic increase in protein engineering, reported excellent multiple contact of enzymes with substrates, with increased selectivity.

The development of biocatalysts that, in comparison with chemical catalysts, are faster, less expensive, and more versatile in their selection and preference for substrates, that can catalyze an increased range of reactions, and that have higher temperature stability and improved solvent compatibility is promising for the sustainability of various products/processes in the market today.

16.9 Agricultural Chemicals

Agricultural chemicals, mainly fertilizers and pesticides, are used in massive amounts worldwide to sustain the productivity of land. Because of their widespread use, agrochemicals are an important source of pollution, pose health risks, and consume large amounts of resources in their production. Enzymology can present useful products that can replace conventional agrochemicals and the methods used to degrade agrochemicals (Malik et al. 2017). In addition, biotechnology can provide animal feed with enhanced nutritional and storage characteristics, to improve the sustainability of animal production.

16.10 Fiber, Textiles, Pulp, and Paper Processing and Other Applications

Through biotechnology and improved silviculture, trees and other bioresources used in papermaking can be specifically tailored to match the properties required in cellulose fibers for different product applications, thus showing potential to increase the paper yield and product quality. Producing optimal fibers for papermaking through genetic engineering is an important long-term objective that requires a better understanding than we have at present of fiber biosynthesis in plants. Furthermore, the use of engineered microorganisms and enzymes can displace many of the environmentally adverse practices used in pulp processing. Some of these developments are discussed next.

In kraft pulping, bleaching of the pulp remains one of the most expensive operations and is a prime target for cost reduction. Because of the polluting potential of chlorine bleach, pulp mills are mostly changing to bleaching methods that do not require elemental chlorine. The use of low-molecular weight xylanase from *Trichoderma viride* VKF3 has recently been reported for the bio-bleaching of newspaper pulp (Nathan et al. 2017). Oxidative enzymes such as laccase provide other promising options for reducing costs in pulp mills. Other processing improvements have been achieved by using lipases to control pitch deposits; cellulases to improve rates of pulp dewatering; and pectinases for digesting pectins. Ongoing developments will provide engineered enzymes that are better suited to the needs of pulp processing and cost less than the enzymes used at present. In future, it may be possible to manufacture unique paper products by developing enzymes that can be used to control the properties of the pulp fibers and, therefore, the end product.

The production of paper consumes huge amounts of water. Extensive research is underway for the treatment of wastewater from paper mills, the aim being total recycling. Pulp and paper mills in Canada are aiming for total effluent reuse after secondary and tertiary biotreatment. Wastewater recycling potentially saves on the expense of treating any freshwater entering the mill and greatly reduces the environmental impact of effluent disposal.

In the processing of textiles, cellulose pulp is usually bleached with hydrogen peroxide, which must be removed before the fibers are colored. The traditional method for the removal of hydrogen peroxide relied on extensive washing in hot water and the use of inorganic salts. The enzymatic process saves water and energy and the effluent is ecologically harmless. Of note, *Aspergillus oryzae* lipase is capable of modifying polyethylene terephthalate fabrics, improving their hydrophilicity and anti-static capacity, while the immobilization of porcine-pancreas lipase on zirconia-coated alkylamine glass beads by glutaraldehyde coupling improved washing properties in cotton cloth. Fungi such as *M. miehei* has been reported for esterification in the presence of pentane (Bloomer et al. 1992). Other fungi, such as *C. rugosa*, *Penicillium roqueforti*, and *Humicola lanuginosa* have the ability to grow in medium supplemented with organic solvents such as cyclohexane and hexane. These fungi have industrial applications in the field of organic synthesis as well as in the textile industry (Mehta et al. 2017).

Enzyme options in the textile industry range from lipases to amylases. Lipases, together with alpha amylase, are used for the desizing of denim and other cotton fabrics on a commercial scale. Nippon Paper Industries (Tokyo, Japan) developed a pitch control method that used a fungal lipase from *C. rugosa* to hydrolyse up to 90% of wood triglycerides. *Rhizomucor meihei* lipase is used as a biocatalyst in personal care products such as skin and sun-tan creams and bath oils. *C. antarctica* lipase B-synthesized amphiphilic compounds are important in the cosmetic industry as they have a range of beneficial properties for the skin. The lipase component increases detergency and prevents scaling. Recently, lipase from *Rhizopus nigricans* showed maximum lipolytic activity, as well as bio-emulsification activity, indicating high bio-surfactant production in kerosene A lipase obtained from *C. cylindracea* considerably reduced pitch problems and talc consumption of triglyceride in groundwood pulp. *C. antarctica* lipase A was also used in pitch control in the paper industry (Mehta et al. 2017).

Bacillus subtilis is one of the most widely used bacteria for the production of industrial enzymes. *Bacillus* spp., especially *B. subtilis* and *B. licheniformis*, are the sources of most extracellular proteases (Kamal et al. 2016). These enzymes sourced from *B. subtilis* (6381.75 U/mg), *B. altitudinis* (MCCB0014) (7407.5 U/mg), *B. circulans* MTCC 7906 (3147.33 U/mL), and *B. alcalophilus* ATCC 21522 (18,000 U/mg) were reported to exhibit high activity (Kamal et al. 2016).

Enzymes such as alcalase and savinase from *B. licheniformis* and other *Bacillus* spp. are also used in the detergent industry and the textile industry. Enzymatic degradation using alkaline proteases with keratinolytic activity (keratinases) is an attractive method for hydrolysis of proteins and keratins and also helps to reduce the biological oxygen demand (BOD) for aquatic macro and micro flora. *Bacillus* spp. are extensively reported as the bacterial source of keratinases for the degradation of feathers (Kamal et al. 2016). A novel *Chryseobacterium* sp. was screened for cold-active protease production in the presence of a high concentration of NaCl, and its tolerance to several organic solvents, surfactants, and detergents was reported. Classical optimization for enhanced protease production, of 18 U/mg to 26 U/mg, was studied and reported (Mageswari et al. 2017). A protease from *B. licheniformis* K-3 showed remarkable tolerance to detergents such as cetrimonium bromide, sodium dodecyl sulfate, and Tween-20, suggesting its industrial applications for the de-gelatinization of X-ray films and the dehairing of animal hide (Singh and Bajaj 2017). Studies also suggest the application of a thermostable and pH-stable protease from *B. licheniformis* K-3, using agroindustrial/forestry residues as an inexpensive substrate for cost-effective enzyme production. A serine protease from a newly isolated *Bacillus* sp. was reported to show efficient silk-degumming, sericin-degrading, and color-bleaching activities (Suwannaphana et al. 2017).

Cellulose, hemicellulose, pectin (carbohydrate), and lignin (noncarbohydrate) polymers are the main substrates of lignocellulose-degrading enzymes. These polymers are present in large amounts in the primary cell walls and dietary fibers of major fruits and vegetables. During the processing of fruits and vegetables to the corresponding final food products, lignocellulosic substrates are hydrolyzed by different lignocellulolytic enzymes (Toushik et al. 2017).

The biological treatment of textile wastewater varies widely, ranging from bacterial, fungal culture (*Armillaria* sp. F022), yeast to any consortia. Reports of the use of enzymes in textile wastewater treatment started in 1970 with the isolation of three microbial strains, viz. *B. subtilis, Aeromonas hydrophila*, and *B. cereus*. A wide range of aerobic and anaerobic bacteria, such as *Pseudomonas sp.*, *B. subtilis*, *Geobacillus sp.*, *Escherichia coli*, *Rhabdobacter sp.*, *Enterococcus sp.*, *Staphylococcus sp.*, *Corneybacterium sp.*, *Lactobacillus sp.*, *Xenophilus sp.*, *Clostridium sp.*, *Acinetobacter sp.*, *Micrococcus sp.*, *Dermacoccus sp.*, *Rhizobium sp.*, *Proteus sp.*, *Morganella sp.*, *Aeromonas sp.*, *Alcaligenes sp.*, *Klebsiellla sp.*, *Shewanella sp.*, and *Alishewanella sp.*, have been extensively reported to show good, nonspecific biodegradation of azo dyes. *Pseudomonas sp.* is widely used in decolorization studies because of its capacity to degrade a variety of azo dyes (Red HE7B, Reactive Blue 172, Reactive Red 22, Reactive Red 2, and orange I and II). *Pseudomonas sp.* has shown its potential for the degradation of commercial azo dyes used in textile wastewaters (Sarkar et al. 2017). The use of microbial enzymes in the degradation of synthetic azo dyes in the textile industry is a sustainable methodology that can be employed by industry on a large scale. An alkali-tolerant EG gene of *B. subtilis* Y106 was homologously overexpressed to obtain a suitable enzyme for pulp modification (Wang et al. 2017). For eliminating textile waste from the environment, the co-plantation of *Typha angustifolia* and *Paspalum scrobiculatum* has shown enhanced removal of dye such as Congo red from textile effluent (Chandanshive et al. 2017).

Enzymes have been strongly accepted as a green alternative for use in many textile processes. These biocatalysts are not consumed and immobilization has been adopted as the most promising tool for their efficient recovery and reuse. Smart polymers and nanoparticle materials have been used for textile applications (Madhu and Chakraborty 2017).

In regard to efficient process techniques, the immobilization of lignin-modifying enzymes (LMEs), including lignin peroxidase, manganese peroxidase, and white-rot fungi laccase, has also been studied. The successful use of immobilized LMEs in the decolorization and/or detoxification of industrial dyes and dye-based industrial wastewater effluents has also been reported (Bilal et al. 2017).

16.11 Environmental Biotransformation and Bioremediation

Historically, the treatment of municipal wastewater by the activated sludge method has represented a major use of microorganisms in environmental care and bioremediation applications. The use of microbial extremozymes has made its mark in the field of environmental biotransformation and biodegradation, with a long history of applications. From the mid-1990s to 2017, many applications and reports have suggested the use of microorganisms in bioremediation and degradation—from environmentally friendly biotransformation in the pharmaceutical industry to the employment of microbial enzymes such as lipases in biosensors for the detection of specific pollutant levels; these applications have shown high efficiency, with wide diversity in the field, including the removal of nitrates and phosphates from wastewaters. A very interesting study at the laboratory-scale batch level used *B. cereus* AKG1 and AKG2 to treat wastewater, investigating BOD, chemical oxygen demand, and total organic carbon (Nikama et al. 2017). Hyper phenol-tolerant microorganisms from oil refineries and oil exploration sites were investigated for their potential to biotransform phenol by Sarkar et al. (2017). Biodegradation and detoxification of dyes is also possible through consotia of *Providencia rettgeri* strain HSL1 and *Pseudomonas* sp. SUK1 (Lade et al. 2015). *Pseudo Cochliobolus verrucosus* NFCC 3818, an ethyl glycol-tolerant lignolytic *Ascomycete* strain, has shown capacity for the detoxification and degradation of azo dyes.

Enhanced solvent tolerance of a psychrophilic phthalate esterase in an arctic bacterium, *Sphingomonas glacialis* PAMC 26605, was seen after the cloning and characterization of this esterase (Hong et al. 2017). For soil bioremediation, bacterial associations with plants have been observed for *Azotobacter* and *Lepidium sativum*, with tolerance for heavy metals being observed (Sobariu et al. 2017).

16.12 Conclusion

Economic and biotechnological benefits with respect to cost, productivity, reduction of environmental hazards, and sustainability are the well reported and evident advantages of enzymology. In the production of critical and chiral molecules, microbial enzymes have unrivalled precision, owing to their enantio-, regio-, and substrate selectivity, and this selectivity has supported the use of nitrile-hydrolyzing enzymes in industry today (Xue et al. 2015). Industrial-friendly enzyme-producing organisms, including bacteria, filamentous fungi, yeasts, and plants have been well studied for their possible use in the commercial production of carboxylic acids and amides on an industrial scale. Key properties of such organisms, such as enantioselectivity and enantio-retentivity, come with supporting traits such as thermostability, halostability, pH stability, and organic solvent tolerance. These characteristics occur in various microbial enzymes that assist in the production of specific molecules at lower cost and better yield than what is seen with conventional methods. The use of these enzymes has reduced hazardous impacts on the environment. It is time for us to focus on the commercial-scale production of such enzymes and their products in order to increase the overall bioeconomy.

References

Bayat M, Mehrnia MR, Hosseinzadeh M, Sheikh-Sofla R (2015) Petrochemical wastewater treatment and reuse by MBR: a pilot study for ethylene oxide/ethylene glycol and olefin units. J Ind Eng Chem 25:265–271. https://doi.org/10.1016/j.jiec.2014.11.003

Bernal C, Guzman F, Illane A, Wilson L (2018) Selective and eco-friendly synthesis of lipoaminoacid-based surfactants for food, using immobilized lipase and protease biocatalysts. Food Chem 239:189–195. https://doi.org/10.1016/j.foodchem.2017.06.105

Bilal M, Asgher M, Parra-Saldivar R, Hu H, Wang W, Zhang X, Iqbal H (2017) Immobilized ligninolytic enzymes: an innovative and environmental responsive technology to tackle dye-based industrial pollutants – a review. Sci Total Environ 576:646–659

Bloomer S, Adlercreutz P, Mattiasson B (1992) Facile synthesis of fatty acid esters in High Yields. Enzym Microb Technol 14:546–552

Cai WW, Su EZ, Zhu SJ, Ren YH, Wei DZ (2014) Characterization of a nitrilase from *Arthrobacter aurescens* CYC705 for synthesis of iminodiacetic acid. J Gen Appl Microbiol 60(6):207–214. https://doi.org/10.2323/jgam.60.207

Chandanshive VV, Rane NR, Tamboli AS, Gholavec AR, Khandare RV, Govindwar SP (2017) Co-plantation of aquatic macrophytes Typha angustifolia and Paspalum scrobiculatum for effective treatment of textile industry effluent. J Hazard Mater 338:47–56

Choi HJ, Lim BR, Park YJ, Joo WH (2017) Improvement in solvent tolerance by exogenous glycerol in *Pseudomonas* sp. BCNU 106. Lett Appl Microbiol 65(2):147–152

Choudhury P, Bhunia B (2015) Industrial application of lipase: a review. Biopharm J 1(2):41–47

Classen T, Pietruszka J (2017) Complex molecules, clever solutions – enzymatic approaches towards natural product and active agent syntheses. Bioorg Med Chem 26(7):2185–1303. https://doi.org/10.1016/j.bmc.2017.06.045

Contesini FJ, Calzado F, Madeira JS, Rubio MV, Zubieta MP, de Melo RR, Gonçalves TA (2016) *Aspergillus* lipases: biotechnological and industrial application. Fungal Metab:639–666

de Carvalho CC (2017) Whole cell biocatalysts: essential workers from nature to the industry. Microb Biotechnol 10(2):250–263

Eun Bae J, Kim IJ, Nam KH (2017) Crystal structure of glucose isomerase in complex with xylitol inhibitor in one metal binding mode. Biochem Biophys Res Commun 493(1):666–670

Fernandes P, Ferreira BS et al (2003) Solvent tolerance in bacteria: role of efflux pumps and cross-resistance with antibiotics. Int J Antimicrob Agents 22:211–216

Gong SJ, Ming Z, Song J, Min Z, Hong Z (2012) Nitrilases in nitrile biocatalysis: recent progress and forthcoming research. Microb Cell Factories 11:142. https://doi.org/10.1186/1475-2859-11-142

Gurung N, Ray S, Bose S, Rai V (2013) A broader view: microbial enzymes and their relevance in industries, medicine, and beyond. BioMed Res Int 2013:1–18. https://doi.org/10.1155/2013/329121

Halan B (2017) Growth of *Pseudomonas taiwanensis* VLB120ΔC biofilms in the presence of *n*-butanol. Microb Biotechnol 10(4):745–755

Han C, Yao P, Yuan J, Duan Y, Feng J, Wang M, Wub Q, Zhu D (2015) Nitrilase-catalyzed hyd: rolysis of 3-aminopropionitrile at high concentration with a tandem reaction strategy for shifting the reaction to β-alanine formation. J Mol Catal B Enzym 115:113–118. https://doi.org/10.1016/j.molcatb.2015.02.007

Hemamalini R, Khare SK (2016) Purification and characterization of active aggregates of an organic solvent tolerant lipase from *Marinobacter* sp. EMB5 Insights Enzyme Res 1(1):3

Hong DK, Jang SH, Lee CW (2017) Gene cloning and characterization of a psychrophilic phthalate esterase with organic solvent tolerance from an Arctic bacterium *Sphingomonas glacialis* PAMC 26605. J Mol Catal B: Enzym 133:S337–S345. https://doi.org/10.1016/j.molcatb.2017.02.004

Hunt RW, Chinnasamy S, Bhatnagar A, Das KC (2010) Effect of biochemical stimulants on biomass productivity and metabolite content of the microalga, Chlorella sorokiniana. Appl Biochem Biotechnol 162(8):2400–2414

Jérôme F, Luque R, Liao Y, Dewaele A, Verboekend D, Sels BF (2017) Alkylphenols as bio-based solvents: properties, manufacture and applications. Wiley. https://doi.org/10.1002/9781119065357

Kamal S, Rehman S, Iqbal HMN (2016) Biotechnological valorization of proteases: from hyperproduction to industrial exploitation—a review. Environ Prog Sustain Energy 36(2):511–522. https://doi.org/10.1002/ep.12447

Kang LJ, Meng ZT, Hu C, Zhang Y, Guo HL, Li Q, Li M (2017) Screening, purification, and characterization of a novel organic solvent-tolerant esterase, Lip2, from *Monascus purpureus* strain M7. Extremophiles 21:345–355

Koutinasa M, Yiangoua C, Natália M, Katerina O, Albert L, Francisco C, Ferreira VS (2018) Application of commercial and non-commercial immobilized lipases for biocatalytic production of ethyl lactate in organic solvents. Bioresour Technol 247:496–503. https://doi.org/10.1016/j.biortech.2017.09.130

Lade H, Kadam A, Paul D, Govindwar S (2015) Biodegradation and detoxification of textile azo dyes by bacterial consortium under sequential microaerophilic/aerobic processes. EXCLI J 14:158–174. https://doi.org/10.17179/excli2014-642

Li XZ, Zhang L et al (1998) Role of the multidrug efflux systems of Pseudomonas aeruginosa in organic solvent tolerance. J Bacteriol 180(11):2987–2991

Li G, Wang L, Reetz MT (2017) Biocatalysts for the pharmaceutical industry created by structure-guided directed evolution of stereoselective enzymes. Bioorg Med Chem 26:1241–1251. https://doi.org/10.1016/j.bmc.2017.05.021

Li S, Yang X, Yang S, Zhu M, Wang X (2012) Technology prospecting on enzymes: application, marketing and engineering. Comput Struct Biotechnol J 2(3):e201209017. https://doi.org/10.5936/csbj.201209017

Lidija T, Živkovića I, Živković LS, Beškoskic VP, Gopčevića KR, JokićdDragoslav BM, Radosavljeviće S, Karadžića IM (2017) The Candida rugosa lipase adsorbed onto titania as nano biocatalyst with improved thermostability and reuse potential in aqueous and organic media. J Mol Catal B Enzym 133(1):533–542. https://doi.org/10.1016/j.molcatb.2017.06.001

Liu ZQ, Baker PJ, Cheng F, Xue YP, Zheng Y-G, Shen Y-C (2013) Screening and improving the recombinant nitrilases and application in biotransformation of iminodiacetonitrile to iminodiacetic acid. PLoS One 8(6):67197. https://doi.org/10.1371/journal.pone.0067197

Ludmila M, Křen V (2010) Biotransformations with nitrilases. Curr Opin Chem Biol 14(2):130–137. https://doi.org/10.1016/j.cbpa.2009.11.018

Madhu A, Chakraborty JN (2017) Developments in application of enzymes for textile processing. J Clean Prod 145:114–133

Mageswari A, Subramanian P, Chandrasekaran S, Karthikeyan S, Gothandam KM (2017) Systematic functional analysis and application of a cold-active serine protease from a novel *Chryseobacterium* sp. Food Chem 217:18–27

Maksimova YG, Gorbunova AN, Demakov VA (2017) Stereoselective biotransformation of phenylglycine nitrile by heterogeneous biocatalyst based on immobilized bacterial cells and enzyme preparation. Dokl Biochem Biophys 474(1):183–185. https://doi.org/10.1134/S1607672917030139

Malik Z, Ahmad M, Abassi GH, Dawood M, Hussain A, Jamil M (2017) Agrochemicals and soil microbes: interaction for soil health. Soil Biol Biochem 49:139–152

Manefield M, Lee M, Koenig J (2017) The nature and relevance of solvent stress in microbes and mechanisms of tolerance. In: Microbial ecology of extreme environments, pp 201–213. https://doi.org/10.1007/978-3-319-51686-8_9

Martínez AT, Ruiz-Dueñas FJ, Camarero S, Serrano A, Linde D, Lund H, Vind J, Tovborg M, Herold-Majumdar OM, Hofrichter M, Liers C, Ullrich R, Scheibner K, Sannia G, Piscitelli A, Pezzella C, Sener ME, Kılıç S, van WJH B, Guallar V, Lucas MF, Zuhse R, Ludwig R, Hollmann F, Fernández-Fueyo E, Record E, Faulds CB, Tortajada M, Winckelmann I, Rasmussen JA, Gelo-Pujic M, Gutiérrez, Del Río JC, Rencoret J, Alcalde M (2017) Oxidoreductases on their way to industrial biotransformations. Biotechnol Adv 35(6):815–831. https://doi.org/10.1016/j.biotechadv.2017.06.003

Martınkova L, Rucka L, Nesvera J, Patek M (2017) Recent advances and challenges in the heterologous production of microbial nitrilases for biocatalytic applications. World J Microbiol Biotechnol 33:8. https://doi.org/10.1007/s11274-016-2173-6

Matsumae H, Furui M, Shibatani T (1993) Lipase-catalyzed asymmetric hydrolysis of 3-phenylglycidic acid ester, the key intermediate in the synthesis of diltiazem hydrochloride. J Ferment Bioeng 75(2):93–98. https://doi.org/10.1016/0922-338X(93)90216-U

Mazmouza R, Essadika I, Hamdaneb D, Méjeana A, Plouxac O (2018) Characterization of CyrI, the hydroxylase involved in the last step of cylindrospermopsin biosynthesis: binding studies, site-directed mutagenesis and stereoselectivity. Arch Biochem Biophys 647:1–9. https://doi.org/10.1016/j.abb.2018.04.007

Mehta A, Bodh U, Gupta R (2017) Fungal lipases: a review. J Biotech Res 8:58–77

Miazek K, Kratky L, Sulc R, Jirout T, Aguedo M, Richel A, Goffin D (2017) Effect of organic solvents on microalgae growth, metabolism and industrial bioproducts extraction: a review. Int J Mol Sci 18(7):1429

Molina-Santiago C, Udaondo Z (2017) Global transcriptional response of solvent-sensitive and solvent-tolerant *Pseudomonas putida* strains exposed to toluene. Environ Microbiol 19(2):645–658

Nathan VK, Rani ME, Rathinasamy G, Dhiraviam KN (2017) Low molecular weight xylanase from *Trichoderma viride* VKF3 for bio-bleaching of newspaper pulp. Bioresources 12(3):5264–5278

Ni K, Wang H, Zhang LZM, Zhang S, Ren Y, Wei D (2013) Efficient production of (R)-(−)-mandelic acid in biphasic system by immobilized recombinant E. coli. J Biotechnol 167(4):433–440. https://doi.org/10.1016/j.jbiotec.2013.07.024

Nikama M, Patil S, Chaudhari A, Patil U, Khandare R, Govindwar S (2017) Biodegradation and detoxification of azo solvent dye by ethylene glycol tolerant ligninolytic ascomycete strain of *Pseudocochliobolus verruculosus* NFCCI 3818. Biocatal Agric Biotechnol 9:209–217. https://doi.org/10.1016/j.bcab.2017.01.004

Sarkar S, Banerjee A, Halder U, Biswas R, Bandopadhyay R (2017) Degradation of synthetic azo dyes of textile industry: a sustainable approach using microbial enzymes. Water Conserv Sci Eng 2(4):121–131

Sharma R, Vashist S (2017) Mutated microbe for low cost production of high value commercial/s, TEMP/E-1/28366/2017-DEL, August 5, 2017

Sikora A, Sroka WD, Siódmiak T, Marszall MR (2017) Kinetic resolution of (R,S)-atenolol with the use of lipases in various organic solvents. Curr Org Synth 14(5):747–754

Singh S, Banerjee UC (2005) Enantioselective hydrolysis of methoxyphenyl glycidic acid methyl ester [(±)-MPGM] by a thermostable and alkalostable lipase from *Pseudomonas aeruginosa.* J Mol Catal B Enzym 36(1–6):30–35. https://doi.org/10.1016/j.molcatb.2005.07.005

Singh S, Bajaj BK (2017) Agroindustrial/forestry residues as substrates for production of thermo-active alkaline protease from *Bacillus licheniformis* K-3 having multifaceted hydrolytic potential. Waste Biomass Valoriz 8(2):453–462

Sobariu DL, DIT F, Diaconu M, Pavel LV, Hlihor RM, Drăgoi EN, Curteanu S, Lenz M, Corvini PF, Gavrilescu MN (2017) Rhizobacteria and plant symbiosis in heavy metal uptake and its implications for soil bioremediation. New Biotechnol 39(Pt A):125–134. https://doi.org/10.1016/j.nbt.2016.09.002

Sun H, Zhang H, Ang EL, Zhao H (2018) Biocatalysis for the synthesis of pharmaceuticals and pharmaceutical intermediates. Bioorg Med Chem 26(7):1275–1284. https://doi.org/10.1016/j.bmc.2017.06.043

Suwannaphana S, Fufeungsombut E, Chim-anage APP (2017) A serine protease from newly isolated Bacillus sp. for efficient silk degumming, sericin degrading and colour bleaching activities. Int Biodeterior Biodegrad 117:141–149. https://doi.org/10.1016/j.ibiod.2016.12.009

Toushik SH, Lee KT, Lee JS, Kim KS (2017) Functional applications of lignocellulolytic enzymes in the fruit and vegetable processing industries. J Food Sci 82(3):585–593

Truppo MD (2017) Biocatalysis in the pharmaceutical industry: the need for speed. Med Chem Lett 8(5):476–480

Tseng WC, Wu TJ, Chang YJ, Cheng HW, Fang TY (2017) Overexpression and characterization of a recombinant l-ribose isomerase from *Actinotalea fermentans* ATCC 43279. J Biotechnol 259:168–174

Qiu J, Su EZ, Wang W, Wei DZ (2014a) High yield synthesis of D-phenylglycine and its derivatives by nitrilase mediated dynamic kinetic resolution in aqueous-1-octanol biphasic system. Tetrahedron Lett 55(8):1448–1451. https://doi.org/10.1016/j.tetlet.2014.01.044

Qiu J, Su EZ, Wang HL, Cai WW, Wang W, Wei DZ (2014b) Cloning, overexpression, and characterization of a high enantioselective nitrilase from Sphingomonas wittichii RW1 for asymmetric synthesis of (R)-phenylglycine. Appl Biochem Biotechnol 173(2):365–377. https://doi.org/10.1007/s12010-014-0845-y

Vafaeezadeh M, Alinezhad H (2016) Bronsted acidic ionic liquids: green catalysts for essential organic reactions. J Mol Liq 218:95–105. https://doi.org/10.1016/j.molliq.2016.02.017

Vekariya RL (2017) A review of ionic liquids: applications towards catalytic organic transformations. J Mol Liq 227:44–60. https://doi.org/10.1016/j.molliq.2016.11.123

Vesela AB, Krenkova A, Martınkova L (2015) Exploring the potential of fungal arylacetonitrilases in mandelic acid synthesis. Mol Biotechnol 57(5):466–474. https://doi.org/10.1007/s12033-015-9840-y

Wang HL, Gao WY, Sun HH, Chen LF, Zhang LJ, Wang XD, Wei DZ (2015a) Protein engineering of a nitrilase from *Burkholderia cenocepacia* J2315 for efficient and enantioselective production of (R)-o-chloromandelic acid. Appl Environ Microbiol 81(24):8469–8477. https://doi.org/10.1128/AEM.02688-15

Wang HL, Fan HY, Sun HH, Zhao L, Wei DZ (2015b) Process development for the production of (R)-(−)-mandelic acid by recombinant *Escherichia coli* cells harboring nitrilase from *Burkholderia cenocepacia* J2315. Org Process Res Dev 19(12):2012–2016. https://doi.org/10.1021/acs.oprd.5b00269

Wang M, Du J, Zhang D, Li X, Zhao J (2017) Modification of different pulps by homologous overexpression alkali-tolerant endoglucanase in *Bacillus subtilis* Y106. Sci Rep 7:3321. https://doi.org/10.1038/s41598-017-03215-9

Wei Y, Xia S, He C, Xiong W, Xu H (2016) Highly enantioselective production of a chiral intermediate of sitagliptin by a novel isolate of *Pseudomonas pseudoalcaligenes*. Biotechnol Lett 38:841–846

Xue YP, Wang YP, Xu Z, Liu ZQ, Shu XR, Jia DX, Zheng YG, Shen YC (2015) Chemoenzymatic synthesis of gabapentin by combining nitrilase-mediated hydrolysis with hydrogenation over Raney-nickel. Catal Commun 66:121–125. https://doi.org/10.1016/j.catcom.2015.03.035

Yamada H, Kobayashi M (2014) Nitrile hydratase and its application to industrial production of acrylamide. Biosci Biotechnol Biochem 60:1391–1400. https://doi.org/10.1271/bbb.60.1391

Yoshida T, Mitsukura K, Mizutani T, Nakashima R, Shimizu Y, Kawabata H, Nagasawa T (2013) Enantioselective synthesis of (S)-2-cyano-2-methylpentanoic acid by nitrilase. Biotechnol Lett 35(5):685–688. https://doi.org/10.1007/s10529-012-1131-0

Use of Acidophilic or Acidotolerant *Actinobacteria* for Sustainable Agricultural Production in Acidic Soils

17

K. Tamreihao, Nimaichand Salam,
and Debananda S. Ningthoujam

Abstract

The quest for increasing agricultural production for the burgeoning human population had been effective with the use of nitrogen-based fertilizers. However, its prolong use and occurrence of acid rain resulted in dropping the soil pH below 5.0, whose environmental conditions considerably decreased the beneficial effects of soil neutrophilic bacteria while increasing the abundance of pathogenic fungi. Furthermore, the use of pesticide and synthetic fertilizers had adverse effect on human health and environment. An alternative method will therefore rely on minor groups of bacteria that can sustain its growth under extreme condition. And particularly for designing products to be applied in acidic soil, acidophilic and/or acidotolerant *Actinobacteria* having antifungal and/or plant growth-promoting activities had tremendous potential for developments as novel biocontrol and/or biofertilizer products. As *Actinobacteria* can survive under many adverse environment conditions by forming spores, they can be promising bio-agents for sustainable agricultural production. *Actinobacteria* may help in the degradation of organic matter into humus and release of nitrogen, carbon, and ammonia, in turn supplying the nutrients to agricultural crops in acidic soil. Release of ammonia due to decomposition of chitin by chitinase-producing *Actinobacteria* may raise the pH of soils, paving a way for other neutrophilic plant growth-promoting bacteria.

K. Tamreihao (✉) · D. S. Ningthoujam
Microbial Biotechnology Research Laboratory, Department of Biochemistry, Manipur University, Imphal, Manipur, India

N. Salam
State Key Laboratory of Biocontrol and Guangdong Provincial Key Laboratory of Plant Resources, School of Life Sciences, Sun Yat-Sen University, Guangzhou, Guangdong, China

D. Egamberdieva et al. (eds.), *Extremophiles in Eurasian Ecosystems: Ecology, Diversity, and Applications*, Microorganisms for Sustainability 8,
https://doi.org/10.1007/978-981-13-0329-6_17

Keywords
Actinobacteria · Acidophilic · Acidotolerant · Biocontrol · Biofertilizer · *Streptomyces* · Agricultural crops

17.1 Introduction

The ever-increasing world population necessitates the production levels of agriculture to raise by 60% in 2030–2050 relative to its production in 2005–2007 in order to meet the demands of the population that is expected to increase from 7 to 9 billion in 2050 (Schröder 2014). Most of this population growth is projected to occur within the developing countries, especially in Africa and Asia. However, the world's average agricultural production has not kept up since 2000 to meet the demand of the increasing population, and demand has outstripped its production (FAO 2000). In order to feed the burgeoning human population, there is an imperative need for 70–100% increase in global agricultural productivity by 2050 (Godfray et al. 2010).

Furthermore, with the increase in population and a need for more urbanization and industrialization, the loss of agricultural lands is predicted to increase rapidly in the coming decades. This means that increasing food demand must be met using ever-decreasing area of arable lands (Godfray et al. 2010). As agricultural production intensified with increase in human populations over the past few decades, producers became more and more dependent on agrochemicals for crop protection. However, the massive use of chemical pesticides and fertilizers causes the development of resistant pathogens and poses serious risks to the environment and human health (de Weger et al. 1995; Leach and Mumford 2008). Extensive use of ammonia-based fertilizers and environmental factors like acid rain cause pH of soil to drop below 5.0 which contributes to significant decline in neutrophilic bacterial population and abundance of pathogenic fungi comparatively. So, for better plant growth to maintain the productivity levels and to curb the plant diseases caused by fungi, there is a dire/pressing need for exploration of acidophilic/acidotolerant bacteria for application as efficient biocontrol and biofertilizing agents (Ventura 2000; Haney et al. 2000; Tamreihao et al. 2016b).

Actinobacteria are major components of the bacterial populations present in the soil. They are aerobic, Gram-positive and grow as branching filaments consisting of vegetative mycelia and aerial hyphae that play important ecological roles in soil nutrient cycling (Franco-Correa et al. 2010; de Jesus Sousa and Olivares 2016). The most extensively studied *Actinobacteria* belong to genera *Streptomyces*. *Actinobacteria* especially *Streptomyces* have been reported to produce several important bioactive secondary metabolites that can be used as biocontrol and biofertilizing agents for application in agriculture (Goodfellow and Williams 1983; Nimaichand et al. 2013; Tamreihao et al. 2016a). *Actinobacteria* can promote plant growth directly or indirectly. The main mechanisms by which they directly contribute to the plant growth are production of phytohormones such as indole-3-acetic acid (IAA), cytokinins, and gibberellins; enhancement of plant nutrition by solubilization of minerals such as inorganic phosphate; and production of siderophores that chelate iron and 1-aminocyclopropane-1-carboxylic acid deaminase (Bhattacharyya and Jha 2012). *Actinobacteria* indirectly promote the plant by biocontrol of phytopathogens through the production of antibiotics and volatile compounds, synthesis of fungal cell wall-degrading extracellular enzymes, induction of systemic resistance, and competition for nutrients and niches within the rhizosphere (Bhattacharyya and Jha 2012; Podile and Kishore 2006).

In the last decade, intensive research has been focused on minor groups of *Actinobacteria*, including those that are difficult to isolate and cultivate and those that grow under extreme conditions. However, majority of the soil *Actinobacteria* grow in neutral and slightly alkaline conditions; thus isolation has been mostly based on neutrophilic strain (Franco-Correa et al. 2010). Until the investigations of Corke and Chase (1964) and Khan and Williams (1975), all beneficial *Actinobacteria* were believed to be neutrophilic. Acidophilic bacteria grow in the pH range 3.5–6.5, with optimum growth between pH 4.5 and 5.5 (Khan and Williams 1975). Acidophilic *Actinobacteria* can be assigned to two groups, namely, a moderately acidophilic group which grow from pH 4.5 to 7.5, with an optimal pH of around 6.0, and a group of obligately acidophilic *Actinobacteria* which grow from pH 3.5 to 6.5, with an optimal pH of around 4.5 (Bull 2011; Xu et al. 2006; Guo et al. 2015). The most frequently encountered acidophilic/acidotolerant *Actinobacteria* belong to the genus *Streptomyces* (Hagedorn 1976; Guo et al. 2015; Poomthongdee et al. 2015). Unlike *Streptomyces*, members of the genus *Streptacidiphilus* are strictly acidophilic (Cho et al. 2008; Golinska et al. 2013; Huang et al. 2004; Kim et al. 2003; Wang et al. 2006).

17.2 General Mechanisms for Acid Tolerance in Bacteria

In order to survive in potentially stressful environmental conditions such as acids, antibiotics, heat, etc., bacteria possess different effective mechanisms to cope and survive. And among these, an acidic condition is the most common condition encountered by several bacteria (Winfield and Groisman 2003; Liu et al. 2015). The pH levels inside the stomach of mammals can drop drastically to a relative value of 2.0. However, bacteria such as *Escherichia coli*, *Salmonella enteric*, and *Shigella flexneri* remain unaffected and show capability to withstand and survive in extreme acidic conditions (Kanjee and Houry 2013; Spector and Kenyon 2012).

There are several mechanisms used by acidophilic and/or acidotolerant bacteria for resistance or tolerance against acidic environments. The general acid-resistant mechanisms used by acidophilic/acidotolerant bacteria include the use of H^+ antiport systems such as H^+-ATPase activity, acid end-product efflux, decreased proton permeability to maintain a low intracellular concentration of protons, and synthesis of alkali products to neutralize acid generated during extracellular metabolism (Liu et al. 2015). Bacteria such as *Pseudomonas aeruginosa* (Williams and Camara 2009) and *Streptococcus mutans* (Li et al. 2001) form dense biofilms to protect cells against extracellular acid shock (Liu et al. 2015).

17.3 Biocontrol Tools Against Fungal Pathogens

17.3.1 Antibiotic Production

Actinobacteria especially members of the genus *Streptomyces* are famous for their ability to produce several important secondary metabolites having antimicrobial and plant growth-promoting traits for application in medicine and agriculture (Goodfellow and Williams 1983; Nimaichand et al. 2013; Barka et al. 2015; Viaene et al. 2016; Tamreihao et al. 2016b). Approximately 60% of agricultural antibiotics are derived from the genus *Streptomyces* (Ilic et al. 2007). Acidophilic and/acidotolerant bacteria have been reported to exhibit higher ratio of antifungal activity over neutrophilic *Actinobacteria* (Khan and Williams 1975; Zakalyukina and Zenova 2007; Guo et al. 2015; Poomthongdee et al. 2015). And among *Actinobacteria*, strains classified within the genus *Streptomyces* showed the higher rate and broader spectrum of antagonistic activities (Guo et al. 2015; Poomthongdee et al. 2015).

Acidophilic *Actinobacteria* isolated from acidic soil exhibited antifungal activity against *Fusarium* sp., *Curvularia* sp., and *Colletotrichum gloeosporioides* (Niyasom et al. 2015). Similarly, acidophilic *Streptomyces* spp. isolated from acidic soils inhibited the growth of *Fusarium* sp. and *F. oxysporum* (Zakalyukina and Zenova 2007; Guo et al. 2015). Majority of 212 *Actinobacteria* obtained using acidified media (pH 5.5) showed antagonistic activity against rice fungal pathogens such as *F. moniliforme*, *Helminthosporium oryzae*, and *Rhizoctonia solani* (Poomthongdee et al. 2015). Similarly, acidotolerant *Streptomyces* sp. MBRL 10 inhibited the growth of rice fungal pathogens such as *R. solani*, *R. oryzae-sativae*, *H. oryzae*, *Pyricularia oryzae*, *F. oxysporum*, and *Curvularia oryzae* (Tamreihao et al. 2016b). The volatile compound(s) produced by the strain also exhibited antifungal activity. The metabolites present in the culture filtrate also exhibited significant inhibition zone (Tamreihao et al. 2016b). Boukaew et al. (2013) reported that the volatile compound; 3,7-dimethylocta-1,6-dien-3-ol (L-linalool), produced by *Streptomyces philanthi* inhibited the growth of rice pathogenic fungi *R. solani*, *P. grisea*, *H. oryzae*, and *F. fujikuroi*. It could effectively suppress the growth of *R. solani* and reduce the incidence and/or severity of leaf blight in rice plant.

The crude extract from acidophilic *Streptomyces* sp. inhibited mycelial growth of 22 species of fungi (Lyu et al. 2017). The crude extract also effectively inhibited spore germination of *Botrytis cinerea* and *Rhizopus stolonifer*. The two antibiotics reveromycins A and B extracted from *Streptomyces* sp. effectively suppressed mycelial growth of *Botrytis cinerea*, *Mucor hiemalis*, *Rhizopus stolonifer*and *Sclerotinia sclerotiorum*, and spore germination of *B. cinerea*, *M. hiemalis*, and *R. stolonifer* under acidic pH conditions. The suppressive efficacies of the antibiotics were greatly affected by ambient pH. The antifungal activity of reveromycins A and B was higher at pH 4.5 and 5.5, whereas the activity decreased at pH 7.0, implying that application of *Streptomyces* sp. in acidic soil may achieve higher antifungal efficacy in the suppression of pathogenic fungi. Treatment of strawberries with the crude extract and reveromycin A significantly suppressed the strawberry fruit rot caused by *B. cinerea*, *M. hiemalis*, *R. stolonifer*, and *S. sclerotiorum* (Lyu et al. 2017). A list of antibiotics produced by acidophilic *Actinobacteria* is provided in Table 17.1.

Table 17.1 Antibiotic and plant growth-promoting compounds produced by acidophilic and/acidotolerant *Actinobacteria*

Actinobacteria genus and strain	Antibiotics/ compound produced	Bioactivity	References
Streptomyces sp. 3–10	Reveromycin A	Antifungal	Lyu et al. (2017)
Streptomyces sp. 3–10	Reveromycin B	Antifungal	Lyu et al. (2017)
Streptomyces sp. FXJ1.532	Macrolide	Antifungal	Guo et al. (2015) and Kim et al. (2011)
Streptomyces sp. FXJi.408	Polyene macrolide	Antifungal	Guo et al. (2015) and Bhat and Narayanan (1996)
Streptomyces sp. FXJ1.535	Angucycline-type compound	Antifungal	Guo et al. (2015) and Zhang et al. (2012)
Streptomyces sp. GTVL2G15	Actinomycin-like compound	Antifungal	Guo et al. (2015) and Xiong et al. (2012)
Streptomyces sp. FXJ1.275	Peptide compound	Antifungal	Guo et al. (2015) and Seo et al. (2012)
Saccharothrix sp. FXJ1.021	Anthracycline-related compound	Antifungal	Guo et al. (2015) and Kim et al. (2000)
Actinomadura sp. FXJ1.344	Angucycline-type compound	Antifungal	Guo et al. (2015) and Xiong et al. (2012)
Amycolatopsis sp. FXJ1.406	Peptide compound	Antifungal	Guo et al. (2015) and Seo et al. (2012)
Streptomyces sp. MBRL 10	Volatile compound	Antifungal	Tamreihao et al. (2016b)
Streptomyces sp. FXJ1.172	Thienodolin analogue	Plant growth regulator	Guo et al. (2015) and Kanbe et al. (1993)
Streptomyces sp. MBRL 10	IAA, siderophore	Plant growth promotion (PGP)	Tamreihao et al. (2016b)
Streptomyces sp.	Siderophore	PGP	Poomthongdee et al. (2015)
Streptomyces sp. FXJ1.066	Siderophore	PGP	Guo et al. (2015)
Streptomyces sp. FXJ23y	Siderophore	PGP	Guo et al. (2015)
Lentzea jiangxiensis FXJ1.034	Siderophore	PGP	Guo et al. (2015)
Streptosporangium sp. FXJ1.1111	Siderophore	PGP	Guo et al. (2015)

17.3.2 Fungal Cell Wall-Degrading Enzyme Production

Actinobacteria are of enormous importance since they possess a capacity to produce and secrete a variety of extracellular hydrolytic enzymes. *Actinobacteria* help in the decomposition of organic matter into humus and release of nitrogen, carbon, and ammonia, in turn supplying the nutrients to agricultural crops. *Streptomyces* spp. play an important role in degrading complex organics such as cellulose, lignin, and chitin (Crawford et al. 1983; Prasad et al. 2013; Brzezinska et al. 2013). Acidophilic *Actinobacteria* are probably active in decomposition processes in acidic soils, and their exoenzymes, chitinases, and diastases are adapted to function at a lower pH than those from neutrophilic *Actinobacteria* (Williams and Flowers 1978; Williams and Robinson 1981).

Acidophilic/acidotolerant *Actinobacteria* especially *Streptomyces* sp. have been reported to play significant role in the degradation of chitin in acidic soil and litter where fungi are major colonizers, and release of ammonia by the deacetylation and deamination of N-acetylglucosamine residues increases the pH of the soil (Williams and Robinson 1981). This paves way for other neutrophilic PGPB to populate and compete with the pathogens (Tamreihao et al. 2016b).

Acidotolerant *Streptomyces* sp. showing antagonistic activity against important rice fungal pathogens in plate assay could produce fungal cell wall-degrading enzymes such as chitinase, β-1,3-glucanase, lipase, and protease (Tamreihao et al. 2016b). Similarly, *Streptomyces* sp. producing cell wall-degrading enzymes such as chitinase, β-1,3-glucanase, lipase, and protease inhibited the mycelial growth of *R. solani*, *Colletotrichum gloeosporioides*, *Alternaria brassicae*, and *Phytophthora capsici* (Srividya et al. 2012). A higher proportion of the acidophilic/acidotolerant *Actinobacteria* exhibited chitinolytic activity over neutrophilic *Actinobacteria.* And among the acidic actinobacterial isolates, *Streptomyces* spp. showed higher percentage of chitinolytic activity (Guo et al. 2015). Many plant pathogenic fungi such as *Botrytis cinerea* and *Sclerotinia sclerotiorum* secrete oxalic acid to facilitate their infection and colonization of plant tissues (Choquer et al. 2007; Williams et al. 2011). Oxalic acid can acidify the surrounding environment. Application of chitinase-producing acidophilic/acidotolerant bioactive actinobacterial strains can be a source of new effective agents for controlling fungal plant diseases for sustainable agricultural product alternative to synthetic chemicals in acidic soils.

17.4 Biofertilizing Tools for Growth Promotion

Actinobacteria directly stimulate the growth of plants by production of phytohormones such as IAA, cytokinins, and gibberellins. IAA is a phytohormone essential for the growth and development of plants including cell plasticity, tissue elongation, embryogenesis, tip dominance and emergence of lateral roots (Teale et al. 2006). Acidotolerant *Streptomyces* sp. could produce 25 μg/ml of IAA in the presence of 2 mg/ml of tryptophan. *Streptomyces* spp. have been reported to produce IAA in the range of 11–144 μg/ml when the production medium was supplemented with 2 mg/ml of tryptophan (Khamna et al. 2010). IAA producing *Streptomyces* spp. promoted seed germination and plant growth in maize and cowpea (Khamna et al. 2010). IAA produced by *Streptomyces hygroscopicus* stimulated root elongation and induced the formation of adventitious roots in kidney beans (Lin and Xu 2013).

Despite the abundance of phosphorus in the soil (often as high as 400–1200 mg/kg of soil), only 0.1% of the total phosphates exists in a soluble form available for uptake by plants. Most of the soil phosphorus is in insoluble form and, therefore, not available to support plant growth. The insoluble phosphorus is present as inorganic minerals such as apatite or as one of several organic forms including inositol phosphate (soil phytate), phosphomonoesters, and phosphotriesters (Khan et al. 2007; Zhou et al. 1992). In addition, phosphate fertilizers applied in agricultural fields are quickly immobilized in the soil or quickly washed away by rain waters, polluting rivers and ground waters (Hamdali et al. 2008a), and only 10–30% of applied phosphate fertilizer is taken up by plants in the year of application (Mclaughlin et al. 1988). The major mechanisms used by bacteria for solubilization of inorganic phosphate include the synthesis of low molecular weight organic acids such as gluconic acid, citric acid, succinic acid and oxalic acid (Rodriguez et al. 2004; Rajput et al. 2013). Acidophilic/acidotolerant *Streptomyces* spp. have been reported to solubilize inorganic phosphate (Poomthongdee et al. 2015; Tamreihao et al. 2016b). Hamdali et al. (2008a) reported that rock phosphate (RP) solubilizing *Streptomyces griseus*, *Streptomyces cavourensis*, and *Micromonospora aurantiaca* could stimulate the growth of wheat plants in soil supplemented with RP. Growth promotion was correlated with significant increase in nitrogen and phosphorus contents of the plant tissues (Hamdali et al. 2008b). Phosphate-solubilizing *Streptomyces corchorusii* enhance seed germination, growth promotion, and grain yield production of rice plants under pot and field conditions (Tamreihao et al. 2016a).

Production of siderophore, a compound that can chelate iron and make the bound iron available to the plants (Burd et al. 1998; Dimpka et al. 2008), by acidophilic/acidotolerant *Actinobacteria* especially *Streptomyces* has been reported (Poomthongdee et al. 2015; Tamreihao et al. 2016b). Genome sequence of acidophilic *Streptacidiphilus oryzae* contains genes related to siderophore production (Kim et al. 2015). Siderophore-producing *Actinobacteria* have been reported to enhance plant growth in cowpea (Dimpka et al. 2008), chickpea (Misk and Franco 2011), and wheat (Sadeghi et al. 2012). Siderophore-producing *Streptomyces* spp. have been reported to exhibit biocontrol activities against plant pathogens (Dimpka et al. 2008; Sadeghi et al. 2012).

Treatment of rice seeds with acidotolerant *Streptomyces* sp. having plant growth-promoting traits such as IAA, siderophore production and phosphate solubilization could enhance seed germination and growth of rice seedlings under gnotobiotic conditions. The strain also enhanced the growth of rice plants under greenhouse conditions (Tamreihao et al. 2016a, b).

17.5 Conclusions and Future Perspectives

The greatest challenge for agricultural crop production in the current century is to meet the increasing world population and to reduce the use of synthetic chemical as the latter pose serious risk to human health and environment. The use of plant growth-promoting bacteria (PGPB) to enhance agricultural crop production has emerged as sustainable and alternative tools to meet this challenge. As the soil pH drop below 5.0 due to massive and prolonged use of synthetic nitrogenous fertilizers and environmental factor such as acid rain, intensive search for a group of PGPB that can survive under extreme conditions (i.e., alkaline and acidic conditions) has become a prime importance for sustainable production of agricultural crops

(Goodfellow and O'Donnell 1989; Lazzarini et al. 2000; Zakalyukina and Zenova 2007; Tamreihao et al. 2016b). Since many *Actinobacteria* show resistance to many extreme environmental conditions due to formation of spores (Chater 1993), they can be explored and used for development of promising plant growth-promoting agents for sustainable production of agricultural crops in acidic soils.

There is scanty report on acidophilic and/or acidotolerant *Actinobacteria* for their potential as biocontrol and plant growth-promoting activities. There is an urgent need for further exploration of acidophilic/acidotolerant *Actinobacteria* especially *Streptomyces*, having biocontrol and biofertilizer potential for application in an environment, where the pH of the soil drop below 5.0 due to excessive use of nitrogenous fertilizers, acid rain and acid mine drainage for sustainable development of agricultural crops. Acidophilic and/or acidotolerant *Actinobacteria* can find great utilization in industrial bioprocess such as lactic acid fermentation during industrial production. It can also find application in phytoremediation of environmental pollutants where soil can be contaminated with polycyclic aromatic hydrocarbons due to acid mine drainage (Liu et al. 2015).

As acidophilic and/or acidotolerant *Actinobacteria* especially *Streptomyces* can survive and propagate in acidic environment, they can enhance soil productivity and soil health, prevent plant diseases, and ultimately increase the production of agricultural crops. Release of ammonia by the deacetylation and deamination of N-acetylglucosamine residues by chitinase-producing *Streptomyces* may raise the pH of the soil, paving the way for the other neutrophilic PGPB to colonize and compete with the pathogens.

Acknowledgments Authors acknowledge the grant from the Department of Biotechnology (DBT), Government of India, given to the Advanced Level State Biotech Hub (BT/04/NE/2009). KT thanks Phungmila Vashum for critical review and proofreading.

References

Barka EA, Vatsa P, Sanchez L, Gaveau-Vaillant N, Jacquard C, Klenk HP, Clément C, Ouhdouch Y, van Wezel GP (2015) Taxonomy, physiology, and natural products of Actinobacteria. Microbiol Mol Biol Rev 80:1–43

Bhat JM, Narayanan MS (1996) Antifungal activity (in vitro) of certain polyene macrolide antibiotics against various plant pathogens. Hindustan Antibiot Bull 38:32–36

Bhattacharyya PN, Jha DK (2012) Plant growth-promoting rhizobacteria (PGPR): emergence in agriculture. World J Microbiol Biotechnol 28:1327–1350

Boukaew S, Plubrukam A, Prasertsan P (2013) Effect of volatile substances from *Streptomyces philanthi* RM-1-138 on growth of *Rhizoctonia solani* on rice leaf. Biol Control 58:471–482

Brzezinska MS, Jankiewicz U, Walczak M (2013) Biodegradation of chitinous substances and chitinase production by the soil actinomycete *Streptomyces rimosus*. Int Biodeterior Biodegrad 201:104–110

Bull AT (2011) Actinobacteria of the extremobiosphere. In: Horikoshi K, Antranikian G, Bull AT, Robb FT, Stetter KO (eds) Extremophiles handbook. Springer Verlag, Tokyo, pp 1203–1240

Burd GI, Dixon DG, Glick BR (1998) A plant growth-promoting bacterium that decreases nickel toxicity in seedlings. Appl Environ Microbiol 64:3663–3668

Chater KF (1993) Genetics of differentiation in *Streptomyces*. Annu Rev Microbiol 47:685–713

Cho S-H, Han J-H, Ko H-Y, Kim SB (2008) Streptacidiphilus anmyonensis sp. nov., Streptacidiphilus rugosus sp. nov. and Streptacidiphilus melanogenes sp. nov., acidophilic actinobacteria isolated from Pinus soils. Int J Syst Evol Microbiol 58:1566–1570

Choquer M, Fournier E, Kunz C, Levis C, Pradier JM, Simon A, Viaud M (2007) *Botrytis cinerea* virulence factors: new insights into a necrotrophic and polyphageous pathogen. FEMS Microbiol Lett 277:1–10. https://doi.org/10.1111/j.1574-6968.2007.00930.x

Corke CT, Chase FE (1964) Comparative studies of actinomycetes populations in acid podzolic and neutral mull forest soil. Proc Soil Sci Am 28:68–69

Crawford DL, Pometto AL III, Crawford RL (1983) Lignin degradation by *Streptomyces viridosporus*: isolation and characterization of a new polymeric lignin degradation intermediate. Appl Environ Microbiol 45:898–904

de Jesus Sousa JA, Olivares FL (2016) Plant growth promotion by streptomycetes: ecophysiology, mechanisms and applications. Chem Biol Technol Agric 3:24. https://doi.org/10.1186/s40538-016-0073-5

de Weger LA, van der Bij AJ, Dekkers LC, Simons M, Wijffelman CA, Lugtenberg BJJ (1995) Colonization of the rhizosphere of crop plants by plant-beneficial pseudomonads. FEMS Microbiol Ecol 17:221–228

Dimpka C, Svatoš A, Merten D, Büchel G, Kothe E (2008) Hydroxamate siderophore produced by *Streptomyces acidiscbies* E13 bind nickel and promote growth in cowpea (*Vigna unguiculata* L.) under nickel stress. Can J Microbiol 54:163–172

FAO (2000) Bridging the gap in rice production. http://www.fao.org/News/2000/001203-e.htm (17/6/2008)

Franco-Correa M, Quintana A, Duque C, Saurez C, Rodríguez MX, Barea J-M (2010) Evaluation of actinomycete strains for key traits related with plant growth promotion and mycorrhiza helping activities. Appl Soil Ecol 45:209–217

Godfray HCJ, Beddington JR, Crute IR, Haddad L, Lawrence D, Muir JF, Pretty J, Robinson S, Thomas SM, Toulmin C (2010) Food security: the challenge of feeding 9 billion people. Science 327:812–818

Goodfellow M, O'Donnell AG (1989) Search and discovery of industrially significant actinomycetes. In: Baumberg S, Hunter IS, Rhodes PM (eds) Microbial products: new approaches, Society for General Microbiology Symposium No. 44. Cambridge University Press, Cambridge, pp 343–383

Goodfellow M, Williams ST (1983) Ecology of actinomycetes. Annu Rev Microbiol 29:319–322

Golinska P, Ahmed L, Wang D, Goodfellow M (2013) Streptacidiphilus durhamensis sp. nov., isolated from a spruce forest soil. Antonia van Leeuwenhoek 104:199–206

Guo X, Liu N, Li X, Ding Y, Shang F, Gao Y, Ruan J, Huang Y (2015) Red soils harbour diverse culturable actinomycetes that are promising sources of novel secondary metabolites. Appl Environ Microbiol 81:3086–3103

Hagedorn C (1976) Influence of soil acidity on Streptomyces population inhabiting forest soils. Appl Environ Microbiol 32:368–375

Hamdali H, Hafidi M, Virolle MJ, Ouhdouch Y (2008a) Rock phosphate-solubilizing actinomycetes: screening for plant growth-promoting activities. World J Microbiol Biotechnol 24:2565–2575

Hamdali H, Hafidi M, Virolle MJ, Ouhdouch Y (2008b) Growth promotion and protection against damping-off of wheat by two rock phosphate solubilising actinomycetes in a P-deficient soil under greenhouse conditions. Appl Soil Ecol 40:510–517

Haney RL, Senseman SA, Hons FM, Zuberer DA (2000) Effect of glyphosate on soil microbial activity and biomass. Weed Sci 48:89–93

Huang Y, Cui Q, Wang L, Rodriguez C, Quintana E, Goodfellow M, Liu Z (2004) Streptacidiphilus jiangxiensis sp. nov., a novel actinomycete isolated from acidic rhizosphere soil in China. Antonie Van Leeuwenhoek 86:159–165

Ilic SB, Konstantinovic SS, Todorovic ZB, Lazic ML, Veljkovic VB, Jokovic N, Radovanovic BC (2007) Characterization and antimicrobial activity of the bioactive metabolites in Streptomycete isolates. Microbiology 76:421–428

Kanbe K, Okamura M, Hattori S, Naganawa H, Hamada M, Okami Y, Takeuchi T (1993) Thienodolin, a new plant growth-regulating substance produced by a Streptomycete strain: I. Taxonomy and fermentation of the producing strain, and the isolation and characterization of thienodolin. Biosci Biotech Biochem 57:632–635

Kanjee U, Houry WA (2013) Mechanisms of acid resistance in *Escherichia coli*. Annu Rev Microbiol 67:65–81

Khamna S, Yokota A, Peberdy J, Lumyong S (2010) Indole-3-acetic acid production by *Streptomyces* sp. isolated from Thai medicinal rhizosphere soils. Eur Asian J Bio Sci 4:23–32

Khan MR, Williams ST (1975) Studies on the ecology of actinomycetes in soil: VIII. Distribution and characteristics of acidophilic actinomycetes. Soil Biol Biochem 7:345–348

Khan MS, Zaidi A, Wani PA (2007) Role of phosphate solubilizing microorganisms in sustainable agriculture – a review. Agron Sustain Dev 27:29–43

Kim BS, Moon SS, Hwang BK (2000) Structure elucidation and antifungal activity of an anthracycline antibiotic, daunomycin, isolated from *Actinomadura roseola*. J Agric Food Chem 48:1875–1881

Kim BS, Moon SS, Hwang BK (2011) Isolation, identification, and antifungal activity of a macrolide antibiotic, oligomycin A, produces by *Streptomyces libani*. Can J Bot 77:850–858

Kim SB, Lonsdale J, Seong CN, Goodfellow M (2003) Streptacidiphilus gen. nov., acidophilic actinomycetes with wall chemotype I and emendation of the family Streptomycetaceae (Waksman and Henrici 1943 AL) emend. Rainey et al. 1997.Antonie Van Leeuwenhoek 83:107–116

Kim YR, Park S, Kim T-S, Kim M-K, Han J-H, Joung Y, Kim B (2015) Draft genome sequence of *Streptacidiphilus oryzae* TH49^{T}, an acidophilic actinobacterium isolated from soil. Genome Announc 3:e00703-15. https://doi.org/10.1128/genomeA.00703-15

Lazzarini A, Caveletti L, Toppo G, Marinelli F (2000) Rare genera of actinomycetes as potential producers of new antibiotics. Antonie Van Leeuwenhoek 78:399–405

Leach AW, Mumford JD (2008) Pesticide environmental accounting: a method for assessing the external costs of individual pesticide applications. Environ Pollut 151:139–147

Li YH, Hanna MN, Svensater G, Ellen RP, Cvitkovitch DG (2001) Cell density modulates acid adaptation in *Streptococcus mutans*: implications for survival in biofilms. J Bacteriol 183:6875–6884

Lin L, Xu X (2013) Indole-3-acetic acid production by endophytic *Streptomyces* sp. En-1 isolated from medicinal plants. Curr Microbiol 67:209–217

Liu Y, Tang H, Lin Z, Xu P (2015) Mechanisms of acid tolerance in bacteria and prospects in biotechnology and bioremediation. Biotechnol Adv 33:1484–1492

Lyu A, Liu H, Che H, Yang L, Zhang J, Wu M, Chen W, Li G (2017) Reveromycins A and B from *Streptomyces* sp. 3–10: antifungal activity against plant pathogenic fungi *In vitro* and in a Strawberry food model system. Front Microbiol 8:550. https://doi.org/10.3389/fmicb.2017.00550

Mclaughlin BR, Russell PR, Carroll DJ (1988) Cogen plant includes 12-acre greenhouse, oyster farm. Power Eng 92:40–42

Misk A, Franco C (2011) Biocontrol of chickpea root rot using endophytic actinobacteria. Biol Control 56:811–822

Nimaichand S, Tamrihao K, Yang L-L, Zhu W-Y, Zhang Y-G, Li L, Tang S-K, Ningthoujam DS, Li W-J (2013) *Streptomyces hundungensis* sp. nov., a novel actinomycete with antifungal activity and plant growth promoting traits. J Antibiot 66:205–209

Niyasom C, Boonmak S, Meesri N (2015) Antimicrobial activity of acidophilic actinomycetes isolated from acidic soil. KTMITL Sci Tech J 15:62–69

Podile AR, Kishore GK (2006) Plant growth-promoting rhizobacteria. In: Gnanamanickam SS (ed) Plant-associated bacteria. Springer, Dordrecht, pp 195–230

Poomthongdee N, Duangmal K, Pathom-aree W (2015) Acidophilic actinomycetes from rhizospheric soil: diversity and properties beneficial to plants. J Antibiot 68:106–114

Prasad P, Singh T, Bedi S (2013) Characterization of the cellulolytic enzyme produced by *Streptomyces griseorubens* (Accession No. AB184139) isolated from Indian soil. J King Saud Univ-Sci 25:245–250

Rajput MS, Kumar G, Rajkumar S (2013) Repression of oxalic acid-mediated mineral phosphate solubilization in rhizospheric isolates of *Klebsiella pneumoniae* by succinate. Arch Microbiol 195:81–88

Rodriguez H, Gonzalez T, Goire I, Bashan Y (2004) Gluconic acid production and phosphate solubilization by the plant growth-promoting bacterium *Azospirillum* spp. Naturwissenschaften 91:552–555

Sadeghi A, Karimi E, Dahaji PA, Javid MG, Dalvand Y, Askari H (2012) Plant growth promoting activity of an auxin and siderophore producing isolate of *Streptomyces* under saline soil conditions. World J Microbiol Biotechnol 28:1503–1509

Schröder JJ (2014) The position of mineral nitrogen fertilizer in efficient use of nitrogen and land: a review. Nat Resour 5:936–948

Seo M-D, Won H-S, Kim J-H, Mishig-Ochir T, Lee B-J (2012) Antimicrobial peptides for therapeutic applications: a review. Molecules 17:12276–12286

Spector MP, Kenyon WJ (2012) Resistance and survival strategies of *Salmonella enteric* to environmental stresses. Food Res Int 45:455–481

Srividya S, Thapa A, Bhat DV, Golmei K, Dey N (2012) *Streptomyces* sp. 9p as effective biocontrol against chilli soilborne fungal pathogens. Euro J Exp Biol 2:163–173

Tamreihao K, Ningthoujam DS, Nimaichand S, Singh ES, Reena P, Singh SH, Nongthomba U (2016a) Biocontrol and plant growth promoting activities of *Streptomyces corchorusii* strain UCR3-16 and preparation of powder formulation for application as biofertilizer agents for rice plant. Microbiol Res 192:260–270

Tamreihao K, Nimaichand S, Chanu SB, Devi KA, Lynda R, Jeeniita N, Ningthoujam DS (2016b) Acidotolerant *Streptomyces* sp. MBRL 10 from limestone quarry site antagonism against fungal pathogens and growth promotion in rice plants. J King Saud Univ-Sci 30(2):143–292. https://doi.org/10.1016/j.jksus.2016.10.003

Teale WD, Paponov IA, Palme K (2006) Auxin in action: signalling, transport and the control of plant growth and development. Nat Rev Mol Cell Biol 7:847–859

Ventura LA (2000) The effect of soil pH on plant growth. Science Experiments on File Revised Facts on the File, Inc 4:1–5

Viaene T, Langendries S, Beirinckx S, Maes M, Goormachtig (2016) Streptomyces as a plant's best friend ? FEMS Microbiol Eco 92:1–10

Wang L, Huang Y, Liu Z, Goodfellow M, Rodriguez C (2006) Streptacidiphilus oryzae sp. nov., an actinomycete isolated from rice-field soil in Thailand. Int J Syst Evol Microbiol 56:1257–1261

Williams P, Camara M (2009) Quorum sensing and environmental adaptation in *Pseudomonas aeruginosa*: a tale of regulatory networks and multifunctional signal molecules. Curr Opin Microbiol 12:182–191

Williams ST, Flowers TH (1978) The influence of pH on starch hydrolysis of neutrophilic and acidophilic streptomycetes. Microbios 20:99–406

Williams ST, Robinson CS (1981) The role of Streptomycetes in decomposition of chitin in acidic soils. J Gen Microbiol 127:55–63

Williams B, Kabbage M, Min JY, Britt R, Dickman MB (2011) Tipping the balance: *Sclerotinia sclerotiorum* secreted oxalic acid suppresses host defenses by manipulating the host redox environment. PLoS Pathog 7:e1002107. https://doi.org/10.1371/journal.ppat.1002107

Winfield MD, Groisman EA (2003) Role of nonhost environment in the lifestyles of *Salmonella* and *Escherichia coli*. Appl Environ Microbiol 69:3687–3694

Xiong Z-Q, Zhang Z-P, Li J-H, Wei S-J, Tu G-Q (2012) Characterization of *Streptomyces padanus* JAU4234, a producer of actinomycin X2, fungichromin, and a new polyene macrolide antibiotic. Appl Environ Microbiol 78:589–592

Xu C, Wang L, Cui Q, Huang Y, Liu Z, Zheng G, Goodfellow M (2006) Neutrotolerant acidophilic *Streptomyces* species isolated from acidic soils in China: *Streptomyces guanduensis* sp. no., *Streptomyces paucisporeus* sp. nov., *Streptomyces rubidus* sp. nov. and *Streptomyces yanglinensis* sp. nov. Intl J Syst Evol Microbiol 56:1109–1115

Zakalyukina YV, Zenova GM (2007) Antagonistic activity of soil acidophilic actinomycetes. Biol Bull 34:329–332

Zhang H, Wang H, Wang Y, Cui H, Xie Z, Pu Y, Pei S, Li F, Qin S (2012) Genomic sequence-based discovery of novel angucyclinone antibiotics from marine *Streptomyces* sp. W007. FEMS Microbiol Lett 332:105–112

Zhou K, Binkley D, Doxtader KG (1992) A new method for estimating gross phosphorus mineralization and immobilization rates in soils. Plant Soil 147:243–250

Printed by Printforce, the Netherlands